Electric Machinery

Sixth Edition

A. E. Fitzgerald
*Late Vice President for Academic Affairs
and Dean of the Faculty
Northeastern University*

Charles Kingsley, Jr.
*Late Associate Professor of Electrical
Engineering, Emeritus
Massachusetts Institute of Technology*

Stephen D. Umans
*Principal Research Engineer
Department of Electrical Engineering and
Computer Science
Laboratory for Electromagnetic and
Electronic Systems
Massachusetts Institute of Technology*

Mc Graw Hill

Boston   Burr Ridge, IL   Dubuque, IA   Madison, WI   New York   San Francisco   St. Louis
Bangkok   Bogotá   Caracas   Kuala Lumpur   Lisbon   London   Madrid   Mexico City
Milan   Montreal   New Delhi   Santiago   Seoul   Singapore   Sydney   Taipei   Toronto

McGraw-Hill Higher Education ⚛

*A Division of The **McGraw-Hill** Companies*

ELECTRIC MACHINERY, SIXTH EDITION

Published by McGraw-Hill, a business unit of The McGraw-Hill Companies, Inc., 1221 Avenue of the Americas, New York, NY 10020. Copyright © 2003, 1990, 1983, 1971, 1961, 1952 by The McGraw-Hill Companies, Inc. All rights reserved. Copyright renewed 1980 by Rosemary Fitzgerald and Charles Kingsley, Jr. All rights reserved. No part of this publication may be reproduced or distributed in any form or by any means, or stored in a database or retrieval system, without the prior written consent of The McGraw-Hill Companies, Inc., including, but not limited to, in any network or other electronic storage or transmission, or broadcast for distance learning.

Some ancillaries, including electronic and print components, may not be available to customers outside the United States.

This book is printed on acid-free paper.

International   1 2 3 4 5 6 7 8 9 0 DOC/DOC 0 9 8 7 6 5 4 3 2
Domestic   1 2 3 4 5 6 7 8 9 0 DOC/DOC 0 9 8 7 6 5 4 3 2

ISBN 0-07-366009-4
ISBN 0-07-112193-5 (ISE)

Publisher: *Elizabeth A. Jones*
Developmental editor: *Michelle L. Flomenhoft*
Executive marketing manager: *John Wannemacher*
Project manager: *Rose Koos*
Production supervisor: *Sherry L. Kane*
Media project manager: *Jodi K. Banowetz*
Senior media technology producer: *Phillip Meek*
Coordinator of freelance design: *Rick D. Noel*
Cover designer: *Rick D. Noel*
Cover image courtesy of: *Rockwell Automation/Reliance Electric*
Lead photo research coordinator: *Carrie K. Burger*
Compositor: *Interactive Composition Corporation*
Typeface: *10/12 Times Roman*
Printer: *R. R. Donnelley & Sons Company/Crawfordsville, IN*

Library of Congress Cataloging-in-Publication Data

Fitzgerald, A. E. (Arthur Eugene), 1909–
    Electric machinery / A. E. Fitzgerald, Charles Kingsley, Jr., Stephen D. Umans. —6th ed.
      p.   cm. —(McGraw-Hill series in electrical engineering. Power and energy)
    Includes index.
    ISBN 0-07-366009-4—ISBN 0-07-112193-5
    1. Electric machinery. I. Kingsley, Charles, 1904–. II. Umans, Stephen D. III. Title.
  IV. Series.

  TK2181 .F5   2003
  621.31′042—dc21
                               2002070988
                               CIP

INTERNATIONAL EDITION ISBN 0-07-112193-5
Copyright © 2003. Exclusive rights by The McGraw-Hill Companies, Inc., for manufacture and export. This book cannot be re-exported from the country to which it is sold by McGraw-Hill. The International Edition is not available in North America.

www.mhhe.com

This book is dedicated to my mom, Nettie Umans, and my aunt, Mae Hoffman, and in memory of my dad, Samuel Umans.

ABOUT THE AUTHORS

The late **Arthur E. Fitzgerald** was Vice President for Academic Affairs at Northeastern University, a post to which he was appointed after serving first as Professor and Chairman of the Electrical Engineering Department, followed by being named Dean of Faculty. Prior to his time at Northeastern University, Professor Fitzgerald spent more than 20 years at the Massachusetts Institute of Technology, from which he received the S.M. and Sc.D., and where he rose to the rank of Professor of Electrical Engineering. Besides *Electric Machinery*, Professor Fitzgerald was one of the authors of *Basic Electrical Engineering,* also published by McGraw-Hill. Throughout his career, Professor Fitzgerald was at the forefront in the field of long-range power system planning, working as a consulting engineer in industry both before and after his academic career. Professor Fitzgerald was a member of several professional societies, including Sigma Xi, Tau Beta Pi, and Eta Kappa Nu, and he was a Fellow of the IEEE.

The late **Charles Kingsley, Jr.** was Professor in the Department of Electrical Engineering and Computer Science at the Massachusetts Institute of Technology, from which he received the S.B. and S.M. degrees. During his career, he spent time at General Electric, Boeing, and Dartmouth College. In addition to *Electric Machinery,* Professor Kingsley was co-author of the textbook *Magnetic Circuits and Transformers*. After his retirement, he continued to participate in research activities at M.I.T. He was an active member and Fellow of the IEEE, as well as its predecessor society, the American Institute of Electrical Engineers.

Stephen D. Umans is Principal Research Engineer in the Electromechanical Systems Laboratory and the Department of Electrical Engineering and Computer Science at the Massachusetts Institute of Technology, from which he received the S.B., S.M., E.E., and Sc.D. degrees, all in electrical engineering. His professional interests include electromechanics, electric machinery, and electric power systems. At MIT, he has taught a wide range of courses including electromechanics, electromagnetics, electric power systems, circuit theory, and analog electronics. He is a Fellow of the IEEE and an active member of the Power Engineering Society.

BRIEF CONTENTS

CONTENTS

PREFACE

T he chief objective of *Electric Machinery* continues to be to build a strong foundation in the basic principles of electromechanics and electric machinery. Through all of its editions, the emphasis of *Electric Machinery* has been on both physical insight and analytical techniques. Mastery of the material covered will provide both the basis for understanding many real-world electric-machinery applications as well as the foundation for proceeding on to more advanced courses in electric machinery design and control.

Although much of the material from the previous editions has been retained in this edition, there have been some significant changes. These include:

■ A chapter has been added which introduces the basic concepts of power electronics as applicable to motor drives.

■ Topics related to machine control, which were scattered in various chapters in the previous edition, have been consolidated in a single chapter on speed and torque control. In addition, the coverage of this topic has been expanded significantly and now includes field-oriented control of both synchronous and induction machines.

■ MATLAB[®1] examples, practice problems, and end-of-chapter problems have been included in the new edition.

■ The analysis of single-phase induction motors has been expanded to cover the general case in which the motor is running off both its main winding and its auxiliary winding (supplied with a series capacitor).

Power electronics are a significant component of many contemporary electric-machine applications. This topic is included in Chapter 10 of this edition of *Electric Machinery* in recognition of the fact that many electric-machinery courses now include a discussion of power electronics and drive systems. However, it must be emphasized that the single chapter found here is introductory at best. One chapter cannot begin to do justice this complex topic any more than a single chapter in a power-electronics text could adequately introduce the topic of electric machinery.

The approach taken here is to discuss the basic properties of common power electronic components such as diodes, SCRs, MOSFETs, and IGBTs and to introduce simple models for these components. The chapter then illustrates how these components can be used to achieve two primary functions of power-electronic circuits in drive applications: rectification (conversion of ac to dc) and inversion (conversion of dc to ac). Phase-controlled rectification is discussed as a technique for controlling the dc voltage produced from a fixed ac source. Phase-controlled rectification can be used

---

[1] MATLAB is a registered trademark of The MathWorks, Inc.

to drive dc machines as well as to provide a controllable dc input to inverters in ac drives. Similarly, techniques for producing stepped and pulse-width-modulated waveforms of variable amplitudes and frequency are discussed. These techniques are at the heart of variable-speed drive systems which are commonly found in variable-speed ac drives.

Drive-systems based upon power electronics permit a great deal of flexibility in the control of electric machines. This is especially true in the case of ac machines which used to be found almost exclusively in applications where they were supplied from the fixed-frequency, fixed-voltage power system. Thus, the introduction to power electronics in Chapter 10 is followed by a chapter on the control of electric machines.

Chapter 11 brings together material that was distributed in various chapters in the previous edition. It is now divided into three main sections: control of dc motors, control of synchronous motors, and control of induction motors. A brief fourth section discusses the control of variable-reluctance motors. Each of these main sections begins with a discussion of speed control followed by a discussion of torque control.

Many motor-drive systems are based upon the technique of field-oriented control (also known as vector control). A significant addition to this new edition is the discussion of field-oriented control which now appears in Chapter 11. This is somewhat advanced material which is not typically found in introductory presentations of electric machinery. As a result, the chapter is structured so that this material can be omitted or included at the discretion of the instructor. It first appears in the section on torque control of synchronous motors, in which the basic equations are derived and the analogy with the control of dc machines is discussed. It appears again in its most commonly used form in the section on the torque control of induction motors.

The instructor should note that a complete presentation of field-oriented control requires the use of the dq0 transformation. This transformation, which appeared for synchronous machines in Chapter 6 of the previous edition, is now found in Appendix C of this edition. In addition, the discussion in this appendix has been expanded to include a derivation of the dq0 transformation for induction machines in which both stator and rotor quantities must be transformed.

Although very little in the way of sophisticated mathematics is required of the reader of this book, the mathematics can get somewhat messy and tedious. This is especially true in the analyis of ac machines in which there is a significant amount of algebra involving complex numbers. One of the significant positive developments in the last decade or so is the widespread availability of programs such as MATLAB which greatly facilitate the solution of such problems. MATLAB is widely used in many universities and is available in a student version.[2]

In recognition of this development, this edition incorporates MATLAB in examples and practice problems as well as in end-of-chapter problems. It should be emphasized, though, that the use of MATLAB is not in any way a requirement for the adoption or use of *Electric Machinery*. Rather, it is an enhancement. The book

---

[2] The MATLAB Student Version is published and distributed by The MathWorks, Inc. (http://www.mathworks.com).

now includes interesting examples which would have otherwise been too mathematically tedious. Similarly, there are now end-of-chapter problems which are relatively straightforward when done with MATLAB but which would be quite impractical if done by hand. Note that each MATLAB example and practice problem has been notated with the symbol 🔲, found in the margin of the book. End-of-chapter problems which suggest or require MATLAB are similarly notatated.

It should be emphasized that, in addition to MATLAB, a number of other numerical-analysis packages, including various spread-sheet packages, are available which can be used to perform calculations and to plot in a fashion similar to that done with MATLAB. If MATLAB is not available or is not the package of preference at your institution, instructors and students are encouraged to select any package with which they are comfortable. Any package that simplifies complex calculations and which enables the student to focus on the concepts as opposed to the mathematics will do just fine.

In addition, it should be noted that even in cases where it is not specifically suggested, most of the end-of-chapter problems in the book can be worked using MATLAB or an equivalent program. Thus, students who are comfortable using such tools should be encouraged to do so to save themselves the need to grind through messy calculations by hand. This approach is a logical extension to the use of calculators to facilitate computation. When solving homework problems, the students should still, of course, be required to show on paper how they formulated their solution, since it is the formulation of the solution that is key to understanding the material. However, once a problem is properly formulated, there is typically little additional to be learned from the number crunching itself. The learning process then continues with an examination of the results, both in terms of understanding what they mean with regard to the topic being studied as well as seeing if they make physical sense.

One additional benefit is derived from the introduction of MATLAB into this edition of *Electric Machinery*. As readers of previous editions will be aware, the treatment of single-phase induction motors was never complete in that an analytical treatment of the general case of a single-phase motor running with both its main and auxiliary windings excited (with a capacitor in series with the auxiliary winding) was never considered. In fact, such a treatment of single-phase induction motors is not found in any other introductory electric-machinery textbook of which the author is aware.

The problem is quite simple: this general treatment is mathematically complex, requiring the solution of a number of simultaneous, complex algebraic equations. This, however, is just the sort of problem at which programs such as MATLAB excel. Thus, this new edition of *Electric Machinery* includes this general treatment of single-phase induction machines, complete with a worked out quantitative example and end-of-chapter problems.

It is highly likely that there is simply too much material in this edition of *Electric Machinery* for a single introductory course. However, the material in this edition has been organized so that instructors can pick and choose material appropriate to the topics which they wish to cover. As in the fifth edition, the first two chapters introduce basic concepts of magnetic circuits, magnetic materials, and transformers. The third chapter introduces the basic concept of electromechanical energy conversion. The fourth chapter then provides an overview of and on introduction to the various machine types. Some instructors choose to omit all or most of the material in Chapter 3 from an introductory course. This can be done without a significant impact to the understanding of much of the material in the remainder of the book.

The next five chapters provide a more in-depth discussion of the various machine types: synchronous machines in Chapter 5, induction machines in Chapter 6, dc machines in Chapter 7, variable-reluctance machines in Chapter 8, and single/two-phase machines in Chapter 9. Since the chapters are pretty much independent (with the exception of the material in Chapter 9 which builds upon the polyphase-induction-motor discussion of Chapter 6), the order of these chapters can be changed and/or an instructor can choose to focus on one or two machine types and not to cover the material in all five of these chapters.

The introductory power-electronics discussion of Chapter 10 is pretty much stand-alone. Instructors who wish to introduce this material should be able to do so at their discretion; there is no need to present it in a course in the order that it is found in the book. In addition, it is not required for an understanding of the electric-machinery material presented in the book, and instructors who elect to cover this material in a separate course will not find themselves handicapped in any way by doing so.

Finally, instructors may wish to select topics from the control material of Chapter 11 rather than include it all. The material on speed control is essentially a relatively straightforward extension of the material found in earlier chapters on the individual machine types. The material on field-oriented control requires a somewhat more sophisticated understanding and builds upon the dq0 transformation found in Appendix C. It would certainly be reasonable to omit this material in an introductory course and to delay it for a more advanced course where sufficient time is available to devote to it.

McGraw-Hill has set up a website, **www.mhhe.com/umans**, to support this new edition of *Electric Machinery*. The website will include a downloadable version of the solutions manual (for instructors only) as well as PowerPoint slides of figures from the book. This being a new feature of *Electric Machinery,* we are, to a great extent, starting with a blank slate and will be exploring different options for supplementing and enhancing the text. For example, in recognition of the fact that instructors are always looking for new examples and problems, we will set up a mechanism so that instructors can submit examples and problems for publication on the website (with credit given to their authors) which then can be shared with other instructors.

We are also considering setting up a section of the website devoted to MATLAB and other numerical analysis packages. For users of MATLAB, the site might contain hints and suggestions for applying MATLAB to *Electric Machinery* as well as perhaps some Simulink[3] examples for instructors who wish to introduce simulations into their courses. Similarly, instructors who use packages other than MATLAB might

[3] Simulink is a registered trademark of The MathWorks, Inc.

want to submit their suggestions and experiences to share with other users. In this context, the website would appear again to be an ideal resource for enhancing interaction between instructors.

Clearly, the website will be a living document which will evolve in response to input from users. I strongly urge each of you to visit it frequently and to send in suggestions, problems, and examples, and comments. I fully expect it to become a valuable resource for users of *Electric Machinery* around the world.

Professor Kingsley first asked this author to participate in the fourth edition of *Electric Machinery;* the professor was actively involved in that edition. He participated in an advisory capacity for the fifth edition. Unfortunately, Professor Kingsley passed away since the publication of the fifth edition and did not live to see the start of the work on this edition. He was a fine gentleman, a valued teacher and friend, and he is missed.

I wish to thank a number of my colleagues for their insight and helpful discussions during the production of this edition. My friend, Professor Jeffrey Lang, who also provided invaluable insight and advice in the discussion of variable-reluctance machines which first appeared in the fifth edition, was extremely helpful in formulating the presentations of power electronics and field-oriented control which appear in this edition. Similarly, Professor Gerald Wilson, who served as my graduate thesis advisor, has been a friend and colleague throughout my career and has been a constant source of valuable advice and insight.

On a more personal note, I would like to express my love for my wife Denise and our children Dalya and Ari and to thank them for putting up with the many hours of my otherwise spare time that this edition required. I promised the kids that I would read the Harry Potter books when work on this edition of *Electric Machinery* was completed and I had better get to it! In addition, I would like to recognize my life-long friend David Gardner who watched the work on this edition with interest but who did not live to see it completed. A remarkable man, he passed away due to complications from muscular dystrophy just a short while before the final draft was completed.

Finally, I wish to thank the reviewers who participated in this project and whose comments and suggestions played a valuable role in the final form of this edition. These include Professors:

Ravel F. Ammerman, *Colorado School of Mines*
Juan Carlos Balda, *University of Arkansas, Fayetteville*
Miroslav Begovic, *Georgia Institute of Technology*
Prasad Enjeti, *Texas A&M University*
Vernold K. Feiste, *Southern Illinois University*
Thomas G. Habetler, *Georgia Institute of Technology*
Steven Hietpas, *South Dakota State University*
Heath Hofmann, *Pennsylvania State University*
Daniel Hutchins, *U.S. Naval Academy*
Roger King, *University of Toledo*

Alexander E. Koutras, *California Polytechnic State University, Pomona*

Bruno Osorno, *California State University, Northridge*

Henk Polinder, *Delft University of Technology*

Gill Richards, *Arkansas Tech University*

Duane F. Rost, *Youngstown State University*

Melvin Sandler, *The Cooper Union*

Ali O. Shaban, *California Polytechnic State University, San Luis Obispo*

Alan Wallace, *Oregon State University*

I would like to specifically acknowledge Professor Ibrahim Abdel-Moneim Abdel-Halim of Zagazig University, whose considerable effort found numerous typos and numerical errors in the draft document.

Stephen D. Umans
Cambridge, MA
March 5, 2002

1

Magnetic Circuits and Magnetic Materials

T
he objective of this book is to study the devices used in the interconversion of electric and mechanical energy. Emphasis is placed on electromagnetic rotating machinery, by means of which the bulk of this energy conversion takes place. However, the techniques developed are generally applicable to a wide range of additional devices including linear machines, actuators, and sensors.

Although not an electromechanical-energy-conversion device, the transformer is an important component of the overall energy-conversion process and is discussed in Chapter 2. The techniques developed for transformer analysis form the basis for the ensuing discussion of electric machinery.

Practically all transformers and electric machinery use ferro-magnetic material for shaping and directing the magnetic fields which act as the medium for transferring and converting energy. Permanent-magnet materials are also widely used. Without these materials, practical implementations of most familiar electromechanical-energy-conversion devices would not be possible. The ability to analyze and describe systems containing these materials is essential for designing and understanding these devices.

This chapter will develop some basic tools for the analysis of magnetic field systems and will provide a brief introduction to the properties of practical magnetic materials. In Chapter 2, these results will then be applied to the analysis of transformers. In later chapters they will be used in the analysis of rotating machinery.

In this book it is assumed that the reader has basic knowledge of magnetic and electric field theory such as given in a basic physics course for engineering students. Some readers may have had a course on electromagnetic field theory based on Maxwell's equations, but an in-depth understanding of Maxwell's equations is not a prerequisite for study of this book. The techniques of magnetic-circuit analysis, which represent algebraic approximations to exact field-theory solutions, are widely used in the study of electromechanical-energy-conversion devices and form the basis for most of the analyses presented here.

1.1   INTRODUCTION TO MAGNETIC CIRCUITS

The complete, detailed solution for magnetic fields in most situations of practical engineering interest involves the solution of Maxwell's equations along with various constitutive relationships which describe material properties. Although in practice exact solutions are often unattainable, various simplifying assumptions permit the attainment of useful engineering solutions.[1]

We begin with the assumption that, for the systems treated in this book, the frequencies and sizes involved are such that the displacement-current term in Maxwell's equations can be neglected. This term accounts for magnetic fields being produced in space by time-varying electric fields and is associated with electromagnetic radiation. Neglecting this term results in the magneto-quasistatic form of the relevant Maxwell's equations which relate magnetic fields to the currents which produce them.

$$\oint_C \mathbf{H} \, \mathbf{dl} = \int_S \mathbf{J} \cdot \mathbf{da} \tag{1.1}$$

$$\oint_S \mathbf{B} \cdot \mathbf{da} = 0 \tag{1.2}$$

Equation 1.1 states that the line integral of the tangential component of the *magnetic field intensity* $\mathbf{H}$ around a closed contour C is equal to the total current passing through any surface S linking that contour. From Eq. 1.1 we see that the source of $\mathbf{H}$ is the *current density* $\mathbf{J}$. Equation 1.2 states that the *magnetic flux density* $\mathbf{B}$ is conserved, i.e., that no net flux enters or leaves a closed surface (this is equivalent to saying that there exist no monopole charge sources of magnetic fields). From these equations we see that the magnetic field quantities can be determined solely from the instantaneous values of the source currents and that time variations of the magnetic fields follow directly from time variations of the sources.

A second simplifying assumption involves the concept of the *magnetic circuit*. The general solution for the magnetic field intensity $\mathbf{H}$ and the magnetic flux density $\mathbf{B}$ in a structure of complex geometry is extremely difficult. However, a three-dimensional field problem can often be reduced to what is essentially a one-dimensional circuit equivalent, yielding solutions of acceptable engineering accuracy.

A magnetic circuit consists of a structure composed for the most part of high-permeability magnetic material. The presence of high-permeability material tends to cause magnetic flux to be confined to the paths defined by the structure, much as currents are confined to the conductors of an electric circuit. Use of this concept of

---

[1] Although exact analytical solutions cannot be obtained, computer-based numerical solutions (the finite-element and boundary-element methods form the basis for a number of commercial programs) are quite common and have become indespensible tools for analysis and design. However, such techniques are best used to refine analyses based upon analytical techniques such as are found in this book. Their use contributes little to a fundamental understanding of the principles and basic performance of electric machines and as a result they will not be discussed in this book.

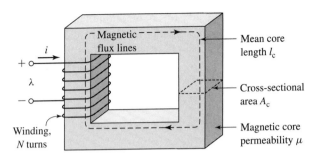

Figure 1.1 Simple magnetic circuit.

the magnetic circuit is illustrated in this section and will be seen to apply quite well to many situations in this book.[2]

A simple example of a magnetic circuit is shown in Fig. 1.1. The core is assumed to be composed of magnetic material whose permeability is much greater than that of the surrounding air ($\mu \gg \mu_0$). The core is of uniform cross section and is excited by a winding of N turns carrying a current of i amperes. This winding produces a magnetic field in the core, as shown in the figure.

Because of the high permeability of the magnetic core, an exact solution would show that the magnetic flux is confined almost entirely to the core, the field lines follow the path defined by the core, and the flux density is essentially uniform over a cross section because the cross-sectional area is uniform. The magnetic field can be visualized in terms of flux lines which form closed loops interlinked with the winding.

As applied to the magnetic circuit of Fig. 1.1, the source of the magnetic field in the core is the ampere-turn product Ni. In magnetic circuit terminology Ni is the *magnetomotive force* (mmf) $\mathcal{F}$ acting on the magnetic circuit. Although Fig. 1.1 shows only a single coil, transformers and most rotating machines have at least two windings, and Ni must be replaced by the algebraic sum of the ampere-turns of all the windings.

The *magnetic flux* ϕ crossing a surface S is the surface integral of the normal component of **B**; thus

$$\phi = \int_S \mathbf{B} \cdot \mathbf{da} \tag{1.3}$$

In SI units, the unit of ϕ is the *weber* (Wb).

Equation 1.2 states that the net magnetic flux entering or leaving a closed surface (equal to the surface integral of **B** over that closed surface) is zero. This is equivalent to saying that all the flux which enters the surface enclosing a volume must leave that volume over some other portion of that surface because magnetic flux lines form closed loops.

---

[2] For a more extensive treatment of magnetic circuits see A. E. Fitzgerald, D. E. Higgenbotham, and A. Grabel, *Basic Electrical Engineering*, 5th ed., McGraw-Hill, 1981, chap. 13; also E. E. Staff, M.I.T., *Magnetic Circuits and Transformers*, M.I.T. Press, 1965, chaps. 1 to 3.

These facts can be used to justify the assumption that the magnetic flux density is uniform across the cross section of a magnetic circuit such as the core of Fig. 1.1. In this case Eq. 1.3 reduces to the simple scalar equation

$$\phi_c = B_c A_c \tag{1.4}$$

where ϕ_c = flux in core

$\quad B_c$ = flux density in core

$\quad A_c$ = cross-sectional area of core

From Eq. 1.1, the relationship between the mmf acting on a magnetic circuit and the magnetic field intensity in that circuit is.[3]

$$\mathcal{F} = Ni = \oint \mathbf{H} \mathbf{dl} \tag{1.5}$$

The core dimensions are such that the path length of any flux line is close to the mean core length l_c. As a result, the line integral of Eq. 1.5 becomes simply the scalar product $H_c l_c$ of the magnitude of $\mathbf{H}$ and the mean flux path length l_c. Thus, the relationship between the mmf and the magnetic field intensity can be written in magnetic circuit terminology as

$$\mathcal{F} = Ni = H_c l_c \tag{1.6}$$

where H_c is average magnitude of $\mathbf{H}$ in the core.

The direction of H_c in the core can be found from the *right-hand rule,* which can be stated in two equivalent ways. (1) Imagine a current-carrying conductor held in the right hand with the thumb pointing in the direction of current flow; the fingers then point in the direction of the magnetic field created by that current. (2) Equivalently, if the coil in Fig. 1.1 is grasped in the right hand (figuratively speaking) with the fingers pointing in the direction of the current, the thumb will point in the direction of the magnetic fields.

The relationship between the magnetic field intensity $\mathbf{H}$ and the magnetic flux density $\mathbf{B}$ is a property of the material in which the field exists. It is common to assume a linear relationship; thus

$$\mathbf{B} = \mu \mathbf{H} \tag{1.7}$$

where μ is known as the *magnetic permeability*. In SI units, $\mathbf{H}$ is measured in units of *amperes per meter,* $\mathbf{B}$ is in *webers per square meter,* also known as *teslas* (T), and μ is in *webers per ampere-turn-meter,* or equivalently *henrys per meter.* In SI units the permeability of free space is $\mu_0 = 4\pi \times 10^{-7}$ henrys per meter. The permeability of linear magnetic material can be expressed in terms of μ_r, its value relative to that of free space, or $\mu = \mu_r \mu_0$. Typical values of μ_r range from 2000 to 80,000 for materials used

---

[3] In general, the mmf drop across any segment of a magnetic circuit can be calculated as $\int \mathbf{H} \mathbf{dl}$ over that portion of the magnetic circuit.

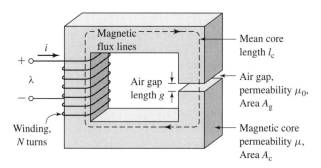

Figure 1.2 Magnetic circuit with air gap.

in transformers and rotating machines. The characteristics of ferromagnetic materials are described in Sections 1.3 and 1.4. For the present we assume that μ_r is a known constant, although it actually varies appreciably with the magnitude of the magnetic flux density.

Transformers are wound on closed cores like that of Fig. 1.1. However, energy conversion devices which incorporate a moving element must have air gaps in their magnetic circuits. A magnetic circuit with an air gap is shown in Fig. 1.2. When the air-gap length g is much smaller than the dimensions of the adjacent core faces, the magnetic flux ϕ will follow the path defined by the core and the air gap and the techniques of magnetic-circuit analysis can be used. If the air-gap length becomes excessively large, the flux will be observed to "leak out" of the sides of the air gap and the techniques of magnetic-circuit analysis will no longer be strictly applicable.

Thus, provided the air-gap length g is sufficiently small, the configuration of Fig. 1.2 can be analyzed as a magnetic circuit with two series components: a magnetic core of permeability μ, cross-sectional area A_c, and mean length l_c, and an air gap of permeability μ_0, cross-sectional area A_g, and length g. In the core the flux density can be assumed uniform; thus

$$B_c = \frac{\phi}{A_c} \tag{1.8}$$

and in the air gap

$$B_g = \frac{\phi}{A_g} \tag{1.9}$$

where ϕ = the flux in the magnetic circuit.

Application of Eq. 1.5 to this magnetic circuit yields

$$\mathcal{F} = H_c l_c + H_g g \tag{1.10}$$

and using the linear B-H relationship of Eq. 1.7 gives

$$\mathcal{F} = \frac{B_c}{\mu} l_c + \frac{B_g}{\mu_0} g \tag{1.11}$$

Here the $\mathcal{F} = Ni$ is the mmf applied to the magnetic circuit. From Eq. 1.10 we see that a portion of the mmf, $\mathcal{F}_c = H_c l_c$, is required to produce magnetic field in the core while the remainder, $\mathcal{F}_g = H_g g$, produces magnetic field in the air gap.

For practical magnetic materials (as is discussed in Sections 1.3 and 1.4), B_c and H_c are not simply related by a known constant permeability μ as described by Eq. 1.7. In fact, B_c is often a nonlinear, multivalued function of H_c. Thus, although Eq. 1.10 continues to hold, it does not lead directly to a simple expression relating the mmf and the flux densities, such as that of Eq. 1.11. Instead the specifics of the nonlinear B_c-H_c relation must be used, either graphically or analytically. However, in many cases, the concept of constant material permeability gives results of acceptable engineering accuracy and is frequently used.

From Eqs. 1.8 and 1.9, Eq. 1.11 can be rewritten in terms of the total flux ϕ as

$$\mathcal{F} = \phi \left(\frac{l_c}{\mu A_c} + \frac{g}{\mu_0 A_g} \right) \tag{1.12}$$

The terms that multiply the flux in this equation are known as the *reluctance* $\mathcal{R}$ of the core and air gap, respectively,

$$\mathcal{R}_c = \frac{l_c}{\mu A_c} \tag{1.13}$$

$$\mathcal{R}_g = \frac{g}{\mu_0 A_g} \tag{1.14}$$

and thus

$$\mathcal{F} = \phi(\mathcal{R}_c + \mathcal{R}_g) \tag{1.15}$$

Finally, Eq. 1.15 can be inverted to solve for the flux

$$\phi = \frac{\mathcal{F}}{\mathcal{R}_c + \mathcal{R}_g} \tag{1.16}$$

or

$$\phi = \frac{\mathcal{F}}{\frac{l_c}{\mu A_c} + \frac{g}{\mu_0 A_g}} \tag{1.17}$$

In general, for any magnetic circuit of total reluctance $\mathcal{R}_{tot}$, the flux can be found as

$$\phi = \frac{\mathcal{F}}{\mathcal{R}_{tot}} \tag{1.18}$$

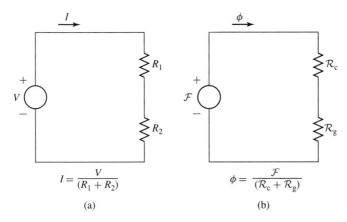

Figure 1.3 Analogy between electric and magnetic circuits.
(a) Electric circuit, (b) magnetic circuit.

The term which multiplies the mmf is known as the *permeance* P and is the inverse of the reluctance; thus, for example, the total permeance of a magnetic circuit is

$$P_{\text{tot}} = \frac{1}{\mathcal{R}_{\text{tot}}} \tag{1.19}$$

Note that Eqs. 1.15 and 1.16 are analogous to the relationships between the current and voltage in an electric circuit. This analogy is illustrated in Fig. 1.3. Figure 1.3a shows an electric circuit in which a voltage V drives a current I through resistors R_1 and R_2. Figure 1.3b shows the schematic equivalent representation of the magnetic circuit of Fig. 1.2. Here we see that the mmf $\mathcal{F}$ (analogous to voltage in the electric circuit) drives a flux ϕ (analogous to the current in the electric circuit) through the combination of the reluctances of the core $\mathcal{R}_c$ and the air gap $\mathcal{R}_g$. This analogy between the solution of electric and magnetic circuits can often be exploited to produce simple solutions for the fluxes in magnetic circuits of considerable complexity.

The fraction of the mmf required to drive flux through each portion of the magnetic circuit, commonly referred to as the *mmf drop* across that portion of the magnetic circuit, varies in proportion to its reluctance (directly analogous to the voltage drop across a resistive element in an electric circuit). From Eq. 1.13 we see that high material permeability can result in low core reluctance, which can often be made much smaller than that of the air gap; i.e., for $(\mu A_c / l_c) \gg (\mu_0 A_g / g)$, $\mathcal{R}_c \ll \mathcal{R}_g$ and thus $\mathcal{R}_{\text{tot}} \approx \mathcal{R}_g$. In this case, the reluctance of the core can be neglected and the flux and hence B can be found from Eq. 1.16 in terms of $\mathcal{F}$ and the air-gap properties alone:

$$\phi \approx \frac{\mathcal{F}}{\mathcal{R}_g} = \frac{\mathcal{F}\mu_0 A_g}{g} = Ni \frac{\mu_0 A_g}{g} \tag{1.20}$$

As will be seen in Section 1.3, practical magnetic materials have permeabilities which are not constant but vary with the flux level. From Eqs. 1.13 to 1.16 we see that as

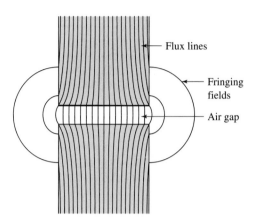

Figure 1.4   Air-gap fringing fields.

long as this permeability remains sufficiently large, its variation will not significantly affect the performance of the magnetic circuit.

In practical systems, the magnetic field lines "fringe" outward somewhat as they cross the air gap, as illustrated in Fig. 1.4. Provided this fringing effect is not excessive, the magnetic-circuit concept remains applicable. The effect of these *fringing fields* is to increase the effective cross-sectional area A_g of the air gap. Various empirical methods have been developed to account for this effect. A correction for such fringing fields in short air gaps can be made by adding the gap length to each of the two dimensions making up its cross-sectional area. In this book the effect of fringing fields is usually ignored. If fringing is neglected, $A_g = A_c$.

In general, magnetic circuits can consist of multiple elements in series and parallel. To complete the analogy between electric and magnetic circuits, we can generalize Eq. 1.5 as

$$\mathcal{F} = \oint \mathbf{H} \mathbf{dl} = \sum_k \mathcal{F}_k = \sum_k H_k l_k \tag{1.21}$$

where $\mathcal{F}$ is the mmf (total ampere-turns) acting to drive flux through a closed loop of a magnetic circuit,

$$\mathcal{F} = \int_S \mathbf{J} \cdot \mathbf{da} \tag{1.22}$$

and $\mathcal{F}_k = H_k l_k$ is the *mmf drop* across the k'th element of that loop. This is directly analogous to Kirchoff's voltage law for electric circuits consisting of voltage sources and resistors

$$V = \sum_k R_k i_k \tag{1.23}$$

where V is the source voltage driving current around a loop and $R_k i_k$ is the voltage drop across the k'th resistive element of that loop.

Similarly, the analogy to Kirchoff's current law

$$\sum_n i_n = 0 \tag{1.24}$$

which says that the sum of currents into a node in an electric circuit equals zero is

$$\sum_n \phi_n = 0 \tag{1.25}$$

which states that the sum of the flux into a node in a magnetic circuit is zero.

We have now described the basic principles for reducing a magneto-quasistatic field problem with simple geometry to a *magnetic circuit model*. Our limited purpose in this section is to introduce some of the concepts and terminology used by engineers in solving practical design problems. We must emphasize that this type of thinking depends quite heavily on engineering judgment and intuition. For example, we have tacitly assumed that the permeability of the "iron" parts of the magnetic circuit is a constant known quantity, although this is not true in general (see Section 1.3), and that the magnetic field is confined soley to the core and its air gaps. Although this is a good assumption in many situations, it is also true that the winding currents produce magnetic fields outside the core. As we shall see, when two or more windings are placed on a magnetic circuit, as happens in the case of both transformers and rotating machines, these fields outside the core, which are referred to as *leakage fields,* cannot be ignored and significantly affect the performance of the device.

EXAMPLE 1.1

The magnetic circuit shown in Fig. 1.2 has dimensions $A_c = A_g = 9 \text{ cm}^2$, $g = 0.050 \text{ cm}$, $l_c = 30 \text{ cm}$, and $N = 500$ turns. Assume the value $\mu_r = 70{,}000$ for core material. (*a*) Find the reluctances $\mathcal{R}_c$ and $\mathcal{R}_g$. For the condition that the magnetic circuit is operating with $B_c = 1.0 \text{ T}$, find (*b*) the flux ϕ and (*c*) the current i.

■ **Solution**

a. The reluctances can be found from Eqs. 1.13 and 1.14:

$$\mathcal{R}_c = \frac{l_c}{\mu_r \mu_0 A_c} = \frac{0.3}{70{,}000 \, (4\pi \times 10^{-7})(9 \times 10^{-4})} = 3.79 \times 10^3 \quad \frac{\text{A} \cdot \text{turns}}{\text{Wb}}$$

$$\mathcal{R}_g = \frac{g}{\mu_0 A_g} = \frac{5 \times 10^{-4}}{(4\pi \times 10^{-7})(9 \times 10^{-4})} = 4.42 \times 10^5 \quad \frac{\text{A} \cdot \text{turns}}{\text{Wb}}$$

b. From Eq. 1.4,

$$\phi = B_c A_c = 1.0(9 \times 10^{-4}) = 9 \times 10^{-4} \text{ Wb}$$

c. From Eqs. 1.6 and 1.15,

$$i = \frac{\mathcal{F}}{N} = \frac{\phi(\mathcal{R}_c + \mathcal{R}_g)}{N} = \frac{9 \times 10^{-4}(4.46 \times 10^5)}{500} = 0.80 \text{ A}$$

Practice Problem 1.1

Find the flux ϕ and current for Example 1.1 if (a) the number of turns is doubled to $N = 1000$ turns while the circuit dimensions remain the same and (b) if the number of turns is equal to $N = 500$ and the gap is reduced to 0.040 cm.

Solution

    a. $\phi = 9 \times 10^{-4}$ Wb and $i = 0.40$ A
    b. $\phi = 9 \times 10^{-4}$ Wb and $i = 0.64$ A

EXAMPLE 1.2

The magnetic structure of a synchronous machine is shown schematically in Fig. 1.5. Assuming that rotor and stator iron have infinite permeability ($\mu \rightarrow \infty$), find the air-gap flux ϕ and flux density B_g. For this example $I = 10$ A, $N = 1000$ turns, $g = 1$ cm, and $A_g = 2000$ cm^2.

■ Solution

Notice that there are two air gaps in series, of total length $2g$, and that by symmetry the flux density in each is equal. Since the iron permeability here is assumed to be infinite, its reluctance is negligible and Eq. 1.20 (with g replaced by the total gap length $2g$) can be used to find the flux

$$\phi = \frac{NI\mu_0 A_g}{2g} = \frac{1000(10)(4\pi \times 10^{-7})(0.2)}{0.02} = 0.13 \text{ Wb}$$

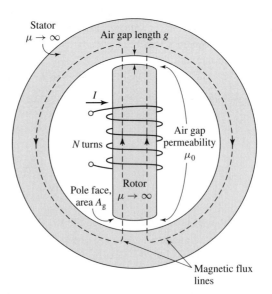

Stator
$\mu \rightarrow \infty$

Air gap length g

I

N turns

Air gap permeability
μ_0

Pole face, area A_g

Rotor
$\mu \rightarrow \infty$

Magnetic flux lines

Figure 1.5   Simple synchronous machine.

and

$$B_g = \frac{\phi}{A_g} = \frac{0.13}{0.2} = 0.65 \text{ T}$$

For the magnetic structure of Fig. 1.5 with the dimensions as given in Example 1.2, the air-gap flux density is observed to be $B_g = 0.9$ T. Find the air-gap flux ϕ and, for a coil of $N = 500$ turns, the current required to produce this level of air-gap flux.

Solution

$\phi = 0.18$ Wb and $i = 28.6$ A.

1.2 FLUX LINKAGE, INDUCTANCE, AND ENERGY

When a magnetic field varies with time, an electric field is produced in space as determined by *Faraday's law:*

$$\oint_C \mathbf{E} \cdot \mathbf{ds} = -\frac{d}{dt} \int_S \mathbf{B} \cdot \mathbf{da} \tag{1.26}$$

Equation 1.26 states that the line integral of the *electric field intensity* **E** around a closed contour C is equal to the time rate of change of the magnetic flux linking (i.e. passing through) that contour. In magnetic structures with windings of high electrical conductivity, such as in Fig. 1.2, it can be shown that the **E** field in the wire is extremely small and can be neglected, so that the left-hand side of Eq. 1.26 reduces to the negative of the *induced voltage*[4] e at the winding terminals. In addition, the flux on the right-hand side of Eq. 1.26 is dominated by the core flux ϕ. Since the winding (and hence the contour C) links the core flux N times, Eq. 1.26 reduces to

$$e = N\frac{d\varphi}{dt} = \frac{d\lambda}{dt} \tag{1.27}$$

where λ is the *flux linkage* of the winding and is defined as

$$\lambda = N\varphi \tag{1.28}$$

Flux linkage is measured in units of webers (or equivalently weber-turns). The symbol φ is used to indicate the instantaneous value of a time-varying flux.

In general the flux linkage of a coil is equal to the surface integral of the normal component of the magnetic flux density integrated over any surface spanned by that coil. Note that the direction of the induced voltage e is defined by Eq. 1.26 so that if

---

[4] The term *electromotive force* (emf) is often used instead of *induced voltage* to represent that component of voltage due to a time-varying flux linkage.

the winding terminals were short-circuited, a current would flow in such a direction as to oppose the change of flux linkage.

For a magnetic circuit composed of magnetic material of constant magnetic permeability or which includes a dominating air gap, the relationship between ϕ and i will be linear and we can define the *inductance L* as

$$L = \frac{\lambda}{i} \tag{1.29}$$

Substitution of Eqs. 1.5, 1.18 and 1.28 into Eq. 1.29 gives

$$L = \frac{N^2}{\mathcal{R}_{\text{tot}}} \tag{1.30}$$

From which we see that the inductance of a winding in a magnetic circuit is proportional to the square of the turns and inversely proportional to the reluctance of the magnetic circuit associated with that winding.

For example, from Eq. 1.20, under the assumption that the reluctance of the core is negligible as compared to that of the air gap, the inductance of the winding in Fig. 1.2 is equal to

$$L = \frac{N^2}{(g/\mu_0 A_g)} = \frac{N^2 \mu_0 A_g}{g} \tag{1.31}$$

Inductance is measured in *henrys* (H) or *weber-turns per ampere*. Equation 1.31 shows the dimensional form of expressions for inductance; inductance is proportional to the square of the number of turns, to a magnetic permeability, and to a cross-sectional area and is inversely proportional to a length. It must be emphasized that, strictly speaking, the concept of inductance requires a linear relationship between flux and mmf. Thus, it cannot be rigorously applied in situations where the nonlinear characteristics of magnetic materials, as is discussed in Sections 1.3 and 1.4, dominate the performance of the magnetic system. However, in many situations of practical interest, the reluctance of the system is dominated by that of an air gap (which is of course linear) and the nonlinear effects of the magnetic material can be ignored. In other cases it may be perfectly acceptable to assume an average value of magnetic permeability for the core material and to calculate a corresponding average inductance which can be used for calculations of reasonable engineering accuracy. Example 1.3 illustrates the former situation and Example 1.4 the latter.

EXAMPLE 1.3

The magnetic circuit of Fig. 1.6a consists of an N-turn winding on a magnetic core of infinite permeability with two parallel air gaps of lengths g_1 and g_2 and areas A_1 and A_2, respectively.

Find (*a*) the inductance of the winding and (*b*) the flux density B_1 in gap 1 when the winding is carrying a current i. Neglect fringing effects at the air gap.

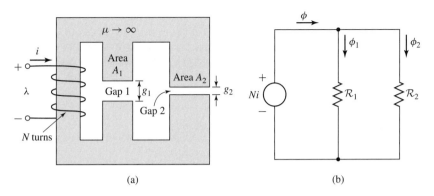

Figure 1.6 (a) Magnetic circuit and (b) equivalent circuit for Example 1.3.

■ **Solution**

a. The equivalent circuit of Fig. 1.6b shows that the total reluctance is equal to the parallel combination of the two gap reluctances. Thus

$$\phi = \frac{Ni}{\frac{\mathcal{R}_1 \mathcal{R}_2}{\mathcal{R}_1 + \mathcal{R}_2}}$$

where

$$\mathcal{R}_1 = \frac{g_1}{\mu_0 A_1} \qquad\qquad \mathcal{R}_2 = \frac{g_2}{\mu_0 A_2}$$

From Eq. 1.29,

$$L = \frac{\lambda}{i} = \frac{N\phi}{i} = \frac{N^2(\mathcal{R}_1 + \mathcal{R}_2)}{\mathcal{R}_1 \mathcal{R}_2}$$

$$= \mu_0 N^2 \left(\frac{A_1}{g_1} + \frac{A_2}{g_2} \right)$$

b. From the equivalent circuit, one can see that

$$\phi_1 = \frac{Ni}{\mathcal{R}_1} = \frac{\mu_0 A_1 Ni}{g_1}$$

and thus

$$B_1 = \frac{\phi_1}{A_1} = \frac{\mu_0 Ni}{g_1}$$

EXAMPLE 1.4

In Example 1.1, the relative permeability of the core material for the magnetic circuit of Fig. 1.2 is assumed to be $\mu_r = 70,000$ at a flux density of 1.0 T.

a. For this value of μ_r calculate the inductance of the winding.
b. In a practical device, the core would be constructed from electrical steel such as M-5

electrical steel which is discussed in Section 1.3. This material is highly nonlinear and its relative permeability (defined for the purposes of this example as the ratio B/H) varies from a value of approximately $\mu_r = 72{,}300$ at a flux density of $B = 1.0$ T to a value of on the order of $\mu_r = 2900$ as the flux density is raised to 1.8 T. (a) Calculate the inductance under the assumption that the relative permeability of the core steel is 72,300. (b) Calculate the inductance under the assumption that the relative permeability is equal to 2900.

■ Solution

a. From Eqs. 1.13 and 1.14 and based upon the dimensions given in Example 1.1,

$$\mathcal{R}_c = \frac{l_c}{\mu_r \mu_0 A_c} = \frac{0.3}{72{,}300\,(4\pi \times 10^{-7})(9 \times 10^{-4})} = 3.67 \times 10^3 \;\frac{\text{A} \cdot \text{turns}}{\text{Wb}}$$

while $\mathcal{R}_g$ remains unchanged from the value calculated in Example 1.1 as $\mathcal{R}_g = 4.42 \times 10^5$ A · turns / Wb.

Thus the total reluctance of the core and gap is

$$\mathcal{R}_{tot} = \mathcal{R}_c + \mathcal{R}_g = 4.46 \times 10^5 \;\frac{\text{A} \cdot \text{turns}}{\text{Wb}}$$

and hence from Eq. 1.30

$$L = \frac{N^2}{\mathcal{R}_{tot}} = \frac{500^2}{4.46 \times 10^5} = 0.561\ \text{H}$$

b. For $\mu_r = 2900$, the reluctance of the core increases from a value of 3.79×10^3 A · turns / Wb to a value of

$$\mathcal{R}_c = \frac{l_c}{\mu_r \mu_0 A_c} = \frac{0.3}{2900\,(4\pi \times 10^{-7})(9 \times 10^{-4})} = 9.15 \times 10^4 \;\frac{\text{A} \cdot \text{turns}}{\text{Wb}}$$

and hence the total reluctance increases from 4.46×10^5 A · turns / Wb to 5.34×10^5 A · turns / Wb. Thus from Eq. 1.30 the inductance decreases from 0.561 H to

$$L = \frac{N^2}{\mathcal{R}_{tot}} = \frac{500^2}{5.34 \times 10^5} = 0.468\ \text{H}$$

This example illustrates the linearizing effect of a dominating air gap in a magnetic circuit. In spite of a reduction in the permeablity of the iron by a factor of $72{,}300/2900 = 25$, the inductance decreases only by a factor of $0.468/0.561 = 0.83$ simply because the reluctance of the air gap is significantly larger than that of the core. In many situations, it is common to assume the inductance to be constant at a value corresponding to a finite, constant value of core permeability (or in many cases it is assumed simply that $\mu_r \to \infty$). Analyses based upon such a representation for the inductor will often lead to results which are well within the range of acceptable engineering accuracy and which avoid the immense complication associated with modeling the nonlinearity of the core material.

Repeat the inductance calculation of Example 1.4 for a relative permeability $\mu_r = 30,000$.

Solution

$L = 0.554$ H

EXAMPLE 1.5

Using MATLAB,[†] plot the inductance of the magnetic circuit of Example 1.1 and Fig. 1.2 as a function of core permeability over the range $100 \le \mu_r \le 100,000$.

■ Solution

Here is the MATLAB script:

```
clc
clear

% Permeability of free space
mu0 = pi*4.e-7;

%All dimensions expressed in meters
Ac = 9e-4; Ag = 9e-4; g = 5e-4; lc = 0.3;
N = 500;

%Reluctance of air gap
Rg = g/(mu0*Ag);

for n = 1:101
mur(n) = 100 + (100000 - 100)*(n-1)/100;
%Reluctance of core
Rc(n) = lc/(mur(n)*mu0*Ac);
Rtot = Rg+Rc(n);
%Inductance
L(n) = N^2/Rtot;
end

plot(mur,L)
xlabel('Core relative permeability')
ylabel('Inductance [H]')
```

The resultant plot is shown in Fig. 1.7. Note that the figure clearly confirms that, for the magnetic circuit of this example, the inductance is quite insensitive to relative permeability until the relative permeability drops to on the order of 1000. Thus, as long as the effective relative permeability of the core is "large" (in this case greater than 1000), any nonlinearities in the properties of the core material will have little effect on the terminal properties of the inductor.

[†] MATLAB is a registered trademark of The MathWorks, Inc.

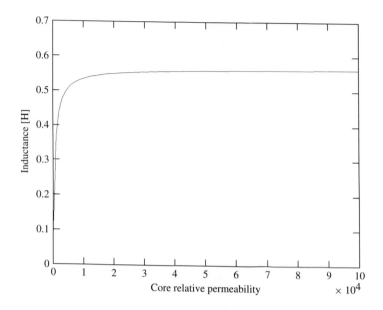

Figure 1.7   MATLAB plot of inductance vs. relative permeability for Example 1.5.

Practice Problem 1.4

Write a MATLAB script to plot the inductance of the magnetic circuit of Example 1.1 with $\mu_r = 70{,}000$ as a function of air-gap length as the the air-gap is varied from 0.01 cm to 0.10 cm.

Figure 1.8 shows a magnetic circuit with an air gap and two windings. In this case note that the mmf acting on the magnetic circuit is given by the *total ampere-turns* acting on the magnetic circuit (i.e., the net ampere turns of both windings) and that the reference directions for the currents have been chosen to produce flux in the same direction. The total mmf is therefore

$$\mathcal{F} = N_1 i_1 + N_2 i_2 \tag{1.32}$$

and from Eq. 1.20, with the reluctance of the core neglected and assuming that $A_c = A_g$, the core flux ϕ is

$$\phi = (N_1 i_1 + N_2 i_2)\frac{\mu_0 A_c}{g} \tag{1.33}$$

In Eq. 1.33, ϕ is the *resultant core flux* produced by the total mmf of the two windings. It is this resultant ϕ which determines the operating point of the core material.

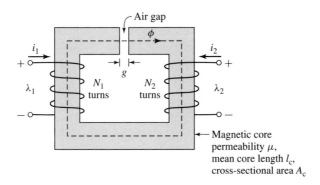

Figure 1.8 Magnetic circuit with two windings.

If Eq. 1.33 is broken up into terms attributable to the individual currents, the resultant flux linkages of coil 1 can be expressed as

$$\lambda_1 = N_1\phi = N_1{}^2\left(\frac{\mu_0 A_c}{g}\right)i_1 + N_1 N_2\left(\frac{\mu_0 A_c}{g}\right)i_2 \qquad (1.34)$$

which can be written

$$\lambda_1 = L_{11}i_1 + L_{12}i_2 \qquad (1.35)$$

where

$$L_{11} = N_1{}^2\frac{\mu_0 A_c}{g} \qquad (1.36)$$

is the *self-inductance* of coil 1 and $L_{11}i_1$ is the flux linkage of coil 1 due to its own current i_1. The *mutual inductance* between coils 1 and 2 is

$$L_{12} = N_1 N_2\frac{\mu_0 A_c}{g} \qquad (1.37)$$

and $L_{12}i_2$ is the flux linkage of coil 1 due to current i_2 in the other coil. Similarly, the flux linkage of coil 2 is

$$\lambda_2 = N_2\phi = N_1 N_2\left(\frac{\mu_0 A_c}{g}\right)i_1 + N_2{}^2\left(\frac{\mu_0 A_c}{g}\right)i_2 \qquad (1.38)$$

or

$$\lambda_2 = L_{21}i_1 + L_{22}i_2 \qquad (1.39)$$

where $L_{21} = L_{12}$ is the mutual inductance and

$$L_{22} = N_2{}^2\frac{\mu_0 A_c}{g} \qquad (1.40)$$

is the self-inductance of coil 2.

It is important to note that the resolution of the resultant flux linkages into the components produced by i_1 and i_2 is based on superposition of the individual effects and therefore implies a linear flux-mmf relationship (characteristic of materials of constant permeability).

Substitution of Eq. 1.29 in Eq. 1.27 yields

$$e = \frac{d}{dt}(Li) \tag{1.41}$$

for a magnetic circuit with a single winding. For a static magnetic circuit, the inductance is fixed (assuming that material nonlinearities do not cause the inductance to vary), and this equation reduces to the familiar circuit-theory form

$$e = L\frac{di}{dt} \tag{1.42}$$

However, in electromechanical energy conversion devices, inductances are often time-varying, and Eq. 1.41 must be written as

$$e = L\frac{di}{dt} + i\frac{dL}{dt} \tag{1.43}$$

Note that in situations with multiple windings, the total flux linkage of each winding must be used in Eq. 1.27 to find the winding-terminal voltage.

The power at the terminals of a winding on a magnetic circuit is a measure of the rate of energy flow into the circuit through that particular winding. The *power, p,* is determined from the product of the voltage and the current

$$p = ie = i\frac{d\lambda}{dt} \tag{1.44}$$

and its unit is *watts* (W), or *joules per second*. Thus the change in *magnetic stored energy* ΔW in the magnetic circuit in the time interval t_1 to t_2 is

$$\Delta W = \int_{t_1}^{t_2} p \, dt = \int_{\lambda_1}^{\lambda_2} i \, d\lambda \tag{1.45}$$

In SI units, the magnetic stored energy W is measured in *joules* (J).

For a single-winding system of constant inductance, the change in magnetic stored energy as the flux level is changed from λ_1 to λ_2 can be written as

$$\Delta W = \int_{\lambda_1}^{\lambda_2} i \, d\lambda = \int_{\lambda_1}^{\lambda_2} \frac{\lambda}{L} \, d\lambda = \frac{1}{2L}(\lambda_2^2 - \lambda_1^2) \tag{1.46}$$

The total magnetic stored energy at any given value of λ can be found from setting λ_1 equal to zero:

$$W = \frac{1}{2L}\lambda^2 = \frac{L}{2}i^2 \tag{1.47}$$

EXAMPLE 1.6

For the magnetic circuit of Example 1.1 (Fig. 1.2), find (*a*) the inductance L, (*b*) the magnetic stored energy W for $B_c = 1.0$ T, and (*c*) the induced voltage e for a 60-Hz time-varying core flux of the form $B_c = 1.0 \sin \omega t$ T where $\omega = (2\pi)(60) = 377$.

■ **Solution**

a. From Eqs. 1.16 and 1.29 and Example 1.1,

$$L = \frac{\lambda}{i} = \frac{N\phi}{i} = \frac{N^2}{\mathcal{R}_c + \mathcal{R}_g}$$

$$= \frac{500^2}{4.46 \times 10^5} = 0.56 \text{ H}$$

Note that the core reluctance is much smaller than that of the gap ($\mathcal{R}_c \ll \mathcal{R}_g$). Thus to a good approximation the inductance is dominated by the gap reluctance, i.e.,

$$L \approx \frac{N^2}{\mathcal{R}_g} = 0.57 \text{ H}$$

b. In Example 1.1 we found that when $B_c = 1.0$ T, $i = 0.80$A. Thus from Eq. 1.47,

$$W = \frac{1}{2}Li^2 = \frac{1}{2}(0.56)(0.80)^2 = 0.18 \text{ J}$$

c. From Eq. 1.27 and Example 1.1,

$$e = \frac{d\lambda}{dt} = N\frac{d\varphi}{dt} = NA_c\frac{dB_c}{dt}$$

$$= 500 \times (9 \times 10^{-4}) \times (377 \times 1.0 \cos(377t))$$

$$= 170 \cos(377t) \quad V$$

Repeat Example 1.6 for $B_c = 0.8$ T, assuming the core flux varies at 50 Hz instead of 60 Hz.

Solution

    a. The inductance L is unchanged.
    b. $W = 0.115$ J
    c. $e = 113 \cos(314t)$ V

1.3 PROPERTIES OF MAGNETIC MATERIALS

In the context of electromechanical energy conversion devices, the importance of magnetic materials is twofold. Through their use it is possible to obtain large magnetic flux densities with relatively low levels of magnetizing force. Since magnetic forces and energy density increase with increasing flux density, this effect plays a large role in the performance of energy-conversion devices.

In addition, magnetic materials can be used to constrain and direct magnetic fields in well-defined paths. In a transformer they are used to maximize the coupling between the windings as well as to lower the excitation current required for transformer operation. In electric machinery, magnetic materials are used to shape the fields to obtain desired torque-production and electrical terminal characteristics. Thus a knowledgeable designer can use magnetic materials to achieve specific desirable device characteristics.

Ferromagnetic materials, typically composed of iron and alloys of iron with cobalt, tungsten, nickel, aluminum, and other metals, are by far the most common magnetic materials. Although these materials are characterized by a wide range of properties, the basic phenomena responsible for their properties are common to them all.

Ferromagnetic materials are found to be composed of a large number of domains, i.e., regions in which the magnetic moments of all the atoms are parallel, giving rise to a net magnetic moment for that domain. In an unmagnetized sample of material, the domain magnetic moments are randomly oriented, and the net resulting magnetic flux in the material is zero.

When an external magnetizing force is applied to this material, the domain magnetic moments tend to align with the applied magnetic field. As a result, the domain magnetic moments add to the applied field, producing a much larger value of flux density than would exist due to the magnetizing force alone. Thus the *effective permeability* μ, equal to the ratio of the total magnetic flux density to the applied magnetic-field intensity, is large compared with the permeability of free space μ_0. As the magnetizing force is increased, this behavior continues until all the magnetic moments are aligned with the applied field; at this point they can no longer contribute to increasing the magnetic flux density, and the material is said to be fully *saturated*.

In the absence of an externally applied magnetizing force, the domain magnetic moments naturally align along certain directions associated with the crystal structure of the domain, known as *axes of easy magnetization*. Thus if the applied magnetizing force is reduced, the domain magnetic moments relax to the direction of easy magnetism nearest to that of the applied field. As a result, when the applied field is reduced to zero, although they will tend to relax towards their initial orientation, the magnetic dipole moments will no longer be totally random in their orientation; they will retain a net magnetization component along the applied field direction. It is this effect which is responsible for the phenomenon known as *magnetic hysteresis*.

Due to this hystersis effect, the relationship between B and H for a ferromagnetic material is both nonlinear and multivalued. In general, the characteristics of the material cannot be described analytically. They are commonly presented in graphical form as a set of empirically determined curves based on test samples of the material using methods prescribed by the American Society for Testing and Materials (ASTM).[5]

---

[5] Numerical data on a wide variety of magnetic materials are available from material manufacturers. One problem in using such data arises from the various systems of units employed. For example, magnetization may be given in oersteds or in ampere-turns per meter and the magnetic flux density in gauss, kilogauss, or teslas. A few useful conversion factors are given in Appendix E. The reader is reminded that the equations in this book are based upon SI units.

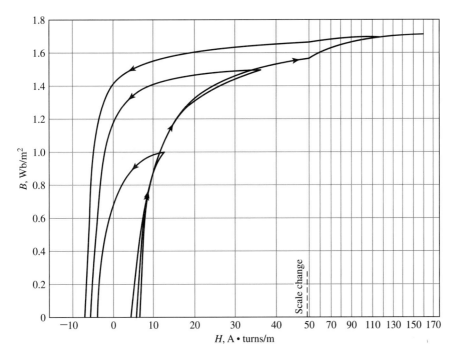

Figure 1.9 *B-H* loops for M-5 grain-oriented electrical steel 0.012 in thick. Only the top halves of the loops are shown here. (*Armco Inc.*)

The most common curve used to describe a magnetic material is the *B-H curve* or *hysteresis loop*. The first and second quadrants (corresponding to $B \geq 0$) of a set of hysteresis loops are shown in Fig. 1.9 for M-5 steel, a typical grain-oriented electrical steel used in electric equipment. These loops show the relationship between the magnetic flux density B and the magnetizing force H. Each curve is obtained while cyclically varying the applied magnetizing force between equal positive and negative values of fixed magnitude. Hysteresis causes these curves to be multivalued. After several cycles the *B-H* curves form closed loops as shown. The arrows show the paths followed by B with increasing and decreasing H. Notice that with increasing magnitude of H the curves begin to flatten out as the material tends toward saturation. At a flux density of about 1.7 T, this material can be seen to be heavily saturated.

Notice that as H is decreased from its maximum value to zero, the flux density decreases but not to zero. This is the result of the relaxation of the orientation of the magnetic moments of the domains as described above. The result is that there remains a *remanant magnetization* when H is zero.

Fortunately, for many engineering applications, it is sufficient to describe the material by a single-valued curve obtained by plotting the locus of the maximum values of B and H at the tips of the hysteresis loops; this is known as a *dc* or *normal magnetization curve*. A dc magnetization curve for M-5 grain-oriented electrical steel

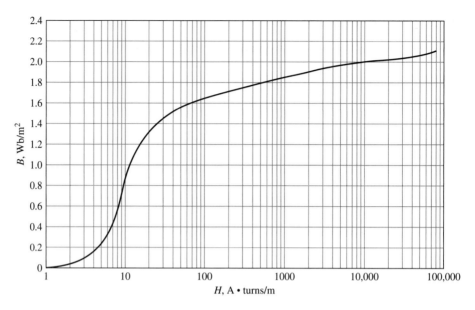

Figure 1.10 Dc magnetization curve for M-5 grain-oriented electrical steel 0.012 in thick. (*Armco Inc.*)

is shown in Fig. 1.10. The dc magnetization curve neglects the hysteretic nature of the material but clearly displays its nonlinear characteristics.

EXAMPLE 1.7

Assume that the core material in Example 1.1 is M-5 electrical steel, which has the dc magnetization curve of Fig. 1.10. Find the current i required to produce $B_c = 1$ T.

■ **Solution**

The value of H_c for $B_c = 1$ T is read from Fig. 1.10 as

$$H_c = 11 \text{ A} \cdot \text{turns/m}$$

The mmf drop for the core path is

$$\mathcal{F}_c = H_c l_c = 11(0.3) = 3.3 \text{ A} \cdot \text{turns}$$

The mmf drop across the air gap is

$$\mathcal{F}_g = H_g g = \frac{B_g g}{\mu_0} = \frac{5 \times 10^{-4}}{4\pi \times 10^{-7}} = 396 \text{ A} \cdot \text{turns}$$

The required current is

$$i = \frac{\mathcal{F}_c + \mathcal{F}_g}{N} = \frac{399}{500} = 0.80 \text{ A}$$

Repeat Example 1.7 but find the current i for $B_c = 1.6$ T. By what factor does the current have to be increased to result in this factor of 1.6 increase in flux density?

Solution

The current i can be shown to be 1.302 A. Thus, the current must be increased by a factor of $1.302/0.8 = 1.63$. Because of the dominance of the air-gap reluctance, this is just slightly in excess of the fractional increase in flux density in spite of the fact that the core is beginning to significantly saturate at a flux density of 1.6 T.

1.4 AC EXCITATION

In ac power systems, the waveforms of voltage and flux closely approximate sinusoidal functions of time. This section describes the excitation characteristics and losses associated with steady-state ac operation of magnetic materials under such operating conditions. We use as our model a closed-core magnetic circuit, i.e., with no air gap, such as that shown in Fig. 1.1 or the transformer of Fig. 2.4. The magnetic path length is l_c, and the cross-sectional area is A_c throughout the length of the core. We further assume a sinusoidal variation of the core flux $\varphi(t)$; thus

$$\varphi(t) = \phi_{max} \sin \omega t = A_c B_{max} \sin \omega t \tag{1.48}$$

where $\phi_{max} =$ amplitude of core flux φ in webers

$\quad B_{max} =$ amplitude of flux density B_c in teslas

$\quad\quad \omega =$ angular frequency $= 2\pi f$

$\quad\quad\; f =$ frequency in Hz

From Eq. 1.27, the voltage induced in the N-turn winding is

$$e(t) = \omega N \phi_{max} \cos(\omega t) = E_{max} \cos \omega t \tag{1.49}$$

where

$$E_{max} = \omega N \phi_{max} = 2\pi f N A_c B_{max} \tag{1.50}$$

In steady-state ac operation, we are usually more interested in the *root-mean-square* or *rms* values of voltages and currents than in instantaneous or maximum values. In general, the rms value of a periodic function of time, $f(t)$, of period T is defined as

$$F_{rms} = \sqrt{\left(\frac{1}{T} \int_0^T f^2(t)\, dt\right)} \tag{1.51}$$

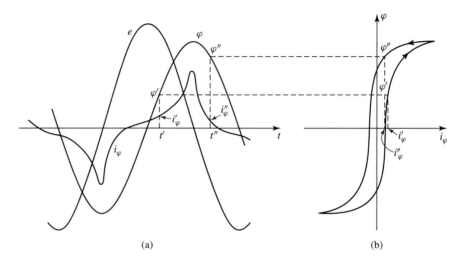

Figure 1.11 Excitation phenomena. (a) Voltage, flux, and exciting current;
(b) corresponding hysteresis loop.

From Eq. 1.51, the rms value of a sine wave can be shown to be $1/\sqrt{2}$ times its peak value. Thus the rms value of the induced voltage is

$$E_{\text{rms}} = \frac{2\pi}{\sqrt{2}} f N A_c B_{\max} = \sqrt{2}\pi f N A_c B_{\max} \tag{1.52}$$

   To produce magnetic flux in the core requires current in the exciting winding known as the *exciting current*, i_φ.[6] The nonlinear magnetic properties of the core require that the waveform of the exciting current differs from the sinusoidal waveform of the flux. A curve of the exciting current as a function of time can be found graphically from the magnetic characteristics of the core material, as illustrated in Fig. 1.11a. Since B_c and H_c are related to φ and i_φ by known geometric constants, the ac hysteresis loop of Fig. 1.11b has been drawn in terms of $\varphi = B_c A_c$ and is $i_\varphi = H_c l_c / N$. Sine waves of induced voltage, e, and flux, φ, in accordance with Eqs. 1.48 and 1.49, are shown in Fig. 1.11a.

   At any given time, the value of i_φ corresponding to the given value of flux can be found directly from the hysteresis loop. For example, at time t' the flux is φ' and the current is i_φ'; at time t'' the corresponding values are φ'' and i_φ''. Notice that since the hysteresis loop is multivalued, it is necessary to be careful to pick the rising-flux values (φ' in the figure) from the rising-flux portion of the hysteresis loop; similarly the falling-flux portion of the hysteresis loop must be selected for the falling-flux values (φ'' in the figure).

---

[6] More generally, for a system with multiple windings, the exciting mmf is the net ampere-turns acting to produce flux in the magnetic circuit.

Notice that, because the hysteresis loop "flattens out" due to saturation effects, the waveform of the exciting current is sharply peaked. Its rms value $I_{\varphi,\text{rms}}$ is defined by Eq. 1.51, where T is the period of a cycle. It is related to the corresponding rms value $H_{c,\text{rms}}$ of H_c by the relationship

$$I_{\varphi,\text{rms}} = \frac{l_c H_{c,\text{rms}}}{N} \tag{1.53}$$

The ac excitation characteristics of core materials are often described in terms of rms voltamperes rather than a magnetization curve relating B and H. The theory behind this representation can be explained by combining Eqs. 1.52 and 1.53. Thus, from Eqs. 1.52 and 1.53, the rms voltamperes required to excite the core of Fig. 1.1 to a specified flux density is equal to

$$E_{\text{rms}} I_{\varphi,\text{rms}} = \sqrt{2}\pi f N A_c B_{\text{max}} \frac{l_c H_{\text{rms}}}{N}$$

$$= \sqrt{2}\pi f B_{\text{max}} H_{\text{rms}} (A_c l_c) \tag{1.54}$$

In Eq. 1.54, the product $A_c l_c$ can be seen to be equal to the volume of the core and hence the rms exciting voltamperes required to excite the core with sinusoidal can be seen to be proportional to the frequency of excitation, the core volume and the product of the peak flux density and the rms magnetic field intensity. For a magnetic material of mass density ρ_c, the mass of the core is $A_c l_c \rho_c$ and the *exciting rms voltamperes per unit mass*, P_a, can be expressed as

$$P_a = \frac{E_{\text{rms}} I_{\varphi,\text{rms}}}{\text{mass}} = \frac{\sqrt{2}\pi f}{\rho_c} B_{\text{max}} H_{\text{rms}} \tag{1.55}$$

Note that, normalized in this fashion, the rms exciting voltamperes can be seen to be a property of the material alone. In addition, note that they depend only on B_{max} because H_{rms} is a unique function of B_{max} as determined by the shape of the material hysteresis loop at any given frequency f. As a result, the ac excitation requirements for a magnetic material are often supplied by manufacturers in terms of rms voltamperes per unit weight as determined by laboratory tests on closed-core samples of the material. These results are illustrated in Fig. 1.12 for M-5 grain-oriented electrical steel.

The exciting current supplies the mmf required to produce the core flux and the power input associated with the energy in the magnetic field in the core. Part of this energy is dissipated as losses and results in heating of the core. The rest appears as reactive power associated with energy storage in the magnetic field. This reactive power is not dissipated in the core; it is cyclically supplied and absorbed by the excitation source.

Two loss mechanisms are associated with time-varying fluxes in magnetic materials. The first is ohmic $I^2 R$ heating, associated with induced currents in the core material. From Faraday's law (Eq. 1.26) we see that time-varying magnetic fields give rise to electric fields. In magnetic materials these electric fields result in induced

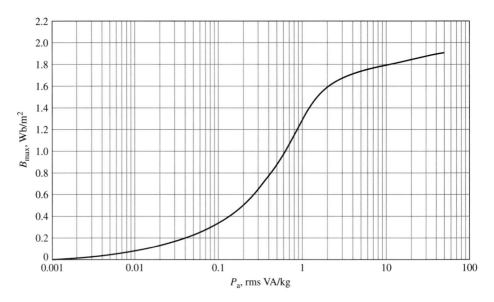

Figure 1.12 Exciting rms voltamperes per kilogram at 60 Hz for M-5 grain-oriented electrical steel 0.012 in thick. (*Armco Inc.*)

currents, commonly referred to as *eddy currents,* which circulate in the core material and oppose changes in flux density in the material. To counteract the corresponding demagnetizing effect, the current in the exciting winding must increase. Thus the resultant "dynamic" *B-H* loop under ac operation is somewhat "fatter" than the hysteresis loop for slowly varying conditions, and this effect increases as the excitation frequency is increased. It is for this reason that the characteristics of electrical steels vary with frequency and hence manufacturers typically supply characteristics over the expected operating frequency range of a particular electrical steel. Note for example that the exciting rms voltamperes of Fig. 1.12 are specified at a frequency of 60 Hz.

To reduce the effects of eddy currents, magnetic structures are usually built of thin sheets of laminations of the magnetic material. These laminations, which are aligned in the direction of the field lines, are insulated from each other by an oxide layer on their surfaces or by a thin coat of insulating enamel or varnish. This greatly reduces the magnitude of the eddy currents since the layers of insulation interrupt the current paths; the thinner the laminations, the lower the losses. In general, eddy-current loss tends to increase as the square of the excitation frequency and also as the square of the peak flux density.

The second loss mechanism is due to the hysteretic nature of magnetic material. In a magnetic circuit like that of Fig. 1.1 or the transformer of Fig. 2.4, a time-varying excitation will cause the magnetic material to undergo a cyclic variation described by a hysteresis loop such as that shown in Fig. 1.13.

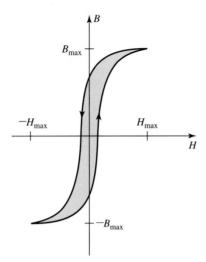

Figure 1.13 Hysteresis loop; hysteresis loss is proportional to the loop area (shaded).

Equation 1.45 can be used to calculate the energy input W to the magnetic core of Fig. 1.1 as the material undergoes a single cycle

$$W = \oint i_\varphi \, d\lambda = \oint \left(\frac{H_c l_c}{N}\right)(A_c N \, dB_c) = A_c l_c \oint H_c \, dB_c \qquad (1.56)$$

Recognizing that $A_c l_c$ is the volume of the core and that the integral is the area of the ac hysteresis loop, we see that each time the magnetic material undergoes a cycle, there is a net energy input into the material. This energy is required to move around the magnetic dipoles in the material and is dissipated as heat in the material. Thus for a given flux level, the corresponding *hysteresis losses* are proportional to the area of the hysteresis loop and to the total volume of material. Since there is an energy loss per cycle, hysteresis power loss is proportional to the frequency of the applied excitation.

In general, these losses depend on the metallurgy of the material as well as the flux density and frequency. Information on core loss is typically presented in graphical form. It is plotted in terms of watts per unit weight as a function of flux density; often a family of curves for different frequencies are given. Figure 1.14 shows the core loss P_c for M-5 grain-oriented electrical steel at 60 Hz.

Nearly all transformers and certain sections of electric machines use sheet-steel material that has highly favorable directions of magnetization along which the core loss is low and the permeability is high. This material is termed *grain-oriented steel*. The reason for this property lies in the atomic structure of a crystal of the silicon-iron

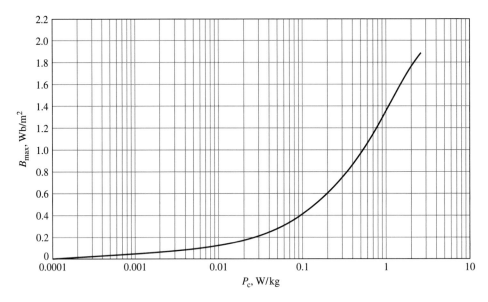

Figure 1.14 Core loss at 60 Hz in watts per kilogram for M-5 grain-oriented electrical steel 0.012 in thick. (*Armco Inc.*)

alloy, which is a body-centered cube; each cube has an atom at each corner as well as one in the center of the cube. In the cube, the easiest axis of magnetization is the cube edge; the diagonal across the cube face is more difficult, and the diagonal through the cube is the most difficult. By suitable manufacturing techniques most of the crystalline cube edges are aligned in the rolling direction to make it the favorable direction of magnetization. The behavior in this direction is superior in core loss and permeability to *nonoriented steels* in which the crystals are randomly oriented to produce a material with characteristics which are uniform in all directions. As a result, oriented steels can be operated at higher flux densities than the nonoriented grades.

Nonoriented electrical steels are used in applications where the flux does not follow a path which can be oriented with the rolling direction or where low cost is of importance. In these steels the losses are somewhat higher and the permeability is very much lower than in grain-oriented steels.

EXAMPLE 1.8

The magnetic core in Fig. 1.15 is made from laminations of M-5 grain-oriented electrical steel. The winding is excited with a 60-Hz voltage to produce a flux density in the steel of $B = 1.5 \sin \omega t$ T, where $\omega = 2\pi 60 \approx 377$ rad/sec. The steel occupies 0.94 of the core cross-sectional area. The mass-density of the steel is 7.65 g/cm^3. Find (*a*) the applied voltage, (*b*) the peak current, (*c*) the rms exciting current, and (*d*) the core loss.

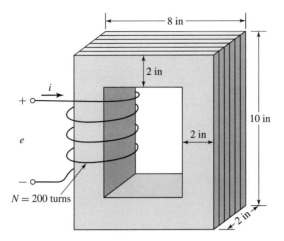

Figure 1.15 Laminated steel core with winding for Example 1.8.

■ Solution

a. From Eq. 1.27 the voltage is

$$e = N\frac{d\varphi}{dt} = NA_c\frac{dB}{dt}$$

$$= 200 \times 4 \text{ in}^2 \times 0.94 \times \left(\frac{1.0 \text{ m}^2}{39.4^2 \text{ in}^2}\right) \times 1.5 \times (377 \cos(377t))$$

$$= 274 \cos(377t) \text{ V}$$

b. The magnetic field intensity corresponding to $B_{max} = 1.5$ T is given in Fig. 1.10 as $H_{max} = 36$ A turns/m. Notice that, as expected, the relative permeability $\mu_r = B_{max}/(\mu_0 H_{max}) = 33{,}000$ at the flux level of 1.5 T is lower than the value of $\mu_r = 72{,}300$ found in Example 1.4 corresponding to a flux level of 1.0 T, yet significantly larger than the value of 2900 corresponding to a flux level of 1.8 T.

$$l_c = (6+6+8+8) \text{ in } \left(\frac{1.0 \text{ m}}{39.4 \text{ in}}\right) = 0.71 \text{ m}$$

The peak current is

$$I = \frac{H_{max}l_c}{N} = \frac{36(0.71)}{200} = 0.13 \text{ A}$$

c. The rms current is obtained from the value of P_a of Fig. 1.12 for $B_{max} = 1.5$ T.

$$P_a = 1.5 \text{ VA/kg}$$

The core volume and weight are

$$V_c = (4 \text{ in}^2)(0.94)(28 \text{ in}) = 105.5 \text{ in}^3$$

$$W_c = (105.5 \text{ in}^3)\left(\frac{2.54 \text{ cm}}{1.0 \text{ in}}\right)^3\left(\frac{7.65 \text{ g}}{1.0 \text{ cm}^3}\right) = 13.2 \text{ kg}$$

The total rms voltamperes and current are

$$P_a = (1.5 \text{ VA/kg})(13.2 \text{ kg}) = 20 \text{ VA}$$

$$I_{\varphi,\text{rms}} = \frac{P_a}{E_{\text{rms}}} = \frac{20}{275(0.707)} = 0.10 \text{ A}$$

d. The core-loss density is obtained from Fig. 1.14 as $P_c = 1.2$ W/kg. The total core loss is

$$P_c = (1.2 \text{ W/kg})(13.2 \text{ kg}) = 16 \text{ W}$$

Practice Problem 1.7

Repeat Example 1.8 for a 60-Hz voltage of $B = 1.0 \sin \omega t$ T.

Solution

    a. $V = 185 \cos 377t$ V
    b. $I = 0.04$ A
    c. $I_\varphi = 0.061$ A
    d. $P_c = 6.7$ W

1.5 PERMANENT MAGNETS

Figure 1.16a shows the second quadrant of a hysteresis loop for Alnico 5, a typical permanent-magnet material, while Fig. 1.16b shows the second quadrant of a hysteresis loop for M-5 steel.[7] Notice that the curves are similar in nature. However, the hysteresis loop of Alnico 5 is characterized by a large value of *residual flux density* or *remanent magnetization, B_r,* (approximately 1.22 T) as well as a large value of *coercivity, H_c,* (approximately -49 kA/m).

The remanent magnetization, B_r, corresponds to the flux density which would remain in a closed magnetic structure, such as that of Fig. 1.1, made of this material, if the applied mmf (and hence the magnetic field intensity H) were reduced to zero. However, although the M-5 electrical steel also has a large value of remanent magnetization (approximately 1.4 T), it has a much smaller value of coercivity (approximately -6 A/m, smaller by a factor of over 7500). The coercivity H_c corresponds to the value of magnetic field intensity (which is proportional to the mmf) required to reduce the material flux density to zero.

The significance of remanent magnetization is that it can produce magnetic flux in a magnetic circuit in the absence of external excitation (such as winding currents). This is a familiar phenomenon to anyone who has afixed notes to a refrigerator with small magnets and is widely used in devices such as loudspeakers and permanent-magnet motors.

---

[7] To obtain the largest value of remanent magnetization, the hysteresis loops of Fig. 1.16 are those which would be obtained if the materials were excited by sufficient mmf to ensure that they were driven heavily into saturation. This is discussed further in Section 1.6.

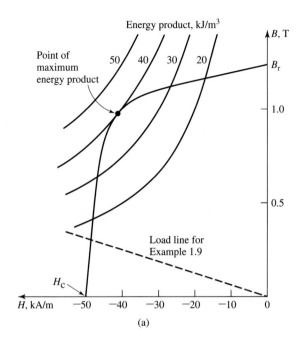

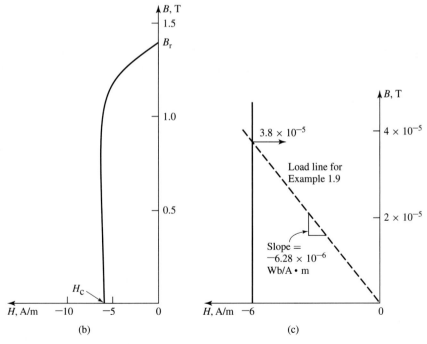

Figure 1.16 (a) Second quadrant of hysteresis loop for Alnico 5; (b) second quadrant of hysteresis loop for M-5 electrical steel; (c) hysteresis loop for M-5 electrical steel expanded for small B. (*Armco Inc.*)

From Fig. 1.16, it would appear that both Alnico 5 and M-5 electrical steel would be useful in producing flux in unexcited magnetic circuits since they both have large values of remanent magnetization. That this is not the case can be best illustrated by an example.

EXAMPLE 1.9

As shown in Fig. 1.17, a magnetic circuit consists of a core of high permeability ($\mu \rightarrow \infty$), an air gap of length $g = 0.2$ cm, and a section of magnetic material of length $l_m = 1.0$ cm. The cross-sectional area of the core and gap is equal to $A_m = A_g = 4$ cm^2. Calculate the flux density B_g in the air gap if the magnetic material is (*a*) Alnico 5 and (*b*) M-5 electrical steel.

■ Solution

a. Since the core permeability is assumed infinite, H in the core is negligible. Recognizing that the mmf acting on the magnetic circuit of Fig. 1.17 is zero, we can write

$$\mathcal{F} = 0 = H_g g + H_m l_m$$

or

$$H_g = -\left(\frac{l_m}{g}\right) H_m$$

where H_g and H_m are the magnetic field intensities in the air gap and the magnetic material, respectively.

Since the flux must be continuous through the magnetic circuit,

$$\phi = A_g B_g = A_m B_m$$

or

$$B_g = \left(\frac{A_m}{A_g}\right) B_m$$

where B_g and B_m are the magnetic flux densities in the air gap and the magnetic material, respectively.

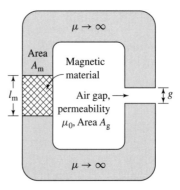

Figure 1.17   Magnetic circuit for Example 1.9.

These equations can be solved to yield a linear relationship for B_m in terms of H_m

$$B_m = -\mu_0 \left(\frac{A_g}{A_m}\right)\left(\frac{l_m}{g}\right) H_m = -5\,\mu_0\,H_m = -6.28 \times 10^{-6} H_m$$

To solve for B_m we recognize that for Alnico 5, B_m and H_m are also related by the curve of Fig. 1.16a. Thus this linear relationship, also known as a *load line*, can be plotted on Fig. 1.16a and the solution obtained graphically, resulting in

$$B_g = B_m = 0.30\ \text{T}$$

b. The solution for M-5 electrical steel proceeds exactly as in part (*a*). The load line is the same as that of part (*a*) because it is determined only by the permeability of the air gap and the geometries of the magnet and the air gap. Hence from Fig. 1.16c

$$B_g = 3.8 \times 10^{-5}\ \text{T} = 0.38\ \text{gauss}$$

which is much less than the value obtained with Alnico 5.

Example 1.9 shows that there is an immense difference between permanent-magnet materials (often referred to as *hard magnetic materials*) such as Alnico 5 and *soft magnetic materials* such as M-5 electrical steel. This difference is characterized in large part by the immense difference in their coercivities H_c. The coercivity can be thought of as a measure of the magnitude of the mmf required to demagnetize the material. As seen from Example 1.9, it is also a measure of the capability of the material to produce flux in a magnetic circuit which includes an air gap. Thus we see that materials which make good permanent magnets are characterized by large values of coercivity H_c (considerably in excess of 1 kA/m).

A useful measure of the capability of permanent-magnet material is known as its *maximum energy product*. This corresponds to the largest B-H product $(B-H)_{max}$, which corresponds to a point on the second quadrant of the hysteresis loop. As can be seen from Eq. 1.56, the product of B and H has the dimensions of energy density (joules per cubic meter). We now show that operation of a given permanent-magnet material at this point will result in the smallest volume of that material required to produce a given flux density in an air gap. As a result, choosing a material with the largest available maximum energy product can result in the smallest required magnet volume.

In Example 1.9, we found an expression for the flux density in the air gap of the magnetic circuit of Fig. 1.17:

$$B_g = \frac{A_m}{A_g} B_m \tag{1.57}$$

We also found that the ratio of the mmf drops across the magnet and the air gap is equal to -1:

$$\frac{H_m l_m}{H_g g} = -1 \tag{1.58}$$

Equation 1.58 can be solved for H_g, and the result can be multiplied by μ_0 to obtain $B_g = \mu_0 H_g$. Multiplying by Eq. 1.57 yields

$$B_g^2 = \mu_0 \left(\frac{l_m A_m}{g A_g} \right)(-H_m B_m)$$

$$= \mu_0 \left(\frac{\text{Vol}_{\text{mag}}}{\text{Vol}_{\text{air gap}}} \right)(-H_m B_m) \tag{1.59}$$

or

$$\text{Vol}_{\text{mag}} = \frac{\text{Vol}_{\text{air gap}} B_g^2}{\mu_0 (-H_m B_m)} \tag{1.60}$$

where Vol_{mag} is the volume of the magnet, $\text{Vol}_{\text{air gap}}$ is the air-gap volume, and the minus sign arises because, at the operating point of the magnetic circuit, H in the magnet (H_m) is negative.

Equation 1.60 is the desired result. It indicates that to achieve a desired flux density in the air gap, the required volume of the magnet can be minimized by operating the magnet at the point of the largest possible value of the B-H product $H_m B_m$, i.e., the point of maximum energy product. Furthermore, the larger the value of this product, the smaller the size of the magnet required to produce the desired flux density. Hence the maximum energy product is a useful performance measure for a magnetic material, and it is often found as a tabulated "figure of merit" on data sheets for permanent-magnet materials.

Note that Eq. 1.59 appears to indicate that one can achieve an arbitrarily large air-gap flux density simply by reducing the air-gap volume. This is not true in practice because as the flux density in the magnetic circuit increases, a point will be reached at which the magnetic core material will begin to saturate and the assumption of infinite permeability will no longer be valid, thus invalidating the derivation leading to Eq. 1.59.

Note also that a curve of constant B-H product is a hyperbola. A set of such hyperbolas for different values of the B-H product is plotted in Fig. 1.16a. From these curves, we see that the maximum energy product for Alnico 5 is 40 kJ/m^3 and that this occurs at the point $B = 1.0$ T and $H = -40$ kA/m.

EXAMPLE 1.10

The magnetic circuit of Fig. 1.17 is modified so that the air-gap area is reduced to $A_g = 2.0\,\text{cm}^2$, as shown in Fig. 1.18. Find the minimum magnet volume required to achieve an air-gap flux density of 0.8 T.

■ **Solution**

The smallest magnet volume will be achieved with the magnet operating at its point of maximum energy product, as shown in Fig. 1.16a. At this operating point, $B_m = 1.0$ T and $H_m = -40$ kA/m.

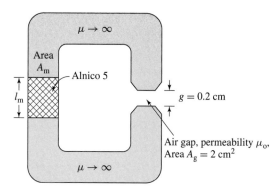

Figure 1.18 Magnetic circuit for Example 1.10.

Thus from Eq. 1.57,

$$A_m = A_g \left(\frac{B_g}{B_m} \right)$$

$$= 2 \text{ cm}^2 \left(\frac{0.8}{1.0} \right) = 1.6 \text{ cm}^2$$

and from Eq. 1.58

$$l_m = -g \left(\frac{H_g}{H_m} \right) = -g \left(\frac{B_g}{\mu_0 H_m} \right)$$

$$= -0.2 \text{ cm} \left(\frac{0.8}{(4\pi \times 10^{-7})(-40 \times 10^3)} \right)$$

$$= 3.18 \text{ cm}$$

Thus the minimum magnet volume is equal to $1.6 \text{ cm}^2 \times 3.18 \text{ cm} = 5.09 \text{ cm}^3$.

Practice Problem 1.8

Repeat Example 1.10 assuming the air-gap area is further reduced to $A_g = 1.8 \text{ cm}^2$ and that the desired air-gap flux density is 0.6 T.

Solution

Minimum magnet volume = 2.58 cm^3.

1.6 APPLICATION OF PERMANENT MAGNET MATERIALS

Examples 1.9 and 1.10 consider the operation of permanent magnetic materials under the assumption that the operating point can be determined simply from a knowledge of the geometry of the magnetic circuit and the properties of the various magnetic

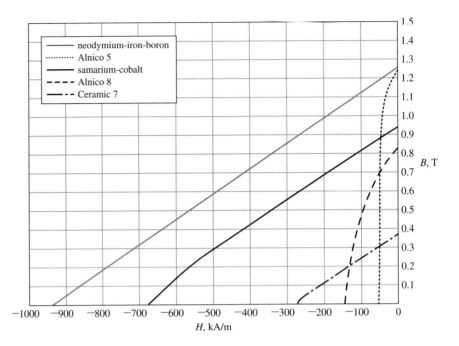

Figure 1.19 Magnetization curves for common permanent-magnet materials.

materials involved. In fact, the situation is more complex.[8] This section will expand upon these issues.

Figure 1.19 shows the magnetization characteristics for a few common permanent magnet materials. Alnico 5 is a widely used alloy of iron, nickel, aluminum, and cobalt, originally discovered in 1931. It has a relatively large residual flux density. Alnico 8 has a lower residual flux density and a higher coercivity than Alnico 5. Hence, it is less subject to demagnetization than Alnico 5. Disadvantages of the Alnico materials are their relatively low coercivity and their mechanical brittleness.

Ceramic permanent magnet materials (also known as *ferrite magnets*) are made from iron-oxide and barium- or strontium-carbonate powders and have lower residual flux densities than Alnico materials but significantly higher coercivities. As a result, they are much less prone to demagnetization. One such material, Ceramic 7, is shown in Fig. 1.19, where its magnetization characteristic is almost a straight line. Ceramic magnets have good mechanical characteristics and are inexpensive to manufacture; as a result, they are the widely used in many permanent magnet applications.

---

[8] For a further discussion of permanent magnets and their application, see P. Campbell, *Permanent Magnet Materials and Their Application,* Cambridge University Press, 1994; R. J. Parker, *Advances in Permanent Magnetism,* John Wiley & Sons, 1990; A. Bosak, *Permanent-Magnet DC Linear Motors,* Clarendon Press·Oxford, 1996; G. R. Slemon and A. Straughen, *Electric Machines,* Addison-Wesley, 1980, Secs 1.20–1.25; and T. J. E. Miller, *Brushless Permanent-Magnet and Reluctance Motor Drives,* Clarendon Press·Oxford, 1989, Chapter 3.

Samarium-cobalt represents a significant advance in permanent magnet technology which began in the 1960s with the discovery of rare earth permanent magnet materials. From Fig. 1.19 it can be seen to have a high residual flux density such as is found with the Alnico materials, while at the same time having a much higher coercivity and maximum energy product. The newest of the rare earth magnetic materials is the neodymium-iron-boron material. It features even larger residual flux density, coercivity, and maximum energy product than does samarium-cobalt.

Consider the magnetic circuit of Fig. 1.20. This includes a section of hard magnetic material in a core of highly permeable soft magnetic material as well as an N-turn excitation winding. With reference to Fig. 1.21, we assume that the hard

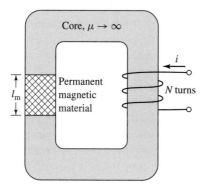

Figure 1.20 Magnetic circuit including both a permanent magnet and an excitation winding.

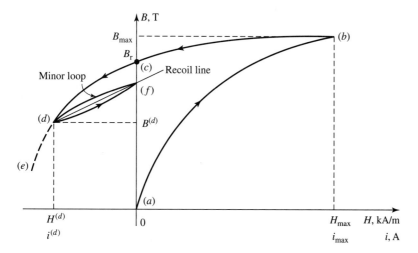

Figure 1.21 Portion of a *B-H* characteristic showing a minor loop and a recoil line.

magnetic material is initially unmagnetized (corresponding to point a of the figure) and consider what happens as current is applied to the excitation winding. Because the core is assumed to be of infinite permeability, the horizontal axis of Fig. 1.21 can be considered to be both a measure of the applied current $i = Hl_m/N$ as well as a measure of H in the magnetic material.

As the current i is increased to its maximum value, the B-H trajectory rises from point a in Fig. 1.21 toward its maximum value at point b. To fully magnetize the material, we assume that the current has been increased to a value i_{max} sufficiently large that the material has been driven well into saturation at point b. When the current is then decreased to zero, the B-H characteristic will begin to form a hysteresis loop, arriving at point c at zero current. At point c, notice that H in the material is zero but B is at its remanent value B_r.

As the current then goes negative, the B-H characteristic continues to trace out a hysteresis loop. In Fig. 1.21, this is seen as the trajectory between points c and d. If the current is then maintained at the value $-i^{(d)}$, the operating point of the magnet will be that of point d. Note that, as in Example 1.9, this same operating point would be reached if the material were to start at point c and, with the excitation held at zero, an air gap of length $g = l_m(A_g/A_m)(-\mu_0 H^{(d)}/B^{(d)})$ were then inserted in the core.

Should the current then be made more negative, the trajectory would continue tracing out the hysteresis loop toward point e. However, if instead the current is returned to zero, the trajectory does not in general retrace the hysteresis loop toward point c. Rather it begins to trace out a *minor hysteresis loop,* reaching point f when the current reaches zero. If the current is then varied between zero and $-i^{(d)}$, the B-H characteristic will trace out the minor loop as shown.

As can be seen from Fig. 1.21, the B-H trajectory between points d and f can be represented by a straight line, known as the *recoil line.* The slope of this line is called the *recoil permeability* μ_R. We see that once this material has been demagnetized to point d, the effective remanent magnetization of the magnetic material is that of point f which is less than the remanent magnetization B_r which would be expected based on the hysteresis loop. Note that should the demagnetization be increased past point d, for example, to point e of Fig. 1.21, a new minor loop will be created, with a new recoil line and recoil permeability.

The demagnetization effects of negative excitation which have just been discussed are equivalent to those of an air gap in the magnetic circuit. For example, clearly the magnetic circuit of Fig. 1.20 could be used as a system to magnetize hard magnetic materials. The process would simply require that a large excitation be applied to the winding and then reduced to zero, leaving the material at a remanent magnetization B_r (point c in Fig. 1.21).

Following this magnetization process, if the material were removed from the core, this would be equivalent to opening a large air gap in the magnetic circuit, demagnetizing the material in a fashion similar to that seen in Example 1.9. At this point, the magnet has been effectively weakened, since if it were again inserted in the magnetic core, it would follow a recoil line and return to a remanent magnetization somewhat less than B_r. As a result, hard magnetic materials, such as the Alnico materials of Fig. 1.19, often do not operate stably in situations with varying mmf and geometry, and there is often the risk that improper operation can significantly demagnetize them. A significant advantage of materials such as Ceramic 7, samarium-cobalt and neodymium-iron-boron is that, because of their "straight-line" characteristic in the second quadrant (with slope close to μ_0), their recoil lines closely match their magnetization characteristic. As a result, demagnetization effects are significantly reduced in these materials and often can be ignored.

At the expense of a reduction in value of the remanent magnetization, hard magnetic materials can be stabilized to operate over a specified region. This procedure, based on the recoil trajectory shown in Fig. 1.21, can best be illustrated by an example.

EXAMPLE 1.11

Figure 1.22 shows a magnetic circuit containing hard magnetic material, a core and plunger of high (assumed infinite) permeability, and a single-turn winding which will be used to magnetize the hard magnetic material. The winding will be removed after the system is magnetized. The plunger moves in the x direction as indicated, with the result that the air-gap area can vary $(2 \text{ cm}^2 \leq A_g \leq 4 \text{ cm}^2)$. Assuming that the hard magnetic material is Alnico 5 and that the system is initially magnetized with $A_g = 2$ cm, (a) find the magnet length l_m such that the system will operate on a recoil line which intersects the maximum B-H product point on the magnetization curve for Alnico 5, (b) devise a procedure for magnetizing the magnet, and (c) calculate the flux density B_g in the air gap as the plunger moves back and forth and the air gap varies between these two limits.

■ **Solution**

a. Figure 1.23a shows the magnetization curve for Alnico 5 and two load lines corresponding to the two extremes of the air gap, $A_g = 2 \text{ cm}^2$ and $A_g = 4 \text{ cm}^2$. We see that the system will operate on the desired recoil line if the load line for $A_g = 2 \text{ cm}^2$ intersects the B-H characteristic at the maximum energy product point (labeled point a in Fig. 1.23a), $B_m^{(a)} = 1.0$ T and $H_m^{(a)} = -40$ kA/m.

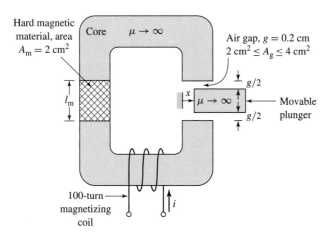

Figure 1.22 Magnetic circuit for Example 1.11.

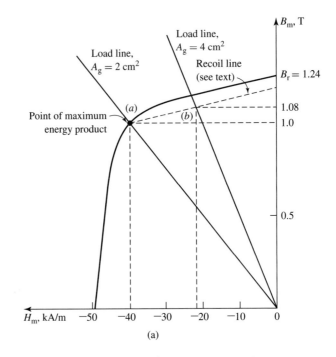

(a)

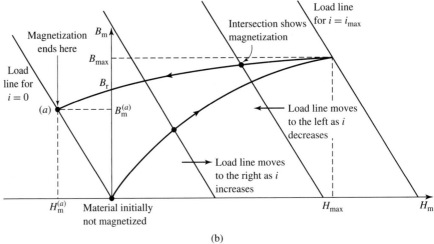

(b)

Figure 1.23   (a) Magnetization curve for Alnico 5 for Example 1.11; (b) series of load lines for $A_g = 2$ cm^2 and varying values of i showing the magnetization procedure for Example 1.11.

From Eqs. 1.57 and 1.58, we see that the slope of the required load line is given by

$$\frac{B_{\mathrm{m}}^{(a)}}{-H_{\mathrm{m}}^{(a)}} = \frac{B_{\mathrm{g}}}{H_{\mathrm{g}}} \frac{A_{\mathrm{g}}}{A_{\mathrm{m}}} \frac{l_{\mathrm{m}}}{g}$$

and thus

$$l_{\mathrm{m}} = g \left(\frac{A_{\mathrm{m}}}{A_{\mathrm{g}}} \right) \left(\frac{B_{\mathrm{m}}^{(a)}}{-\mu_0 H_{\mathrm{m}}^{(a)}} \right)$$

$$= 0.2 \, \mathrm{cm} \left(\frac{2}{2} \right) \left(\frac{1.0}{4\pi \times 10^{-7} \times 4 \times 10^4} \right) = 3.98 \, \mathrm{cm}$$

b. Figure 1.23b shows a series of load lines for the system with $A_{\mathrm{g}} = 2 \, \mathrm{cm}^2$ and with current i applied to the excitation winding. The general equation for these load lines can be readily derived since from Eq. 1.5

$$Ni = H_{\mathrm{m}} l_{\mathrm{m}} + H_{\mathrm{g}} g$$

and from Eqs. 1.3 and 1.7

$$B_{\mathrm{m}} A_{\mathrm{m}} = B_{\mathrm{g}} A_{\mathrm{g}} = \mu_0 H_{\mathrm{g}} A_{\mathrm{g}}$$

Thus

$$B_{\mathrm{m}} = -\mu_0 \left(\frac{A_{\mathrm{g}}}{A_{\mathrm{m}}} \right) \left(\frac{l_{\mathrm{m}}}{g} \right) H_{\mathrm{m}} + \frac{\mu_0 N}{g} \left(\frac{A_{\mathrm{g}}}{A_{\mathrm{m}}} \right) i$$

$$= \mu_0 \left[-\left(\frac{2}{2} \right) \left(\frac{3.98}{0.2} \right) H_{\mathrm{m}} + \frac{100}{2 \times 10^{-3}} \left(\frac{2}{2} \right) i \right]$$

$$= -2.50 \times 10^{-5} H_{\mathrm{m}} + 6.28 \times 10^{-2} i$$

From this equation and Fig. 1.23b, we see that to drive the magnetic material into saturation to the point $B_{\mathrm{max}} - H_{\mathrm{max}}$, the current in the magnetizing winding must be increased to the value i_{max} where

$$i_{\mathrm{max}} = \frac{B_{\mathrm{max}} + 2.50 \times 10^{-5} H_{\mathrm{max}}}{6.28 \times 10^{-2}} \, \mathrm{A}$$

In this case, we do not have a complete hysteresis loop for Alnico 5, and hence we will have to estimate B_{max} and H_{max}. Linearly extrapolating the B-H curve at $H = 0$ back to 4 times the coercivity, that is, $H_{\mathrm{max}} = 4 \times 50 = 200 \, \mathrm{kA/m}$, yields $B_{\mathrm{max}} = 2.1 \, \mathrm{T}$. This value is undoubtedly extreme and will overestimate the required current somewhat. However, using $B_{\mathrm{max}} = 2.1 \, \mathrm{T}$ and $H_{\mathrm{max}} = 200 \, \mathrm{kA/m}$ yields $i_{\mathrm{max}} = 45.2 \, \mathrm{A}$.

Thus with the air-gap area set to $2 \, \mathrm{cm}^2$, increasing the current to 45.2 A and then reducing it to zero will achieve the desired magnetization.

c. Because we do not have specific information about the slope of the recoil line, we shall assume that its slope is the same as that of the B-H characteristic at the point $H = 0$, $B = B_{\mathrm{r}}$. From Fig. 1.23a, with the recoil line drawn with this slope, we see that as the air-gap area varies between 2 and $4 \, \mathrm{cm}^2$, the magnet flux density B_{m} varies between 1.00 and 1.08 T. Since the air-gap flux density equals $A_{\mathrm{m}}/A_{\mathrm{g}}$ times this value, the airgap flux density will equal $(2/2)1.00 = 1.0 \, \mathrm{T}$ when $A_{\mathrm{g}} = 2.0 \, \mathrm{cm}^2$ and $(2/4)1.08 = 0.54 \, \mathrm{T}$ when

$A_g = 4.0 \, \text{cm}^2$. Note from Fig. 1.23a that, when operated with these air-gap variations, the magnet appears to have an effective residual flux density of 1.17 T instead of the initial value of 1.24 T. Note that as long as the air-gap variation is limited to the range considered here, the system will continue to operate on the line labeled "Recoil line" in Fig. 1.23a and the magnet can be said to be *stabilized*.

As has been discussed, hard magnetic materials such as Alnico 5 can be subject to demagnetization, should their operating point be varied excessively. As shown in Example 1.11, these materials can be stabilized with some loss in effective remanent magnetization. However, this procedure does not guarantee absolute stability of operation. For example, if the material in Example 1.11 were subjected to an air-gap area smaller than 2 cm^2 or to excessive demagnetizing current, the effect of the stabilization would be erased and the material would be found to operate on a new recoil line with further reduced magnetization.

However, many materials, such as samarium-cobalt, Ceramic 7, and neodymium-iron-boron (see Fig. 1.19), which have large values of coercivity, tend to have very low values of recoil permeability, and the recoil line is essentially tangent to the *B-H* characteristic for a large portion of the useful operating region. For example, this can be seen in Fig. 1.19, which shows the dc magnetization curve for neodymium-iron-boron, from which we see that this material has a remanent magnetization of 1.25 T and a coercivity of −940 kA/m. The portion of the curve between these points is a straight line with a slope equal to $1.06\mu_0$, which is the same as the slope of its recoil line. As long as these materials are operated on this low-incremental-permeability portion of their *B-H* characteristic, they do not require stabilization, provided they are not excessively demagnetized.

For these materials, it is often convenient to assume that their dc magnetization curve is linear over their useful operating range with a slope equal to the recoil permeability μ_R. Under this assumption, the dc magnetization curve for these materials can be written in the form

$$B = \mu_R(H - H'_c) = B_r + \mu_R H \qquad (1.61)$$

Here, H'_c is the *apparent coercivity* associated with this linear representation. As can be seen from Fig. 1.19, the apparent coercivity is typically somewhat larger in magnitude (i.e. a larger negative value) than the material coercivity H_c because the dc magnetization characteristic tends to bend downward for low values of flux density.

1.7 SUMMARY

Electromechanical devices which employ magnetic fields often use ferromagnetic materials for guiding and concentrating these fields. Because the magnetic permeability of ferromagnetic materials can be large (up to tens of thousands times that of the surrounding space), most of the magnetic flux is confined to fairly well-defined paths determined by the geometry of the magnetic material. In addition, often the frequencies of interest are low enough to permit the magnetic fields to be considered quasi-static, and hence they can be determined simply from a knowledge of the net mmf acting on the magnetic structure.

As a result, the solution for the magnetic fields in these structures can be obtained in a straightforward fashion by using the techniques of magnetic-circuit analysis. These techniques can be used to reduce a complex three-dimensional magnetic field solution to what is essentially a one-dimensional problem. As in all engineering solutions, a certain amount of experience and judgment is required, but the technique gives useful results in many situations of practical engineering interest.

Ferromagnetic materials are available with a wide variety of characteristics. In general, their behavior is nonlinear, and their B-H characteristics are often represented in the form of a family of hysteresis (B-H) loops. Losses, both hysteretic and eddy-current, are functions of the flux level and frequency of operation as well as the material composition and the manufacturing process used. A basic understanding of the nature of these phenomena is extremely useful in the application of these materials in practical devices. Typically, important properties are available in the form of curves supplied by the material manufacturers.

Certain magnetic materials, commonly known as hard or permanent-magnet materials, are characterized by large values of remanent magnetization and coercivity. These materials produce significant magnetic flux even in magnetic circuits with air gaps. With proper design they can be made to operate stably in situations which subject them to a wide range of destabilizing forces and mmf's. Permanent magnets find application in many small devices, including loudspeakers, ac and dc motors, microphones, and analog electric meters.

1.8 PROBLEMS

1.1 A magnetic circuit with a single air gap is shown in Fig. 1.24. The core dimensions are:

Cross-sectional area $A_c = 1.8 \times 10^{-3}$ m^2
Mean core length $l_c = 0.6$ m
Gap length $g = 2.3 \times 10^{-3}$ m
$N = 83$ turns

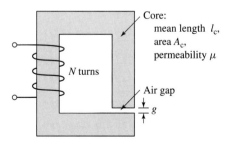

Figure 1.24 Magnetic circuit for Problem 1.1.

Assume that the core is of infinite permeability ($\mu \rightarrow \infty$) and neglect the effects of fringing fields at the air gap and leakage flux. (*a*) Calculate the reluctance of the core $\mathcal{R}_c$ and that of the gap $\mathcal{R}_g$. For a current of $i = 1.5$ A, calculate (*b*) the total flux ϕ, (*c*) the flux linkages λ of the coil, and (*d*) the coil inductance L.

1.2 Repeat Problem 1.1 for a finite core permeability of $\mu = 2500\mu_0$.

1.3 Consider the magnetic circuit of Fig. 1.24 with the dimensions of Problem 1.1. Assuming infinite core permeability, calculate (*a*) the number of turns required to achieve an inductance of 12 mH and (*b*) the inductor current which will result in a core flux density of 1.0 T.

1.4 Repeat Problem 1.3 for a core permeability of $\mu = 1300\mu_0$.

1.5 The magnetic circuit of Problem 1.1 has a nonlinear core material whose permeability as a function of B_m is given by

$$\mu = \mu_0 \left(1 + \frac{3499}{\sqrt{1 + 0.047(B_m)^{7.8}}} \right)$$

where B_m is the material flux density.

a. Using MATLAB, plot a dc magnetization curve for this material (B_m vs. H_m) over the range $0 \le B_m \le 2.2$ T.

b. Find the current required to achieve a flux density of 2.2 T in the core.

c. Again, using MATLAB, plot the coil flux linkages as a function of coil current as the current is varied from 0 to the value found in part (b).

1.6 The magnetic circuit of Fig. 1.25 consists of a core and a moveable plunger of width l_p, each of permeability μ. The core has cross-sectional area A_c and mean length l_c. The overlap area of the two air gaps A_g is a function of the plunger position x and can be assumed to vary as

$$A_g = A_c \left(1 - \frac{x}{X_0} \right)$$

You may neglect any fringing fields at the air gap and use approximations consistent with magnetic-circuit analysis.

a. Assuming that $\mu \rightarrow \infty$, derive an expression for the magnetic flux density in the air gap B_g as a function of the winding current I and as the

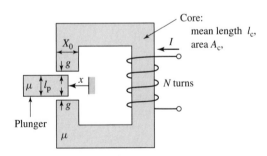

Figure 1.25   Magnetic circuit for Problem 1.6.

plunger position is varied ($0 \le x \le 0.8X_0$). What is the corresponding flux density in the core?

  b. Repeat part (a) for a finite permeability μ.

1.7 The magnetic circuit of Fig. 1.25 and Problem 1.6 has the following dimensions:

$$A_c = 8.2 \text{ cm}^2 \quad l_c = 23 \text{ cm}$$
$$l_p = 2.8 \text{ cm} \quad g = 0.8 \text{ mm}$$
$$X_0 = 2.5 \text{ cm} \quad N = 430 \text{ turns}$$

  a. Assuming a constant permeability of $\mu = 2800\mu_0$, calculate the current required to achieve a flux density of 1.3 T in the air gap when the plunger is fully retracted ($x = 0$).

  b. Repeat the calculation of part (a) for the case in which the core and plunger are composed of a nonlinear material whose permeability is given by

$$\mu = \mu_0 \left(1 + \frac{1199}{\sqrt{1 + 0.05 B_m^8}} \right)$$

where B_m is the magnetic flux density in the material.

  c. For the nonlinear material of part (b), use MATLAB to plot the air-gap flux density as a function of winding current for $x = 0$ and $x = 0.5X_0$.

1.8 An inductor of the form of Fig. 1.24 has dimensions:

    Cross-sectional area $A_c = 3.6 \text{ cm}^2$

    Mean core length $l_c = 15 \text{ cm}$

    $N = 75$ turns

Assuming a core permeability of $\mu = 2100\mu_0$ and neglecting the effects of leakage flux and fringing fields, calculate the air-gap length required to achieve an inductance of 6.0 mH.

1.9 The magnetic circuit of Fig. 1.26 consists of rings of magnetic material in a stack of height h. The rings have inner radius R_i and outer radius R_o. Assume that the iron is of infinite permeability ($\mu \to \infty$) and neglect the effects of magnetic leakage and fringing. For:

    $R_i = 3.4 \text{ cm}$

    $R_o = 4.0 \text{ cm}$

    $h = 2 \text{ cm}$

    $g = 0.2 \text{ cm}$

calculate:

  a. the mean core length l_c and the core cross-sectional area A_c.

  b. the reluctance of the core $\mathcal{R}_c$ and that of the gap $\mathcal{R}_g$.

For $N = 65$ turns, calculate:

  c. the inductance L.

  d. current i required to operate at an air-gap flux density of $B_g = 1.35\text{T}$.

  e. the corresponding flux linkages λ of the coil.

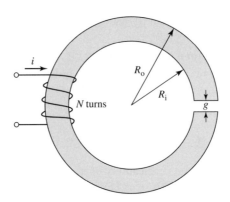

Figure 1.26   Magnetic circuit for
Problem 1.9.

1.10 Repeat Problem 1.9 for a core permeability of $\mu = 750\mu_0$.

1.11 Using MATLAB, plot the inductance of the inductor of Problem 1.9 as a
function of relative core permeability as the core permeability varies for
$\mu_r = 100$ to $\mu_r = 10000$. (Hint: Plot the inductance versus the log of the
relative permeability.) What is the minimum relative core permeability
required to insure that the inductance is within 5 percent of the value
calculated assuming that the core permeability is infinite?

1.12 The inductor of Fig. 1.27 has a core of uniform circular cross-section of area
A_c, mean length l_c and relative permeability μ_r and an N-turn winding. Write
an expression for the inductance L.

1.13 The inductor of Fig. 1.27 has the following dimensions:

$A_c = 1.0 \text{ cm}^2$
$l_c = 15 \text{ cm}$
$g = 0.8 \text{ mm}$
$N = 480 \text{ turns}$

Neglecting leakage and fringing and assuming $\mu_r = 1000$, calculate the
inductance.

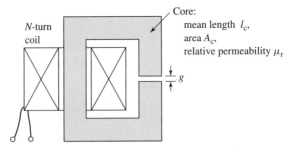

Figure 1.27   Inductor for Problem 1.12.

C/L

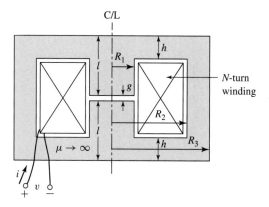

Figure 1.28 Pot-core inductor for Problem 1.15.

1.14 The inductor of Problem 1.13 is to be operated from a 60-Hz voltage source. (*a*) Assuming negligible coil resistance, calculate the rms inductor voltage corresponding to a peak core flux density of 1.5 T. (*b*) Under this operating condition, calculate the rms current and the peak stored energy.

1.15 Consider the magnetic circuit of Fig. 1.28. This structure, known as a *pot-core,* is typically made in two halves. The N-turn coil is wound on a cylindrical bobbin and can be easily inserted over the central post of the core as the two halves are assembled. Because the air gap is internal to the core, provided the core is not driven excessively into saturation, relatively little magnetic flux will "leak" from the core, making this a particularly attractive configuration for a wide variety of applications, both for inductors such as that of Fig. 1.27 and transformers.

Assume the core permeability to be $\mu = 2500\mu_0$ and $N = 200$ turns. The following dimensions are specified:

$$R_1 = 1.5 \text{ cm} \quad R_2 = 4 \text{ cm} \quad l = 2.5 \text{ cm}$$

$$h = 0.75 \text{ cm} \quad g = 0.5 \text{ mm}$$

a. Find the value of R_3 such that the flux density in the outer wall of the core is equal to that within the central cylinder.

b. Although the flux density in the radial sections of the core (the sections of thickness h) actually decreases with radius, assume that the flux density remains uniform. (*i*) Write an expression for the coil inductance and (*ii*) evaluate it for the given dimensions.

c. The core is to be operated at a peak flux density of 0.8 T at a frequency of 60 Hz. Find (*i*) the corresponding rms value of the voltage induced in the winding, (*ii*) the rms coil current, and (*iii*) the peak stored energy.

d. Repeat part (c) for a frequency of 50 Hz.

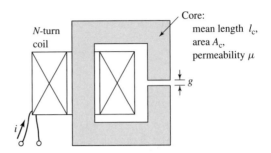

Figure 1.29   Inductor for Problem 1.17.

1.16 A square voltage wave having a fundamental frequency of 60 Hz and equal positive and negative half cycles of amplitude E is applied to a 1000-turn winding surrounding a closed iron core of $1.25 \times 10^{-3}\,\mathrm{m}^2$ cross section. Neglect both the winding resistance and any effects of leakage flux.

a. Sketch the voltage, the winding flux linkage, and the core flux as a function of time.

b. Find the maximum permissible value of E if the maximum flux density is not to exceed 1.15 T.

1.17 An inductor is to be designed using a magnetic core of the form of that of Fig. 1.29. The core is of uniform cross-sectional area $A_c = 5.0\,\mathrm{cm}^2$ and of mean length $l_c = 25$ cm.

a. Calculate the air-gap length g and the number of turns N such that the inductance is 1.4 mH and so that the inductor can operate at peak currents of 6 A without saturating. Assume that saturation occurs when the peak flux density in the core exceeds 1.7 T and that, below saturation, the core has permeability $\mu = 3200\mu_0$.

b. For an inductor current of 6 A, use Eq. 3.21 to calculate (i) the magnetic stored energy in the air gap and (ii) the magnetic stored energy in the core. Show that the total magnetic stored energy is given by Eq. 1.47.

1.18 Consider the inductor of Problem 1.17. Write a simple design program in the form of a MATLAB script to calculate the number of turns and air-gap length as a function of the desired inductance. The script should be written to request a value of inductance (in mH) from the user, with the output being the air-gap length in mm and the number of turns.

The inductor is to be operated with a sinusoidal current at 60 Hz, and it must be designed such that the peak core flux density will be equal to 1.7 T when the inductor current is equal to 4.5 A rms. Write your script to reject any designs for which the gap length is out of the range of 0.05 mm to 5.0 mm or for which the number of turns drops below 5.

Using your program find (a) the minimum and (b) the maximum inductances (to the nearest mH) which will satisfy the given constraints. For each of these values, find the required air-gap length and the number of turns as well as the rms voltage corresponding to the peak core flux.

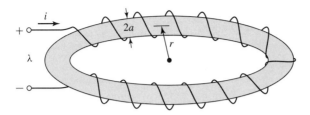

Figure 1.30 Toroidal winding for Problem 1.19.

1.19 A proposed energy storage mechanism consists of an N-turn coil wound around a large nonmagnetic ($\mu = \mu_0$) toroidal form as shown in Fig. 1.30. As can be seen from the figure, the toroidal form has a circular cross section of radius a and toroidal radius r, measured to the center of the cross section. The geometry of this device is such that the magnetic field can be considered to be zero everywhere outside the toroid. Under the assumption that $a \ll r$, the H field inside the toroid can be considered to be directed around the toroid and of uniform magnitude

$$H = \frac{Ni}{2\pi r}$$

For a coil with $N = 1000$ turns, $r = 10$ m, and $a = 0.45$ m:

a. Calculate the coil inductance L.

b. The coil is to be charged to a magnetic flux density of 1.75 T. Calculate the total stored magnetic energy in the torus when this flux density is achieved.

c. If the coil is to be charged at a uniform rate (i.e., $di/dt = $ constant), calculate the terminal voltage required to achieve the required flux density in 30 sec. Assume the coil resistance to be negligible.

1.20 Figure 1.31 shows an inductor wound on a laminated iron core of rectangular cross section. Assume that the permeability of the iron is infinite. Neglect magnetic leakage and fringing in the two air gaps (total gap length $= g$). The N-turn winding is insulated copper wire whose resistivity is ρ Ω·m. Assume

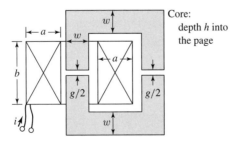

Figure 1.31 Iron-core inductor for Problem 1.20.

that the fraction f_w of the winding space is available for copper; the rest of the space is used for insulation.

a. Calculate the cross-sectional area and volume of the copper in the winding space.

b. Write an expression for the flux density B in the inductor in terms of the current density J_{cu} in the copper winding.

c. Write an expression for the copper current density J_{cu} in terms of the coil current I, the number of turns N, and the coil geometry.

d. Derive an expression for the electric power dissipation in the coil in terms of the current density J_{cu}.

e. Derive an expression for the magnetic stored energy in the inductor in terms of the applied current density J_{cu}.

f. From parts (d) and (e) derive an expression for the L/R time constant of the inductor. Note that this expression is independent of the number of turns in the coil and does not change as the inductance and coil resistance are changed by varying the number of turns.

1.21 The inductor of Fig. 1.31 has the following dimensions:

$$a = h = w = 1.5 \text{ cm} \quad b = 2 \text{ cm} \quad g = 0.2 \text{ cm}$$

The winding factor (i.e., the fraction of the total winding area occupied by conductor) is $f_w = 0.55$. The resistivity of copper is $1.73 \times 10^{-8} \ \Omega\cdot\text{m}$. When the coil is operated with a constant dc applied voltage of 35 V, the air-gap flux density is measured to be 1.4 T. Find the power dissipated in the coil, coil current, number of turns, coil resistance, inductance, time constant, and wire size to the nearest standard size. (Hint: Wire size can be found from the expression

$$\text{AWG} = 36 - 4.312 \ \ln \left(\frac{A_{\text{wire}}}{1.267 \times 10^{-8}} \right)$$

where AWG is the wire size, expressed in terms of the American Wire Gauge, and A_{wire} is the conductor cross-sectional area measured in m^2.)

1.22 The magnetic circuit of Fig. 1.32 has two windings and two air gaps. The core can be assumed to be of infinite permeability. The core dimensions are indicated in the figure.

a. Assuming coil 1 to be carrying a current I_1 and the current in coil 2 to be zero, calculate (*i*) the magnetic flux density in each of the air gaps, (*ii*) the flux linkage of winding 1, and (*iii*) the flux linkage of winding 2.

b. Repeat part (a), assuming zero current in winding 1 and a current I_2 in winding 2.

c. Repeat part (a), assuming the current in winding 1 to be I_1 and the current in winding 2 to be I_2.

d. Find the self-inductances of windings 1 and 2 and the mutual inductance between the windings.

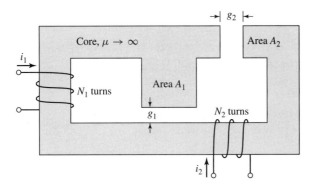

Figure 1.32 Magnetic circuit for Problem 1.22.

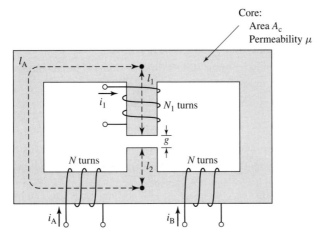

Figure 1.33 Symmetric magnetic circuit for Problem 1.23.

1.23 The symmetric magnetic circuit of Fig. 1.33 has three windings. Windings A and B each have N turns and are wound on the two bottom legs of the core. The core dimensions are indicated in the figure.

a. Find the self-inductances of each of the windings.

b. Find the mutual inductances between the three pairs of windings.

c. Find the voltage induced in winding 1 by time-varying currents $i_A(t)$ and $i_B(t)$ in windings A and B. Show that this voltage can be used to measure the imbalance between two sinusoidal currents of the same frequency.

1.24 The reciprocating generator of Fig. 1.34 has a movable plunger (position x) which is supported so that it can slide in and out of the magnetic yoke while maintaining a constant air gap of length g on each side adjacent to the yoke. Both the yoke and the plunger can be considered to be of infinite permeability. The motion of the plunger is constrained such that its position is limited to $0 \le x \le w$.

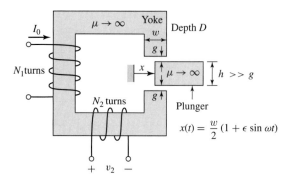

Figure 1.34   Reciprocating generator for Problem 1.24.

There are two windings on this magnetic circuit. The first has N_1 turns and carries a constant dc current I_0. The second, which has N_2 turns, is open-circuited and can be connected to a load.

a. Neglecting any fringing effects, find the mutual inductance between windings 1 and 2 as a function of the plunger position x.

b. The plunger is driven by an external source so that its motion is given by

$$x(t) = \frac{w(1 + \epsilon \sin \omega t)}{2}$$

where $\epsilon \leq 1$. Find an expression for the sinusoidal voltage which is generated as a result of this motion.

1.25 Figure 1.35 shows a configuration that can be used to measure the magnetic characteristics of electrical steel. The material to be tested is cut or punched into circular laminations which are then stacked (with interspersed insulation to avoid eddy-current formation). Two windings are wound over this stack of laminations: the first, with N_l turns, is used to excite a magnetic field in the lamination stack; the second, with N_2 turns, is used to sense the resultant magnetic flux.

The accuracy of the results requires that the magnetic flux density be uniform within the laminations. This can be accomplished if the lamination width $t = R_o - R_i$ is much smaller than the lamination radius and if the excitation winding is wound uniformly around the lamination stack. For the purposes of this analysis, assume there are n laminations, each of thickness Δ. Also assume that winding 1 is excited by a current $i_1 = I_0 \sin \omega t$.

a. Find the relationship between the magnetic field intensity H in the laminations and current i_1 in winding 1.

b. Find the relationship between the voltage v_2 and the time rate of change of the flux density B in the laminations.

c. Find the relationship between the voltage $v_0 = G \int v_2 \, dt$ and the flux density.

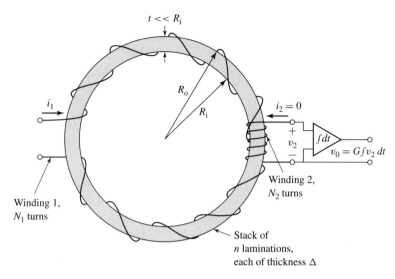

Figure 1.35 Configuration for measurement of magnetic properties of electrical steel.

In this problem, we have shown that the magnetic field intensity H and the magnetic flux density B in the laminations are proportional to the current i_1 and the voltage v_2 by known constants. Thus, B and H in the magnetic steel can be measured directly, and the B-H characteristics as discussed in Sections 1.3 and 1.4 can be determined.

1.26 From the dc magnetization curve of Fig. 1.10 it is possible to calculate the relative permeability $\mu_r = B_c/(\mu_0 H_c)$ for M-5 electrical steel as a function of the flux level B_c. Assuming the core of Fig. 1.2 to be made of M-5 electrical steel with the dimensions given in Example 1.1, calculate the maximum flux density such that the reluctance of the core never exceeds 5 percent of the reluctance of the total magnetic circuit.

1.27 In order to test the properties of a sample of electrical steel, a set of laminations of the form of Fig. 1.35 have been stamped out of a sheet of the electrical steel of thickness 3.0 mm. The radii of the laminations are $R_i = 75$ mm and $R_o = 82$ mm. They have been assembled in a stack of 10 laminations (separated by appropriate insulation to eliminate eddy currents) for the purposes of testing the magnetic properties at a frequency of 100 Hz.

a. The flux in the lamination stack will be excited from a variable-amplitude, 100-Hz voltage source whose peak amplitude is 30 V (peak-to-peak). Calculate the number of turns N_1 for the excitation winding required to insure that the lamination stack can be excited up to a peak flux density of 2.0 T.

b. With a secondary winding of $N_2 = 20$ turns and an integrator gain $G = 1000$, the output of the integrator is observed to be 7.0 V

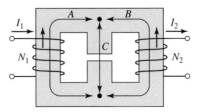

Figure 1.36   Magnetic circuit for Problem 1.28.

peak-to-peak. Calculate (*i*) the corresponding peak flux in the lamination stack and (*ii*) the corresponding amplitude of the voltage applied to the excitation winding.

1.28   The coils of the magnetic circuit shown in Fig. 1.36 are connected in series so that the mmf's of paths A and B both tend to set up flux in the center leg C in the same direction. The coils are wound with equal turns, $N_1 = N_2 = 100$. The dimensions are:

Cross-section area of A and B legs $= 7$ cm^2
Cross-section area of C legs $= 14$ cm^2
Length of A path $= 17$ cm
Length of B path $= 17$ cm
Length of C path $= 5.5$ cm
Air gap $= 0.4$ cm

The material is M-5 grade, 0.012-in steel, with a stacking factor of 0.94. Neglect fringing and leakage.

a.   How many amperes are required to produce a flux density of 1.2 T in the air gap?

b.   Under the condition of part (a), how many joules of energy are stored in the magnetic field in the air gap?

c.   Calculate the inductance.

1.29   The following table includes data for the top half of a symmetric 60-Hz hysteresis loop for a specimen of magnetic steel:

B, T	0	0.2	0.4	0.6	0.7	0.8	0.9	1.0	0.95	0.9	0.8	0.7	0.6	0.4	0.2	0
H, A · turns/m	48	52	58	73	85	103	135	193	80	42	2	−18	−29	−40	−45	−48

Using MATLAB, (*a*) plot this data, (*b*) calculate the area of the hysteresis loop in joules, and (*c*) calculate the corresponding 60-Hz core loss in Watts/kg. The density of M-5 steel is 7.65 g/cm^3.

1.30   Assume the magnetic circuit of Problem 1.1 and Fig. 1.24 to be made up of M-5 electrical steel with the properties described in Figs. 1.10, 1.12, and 1.14. Assume the core to be operating with a 60-Hz sinusoidal flux density of the rms flux density of 1.1 T. Neglect the winding resistance and leakage inductance. Find the winding voltage, rms winding current, and core loss for this operating condition. The density of M-5 steel is 7.65 g/cm³.

1.31 Repeat Example 1.8 under the assumption that all the core dimensions are doubled.

1.32 Using the magnetization characteristics for samarium cobalt given in Fig. 1.19, find the point of maximum energy product and the corresponding flux density and magnetic field intensity. Using these values, repeat Example 1.10 with the Alnico 5 magnet replaced by a samarium-cobalt magnet. By what factor does this reduce the magnet volume required to achieve the desired air-gap flux density?

1.33 Using the magnetization characteristics for neodymium-iron-boron given in Fig. 1.19, find the point of maximum-energy product and the corresponding flux density and magnetic field intensity. Using these values, repeat Example 1.10 with the Alnico 5 magnet replaced by a neodymium-iron-boron magnet. By what factor does this reduce the magnet volume required to achieve the desired air-gap flux density?

1.34 Figure 1.37 shows the magnetic circuit for a permanent-magnet loudspeaker. The voice coil (not shown) is in the form of a circular cylindrical coil which fits in the air gap. A samarium-cobalt magnet is used to create the air-gap dc magnetic field which interacts with the voice coil currents to produce the motion of the voice coil. The designer has determined that the air gap must have radius $R = 1.8$ cm, length $g = 0.1$ cm, and height $h = 0.9$ cm.

Assuming that the yoke and pole piece are of infinite magnetic permeability ($\mu \to \infty$), find the magnet height h_m and the magnet radius R_m that will result in an air-gap magnetic flux density of 1.2 T and require the smallest magnet volume.

(Hint: Refer to Example 1.10 and to Fig. 1.19 to find the point of maximum energy product for samarium cobalt.)

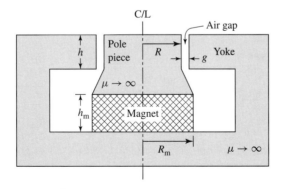

Figure 1.37 Magnetic circuit for the loudspeaker of Problem 1.34 (voice coil not shown).

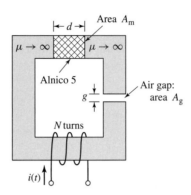

Figure 1.38 Magnetic circuit for
Problem 1.35.

1.35 It is desired to achieve a time-varying magnetic flux density in the air gap of
the magnetic circuit of Fig. 1.38 of the form

$$B_g = B_0 + B_1 \sin \omega t$$

where $B_0 = 0.5$ T and $B_l = 0.25$ T. The dc field B_0 is to be created by a
neodimium-iron-boron magnet, whereas the time-varying field is to be created
by a time-varying current.
   For $A_g = 6$ cm^2, $g = 0.4$ cm, and $N = 200$ turns, find:

a.   the magnet length d and the magnet area A_m that will achieve the desired
dc air-gap flux density and minimize the magnet volume.

b.   the minimum and maximum values of the time-varying current required to
achieve the desired time-varying air-gap flux density. Will this current
vary sinusoidally in time?

CHAPTER 2

Transformers

B efore we proceed with a study of electric machinery, it is desirable to discuss certain aspects of the theory of magnetically-coupled circuits, with emphasis on transformer action. Although the static transformer is not an energy conversion device, it is an indispensable component in many energy conversion systems. A significant component of ac power systems, it makes possible electric generation at the most economical generator voltage, power transfer at the most economical transmission voltage, and power utilization at the most suitable voltage for the particular utilization device. The transformer is also widely used in low-power, low-current electronic and control circuits for performing such functions as matching the impedances of a source and its load for maximum power transfer, isolating one circuit from another, or isolating direct current while maintaining ac continuity between two circuits.

The transformer is one of the simpler devices comprising two or more electric circuits coupled by a common magnetic circuit. Its analysis involves many of the principles essential to the study of electric machinery. Thus, our study of the transformer will serve as a bridge between the introduction to magnetic-circuit analysis of Chapter 1 and the more detailed study of electric machinery to follow.

2.1 INTRODUCTION TO TRANSFORMERS

Essentially, a transformer consists of two or more windings coupled by mutual magnetic flux. If one of these windings, the *primary,* is connected to an alternating-voltage source, an alternating flux will be produced whose amplitude will depend on the primary voltage, the frequency of the applied voltage, and the number of turns. The mutual flux will link the other winding, the *secondary,* [1] and will induce a voltage in it

---

[1] It is conventional to think of the "input" to the transformer as the primary and the "output" as the secondary. However, in many applications, power can flow either way and the concept of primary and secondary windings can become confusing. An alternate terminology, which refers to the windings as "high-voltage" and "low-voltage," is often used and eliminates this confusion.

whose value will depend on the number of secondary turns as well as the magnitude of the mutual flux and the frequency. By properly proportioning the number of primary and secondary turns, almost any desired *voltage ratio,* or *ratio of transformation,* can be obtained.

The essence of transformer action requires only the existence of time-varying mutual flux linking two windings. Such action can occur for two windings coupled through air, but coupling between the windings can be made much more effectively using a core of iron or other ferromagnetic material, because most of the flux is then confined to a definite, high-permeability path linking the windings. Such a transformer is commonly called an *iron-core transformer*. Most transformers are of this type. The following discussion is concerned almost wholly with iron-core transformers.

As discussed in Section 1.4, to reduce the losses caused by eddy currents in the core, the magnetic circuit usually consists of a stack of thin laminations. Two common types of construction are shown schematically in Fig. 2.1. In the *core type* (Fig. 2.1a) the windings are wound around two legs of a rectangular magnetic core; in the *shell type* (Fig. 2.1b) the windings are wound around the center leg of a three-legged core. Silicon-steel laminations 0.014 in thick are generally used for transformers operating at frequencies below a few hundred hertz. Silicon steel has the desirable properties of low cost, low core loss, and high permeability at high flux densities (1.0 to 1.5 T). The cores of small transformers used in communication circuits at high frequencies and low energy levels are sometimes made of compressed powdered ferromagnetic alloys known as *ferrites*.

In each of these configurations, most of the flux is confined to the core and therefore links both windings. The windings also produce additional flux, known as *leakage flux,* which links one winding without linking the other. Although leakage flux is a small fraction of the total flux, it plays an important role in determining the behavior of the transformer. In practical transformers, leakage is reduced by

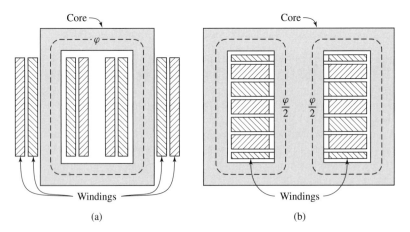

Figure 2.1   Schematic views of (a) core-type and (b) shell-type transformers.

subdividing the windings into sections placed as close together as possible. In the core-type construction, each winding consists of two sections, one section on each of the two legs of the core, the primary and secondary windings being concentric coils. In the shell-type construction, variations of the concentric-winding arrangement may be used, or the windings may consist of a number of thin "pancake" coils assembled in a stack with primary and secondary coils interleaved.

Figure 2.2 illustrates the internal construction of a *distribution transformer* such as is used in public utility systems to provide the appropriate voltage for use by residential consumers. A large power transformer is shown in Fig. 2.3.

Figure 2.2 Cutaway view of self-protected distribution transformer typical of sizes 2 to 25 kVA, 7200:240/120 V. Only one high-voltage insulator and lightning arrester is needed because one side of the 7200-V line and one side of the primary are grounded. (*General Electric Company.*)

Figure 2.3 A 660-MVA three-phase 50-Hz transformer used to step up generator voltage of 20 kV to transmission voltage of 405 kV. (*CEM Le Havre, French Member of the Brown Boveri Corporation.*)

2.2 NO-LOAD CONDITIONS

Figure 2.4 shows in schematic form a transformer with its secondary circuit open and an alternating voltage v_1 applied to its primary terminals. To simplify the drawings, it is common on schematic diagrams of transformers to show the primary and secondary windings as if they were on separate legs of the core, as in Fig. 2.4, even though the windings are actually interleaved in practice. As discussed in Section 1.4, a small steady-state current i_φ, called the *exciting current*, flows in the primary and establishes

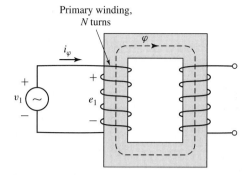

Figure 2.4 Transformer with open secondary.

an alternating flux in the magnetic circuit.[2] This flux induces an emf in the primary equal to

$$e_1 = \frac{d\lambda_1}{dt} = N_1 \frac{d\varphi}{dt} \qquad (2.1)$$

where

λ_1 = flux linkage of the primary winding
φ = flux in the core linking both windings
N_1 = number of turns in the primary winding

The voltage e_1 is in volts when φ is in webers. This emf, together with the voltage drop in the *primary resistance* R_1, must balance the applied voltage v_1; thus

$$v_1 = R_1 i_\varphi + e_1 \qquad (2.2)$$

Note that for the purposes of the current discussion, we are neglecting the effects of primary leakage flux, which will add an additional induced-emf term in Eq. 2.2. In typical transformers, this flux is a small percentage of the core flux, and it is quite justifiable to neglect it for our current purposes. It does however play an important role in the behavior of transformers and is discussed in some detail in Section 2.4.

In most large transformers, the no-load resistance drop is very small indeed, and the induced emf e_1 very nearly equals the applied voltage v_1. Furthermore, the waveforms of voltage and flux are very nearly sinusoidal. The analysis can then be greatly simplified, as we have shown in Section 1.4. Thus, if the instantaneous flux is

$$\varphi = \phi_{max} \sin \omega t \qquad (2.3)$$

---

[2] In general, the exciting current corresponds to the net ampere-turns (mmf) acting on the magnetic circuit, and it is not possible to distinguish whether it flows in the primary or secondary winding or partially in each winding.

the induced voltage is

$$e_1 = N_1 \frac{d\varphi}{dt} = \omega N_1 \phi_{max} \cos \omega t \tag{2.4}$$

where ϕ_{max} is the maximum value of the flux and $\omega = 2\pi f$, the frequency being f Hz. For the current and voltage reference directions shown in Fig. 2.4, the induced emf leads the flux by 90°. The rms value of the induced emf e_1 is

$$E_1 = \frac{2\pi}{\sqrt{2}} f N_1 \phi_{max} = \sqrt{2}\pi f N_1 \phi_{max} \tag{2.5}$$

If the resistive voltage drop is negligible, the counter emf equals the applied voltage. Under these conditions, if a sinusoidal voltage is applied to a winding, a sinusoidally varying core flux must be established whose maximum value ϕ_{max} satisfies the requirement that E_1 in Eq. 2.5 equal the rms value V_1 of the applied voltage; thus

$$\phi_{max} = \frac{V_1}{\sqrt{2}\pi f N_1} \tag{2.6}$$

Under these conditions, the core flux is determined solely by the applied voltage, its frequency, and the number of turns in the winding. This important relation applies not only to transformers but also to any device operated with a sinusoidally-alternating impressed voltage, as long as the resistance and leakage-inductance voltage drops are negligible. The core flux is fixed by the applied voltage, and the required exciting current is determined by the magnetic properties of the core; the exciting current must adjust itself so as to produce the mmf required to create the flux demanded by Eq. 2.6.

Because of the nonlinear magnetic properties of iron, the waveform of the exciting current differs from the waveform of the flux. A curve of the exciting current as a function of time can be found graphically from the ac hysteresis loop, as is discussed in Section 1.4 and shown in Fig. 1.11.

If the exciting current is analyzed by Fourier-series methods, it is found to consist of a fundamental component and a series of odd harmonics. The fundamental component can, in turn, be resolved into two components, one in phase with the counter emf and the other lagging the counter emf by 90°. The in-phase component supplies the power absorbed by hysteresis and eddy-current losses in the core. It is referred to as *core-loss component* of the exciting current. When the core-loss component is subtracted from the total exciting current, the remainder is called the *magnetizing current*. It comprises a fundamental component lagging the counter emf by 90°, together with all the harmonics. The principal harmonic is the third. For typical power transformers, the third harmonic usually is about 40 percent of the exciting current.

Except in problems concerned directly with the effects of harmonic currents, the peculiarities of the exciting-current waveform usually need not be taken into account, because the exciting current itself is small, especially in large transformers. For example, the exciting current of a typical power transformer is about 1 to 2 percent of full-load current. Consequently the effects of the harmonics usually are swamped

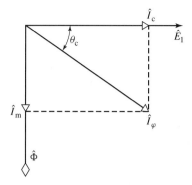

Figure 2.5 No-load phasor diagram.

out by the sinusoidal-currents supplied to other linear elements in the circuit. The exciting current can then be represented by an equivalent sinusoidal current which has the same rms value and frequency and produces the same average power as the actual exciting current. Such representation is essential to the construction of a *phasor diagram,* which represents the phase relationship between the various voltages and currents in a system in vector form. Each signal is represented by a phasor whose length is proportional to the amplitude of the signal and whose angle is equal to the phase angle of that signal as measured with respect to a chosen reference signal.

In Fig. 2.5, the phasors $\hat{E}_1$ and $\hat{\Phi}$ respectively, represent the rms values of the induced emf and the flux. The phasor $\hat{I}_\varphi$ represents the rms value of the equivalent sinusoidal exciting current. It lags the induced emf $\hat{E}_1$ by a phase angle θ_c.

The core loss P_c, equal to the product of the in-phase components of the $\hat{E}_1$ and $\hat{I}_\varphi$, is given by

$$P_c = E_1 I_\varphi \cos \theta_c \qquad (2.7)$$

The component $\hat{I}_c$ in phase with $\hat{E}_1$ is the core-loss current. The component $\hat{I}_m$ in phase with the flux represents an equivalent sine wave current having the same rms value as the magnetizing current. Typical exciting volt-ampere and core-loss characteristics of high-quality silicon steel used for power and distribution transformer laminations are shown in Figs. 1.12 and 1.14.

EXAMPLE 2.1

In Example 1.8 the core loss and exciting voltamperes for the core of Fig. 1.15 at $B_{max} = 1.5$ T and 60 Hz were found to be

$$P_c = 16 \text{ W} \qquad (VI)_{rms} = 20 \text{ VA}$$

and the induced voltage was $274/\sqrt{2} = 194$ V rms when the winding had 200 turns.

Find the power factor, the core-loss current I_c, and the magnetizing current I_m.

■ **Solution**

$$\text{Power factor } \cos\theta_c = \tfrac{16}{20} = 0.80 \text{ (lag)} \quad \text{thus} \quad \theta_c = -36.9°$$

Note that we know that the power factor is lagging because the system is inductive.

Exciting current $I_\varphi = \tfrac{20}{194} = 0.10$ A rms

Core-loss component $I_c = \tfrac{16}{194} = 0.082$ A rms

Magnetizing component $I_m = I_\varphi|\sin\theta_c| = 0.060$ A rms

---

2.3 EFFECT OF SECONDARY CURRENT; IDEAL TRANSFORMER

As a first approximation to a quantitative theory, consider a transformer with a primary winding of N_1 turns and a secondary winding of N_2 turns, as shown schematically in Fig. 2.6. Notice that the secondary current is defined as positive out of the winding; thus positive secondary current produces an mmf in the opposite direction from that created by positive primary current. Let the properties of this transformer be idealized under the assumption that winding resistances are negligible, that all the flux is confined to the core and links both windings (i.e., leakage flux is assumed negligible), that there are no losses in the core, and that the permeability of the core is so high that only a negligible exciting mmf is required to establish the flux. These properties are closely approached but never actually attained in practical transformers. A hypothetical transformer having these properties is often called an *ideal transformer*.

Under the above assumptions, when a time-varying voltage v_1 is impressed on the primary terminals, a core flux φ must be established such that the counter emf e_1 equals the impressed voltage. Thus

$$v_1 = e_1 = N_1 \frac{d\varphi}{dt} \tag{2.8}$$

The core flux also links the secondary and produces an induced emf e_2, and an equal secondary terminal voltage v_2, given by

$$v_2 = e_2 = N_2 \frac{d\varphi}{dt} \tag{2.9}$$

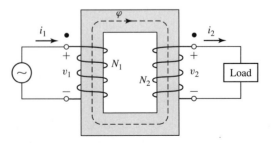

Figure 2.6 Ideal transformer and load.

From the ratio of Eqs. 2.8 and 2.9,

$$\frac{v_1}{v_2} = \frac{N_1}{N_2} \tag{2.10}$$

Thus an ideal transformer transforms voltages in the direct ratio of the turns in its windings.

Now let a load be connected to the secondary. A current i_2 and an mmf $N_2 i_2$ are then present in the secondary. Since the core permeability is assumed very large and since the impressed primary voltage sets the core flux as specfied by Eq. 2.8, the core flux is unchanged by the presence of a load on the secondary, and hence the net exciting mmf acting on the core (equal to $N_1 i_1 - N_2 i_2$) will not change and hence will remain negligible. Thus

$$N_1 i_1 - N_2 i_2 = 0 \tag{2.11}$$

From Eq. 2.11 we see that a compensating primary mmf must result to cancel that of the secondary. Hence

$$N_1 i_1 = N_2 i_2 \tag{2.12}$$

Thus we see that the requirement that the net mmf remain unchanged is the means by which the primary "knows" of the presence of load current in the secondary; any change in mmf flowing in the secondary as the result of a load must be accompanied by a corresponding change in the primary mmf. Note that for the reference directions shown in Fig. 2.6 the mmf's of i_1 and i_2 are in opposite directions and therefore compensate. The net mmf acting on the core therefore is zero, in accordance with the assumption that the exciting current of an ideal transformer is zero.

From Eq. 2.12

$$\frac{i_1}{i_2} = \frac{N_2}{N_1} \tag{2.13}$$

Thus an ideal transformer transforms currents in the inverse ratio of the turns in its windings.

Also notice from Eqs. 2.10 and 2.13 that

$$v_1 i_1 = v_2 i_2 \tag{2.14}$$

i.e., instantaneous power input to the primary equals the instantaneous power output from the secondary, a necessary condition because all dissipative and energy storage mechanisms in the transformer have been neglected.

An additional property of the ideal transformer can be seen by considering the case of a sinusoidal applied voltage and an impedance load. Phasor symbolism can be used. The circuit is shown in simplified form in Fig. 2.7a, in which the dot-marked terminals of the transformer correspond to the similarly marked terminals in Fig. 2.6. The dot markings indicate terminals of corresponding polarity; i.e., if one follows through the primary and secondary windings of Fig. 2.6, beginning at their dot-marked terminals, one will find that both windings encircle the core in the same direction with respect to the flux. Therefore, if one compares the voltages of the two windings, the voltages from a dot-marked to an unmarked terminal will be of the

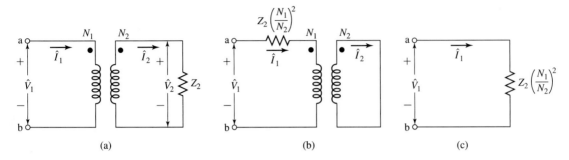

Figure 2.7 Three circuits which are identical at terminals ab when the transformer is ideal.

same instantaneous polarity for primary and secondary. In other words, the voltages $\hat{V}_1$ and $\hat{V}_2$ in Fig. 2.7a are in phase. Also currents $\hat{I}_1$ and $\hat{I}_2$ are in phase as seen from Eq. 2.12. Note again that the polarity of $\hat{I}_1$ is defined as into the dotted terminal and the polarity of $\hat{I}_2$ is defined as out of the dotted terminal.

We next investigate the impedance transformation properties of the ideal transformer. In phasor form, Eqs. 2.10 and 2.13 can be expressed as

$$\hat{V}_1 = \frac{N_1}{N_2}\hat{V}_2 \quad \text{and} \quad \hat{V}_2 = \frac{N_2}{N_1}\hat{V}_1 \tag{2.15}$$

$$\hat{I}_1 = \frac{N_2}{N_1}\hat{I}_2 \quad \text{and} \quad \hat{I}_2 = \frac{N_1}{N_2}\hat{I}_1 \tag{2.16}$$

From these equations

$$\frac{\hat{V}_1}{\hat{I}_1} = \left(\frac{N_1}{N_2}\right)^2 \frac{\hat{V}_2}{\hat{I}_2} \tag{2.17}$$

Noting that the load impedance Z_2 is related to the secondary voltages and currents

$$Z_2 = \frac{\hat{V}_2}{\hat{I}_2} \tag{2.18}$$

where Z_2 is the complex impedance of the load. Consequently, as far as its effect is concerned, an impedance Z_2 in the secondary circuit can be replaced by an equivalent impedance Z_1 in the primary circuit, provided that

$$Z_1 = \left(\frac{N_1}{N_2}\right)^2 Z_2 \tag{2.19}$$

Thus, the three circuits of Fig. 2.7 are indistinguishable as far as their performance viewed from terminals ab is concerned. Transferring an impedance from one side of a transformer to the other in this fashion is called *referring the impedance* to the other side; impedances transform as the square of the turns ratio. In a similar manner, voltages and currents can be *referred* to one side or the other by using Eqs. 2.15 and 2.16 to evaluate the equivalent voltage and current on that side.

To summarize, *in an ideal transformer, voltages are transformed in the direct ratio of turns, currents in the inverse ratio, and impedances in the direct ratio squared; power and voltamperes are unchanged.*

EXAMPLE 2.2

The equivalent circuit of Fig. 2.8a shows an ideal transformer with an impedance $R_2 + jX_2 = 1 + j4$ Ω connected in series with the secondary. The turns ratio $N_1/N_2 = 5{:}1$. (*a*) Draw an equivalent circuit with the series impedance referred to the primary side. (*b*) For a primary voltage of 120 V rms and a short connected across the terminals A-B, calculate the primary current and the current flowing in the short.

■ **Solution**

a. The new equivalent is shown in Fig. 2.8b. The secondary impedance is referred to the primary by the turns ratio squared. Thus

$$R_2' + jX_2' = \left(\frac{N_1}{N_2}\right)^2 (R_2 + jX_2)$$
$$= 25 + j100 \ \Omega$$

b. From Eq. 2.19, a short at terminals A-B will appear as a short at the primary of the ideal transformer in Fig. 2.8b since the zero voltage of the short is reflected by the turns ratio N_1/N_2 to the primary. Hence the primary current will be given by

$$\hat{I}_1 = \frac{\hat{V}_1}{R_2' + jX_2'} = \frac{120}{25 + j100} = 0.28 - j1.13 \text{ A rms}$$

corresponding to a magnitude of 1.16 A rms. From Eq. 2.13, the secondary current will equal $N_1/N_2 = 5$ times that of the current in the primary. Thus the current in the short will have a magnitude of $5(1.16) = 5.8$ A rms.

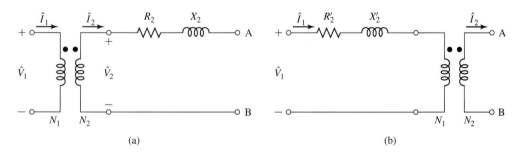

Figure 2.8   Equivalent circuits for Example 2.2. (a) Impedance in series with the secondary. (b) Impedance referred to the primary.

Repeat part (*b*) of Example 2.2 for a series impedance $R_2 + jX_2 = 0.05 + j0.97 \ \Omega$ and a turns ratio of 14:1.

Solution

The primary current is $0.03 - j0.63$ A rms, corresponding to a magnitude of 0.63 A rms. The current in the short will be 14 times larger and thus will be of magnitude 8.82 A rms.

2.4 TRANSFORMER REACTANCES AND EQUIVALENT CIRCUITS

The departures of an actual transformer from those of an ideal transformer must be included to a greater or lesser degree in most analyses of transformer performance. A more complete model must take into account the effects of winding resistances, leakage fluxes, and finite exciting current due to the finite (and indeed nonlinear) permeability of the core. In some cases, the capacitances of the windings also have important effects, notably in problems involving transformer behavior at frequencies above the audio range or during rapidly changing transient conditions such as those encountered in power system transformers as a result of voltage surges caused by lightning or switching transients. The analysis of these high-frequency problems is beyond the scope of the present treatment however, and accordingly the capacitances of the windings will be neglected.

Two methods of analysis by which departures from the ideal can be taken into account are (1) an equivalent-circuit technique based on physical reasoning and (2) a mathematical approach based on the classical theory of magnetically coupled circuits. Both methods are in everyday use, and both have very close parallels in the theories of rotating machines. Because it offers an excellent example of the thought process involved in translating physical concepts to a quantitative theory, the equivalent circuit technique is presented here.

To begin the development of a transformer equivalent circuit, we first consider the primary winding. The total flux linking the primary winding can be divided into two components: the resultant mutual flux, confined essentially to the iron core and produced by the combined effect of the primary and secondary currents, and the primary leakage flux, which links only the primary. These components are identified in the schematic transformer shown in Fig. 2.9, where for simplicity the primary and secondary windings are shown on opposite legs of the core. In an actual transformer with interleaved windings, the details of the flux distribution are more complicated, but the essential features remain the same.

The leakage flux induces voltage in the primary winding which adds to that produced by the mutual flux. Because the leakage path is largely in air, this flux and the voltage induced by it vary linearly with primary current $\hat{I}_1$. It can therefore be represented by a *primary leakage inductance* L_{l_1} (equal to the leakage-flux linkages

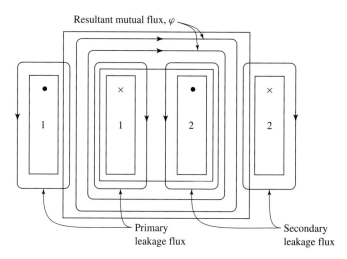

Figure 2.9 Schematic view of mutual and leakage fluxes in a transformer.

with the primary per unit of primary current). The corresponding *primary leakage reactance* X_{1_l} is found as

$$X_{1_l} = 2\pi f L_{1_l} \tag{2.20}$$

In addition, there will be a voltage drop in the primary resistance R_1.

We now see that the primary terminal voltage $\hat{V}_1$ consists of three components: the $\hat{I}_1 R_1$ drop in the primary resistance, the $\hat{I}_1 X_{1_l}$ drop arising from primary leakage flux, and the emf $\hat{E}_1$ induced in the primary by the resultant mutual flux. Fig. 2.10a shows an equivalent circuit for the primary winding which includes each of these voltages.

The resultant mutual flux links both the primary and secondary windings and is created by their combined mmf's. It is convenient to treat these mmf's by considering that the primary current must meet two requirements of the magnetic circuit: It must not only produce the mmf required to produce the resultant mutual flux, but it must also counteract the effect of the secondary mmf which acts to demagnetize the core. An alternative viewpoint is that the primary current must not only magnetize the core, it must also supply current to the load connected to the secondary. According to this picture, it is convenient to resolve the primary current into two components: an exciting component and a load component. The *exciting component* $\hat{I}_\varphi$ is defined as the additional primary current required to produce the resultant mutual flux. It is a nonsinusoidal current of the nature described in Section 2.2.[3] The *load component*

---

[3] In fact, the exciting current corresponds to the net mmf acting on the transformer core and cannot, in general, be considered to flow in the primary alone. However, for the purposes of this discussion, this distinction is not significant.

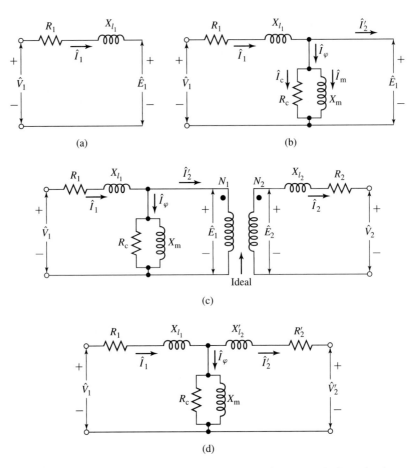

Figure 2.10 Steps in the development of the transformer equivalent circuit.

$\hat{I}'_2$ is defined as the component current in the primary which would exactly counteract the mmf of secondary current $\hat{I}_2$.

Since it is the exciting component which produces the core flux, the net mmf must equal $N_1\hat{I}_\varphi$ and thus we see that

$$N_1\hat{I}_\varphi = N_1\hat{I}_1 - N_2\hat{I}_2$$

$$= N_1(\hat{I}_\varphi + \hat{I}'_2) - N_2\hat{I}_2 \tag{2.21}$$

and from Eq. 2.21 we see that

$$\hat{I}'_2 = \frac{N_2}{N_1}\hat{I}_2 \tag{2.22}$$

From Eq. 2.22, we see that the load component of the primary current equals the secondary current referred to the primary as in an ideal transformer.

The exciting current can be treated as an equivalent sinusoidal current $\hat{I}_\varphi$, in the manner described in Section 2.2, and can be resolved into a core-loss component $\hat{I}_c$ in phase with the emf $\hat{E}_1$ and a magnetizing component $\hat{I}_m$ lagging $\hat{E}_1$ by 90°. In the equivalent circuit (Fig. 2.10b) the equivalent sinusoidal exciting current is accounted for by means of a shunt branch connected across $\hat{E}_1$, comprising a *core-loss resistance* R_c in parallel with a *magnetizing inductance* L_m whose reactance, known as the *magnetizing reactance*, is given by

$$X_m = 2\pi f L_m \qquad (2.23)$$

In the equivalent circuit of (Fig. 2.10b) the power E_1^2/R_c accounts for the core loss due to the resultant mutual flux. R_c is referred to as the *magnetizing resistance* or *core-loss resistance* and together with X_m forms the *excitation branch* of the equivalent circuit, and we will refer to the parallel combination of R_c and X_m as the *exciting impedance* Z_φ. When R_c is assumed constant, the core loss is thereby assumed to vary as E_1^2 or (for sine waves) as $\phi_{max}^2 f^2$, where ϕ_{max} is the maximum value of the resultant mutual flux. Strictly speaking, the magnetizing reactance X_m varies with the saturation of the iron. When X_m is assumed constant, the magnetizing current is thereby assumed to be independent of frequency and directly proportional to the resultant mutual flux. Both R_c and X_m are usually determined at rated voltage and frequency; they are then assumed to remain constant for the small departures from rated values associated with normal operation.

We will next add to our equivalent circuit a representation of the secondary winding. We begin by recognizing that the resultant mutual flux $\hat{\Phi}$ induces an emf $\hat{E}_2$ in the secondary, and since this flux links both windings, the induced-emf ratio must equal the winding turns ratio, i.e.,

$$\frac{\hat{E}_1}{\hat{E}_2} = \frac{N_1}{N_2} \qquad (2.24)$$

just as in an ideal transformer. This voltage transformation and the current transformation of Eq. 2.22 can be accounted for by introducing an ideal transformer in the equivalent circuit, as in Fig. 2.10c. Just as is the case for the primary winding, the emf $\hat{E}_2$ is not the secondary terminal voltage, however, because of the *secondary resistance* R_2 and because the secondary current $\hat{I}_2$ creates secondary leakage flux (see Fig. 2.9). The secondary terminal voltage $\hat{V}_2$ differs from the induced voltage $\hat{E}_2$ by the voltage drops due to secondary resistance R_2 and *secondary leakage reactance* X_{l_2} (corresponding to the *secondary leakage inductance* L_{l_2}), as in the portion of the complete transformer equivalent circuit (Fig. 2.10c) to the right of $\hat{E}_2$.

From the equivalent circuit of Fig. 2.10, the actual transformer therefore can be seen to be equivalent to an ideal transformer plus external impedances. By referring all quantities to the primary or secondary, the ideal transformer in Fig. 2.10c can be moved out to the right or left, respectively, of the equivalent circuit. This is almost invariably done, and the equivalent circuit is usually drawn as in Fig. 2.10d, with the ideal transformer not shown and all voltages, currents, and impedances referred to either the primary or secondary winding. Specifically, for Fig. 2.10d,

$$X'_{l_2} = \left(\frac{N_1}{N_2}\right)^2 X_{l_2} \tag{2.25}$$

$$R'_2 = \left(\frac{N_1}{N_2}\right)^2 R_2 \tag{2.26}$$

and

$$V'_2 = \frac{N_1}{N_2} V_2 \tag{2.27}$$

The circuit of Fig. 2.10d is called the *equivalent-T circuit* for a transformer.

In Fig. 2.10d, in which the secondary quantities are referred to the primary, the referred secondary values are indicated with primes, for example, X'_{l_2} and R'_2, to distinguish them from the actual values of Fig. 2.10c. In the discussion that follows we almost always deal with referred values, and the primes will be omitted. One must simply keep in mind the side of the transformers to which all quantities have been referred.

EXAMPLE 2.3

A 50-kVA 2400:240-V 60-Hz distribution transformer has a leakage impedance of $0.72 + j0.92\ \Omega$ in the high-voltage winding and $0.0070 + j0.0090\ \Omega$ in the low-voltage winding. At rated voltage and frequency, the impedance Z_φ of the shunt branch (equal to the impedance of R_c and jX_m in parallel) accounting for the exciting current is $6.32 + j43.7\ \Omega$ when viewed from the low-voltage side. Draw the equivalent circuit referred to (a) the high-voltage side and (b) the low-voltage side, and label the impedances numerically.

■ Solution

The circuits are given in Fig. 2.11a and b, respectively, with the high-voltage side numbered 1 and the low-voltage side numbered 2. The voltages given on the nameplate of a power system transformer are based on the turns ratio and neglect the small leakage-impedance voltage drops under load. Since this is a 10-to-1 transformer, impedances are referred by multiplying or dividing by 100; for example, the value of an impedance referred to the high-voltage side is greater by a factor of 100 than its value referred to the low-voltage side.

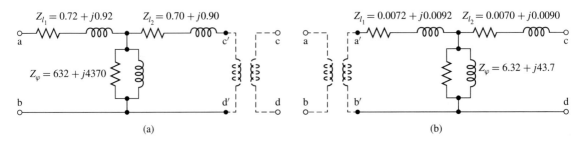

(a)   (b)

Figure 2.11 Equivalent circuits for transformer of Example 2.3 referred to (a) the high-voltage side and (b) the low-voltage side.

The ideal transformer may be explicitly drawn, as shown dotted in Fig. 2.11, or it may be omitted in the diagram and remembered mentally, making the unprimed letters the terminals. If this is done, one must of course remember to refer all connected impedances and sources to be consistent with the omission of the ideal transformer.

If 2400 V rms is applied to the high-voltage side of the transformer of Example 2.3, calculate the magnitude of the current into the magnetizing impedance Z_φ in Figs. 2.11a and b respectively.

Solution

The current through Z_φ is 0.543 A rms when it is referred to the high-voltage side as in Fig. 2.11a and 5.43 A rms when it is referred to the low-voltage side.

2.5 ENGINEERING ASPECTS OF TRANSFORMER ANALYSIS

In engineering analyses involving the transformer as a circuit element, it is customary to adopt one of several approximate forms of the equivalent circuit of Fig. 2.10 rather than the full circuit. The approximations chosen in a particular case depend largely on physical reasoning based on orders of magnitude of the neglected quantities. The more common approximations are presented in this section. In addition, test methods are given for determining the transformer constants.

The approximate equivalent circuits commonly used for constant-frequency power transformer analyses are summarized for comparison in Fig. 2.12. All quantities in these circuits are referred to either the primary or the secondary, and the ideal transformer is not shown.

Computations can often be greatly simplified by moving the shunt branch representing the exciting current out from the middle of the T circuit to either the primary or the secondary terminals, as in Fig. 2.12a and b. These forms of the equivalent circuit are referred to as *cantilever circuits*. The series branch is the combined resistance and leakage reactance of the primary and secondary, referred to the same side. This impedance is sometimes called the *equivalent series impedance* and its components the *equivalent series resistance* R_{eq} and *equivalent series reactance* X_{eq}, as shown in Fig. 2.12a and b.

As compared to the equivalent-T circuit of Fig. 2.10d, the cantilever circuit is in error in that it neglects the voltage drop in the primary or secondary leakage impedance caused by the exciting current. Because the impedance of the exciting branch is typically quite large in large power transformers, the corresponding exciting current is quite small. This error is insignificant in most situations involving large transformers.

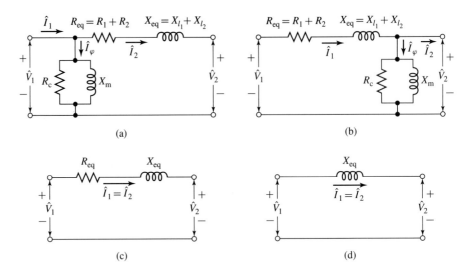

Figure 2.12 Approximate transformer equivalent circuits.

EXAMPLE 2.4

Consider the equivalent-T circuit of Fig. 2.11a of the 50-kVA 2400:240 V distribution transformer of Example 2.3 in which the impedances are referred to the high-voltage side. (*a*) Draw
the cantilever equivalent circuit with the shunt branch at the high-voltage terminal. Calculate
and label R_{eq} and X_{eq}. (*b*) With the low-voltage terminal open-circuit and 2400 V applied to the
high-voltage terminal, calculate the voltage at the low-voltage terminal as predicted by each
equivalent circuit.

■ **Solution**

a. The cantilever equivalent circuit is shown in Fig. 2.13. R_{eq} and X_{eq} are found simply as
   the sum of the high- and low-voltage winding series impedances of Fig. 2.11a

$$R_{eq} = 0.72 + 0.70 = 1.42 \ \Omega$$

$$X_{eq} = 0.92 + 0.90 = 1.82 \ \Omega$$

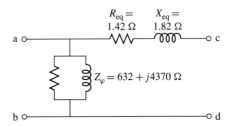

Figure 2.13 Cantilever equivalent
circuit for Example 2.4.

b. For the equivalent-T circuit of Fig. 2.11a, the voltage at the terminal labeled c'-d' will be given by

$$\hat{V}_{c'\text{-}d'} = 2400 \left(\frac{Z_\varphi}{Z_\varphi + Z_{l_1}} \right) = 2399.4 + j0.315 \text{ V}$$

This corresponds to an rms magnitude of 2399.4 V. Reflected to the low-voltage terminals by the low- to high-voltage turns ratio, this in turn corresponds to a voltage of 239.94 V.

Because the exciting impedance is connected directly across the high-voltage terminals in the cantilever equivalent circuit of Fig. 2.13, there will be no voltage drop across any series leakage impedance and the predicted secondary voltage will be 240 V. These two solutions differ by 0.025 percent, well within reasonable engineering accuracy and clearly justifying the use of the cantilever equivalent circuit for analysis of this transformer.

Further analytical simplification results from neglecting the exciting current entirely, as in Fig. 2.12c, in which the transformer is represented as an equivalent series impedance. If the transformer is large (several hundred kilovoltamperes or more), the equivalent resistance R_{eq} is small compared with the equivalent reactance X_{eq} and can frequently be neglected, giving the equivalent circuit of Fig. 2.12d. The circuits of Fig. 2.12c and d are sufficiently accurate for most ordinary power system problems and are used in all but the most detailed analyses. Finally, in situations where the currents and voltages are determined almost wholly by the circuits external to the transformer or when a high degree of accuracy is not required, the entire transformer impedance can be neglected and the transformer considered to be ideal, as in Section 2.3.

The circuits of Fig. 2.12 have the additional advantage that the total equivalent resistance R_{eq} and equivalent reactance X_{eq} can be found from a very simple test in which one terminal is short-circuited. On the other hand, the process of determination of the individual leakage reactances X_{l_1} and X_{l_2} and a complete set of parameters for the equivalent-T circuit of Fig. 2.10c is more difficult. Example 2.4 illustrates that due to the voltage drop across leakage impedances, the ratio of the measured voltages of a transformer will not be indentically equal to the idealized voltage ratio which would be measured if the transformer were ideal. In fact, without some apriori knowledge of the turns ratio (based for example upon knowledge of the internal construction of the transformer), it is not possible to make a set of measurements which uniquely determine the turns ratio, the magnetizing inductance, and the individual leakage impedances.

It can be shown that, simply from terminal measurements, neither the turns ratio, the magnetizing reactance, or the leakage reactances are unique characteristics of a transformer equivalent circuit. For example, the turns ratio can be chosen arbitrarily and for each choice of turns ratio, there will be a corresponding set of values for the leakage and magnetizing reactances which matches the measured characteristic. Each of the resultant equivalent circuits will have the same electrical terminal characteristics, a fact which has the fortunate consequence that any self-consistent set of empirically determined parameters will adequately represent the transformer.

EXAMPLE 2.5

The 50-kVA 2400:240-V transformer whose parameters are given in Example 2.3 is used to step down the voltage at the load end of a feeder whose impedance is $0.30 + j1.60 \ \Omega$. The voltage V_s at the sending end of the feeder is 2400 V.

Find the voltage at the secondary terminals of the transformer when the load connected to its secondary draws rated current from the transformer and the power factor of the load is 0.80 lagging. Neglect the voltage drops in the transformer and feeder caused by the exciting current.

■ Solution

The circuit with all quantities referred to the high-voltage (primary) side of the transformer is shown in Fig. 2.14a, where the transformer is represented by its equivalent impedance, as in Fig. 2.12c. From Fig. 2.11a, the value of the equivalent impedance is $Z_{eq} = 1.42 + j1.82 \ \Omega$ and the combined impedance of the feeder and transformer in series is $Z = 1.72 + j3.42 \ \Omega$. From the transformer rating, the load current referred to the high-voltage side is $I = 50,000/2400 = 20.8$ A.

This solution is most easily obtained with the aid of the phasor diagram referred to the high-voltage side as shown in Fig. 2.14b. Note that the power factor is defined at the load side of the transformer and hence defines the phase angle θ between the load current $\hat{I}$ and the voltage $\hat{V}_2$

$$\theta = -\cos^{-1}(0.80) = -36.87°$$

From the phasor diagram

$$Ob = \sqrt{V_s^2 - (bc)^2} \quad \text{and} \quad V_2 = Ob - ab$$

Note that

$$bc = IX\cos\theta - IR\sin\theta \quad ab = IR\cos\theta + IX\sin\theta$$

where R and X are the combined resistance and reactance, respectively. Thus

$$bc = 20.8(3.42)(0.80) - 20.8(1.72)(0.60) = 35.5 \text{ V}$$

$$ab = 20.8(1.72)(0.80) + 20.8(3.42)(0.60) = 71.4 \text{ V}$$

Substitution of numerical values shows that $V_2 = 2329$ V, referred to the high-voltage side. The actual voltage at the secondary terminals is 2329/10, or

$$V_2 = 233 \text{ V}$$

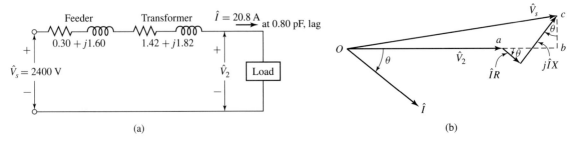

(a)   (b)

Figure 2.14 (a) Equivalent circuit and (b) phasor diagram for Example 2.5.

Repeat Example 2.5 for a load which draws rated current from the transformer with a power factor of 0.8 leading.

Solution

$$V_2 = 239 \text{ V}$$

Two very simple tests serve to determine the parameters of the equivalent circuits of Fig. 2.10 and 2.12. These consist of measuring the input voltage, current, and power to the primary, first with the secondary short-circuited and then with the secondary open-circuited.

Short-Circuit Test   The *short-circuit test* can be used to find the equivalent series impedance $R_{eq} + jX_{eq}$. Although the choice of winding to short-circuit is arbitrary, for the sake of this discussion we will consider the short circuit to be applied to the transformer secondary and voltage applied to primary. For convenience, the high-voltage side is usually taken as the primary in this test. Because the equivalent series impedance in a typical transformer is relatively small, typically an applied primary voltage on the order of 10 to 15 percent or less of the rated value will result in rated current.

Figure 2.15a shows the equivalent circuit with transformer secondary impedance referred to the primary side and a short circuit applied to the secondary. The short-circuit impedance Z_{sc} looking into the primary under these conditions is

$$Z_{sc} = R_1 + jX_{l_1} + \frac{Z_\varphi(R_2 + jX_{l_2})}{Z_\varphi + R_2 + jX_{l_2}} \tag{2.28}$$

Because the impedance Z_φ of the exciting branch is much larger than that of the secondary leakage impedance (which will be true unless the core is heavily saturated by excessive voltage applied to the primary; certainly not the case here), the short-circuit impedance can be approximated as

$$Z_{sc} \approx R_1 + jX_{l_1} + R_2 + jX_{l_2} = R_{eq} + jX_{eq} \tag{2.29}$$

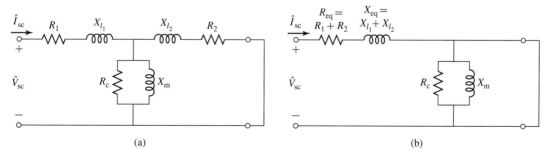

(a)   (b)

Figure 2.15 Equivalent circuit with short-circuited secondary. (a) Complete equivalent circuit. (b) Cantilever equivalent circuit with the exciting branch at the transformer secondary.

Note that the approximation made here is equivalent to the approximation made in reducing the equivalent-T circuit to the cantilever equivalent. This can be seen from Fig. 2.15b; the impedance seen at the input of this equivalent circuit is clearly $Z_{sc} = Z_{eq} = R_{eq} + jX_{eq}$ since the exciting branch is directly shorted out by the short on the secondary.

Typically the instrumentation used for this test will measure the rms magnitude of the applied voltage V_{sc}, the short-circuit current I_{sc}, and the power P_{sc}. Based upon these three measurements, the equivalent resistance and reactance (referred to the primary) can be found from

$$|Z_{eq}| = |Z_{sc}| = \frac{V_{sc}}{I_{sc}} \tag{2.30}$$

$$R_{eq} = R_{sc} = \frac{P_{sc}}{I_{sc}^2} \tag{2.31}$$

$$X_{eq} = X_{sc} = \sqrt{|Z_{sc}|^2 - R_{sc}^2} \tag{2.32}$$

where the symbol $||$ indicates the magnitude of the enclosed complex quantity.

The equivalent impedance can, of course, be referred from one side to the other in the usual manner. On the rare occasions when the equivalent-T circuit in Fig. 2.10d must be resorted to, approximate values of the individual primary and secondary resistances and leakage reactances can be obtained by assuming that $R_1 = R_2 = 0.5R_{eq}$ and $X_{1_1} = X_{1_2} = 0.5X_{eq}$ when all impedances are referred to the same side. Strictly speaking, of course, it is possible to measure R_1 and R_2 directly by a dc resistance measurement on each winding (and then referring one or the other to the other side of the idea transformer). However, as has been discussed, no such simple test exists for the leakage reactances X_{1_1} and X_{1_2}.

Open-Circuit Test The *open-circuit test* is performed with the secondary open-circuited and rated voltage impressed on the primary. Under this condition an exciting current of a few percent of full-load current (less on large transformers and more on smaller ones) is obtained. Rated voltage is chosen to insure that the magnetizing reactance will be operating at a flux level close to that which will exist under normal operating conditions. If the transformer is to be used at other than its rated voltage, the test should be done at that voltage. For convenience, the low-voltage side is usually taken as the primary in this test. If the primary in this test is chosen to be the opposite winding from that of the short-circuit test, one must of course be careful to refer the various measured impedances to the same side of the transformer in order to obtain a self-consistent set of parameter values.

Figure 2.16a shows the equivalent circuit with the transformer secondary impedance referred to the primary side and the secondary open-circuited. The open-circuit impedance Z_{oc} looking into the primary under these conditions is

$$Z_{oc} = R_1 + jX_{1_1} + Z_{\varphi} = R_1 + jX_{1_1} + \frac{R_c\,(jX_m)}{R_c + jX_m} \tag{2.33}$$

Because the impedance of the exciting branch is quite large, the voltage drop in the primary leakage impedance caused by the exciting current is typically negligible, and

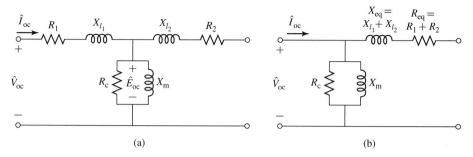

Figure 2.16 Equivalent circuit with open-circuited secondary. (a) Complete equivalent circuit. (b) Cantilever equivalent circuit with the exciting branch at the transformer primary.

the primary impressed voltage $\hat{V}_{oc}$ very nearly equals the emf $\hat{E}_{oc}$ induced by the resultant core flux. Similarly, the primary $I_{oc}^2 R_1$ loss caused by the exciting current is negligible, so that the power input P_{oc} very nearly equals the core loss E_{oc}^2/R_c. As a result, it is common to ignore the primary leakage impedance and to approximate the open-circuit impedance as being equal to the magnetizing impedance

$$Z_{oc} \approx Z_\varphi = \frac{R_c(jX_m)}{R_c + jX_m} \tag{2.34}$$

Note that the approximation made here is equivalent to the approximation made in reducing the equivalent-T circuit to the cantilever equivalent circuit of Fig. 2.16b; the impedance seen at the input of this equivalent circuit is clearly Z_φ since no current will flow in the open-circuited secondary.

As with the short-circuit test, typically the instrumentation used for this test will measure the rms magnitude of the applied voltage, V_{oc}, the open-circuit current I_{oc} and the power P_{oc}. Neglecting the primarly leakage impedance and based upon these three measurements, the magnetizing resistance and reactance (referred to the primary) can be found from

$$R_c = \frac{V_{oc}^2}{P_{oc}} \tag{2.35}$$

$$|Z_\varphi| = \frac{V_{oc}}{Z_{oc}} \tag{2.36}$$

$$X_m = \frac{1}{\sqrt{(1/|Z_\varphi|)^2 - (1/R_c)^2}} \tag{2.37}$$

The values obtained are, of course, referred to the side used as the primary in this test.

The open-circuit test can be used to obtain the core loss for efficiency computations and to check the magnitude of the exciting current. Sometimes the voltage at the terminals of the open-circuited secondary is measured as a check on the turns ratio.

Note that, if desired, a slightly more accurate calculation of X_m and R_c by retaining the measurements of R_1 and X_{1_1} obtained from the short-circuit test (referred to the proper side of the transformer) and basing the derivation on Eq. 2.33. However, such additional effort is rarely necessary for the purposes of engineering accuracy.

EXAMPLE 2.6

With the instruments located on the high-voltage side and the low-voltage side short-circuited, the short-circuit test readings for the 50-kVA 2400:240-V transformer of Example 2.3 are 48 V, 20.8 A, and 617 W. An open-circuit test with the low-voltage side energized gives instrument readings on that side of 240 V, 5.41 A, and 186 W. Determine the efficiency and the voltage regulation at full load, 0.80 power factor lagging.

■ Solution

From the short-circuit test, the magnitude of the equivalent impedance, the equivalent resistance, and the equivalent reactance of the transformer (referred to the high-voltage side as denoted by the subscript H) are

$$|Z_{eq,H}| = \frac{48}{20.8} = 2.31 \ \Omega \quad R_{eq,H} = \frac{617}{20.8^2} = 1.42 \ \Omega$$

$$X_{eq,H} = \sqrt{2.31^2 - 1.42^2} = 1.82 \ \Omega$$

Operation at full-load, 0.80 power factor lagging corresponds to a current of

$$I_H = \frac{50,000}{2400} = 20.8 \ A$$

and an output power

$$P_{output} = P_{load} = (0.8)50,000 = 40,000 \ W$$

The total loss under this operating condition is equal to the sum of the winding loss

$$P_{winding} = I_H^2 R_{eq,H} = 20.8^2(1.42) = 617 \ W$$

and the core loss determined from the open-circuit test

$$P_{core} = 186 \ W$$

Thus

$$P_{loss} = P_{winding} + P_{core} = 803 \ W$$

and the power input to the transformer is

$$P_{input} = P_{output} + P_{loss} = 40,803 \ W$$

The *efficiency* of a power conversion device is defined as

$$efficiency = \frac{P_{output}}{P_{input}} = \frac{P_{input} - P_{loss}}{P_{input}} = 1 - \frac{P_{loss}}{P_{input}}$$

which can be expressed in percent by multiplying by 100 percent. Hence, for this operating condition

$$efficiency = 100\% \left(\frac{P_{output}}{P_{input}} \right) = 100\% \left(\frac{40,000}{40,000 + 803} \right) = 98.0\%$$

The *voltage regulation* of a transformer is defined as the change in secondary terminal voltage from no load to full load and is usually expressed as a percentage of the full-load value. In power systems applications, regulation is one figure of merit for a transformer; a low value indicates that load variations on the secondary of that transformer will not significantly affect the magnitude of the voltage being supplied to the load. It is calculated under the assumption that the primary voltage remains constant as the load is removed from the transformer secondary.

The equivalent circuit of Fig. 2.12c will be used with all quantities referred to the high-voltage side. The primary voltage is assumed to be adjusted so that the secondary terminal voltage has its rated value at full load, or $V_{2H} = 2400$ V. For a load of rated value and 0.8 power factor lagging (corresponding to a power factor angle $\theta = -\cos^{-1}(0.8) = -36.9°$), the load current will be

$$\hat{I}_H = \left(\frac{50 \times 10^3}{2400} \right) e^{-j36.9°} = 20.8(0.8 - j0.6) \text{ A}$$

The required value of the primary voltage V_{1H} can be calculated as

$$\hat{V}_{1H} = \hat{V}_{2H} + \hat{I}_H(R_{eq,H} + jX_{eq,H})$$

$$= 2400 + 20.8(0.80 - j0.60)(1.42 + j1.82)$$

$$= 2446 + j13$$

The magnitude of $\hat{V}_{1H}$ is 2446 V. If this voltage were held constant and the load removed, the secondary voltage on open circuit would rise to 2446 V referred to the high-voltage side. Then

$$\text{Regulation} = \frac{2446 - 2400}{2400}(100\%) = 1.92\%$$

Practice Problem 2.4

Repeat the voltage-regulation calculation of Example 2.6 for a load of 50 kW (rated load, unity power factor).

Solution

$$\text{Regulation} = 1.24\%$$

2.6 AUTOTRANSFORMERS; MULTIWINDING TRANSFORMERS

The principles discussed in previous sections have been developed with specific reference to two-winding transformers. They are also applicable to transformers with other winding configurations. Aspects relating to autotransformers and multiwinding transformers are considered in this section.

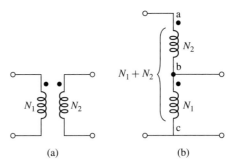

Figure 2.17 (a) Two-winding transformer.
(b) Connection as an autotransformer.

2.6.1 Autotransformers

In Fig. 2.17a, a two-winding transformer is shown with N_1 and N_2 turns on the primary and secondary windings respectively. Substantially the same transformation effect on voltages, currents, and impedances can be obtained when these windings are connected as shown in Fig. 2.17b. Note that, however, in Fig. 2.17b, winding bc is common to both the primary and secondary circuits. This type of transformer is called an *autotransformer*. It is little more than a normal transformer connected in a special way.

   One important difference between the two-winding transformer and the auto-transformer is that the windings of the two-winding transformer are electrically isolated whereas those of the autotransformer are connected directly together. Also, in the autotransformer connection, winding ab must be provided with extra insulation since it must be insulated against the full maximum voltage of the autotransformer. Autotransformers have lower leakage reactances, lower losses, and smaller exciting current and cost less than two-winding transformers when the voltage ratio does not differ too greatly from 1:1.

   The following example illustrates the benefits of an autotransformer for those situations where electrical isolation between the primary and secondary windings is not an important consideration.

EXAMPLE 2.7

The 2400:240-V 50-kVA transformer of Example 2.6 is connected as an autotransformer, as shown in Fig. 2.18a, in which ab is the 240-V winding and bc is the 2400-V winding. (It is assumed that the 240-V winding has enough insulation to withstand a voltage of 2640 V to ground.)

a. Compute the voltage ratings V_H and V_X of the high- and low-voltage sides, respectively, for this autotransformer connection.
b. Compute the kVA rating as an autotransformer.
c. Data with respect to the losses are given in Example 2.6. Compute the full-load efficiency as an autotransformer operating with a rated load of 0.80 power factor lagging.

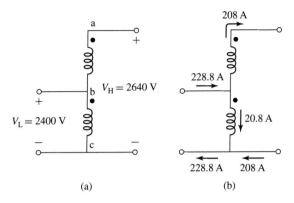

Figure 2.18 (a) Autotransformer connection for Example 2.7. (b) Currents under rated load.

■ Solution

a. Since the 2400-V winding bc is connected to the low-voltage circuit, $V_L = 2400$ V. When $V_{bc} = 2400$ V, a voltage $V_{ab} = 240$ V in phase with V_{bc} will be induced in winding ab (leakage-impedance voltage drops being neglected). The voltage of the high-voltage side therefore is

$$V_H = V_{ab} + V_{bc} = 2640 \text{ V}$$

b. From the rating of 50 kVA as a normal two-winding transformer, the rated current of the 240-V winding is $50,000/240 = 208$ A. Since the high-voltage lead of the autotransformer is connected to the 240-V winding, the rated current I_H at the high-voltage side of the autotransformer is equal to the rated current of the 240-V winding or 208 A. The kVA rating as an autotransformer therefore is

$$\frac{V_H I_H}{1000} = \frac{2640(208)}{1000} = 550 \text{ kVA}$$

Note that, in this connection, the autotransformer has an equivalent turns ratio of 2640/2400. Thus the rated current at the low-voltage winding (the 2400-V winding in this connection) must be

$$I_L = \left(\frac{2640}{2400}\right) 208 \text{ A} = 229 \text{ A}$$

At first, this seems rather unsettling since the 2400-V winding of the transformer has a rated current of 50 kVA/2400 V = 20.8 A. Further puzzling is that fact that this transformer, whose rating as a normal two-winding transformer is 50 kVA, is capable of handling 550 kVA as an autotransformer.

The higher rating as an autotransformer is a consequence of the fact that not all the 550 kVA has to be transformed by electromagnetic induction. In fact, all that the transformer has to do is to boost a current of 208 A through a potential rise of 240 V, corresponding to a power transformation capacity of 50 kVA. This fact is perhaps best illustrated by Fig. 2.18b which shows the currents in the autotransformer under rated conditions. Note that the windings carry only their rated currents in spite of higher rating of the transformer.

c. When it is connected as an autotransformer with the currents and voltages shown in Fig. 2.18, the losses are the same as in Example 2.6, namely, 803 W. But the output as an autotransformer at full load, 0.80 power factor is $0.80(550,000) = 440,000$ W. The efficiency therefore is

$$\left(1 - \frac{803}{440,803} \right) 100\% = 99.82\%$$

The efficiency is so high because the losses are those corresponding to transforming only 50 kVA.

Practice Problem 2.5

A 450-kVA, 460-V:7.97-kV transformer has an efficiency of 97.8 percent when supplying a rated load of unity power factor. If it is connected as a 7.97:8.43-kV autotransformer, calculate its rated terminal currents, rated kVA, and efficiency when supplying a unity-power-factor load.

Solution

The rated current at the 8.43-kV terminal is 978 A, at the 7.97-kV terminal is 1034 A and the transformer rating is 8.25 MVA. Its efficiency supplying a rated, unity-power-factor load is 99.88 percent.

From Example 2.7, we see that when a transformer is connected as an auto-transformer as shown in Fig. 2.17, the rated voltages of the autotransformer can be expressed in terms of those of the two-winding transformer as

Low-voltage:

$$V_{L_{\text{rated}}} = V_{1_{\text{rated}}} \tag{2.38}$$

High-voltage:

$$V_{H_{\text{rated}}} = V_{1_{\text{rated}}} + V_{2_{\text{rated}}} = \left(\frac{N_1 + N_2}{N_1} \right) V_{L_{\text{rated}}} \tag{2.39}$$

The effective turns ratio of the autotransformer is thus $(N_1 + N_2)/N_1$. In addition, the power rating of the autotransformer is equal to $(N_1 + N_2)/N_2$ times that of the two-winding transformer, although the actual power processed by the transformer will not increase over that of the standard two-winding connection.

2.6.2 Multiwinding Transformers

Transformers having three or more windings, known as *multiwinding* or *multicircuit transformers,* are often used to interconnect three or more circuits which may have different voltages. For these purposes a multiwinding transformer costs less and is more efficient than an equivalent number of two-winding transformers. Transformers having a primary and multiple secondaries are frequently found in multiple-output dc power supplies for electronic applications. Distribution transformers used to supply power for domestic purposes usually have two 120-V secondaries connected in series. Circuits for lighting and low-power applications are connected across each of the 120-V windings, while electric ranges, domestic hot-water heaters, clothes-dryers, and other high-power loads are supplied with 240-V power from the series-connected secondaries.

Similarly, a large distribution system may be supplied through a three-phase bank of multiwinding transformers from two or more transmission systems having different voltages. In addition, the three-phase transformer banks used to interconnect two transmission systems of different voltages often have a third, or tertiary, set of windings to provide voltage for auxiliary power purposes in substations or to supply a local distribution system. Static capacitors or synchronous condensers may be connected to the tertiary windings for power factor correction or voltage regulation. Sometimes Δ-connected tertiary windings are put on three-phase banks to provide a low-impedance path for third harmonic components of the exciting current to reduce third-harmonic components of the neutral voltage.

Some of the issues arising in the use of multiwinding transformers are associated with the effects of leakage impedances on voltage regulation, short-circuit currents, and division of load among circuits. These problems can be solved by an equivalent-circuit technique similar to that used in dealing with two-circuit transformers.

The equivalent circuits of multiwinding transformers are more complicated than in the two-winding case because they must take into account the leakage impedances associated with each pair of windings. Typically, in these equivalent circuits, all quantities are referred to a common base, either by use of the appropriate turns ratios as referring factors or by expressing all quantities in per unit. The exciting current usually is neglected.

2.7 TRANSFORMERS IN THREE-PHASE CIRCUITS

Three single-phase transformers can be connected to form a *three-phase transformer bank* in any of the four ways shown in Fig. 2.19. In all four parts of this figure, the windings at the left are the primaries, those at the right are the secondaries, and any primary winding in one transformer corresponds to the secondary winding drawn parallel to it. Also shown are the voltages and currents resulting from balanced impressed primary line-to-line voltages V and line currents I when the ratio of primary-to-secondary turns $N_1/N_2 = a$ and ideal transformers are assumed.[4] Note that the rated voltages and currents at the primary and secondary of the three-phase transformer bank depends upon the connection used but that the rated kVA of the three-phase bank is three times that of the individual single-phase transformers, regardless of the connection.

---

[4] The relationship between three-phase and single-phase quantities is discussed in Appendix A.

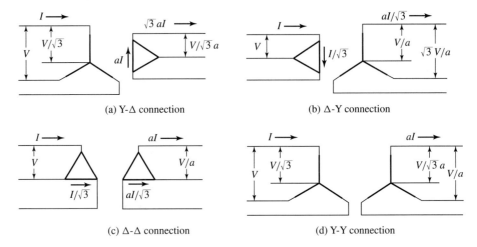

(a) Y-Δ connection   (b) Δ-Y connection

(c) Δ-Δ connection   (d) Y-Y connection

Figure 2.19 Common three-phase transformer connections; the transformer windings are indicated by the heavy lines.

The Y-Δ connection is commonly used in stepping down from a high voltage to a medium or low voltage. One reason is that a neutral is thereby provided for grounding on the high-voltage side, a procedure which can be shown to be desirable in many cases. Conversely, the Δ-Y connection is commonly used for stepping up to a high voltage. The Δ-Δ connection has the advantage that one transformer can be removed for repair or maintenance while the remaining two continue to function as a three-phase bank with the rating reduced to 58 percent of that of the original bank; this is known as the *open-delta, or V,* connection. The Y-Y connection is seldom used because of difficulties with exciting-current phenomena.[5]

Instead of three single-phase transformers, a three-phase bank may consist of one *three-phase transformer* having all six windings on a common multi-legged core and contained in a single tank. Advantages of three-phase transformers over connections of three single-phase transformers are that they cost less, weigh less, require less floor space, and have somewhat higher efficiency. A photograph of the internal parts of a large three-phase transformer is shown in Fig. 2.20.

Circuit computations involving three-phase transformer banks under balanced conditions can be made by dealing with only one of the transformers or phases and recognizing that conditions are the same in the other two phases except for the phase displacements associated with a three-phase system. It is usually convenient to carry out the computations on a single-phase (per-phase-Y, line-to-neutral) basis, since transformer impedances can then be added directly in series with transmission line impedances. The impedances of transmission lines can be referred from one side of the transformer bank to the other by use of the square of the ideal line-to-line voltage

---

[5] Because there is no neutral connection to carry harmonics of the exciting current, harmonic voltages are produced which significantly distort the transformer voltages.

Figure 2.20 A 200-MVA, three-phase, 50-Hz, three-winding, 210/80/10.2-kV transformer removed from its tank. The 210-kV winding has an on-load tap changer for adjustment of the voltage. (*Brown Boveri Corporation.*)

ratio of the bank. In dealing with Y-Δ or Δ-Y banks, all quantities can be referred to the Y-connected side. In dealing with Δ-Δ banks in series with transmission lines, it is convenient to replace the Δ-connected impedances of the transformers by equivalent Y-connected impedances. It can be shown that a balanced Δ-connected circuit of Z_Δ Ω/phase is equivalent to a balanced Y-connected circuit of Z_Y Ω/phase if

$$Z_Y = \frac{1}{3} Z_\Delta \qquad (2.40)$$

EXAMPLE 2.8

Three single-phase, 50-kVA 2400:240-V transformers, each identical with that of Example 2.6, are connected Y-Δ in a three-phase 150-kVA bank to step down the voltage at the load end of a feeder whose impedance is $0.15 + j1.00$ Ω/phase. The voltage at the sending end of the feeder is 4160 V line-to-line. On their secondary sides, the transformers supply a balanced three-phase load through a feeder whose impedance is $0.0005 + j0.0020$ Ω/phase. Find the line-to-line voltage at the load when the load draws rated current from the transformers at a power factor of 0.80 lagging.

■ Solution

The computations can be made on a single-phase basis by referring everything to the high-voltage, Y-connected side of the transformer bank. The voltage at the sending end of the feeder is equivalent to a source voltage V_s of

$$V_s = \frac{4160}{\sqrt{3}} = 2400 \text{ V line-to-neutral}$$

From the transformer rating, the rated current on the high-voltage side is 20.8 A/phase Y. The low-voltage feeder impedance referred to the high voltage side by means of the square of the rated line-to-line voltage ratio of the transformer bank is

$$Z_{\text{lv,H}} = \left(\frac{4160}{240}\right)^2 (0.0005 + j0.0020) = 0.15 + j0.60 \ \Omega$$

and the combined series impedance of the high- and low-voltage feeders referred to the high-voltage side is thus

$$Z_{\text{feeder,H}} = 0.30 + j1.60 \ \Omega/\text{phase Y}$$

Because the transformer bank is Y-connected on its high-voltage side, its equivalent single-phase series impedance is equal to the single-phase series impedance of each single-phase transformer as referred to its high-voltage side. This impedance was originally calculated in Example 2.4 as

$$Z_{\text{eq,H}} = 1.42 + j1.82 \ \Omega/\text{phase Y}$$

Due to the choice of values selected for this example, the single-phase equivalent circuit for the complete system is identical to that of Example 2.5, as can been seen with specific reference to Fig. 2.14a. In fact, the solution on a per-phase basis is exactly the same as the solution to Example 2.5, whence the load voltage referred to the high-voltage side is 2329 V to neutral. The actual line-neutral load voltage can then be calculated by referring this value to the low-voltage side of the transformer bank as

$$V_{\text{load}} = 2329 \left(\frac{240}{4160}\right) = 134 \text{ V line-to-neutral}$$

which can be expressed as a line-to-line voltage by multiplying by $\sqrt{3}$

$$V_{\text{load}} = 134\sqrt{3} = 233 \text{ V line-to-line}$$

Note that this line-line voltage is equal to the line-neutral load voltage calculated in Example 2.5 because in this case the transformers are delta connected on their low-voltage side and hence the line-line voltage on the low-voltage side is equal to the low-voltage terminal voltage of the transformers.

Practice Problem 2.6

Repeat Example 2.8 with the transformers connected Y-Y and all other aspects of the problem statement remaining unchanged.

Solution

405 V line-line

EXAMPLE 2.9

The three transformers of Example 2.8 are reconnected Δ-Δ and supplied with power through a 2400-V (line-to-line) three-phase feeder whose reactance is 0.80 Ω/phase. At its sending end, the feeder is connected to the secondary terminals of a three-phase Y-Δ-connected transformer whose rating is 500 kVA, 24 kV:2400 V (line-to-line). The equivalent series impedance of the sending-end transformer is $0.17 + j0.92$ Ω/phase referred to the 2400-V side. The voltage applied to the primary terminals of the sending-end transformer is 24.0 kV line-to-line.

A three-phase short circuit occurs at the 240-V terminals of the receiving-end transformers. Compute the steady-state short-circuit current in the 2400-V feeder phase wires, in the primary and secondary windings of the receiving-end transformers, and at the 240-V terminals.

■ Solution

The computations will be made on an equivalent line-to-neutral basis with all quantities referred to the 2400-V feeder. The source voltage then is

$$\frac{2400}{\sqrt{3}} = 1385 \text{ V line-to-neutral}$$

From Eq. 2.40, the single-phase-equivalent series impedance of the Δ-Δ transformer seen at its 2400-V side is

$$Z_{eq} = R_{eq} + jX_{eq} = \frac{1.42 + j1.82}{3} = 0.47 + j0.61 \ \Omega/\text{phase}$$

The total series impedance to the short circuit is then the sum of this impedance, that of sending-end transformer and the reactance of the feeder

$$Z_{tot} = (0.47 + j0.61) + (0.17 + j0.92) + j0.80 = 0.64 + j2.33 \ \Omega/\text{phase}$$

which has a magnitude of

$$|Z_{tot}| = 2.42 \ \Omega/\text{phase}$$

The magnitude of the phase current in the 2400-V feeder can now simply be calculated as the line-neutral voltage divided by the series impedance

$$\text{Current in 2400-V feeder} = \frac{1385}{2.42} = 572 \text{ A}$$

and, as is shown in Fig. 2.19c, the winding current in the 2400-V winding of the receiving-end transformer is equal to the phase current divided by $\sqrt{3}$ or

$$\text{Current in 2400-V windings} = \frac{572}{\sqrt{3}} = 330 \text{ A}$$

while the current in the 240-V windings is 10 times this value

$$\text{Current in 240-V windings} = 10 \times 330 = 3300 \text{ A}$$

Finally, again with reference to Fig. 2.19c, the phase current at the 240-V terminals into the short circuit is given by

$$\text{Current at the 240-V terminals} = 3300\sqrt{3} = 5720 \text{ A}$$

Note of course that this same result could have been computed simply by recognizing that the turns ratio of the Δ-Δ transformer bank is equal to 10:1 and hence, under balanced-three-phase conditions, the phase current on the low voltage side will be 10 times that on the high-voltage side.

Practice Problem 2.7

Repeat Example 2.9 under the condition that the three transformers are connected Δ-Y instead of Δ-Δ such that the short low-voltage side of the three-phase transformer is rated 416 V line-to-line.

Solution

$$\text{Current in 2400-V feeder} = 572 \text{ A}$$

$$\text{Current in 2400-V windings} = 330 \text{ A}$$

$$\text{Current in 416-V windings} = 3300 \text{ A}$$

$$\text{Current at the 416-V terminals} = 3300 \text{A}$$

2.8 VOLTAGE AND CURRENT TRANSFORMERS

Transformers are often used in instrumentation applications to match the magnitude of a voltage or current to the range of a meter or other instrument. For example, most 60-Hz power-systems' instrumentation is based upon voltages in the range of 0–120 V rms and currents in the range of 0–5 A rms. Since power system voltages range up to 765-kV line-to-line and currents can be 10's of kA, some method of supplying an accurate, low-level representation of these signals to the instrumentation is required.

One common technique is through the use of specialized transformers known as *potential transformers* or *PT's* and *current transformers* or *CT's*. If constructed with a turns ratio of $N_1:N_2$, an ideal potential transformer would have a secondary voltage equal in magnitude to N_2/N_1 times that of the primary and identical in phase. Similarly, an ideal current transformer would have a secondary output current equal to N_1/N_2 times the current input to the primary, again identical in phase. In other words, potential and current transformers (also referred to as *instrumentation transformers*) are designed to approximate ideal transformers as closely as is practically possible.

The equivalent circuit of Fig. 2.21 shows a transformer loaded with an impedance $Z_b = R_b + j X_b$ at its secondary. For the sake of this discussion, the core-loss resistance R_c has been neglected; if desired, the analysis presented here can be easily expanded to include its effect. Following conventional terminology, the load on an instrumentation transformer is frequently referred to as the *burden* on that transformer, hence the subscript b. To simplify our discussion, we have chosen to refer all the secondary quantities to the primary side of the ideal transformer.

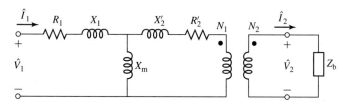

Figure 2.21 Equivalent circuit for an instrumentation transformer.

Consider first a potential transformer. Ideally it should accurately measure voltage while appearing as an open circuit to the system under measurement, i.e., drawing negligible current and power. Thus, its load impedance should be "large" in a sense we will now quantify.

First, let us assume that the transformer secondary is open-circuited (i.e., $|Z_b| = \infty$). In this case we can write that

$$\frac{\hat{V}_2}{\hat{V}_1} = \left(\frac{N_2}{N_1}\right) \frac{jX_m}{R_1 + j(X_1 + X_m)} \qquad (2.41)$$

From this equation, we see that a potential transformer with an open-circuited secondary has an inherent error (in both magnitude and phase) due to the voltage drop of the magnetizing current through the primary resistance and leakage reactance. To the extent that the primary resistance and leakage reactance can be made small compared to the magnetizing reactance, this inherent error can be made quite small.

The situation is worsened by the presence of a finite burden. Including the effect of the burden impedance, Eq. 2.41 becomes

$$\frac{\hat{V}_2}{\hat{V}_1} = \left(\frac{N_2}{N_1}\right) \frac{Z_{eq}Z'_b}{(R_1 + jX_1)(Z_{eq} + Z'_b + R'_2 + jX'_2)} \qquad (2.42)$$

where

$$Z_{eq} = \frac{jX_m(R_1 + jX_1)}{R_1 + j(X_m + X_1)} \qquad (2.43)$$

and

$$Z'_b = \left(\frac{N_1}{N_2}\right)^2 Z_b \qquad (2.44)$$

is the burden impedance referred to the transformer primary.

From these equations, it can be seen that the characterstics of an accurate potential transformer include a large magnetizing reactance (more accurately, a large exciting impedance since the effects of core loss, although neglected in the analysis presented here, must also be minimized) and relatively small winding resistances and leakage reactances. Finally, as will be seen in Example 2.10, the burden impedance must be kept above a minimum value to avoid introducing excessive errors in the magnitude and phase angle of the measured voltage.

EXAMPLE 2.10

A 2400:120-V, 60-Hz potential transformer has the following parameter values (referred to the 2400-V winding):

$$X_1 = 143 \ \Omega \quad X_2' = 164 \ \Omega \quad X_m = 163 \ k\Omega$$

$$R_1 = 128 \ \Omega \quad R_2' = 141 \ \Omega$$

(*a*) Assuming a 2400-V input, which ideally should produce a voltage of 120 V at the low-voltage winding, calculate the magnitude and relative phase-angle errors of the secondary voltage if the secondary winding is open-circuited. (*b*) Assuming the burden impedance to be purely resistive ($Z_b = R_b$), calculate the minimum resistance (maximum burden) that can be applied to the secondary such that the magnitude error is less than 0.5 percent. (*c*) Repeat part (*b*) but find the minimum resistance such that the phase-angle error is less than 1.0 degree.

■ **Solution**

a. This problem is most easily solved using MATLAB.[†] From Eq. 2.41 with $\hat{V}_1 = 2400$ V, the following MATLAB script gives

$$\hat{V}_2 = 119.90 \ \angle 0.045° \ V$$

Here is the MATLAB script:

```
clc
clear

%PT parameters
R1 = 128;
X1 = 143;
Xm = 163e3;

N1 = 2400;
N2 = 120;
N = N1/N2;

%Primary voltage
V1 = 2400;

%Secondary voltage
V2 = V1*(N2/N1)*(j*Xm/(R1+ j*(X1+Xm)));
magV2 = abs(V2);
phaseV2 = 180*angle(V2)/pi;

fprintf('\nMagnitude of V2 = %g [V]',magV2)
fprintf('\n    and angle = %g [degrees]\n\n',phaseV2)
```

b. Here, again, it is relatively straight forward to write a MATLAB script to implement Eq. 2.42 and to calculate the percentage error in the magnitude of voltage $\hat{V}_2$ as compared to the 120 Volts that would be measured if the PT were ideal. The resistive burden R_b

---

[†] MATLAB is a registered trademark of The MathWorks, Inc.

can be initialized to a large value and then reduced until the magnitude error reaches 0.5 percent. The result of such an analysis would show that the minimum resistance is 162.5 Ω, corresponding to a magnitude error of 0.50 percent and a phase angle of 0.22°. (Note that this appears as a resistance of 65 kΩ when referred to the primary.)

c. The MATLAB script of part (b) can be modified to search for the minimum resistive burden that will keep the phase angle error less than 1.0 degrees. The result would show that the minimum resistance is 41.4 Ω, corresponding to a phase angle of 1.00° and a magnitude error of 1.70 percent.

Practice Problem 2.8

Using MATLAB, repeat parts (b) and (c) of Example 2.10 assuming the burden impedance is purely reactive ($Z_b = jX_b$) and finding the corresponding minimum impedance X_b in each case.

Solution

The minimum burden reactance which results in a secondary voltage magnitude within 0.5 percent of the expected 120 V is $X_b = 185.4\ \Omega$, for which the phase angle is 0.25°. The minimum burden reactance which results in a secondary voltage phase-angle of within 1.0° of that of the primary voltage is $X_b = 39.5\ \Omega$, for which the voltage-magnitude error is 2.0 percent.

Consider next a current transformer. An ideal current transformer would accurately measure voltage while appearing as a short circuit to the system under measurement, i.e., developing negligible voltage drop and drawing negligible power. Thus, its load impedance should be "small" in a sense we will now quantify.

Let us begin with the assumption that the transformer secondary is short-circuited (i.e., $|Z_b| = 0$). In this case we can write that

$$\frac{\hat{I}_2}{\hat{I}_1} = \left(\frac{N_1}{N_2}\right) \frac{jX_m}{R_2' + j(X_2' + X_m)} \tag{2.45}$$

In a fashion quite analogous to that of a potential transformer, Eq. 2.45 shows that a current transformer with a shorted secondary has an inherent error (in both magnitude and phase) due to the fact that some of the primary current is shunted through the magnetizing reactance and does not reach the secondary. To the extent that the magnetizing reactance can be made large in comparison to the secondary resistance and leakage reactance, this error can be made quite small.

A finite burden appears in series with the secondary impedance and increases the error. Including the effect of the burden impedance, Eq. 2.45 becomes

$$\frac{\hat{I}_2}{\hat{I}_1} = \left(\frac{N_1}{N_2}\right) \frac{jX_m}{Z_b' + R_2' + j(X_2' + X_m)} \tag{2.46}$$

From these equations, it can be seen that an accurate current transformer has a large magnetizing impedance and relatively small winding resistances and leakage reactances. In addition, as is seen in Example 2.11, the burden impedance on a current transformer must be kept below a maximum value to avoid introducing excessive additional magnitude and phase errors in the measured current.

EXAMPLE 2.11

A 800:5-A, 60-Hz current transformer has the following parameter values (referred to the 800-A winding):

$$X_1 = 44.8 \ \mu\Omega \quad X_2' = 54.3 \ \mu\Omega \quad X_m = 17.7 \ \text{m}\Omega$$

$$R_1 = 10.3 \ \mu\Omega \quad R_2' = 9.6 \ \mu\Omega$$

Assuming that the high-current winding is carrying a current of 800 amperes, calculate the magnitude and relative phase of the current in the low-current winding if the load impedance is purely resistive with $R_b = 2.5 \ \Omega$.

■ Solution

The secondary current can be found from Eq. 2.46 by setting $\hat{I}_1 = 800$ A and $R_b' = (N_1/N_2)^2 R_b = 0.097$ mΩ. The following MATLAB script gives

$$\hat{I}_2 = 4.98 \ \angle 0.346° \ \text{A}$$

Here is the MATLAB script:

```
clc
clear

%CT parameters
R_2p = 9.6e-6;
X_2p = 54.3e-6;
X_m = 17.7e-3;

N_1 = 5;
N_2 = 800;
N = N_1/N_2;

%Load impedance
R_b = 2.5;
X_b = 0;
Z_bp = N^2*(R_b + j * X_b);

% Primary current
I1 = 800;

%Secondary current
I2 = I1*N*j*X_m/(Z_bp + R_2p + j*(X_2p + X_m));

magI2 = abs(I2);
phaseI2 = 180*angle(I2)/pi;

fprintf('\nSecondary current magnitude = %g [A]',magI2)
fprintf('\n    and phase angle = %g [degrees]\n\n',phaseI2)
```

For the current transformer of Example 2.11, find the maximum purely reactive burden $Z_b = jX_b$ such that, for 800 A flowing in the transformer primary, the secondary current will be greater than 4.95 A (i.e., there will be at most a 1.0 percent error in current magnitude).

Solution

X_b must be less than 3.19 Ω

2.9 THE PER-UNIT SYSTEM

Computations relating to machines, transformers, and systems of machines are often carried out in *per-unit* form, i.e., with all pertinent quantities expressed as decimal fractions of appropriately chosen *base values*. All the usual computations are then carried out in these per unit values instead of the familiar volts, amperes, ohms, and so on.

There are a number of advantages to the system. One is that the parameter values of machines and transformers typically fall in a reasonably narrow numerical range when expressed in a per-unit system based upon their rating. The correctness of their values is thus subject to a rapid approximate check. A second advantage is that when transformer equivalent-circuit parameters are converted to their per-unit values, the ideal transformer turns ratio becomes 1:1 and hence the ideal transformer can be eliminated. This greatly simplifies analyses since it eliminates the need to refer impedances to one side or the other of transformers. For complicated systems involving many transformers of different turns ratios, this advantage is a significant one in that a possible cause of serious mistakes is removed.

Quantities such as voltage V, current I, power P, reactive power Q, voltamperes VA, resistance R, reactance X, impedance Z, conductance G, susceptance B, and admittance Y can be translated to and from per-unit form as follows:

$$\text{Quantity in per unit} = \frac{\text{Actual quantity}}{\text{Base value of quantity}} \qquad (2.47)$$

where "Actual quantity" refers to the value in volts, amperes, ohms, and so on. To a certain extent, base values can be chosen arbitrarily, but certain relations between them must be observed for the normal electrical laws to hold in the per-unit system. Thus, for a single-phase system,

$$P_{\text{base}}, Q_{\text{base}}, VA_{\text{base}} = V_{\text{base}} \, I_{\text{base}} \qquad (2.48)$$

$$R_{\text{base}}, X_{\text{base}}, Z_{\text{base}} = \frac{V_{\text{base}}}{I_{\text{base}}} \qquad (2.49)$$

The net result is that *only two independent base quantities can be chosen arbitrarily;* the remaining quantities are determined by the relationships of Eqs. 2.48 and 2.49. In typical usage, values of VA_{base} and V_{base} are chosen first; values of I_{base} and all other quantities in Eqs. 2.48 and 2.49 are then uniquely established.

The value of VA_{base} must be the same over the entire system under analysis. When a transformer is encountered, the values of V_{base} differ on each side and should be chosen in the same ratio as the turns ratio of the transformer. Usually the rated or nominal voltages of the respective sides are chosen. The process of referring quantities to one side of the transformer is then taken care of automatically by using Eqs. 2.48 and 2.49 in finding and interpreting per-unit values.

This can be seen with reference to the equivalent circuit of Fig. 2.10c. If the base voltages of the primary and secondary are chosen to be in the ratio of the turns of the ideal transformer, the per-unit ideal transformer will have a unity turns ratio and hence can be eliminated.

If these rules are followed, the procedure for performing system analyses in per-unit can be summarized as follows:

1. Select a *VA* base and a base voltage at some point in the system.
2. Convert all quantities to per unit on the chosen *VA* base and with a voltage base that transforms as the turns ratio of any transformer which is encountered as one moves through the system.
3. Perform a standard electrical analysis with all quantities in per unit.
4. When the analysis is completed, all quantities can be converted back to real units (e.g., volts, amperes, watts, etc.) by multiplying their per-unit values by their corresponding base values.

When only one electric device, such as a transformer, is involved, the device's own rating is generally used for the volt-ampere base. When expressed in per-unit form on their rating as a base, the characteristics of power and distribution transformers do not vary much over a wide range of ratings. For example, the exciting current is usually between 0.02 and 0.06 per unit, the equivalent resistance is usually between 0.005 and 0.02 per unit (the smaller values applying to large transformers), and the equivalent reactance is usually between 0.015 and 0.10 per unit (the larger values applying to large high-voltage transformers). Similarly, the per-unit values of synchronous- and induction-machine parameters fall within a relatively narrow range. The reason for this is that the physics behind each type of device is the same and, in a crude sense, they can each be considered to be simply scaled versions of the same basic device. As a result, when normalized to their own rating, the effect of the scaling is eliminated and the result is a set of per-unit parameter values which is quite similar over the whole size range of that device.

Often, manufacturers supply device parameters in per unit on the device base. When several devices are involved, however, an arbitrary choice of volt-ampere base must usually be made, and that value must then be used for the overall system. As a result, when performing a system analysis, it may be necessary to convert the supplied per-unit parameter values to per-unit values on the base chosen for the analysis. The following relations can be used to convert per-unit (pu) values from one base to another:

$$(P, Q, VA)_{\text{pu on base 2}} = (P, Q, VA)_{\text{pu on base 1}} \left[\frac{VA_{\text{base 1}}}{VA_{\text{base 2}}} \right] \qquad (2.50)$$

$$(R, X, Z)_{\text{pu on base 2}} = (R, X, Z)_{\text{pu on base 1}} \left[\frac{(V_{\text{base 1}})^2 \, VA_{\text{base 2}}}{(V_{\text{base 2}})^2 \, VA_{\text{base 1}}} \right] \qquad (2.51)$$

$$V_{\text{pu on base 2}} = V_{\text{pu on base 1}} \left[\frac{V_{\text{base 1}}}{V_{\text{base 2}}} \right] \tag{2.52}$$

$$I_{\text{pu on base 2}} = I_{\text{pu on base 1}} \left[\frac{V_{\text{base 2}} VA_{\text{base 1}}}{V_{\text{base 1}} VA_{\text{base 2}}} \right] \tag{2.53}$$

EXAMPLE 2.12

The equivalent circuit for a 100-MVA, 7.97-kV:79.7-kV transformer is shown in Fig. 2.22a. The equivalent-circuit parameters are:

$$X_L = 0.040 \ \Omega \quad X_H = 3.75 \ \Omega \quad X_m = 114 \ \Omega$$

$$R_L = 0.76 \ \text{m}\Omega \quad R_H = 0.085 \ \Omega$$

Note that the magnetizing inductance has been referred to the low-voltage side of the equivalent circuit. Convert the equivalent circuit parameters to per unit using the transformer rating as base.

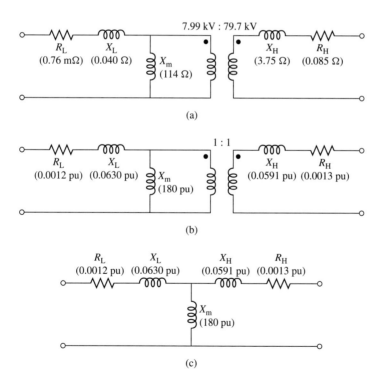

Figure 2.22 Transformer equivalent circuits for Example 2.12. (a) Equivalent circuit in actual units. (b) Per-unit equivalent circuit with 1:1 ideal transformer. (c) Per-unit equivalent circuit following elimination of the ideal transformer.

■ Solution

The base quantities for the transformer are:

Low-voltage side:

$$VA_{base} = 100 \text{ MVA} \quad V_{base} = 7.97 \text{ kV}$$

and from Eqs. 2.48 and 2.49

$$R_{base} = X_{base} = \frac{V_{base}^2}{VA_{base}} = 0.635 \ \Omega$$

High-voltage side:

$$VA_{base} = 100 \text{ MVA} \quad V_{base} = 79.7 \text{ kV}$$

and from Eqs. 2.48 and 2.49

$$R_{base} = X_{base} = \frac{V_{base}^2}{VA_{base}} = 63.5 \ \Omega$$

The per-unit values of the transformer parameters can now be calculated by division by their corresponding base quantities.

$$X_L = \frac{0.040}{0.635} = 0.0630 \text{ per unit}$$

$$X_H = \frac{3.75}{63.5} = 0.0591 \text{ per unit}$$

$$X_m = \frac{114}{0.635} = 180 \text{ per unit}$$

$$R_L = \frac{7.6 \times 10^{-4}}{0.635} = 0.0012 \text{ per unit}$$

$$R_H = \frac{0.085}{63.5} = 0.0013 \text{ per unit}$$

Finally, the voltages representing the turns ratio of the ideal transformer must each be divided by the base voltage on that side of the transformer. Thus the turns ratio of 7.97-kV:79.7-kV becomes in per unit

$$\text{Per-unit turns ratio} = \left(\frac{7.97 \text{ kV}}{7.97 \text{ kV}} \right) : \left(\frac{79.7 \text{ kV}}{79.7 \text{ kV}} \right) = 1 : 1$$

The resultant per-unit equivalent circuit is shown in Fig. 2.22b. Because it has unity turns ratio, there is no need to keep the ideal transformer and hence this equivalent circuit can be reduced to the form of Fig. 2.22c.

EXAMPLE 2.13

The exciting current measured on the low-voltage side of a 50-kVA, 2400:240-V transformer is 5.41 A. Its equivalent impedance referred to the high-voltage side is $1.42 + j1.82 \ \Omega$. Using the transformer rating as the base, express in per unit on the low- and high-voltage sides (a) the exciting current and (b) the equivalent impedance.

■ Solution

The base values of voltages and currents are

$$V_{\text{base,H}} = 2400 \text{ V} \quad V_{\text{base,L}} = 240 \text{ V} \quad I_{\text{base,H}} = 20.8 \text{ A} \quad I_{\text{base,L}} = 208 \text{ A}$$

where subscripts H and L indicate the high- and low-voltage sides, respectively.

From Eq. 2.49

$$Z_{\text{base,H}} = \frac{2400}{20.8} = 115.2 \text{ } \Omega \quad Z_{\text{base,L}} = \frac{240}{208} = 1.152 \text{ } \Omega$$

a. From Eq. 2.47, the exciting current in per unit referred to the low-voltage side can be calculated as:

$$I_{\varphi,\text{L}} = \frac{5.41}{208} = 0.0260 \text{ per unit}$$

The exciting current referred to the high-voltage side is 0.541 A. Its per-unit value is

$$I_{\varphi,\text{H}} = \frac{0.541}{20.8} = 0.0260 \text{ per unit}$$

Note that, as expected, the per-unit values are the same referred to either side, corresponding to a unity turns ratio for the ideal transformer in the per-unit transformer. This is a direct consequence of the choice of base voltages in the ratio of the transformer turns ratio and the choice of a constant volt-ampere base.

b. From Eq. 2.47 and the value for Z_{base}

$$Z_{\text{eq,H}} = \frac{1.42 + j1.82}{115.2} = 0.0123 + j0.0158 \text{ per unit}$$

The equivalent impedance referred to the low-voltage side is $0.0142 + j0.0182\Omega$. Its per-unit value is

$$Z_{\text{eq,L}} = \frac{0.142 + 0.0182}{1.152} = 0.0123 + j0.0158 \text{ per unit}$$

The per-unit values referred to the high- and low-voltage sides are the same, the transformer turns ratio being accounted for in per unit by the base values. Note again that this is consistent with a unity turns ratio of the ideal transformer in the per-unit transformer equivalent circuit.

Practice Problem 2.10

A 15-kVA 120:460-V transformer has an equivalent series impedance of $0.018 + j0.042$ per unit. Calculate the equivalent series impedance in ohms (*a*) referred to the low-voltage side and (*b*) referred to the high-voltage side.

Solution

$$Z_{\text{eq,L}} = 0.017 + j0.040 \text{ } \Omega \quad \text{and} \quad Z_{\text{eq,H}} = 0.25 + j0.60 \text{ } \Omega$$

When they are applied to the analysis of three-phase systems, the base values for the per-unit system are chosen so that the relations for a balanced three-phase system hold between them:

$$(P_{base}, Q_{base}, VA_{base})_{3\text{-phase}} = 3\,VA_{base,\ per\ phase} \tag{2.54}$$

In dealing with three-phase systems, $VA_{base,\ 3\text{-phase}}$, the three-phase volt-ampere base, and $V_{base,\ 3\text{-phase}} = V_{base,\ l\text{-}l}$, the line-to-line voltage base are usually chosen first. The base values for the phase (line-to-neutral) voltage then follows as

$$V_{base,\ l\text{-}n} = \frac{1}{\sqrt{3}} V_{base,\ l\text{-}l} \tag{2.55}$$

Note that the base current for three-phase systems is equal to the phase current, which is the same as the base current for a single-phase (per-phase) analysis. Hence

$$I_{base,\ 3\text{-phase}} = I_{base,\ per\ phase} = \frac{VA_{base,\ 3\text{-phase}}}{\sqrt{3}\ V_{base,\ 3\text{-phase}}} \tag{2.56}$$

Finally, the three-phase base impedance is chosen to the be the single-phase base impedance. Thus

$$\begin{aligned}
Z_{base,\ 3\text{-phase}} &= Z_{base,\ per\ phase} \\
&= \frac{V_{base,\ l\text{-}n}}{I_{base,\ per\ phase}} \\
&= \frac{V_{base,\ 3\text{-phase}}}{\sqrt{3}I_{base,\ 3\text{-phase}}} \\
&= \frac{(V_{base,\ 3\text{-phase}})^2}{VA_{base,\ 3\text{-phase}}}
\end{aligned} \tag{2.57}$$

The equations for conversion from base to base, Eqs. 2.50 through 2.53, apply equally to three-phase base conversion. Note that the factors of $\sqrt{3}$ and 3 relating Δ to Y quantities of volts, amperes, and ohms in a balanced three-phase system are automatically taken care of in per unit by the base values. Three-phase problems can thus be solved in per unit as if they were single-phase problems and the details of transformer (Y vs Δ on the primary and secondary of the transformer) and impedance (Y vs Δ) connections disappear, except in translating volt, ampere, and ohm values into and out of the per-unit system.

EXAMPLE 2.14

Rework Example 2.9 in per unit, specifically calculating the short-circuit phase currents which will flow in the feeder and at the 240-V terminals of the receiving-end transformer bank. Perform the calculations in per unit on the three-phase, 150-kVA, rated-voltage base of the receiving-end transformer.

■ Solution

We start by converting all the impedances to per unit. The impedance of the 500-kVA, 24 kV:2400 V sending end transformer is $0.17 + j0.92$ Ω/phase as referred to the 2400-V side. From Eq. 2.57, the base impedance corresponding to a 2400-V, 150-kVA base is

$$Z_{base} = \frac{2400^2}{150 \times 10^3} = 38.4 \ \Omega$$

From Example 2.9, the total series impedance is equal to $Z_{tot} = 0.64 + j2.33 \ \Omega$/phase and thus in per unit it is equal to

$$Z_{tot} = \frac{0.64 + j2.33}{38.4} = 0.0167 + j0.0607 \text{ per unit}$$

which is of magnitude

$$|Z_{tot}| = 0.0629 \text{ per unit}$$

The voltage applied to the high-voltage side of the sending-end transformer is $V_s = 24.0 \text{ kV} = 1.0$ per unit on a rated-voltage base and hence the short-circuit current will equal

$$I_{sc} = \frac{V_s}{|Z_{tot}|} = \frac{1.0}{0.0629} = 15.9 \text{ per unit}$$

To calculate the phase currents in amperes, it is simply necessary to multiply the per-unit short-circuit current by the appropriate base current. Thus, at the 2400-V feeder the base current is

$$I_{base, \ 2400-V} = \frac{150 \times 10^3}{\sqrt{3} \ 2400} = 36.1 \text{ A}$$

and hence the feeder current will be

$$I_{feeder} = 15.9 \times 36.1 = 574 \text{ A}$$

The base current at the 240-V secondary of the receiving-end transformers is

$$I_{base, \ 240-V} = \frac{150 \times 10^3}{\sqrt{3} \ 240} = 361 \text{ A}$$

and hence the short-circuit current is

$$I_{240-V \ secondary} = 15.9 \times 361 = 5.74 \text{ kA}$$

As expected, these values are equivalent within numerical accuracy to those calculated in Example 2.9.

Practice Problem 2.11

Calculate the magnitude of the short-circuit current in the feeder of Example 2.9 if the 2400-V feeder is replaced by a feeder with an impedance of $0.07 + j0.68 \ \Omega$/phase. Perform this calculation on the 500-kVA, rated-voltage base of the sending-end transformer and express your solution both in per unit and in amperes per phase.

Solution

$$\text{Short-circuit current} = 5.20 \text{ per unit} = 636 \text{ A}$$

EXAMPLE 2.15

A three-phase load is supplied from a 2.4-kV:460-V, 250-kVA transformer whose equivalent series impedance is $0.026 + j0.12$ per unit on its own base. The load voltage is observed to be 438-V line-line, and it is drawing 95 kW at unity power factor. Calculate the voltage at the high-voltage side of the transformer. Perform the calculations on a 460-V, 100-kVA base.

■ **Solution**

The 460-V side base impedance for the transformer is

$$Z_{\text{base, transformer}} = \frac{460^2}{250 \times 10^3} = 0.846 \ \Omega$$

while that based upon a 100-kVA base is

$$Z_{\text{base, 100-kVA}} = \frac{460^2}{100 \times 10^3} = 2.12 \ \Omega$$

Thus, from Eq. 2.51 the per-unit transformer impedance on a 100-kVA base is

$$Z_{\text{transformer}} = (0.026 + j0.12) \left(\frac{0.864}{2.12} \right) = 0.0106 + j.0489 \text{ per unit}$$

The per unit load voltage is

$$\hat{V}_{\text{load}} = \frac{438}{460} = 0.952 \ \angle 0° \text{ per unit}$$

where the load voltage has been chosen as the reference for phase-angle calculations.
The per-unit load power is

$$P_{\text{load}} = \frac{95}{100} = 0.95 \text{ per unit}$$

and hence the per-unit load current, which is in phase with the load voltage because the load is operating at unity power factor, is

$$\hat{I}_{\text{load}} = \frac{P_{\text{load}}}{V_{\text{load}}} = \frac{0.95}{0.952} = 0.998 \ \angle 0° \text{ per unit}$$

Thus we can now calculate the high-side voltage of the transformer

$$\hat{V}_{\text{H}} = \hat{V}_{\text{load}} + \hat{I}_{\text{load}} Z_{\text{transformer}}$$

$$= 0.952 + 0.998(0.0106 + j0.0489)$$

$$= 0.963 + j0.0488 = 0.964 \ \angle 29.0° \text{ per unit}$$

Thus the high-side voltage is equal to $0.964 \times 2400 \text{ V} = 2313 \text{ V}$ (line-line).

Repeat Example 2.15 if the 250-kV three-phase transformer is replaced by a 150-kV transformer also rated at 2.4-kV:460-V and whose equivalent series impedance is $0.038 + j0.135$ per unit on its own base. Perform the calculations on a 460-V, 100-kVA base.

Solution

High-side voltage $= 0.982$ per unit $= 2357$ V (line-line)

2.10 SUMMARY

Although not an electromechanical device, the transformer is a common and indispensable component of ac systems where it is used to transform voltages, currents, and impedances to appropriate levels for optimal use. For the purposes of our study of electromechanical systems, transformers serve as valuable examples of the analysis techniques which must be employed. They offer us opportunities to investigate the properties of magnetic circuits, including the concepts of mmf, magnetizing current, and magnetizing, mutual, and leakage fluxes and their associated inductances.

In both transformers and rotating machines, a magnetic field is created by the combined action of the currents in the windings. In an iron-core transformer, most of this flux is confined to the core and links all the windings. This resultant mutual flux induces voltages in the windings proportional to their number of turns and is responsible for the voltage-changing property of a transformer. In rotating machines, the situation is similar, although there is an air gap which separates the rotating and stationary components of the machine. Directly analogous to the manner in which transformer core flux links the various windings on a transformer core, the mutual flux in rotating machines crosses the air gap, linking the windings on the rotor and stator. As in a transformer, the mutual flux induces voltages in these windings proportional to the number of turns and the time rate of change of the flux.

A significant difference between transformers and rotating machines is that in rotating machines there is relative motion between the windings on the rotor and stator. This relative motion produces an additional component of the time rate of change of the various winding flux linkages. As will be discussed in Chapter 3, the resultant voltage component, known as the *speed voltage,* is characteristic of the process of electromechanical energy conversion. In a static transformer, however, the time variation of flux linkages is caused simply by the time variation of winding currents; no mechanical motion is involved, and no electromechanical energy conversion takes place.

The resultant core flux in a transformer induces a counter emf in the primary which, together with the primary resistance and leakage-reactance voltage drops, must balance the applied voltage. Since the resistance and leakage-reactance voltage drops usually are small, the counter emf must approximately equal the applied voltage and the core flux must adjust itself accordingly. Exactly similar phenomena must take place in the armature windings of an ac motor; the resultant air-gap flux wave must adjust itself to generate a counter emf approximately equal to the applied voltage.

In both transformers and rotating machines, the net mmf of all the currents must accordingly adjust itself to create the resultant flux required by this voltage balance. In any ac electromagnetic device in which the resistance and leakage-reactance voltage drops are small, the resultant flux is very nearly determined by the applied voltage and frequency, and the currents must adjust themselves accordingly to produce the mmf required to create this flux.

In a transformer, the secondary current is determined by the voltage induced in the secondary, the secondary leakage impedance, and the electric load. In an induction motor, the secondary (rotor) current is determined by the voltage induced in the secondary, the secondary leakage impedance, and the mechanical load on its shaft. Essentially the same phenomena take place in the primary winding of the transformer and in the armature (stator) windings of induction and synchronous motors. In all three, the primary, or armature, current must adjust itself so that the combined mmf of all currents creates the flux required by the applied voltage.

In addition to the useful mutual fluxes, in both transformers and rotating machines there are leakage fluxes which link individual windings without linking others. Although the detailed picture of the leakage fluxes in rotating machines is more complicated than that in transformers, their effects are essentially the same. In both, the leakage fluxes induce voltages in ac windings which are accounted for as leakage-reactance voltage drops. In both, the reluctances of the leakage-flux paths are dominated by that of a path through air, and hence the leakage fluxes are nearly linearly proportional to the currents producing them. The leakage reactances therefore are often assumed to be constant, independent of the degree of saturation of the main magnetic circuit.

Further examples of the basic similarities between transformers and rotating machines can be cited. Except for friction and windage, the losses in transformers and rotating machines are essentially the same. Tests for determining the losses and equivalent circuit parameters are similar: an open-circuit, or no-load, test gives information regarding the excitation requirements and core losses (along with friction and windage losses in rotating machines), while a short-circuit test together with dc resistance measurements gives information regarding leakage reactances and winding resistances. Modeling of the effects of magnetic saturation is another example: In both transformers and ac rotating machines, the leakage reactances are usually assumed to be unaffected by saturation, and the saturation of the main magnetic circuit is assumed to be determined by the resultant mutual or air-gap flux.

2.11 PROBLEMS

2.1 A transformer is made up of a 1200-turn primary coil and an open-circuited 75-turn secondary coil wound around a closed core of cross-sectional area 42 cm^2. The core material can be considered to saturate when the rms applied flux density reaches 1.45 T. What maximum 60-Hz rms primary voltage is possible without reaching this saturation level? What is the corresponding secondary voltage? How are these values modified if the applied frequency is lowered to 50 Hz?

2.2 A magnetic circuit with a cross-sectional area of 15 cm^2 is to be operated at 60 Hz from a 120-V rms supply. Calculate the number of turns required to achieve a peak magnetic flux density of 1.8 T in the core.

2.3 A transformer is to be used to transform the impedance of a 8-Ω resistor to an impedance of 75 Ω. Calculate the required turns ratio, assuming the transformer to be ideal.

2.4 A 100-Ω resistor is connected to the secondary of an idea transformer with a turns ratio of 1:4 (primary to secondary). A 10-V rms, 1-kHz voltage source is connected to the primary. Calculate the primary current and the voltage across the 100-Ω resistor.

2.5 A source which can be represented by a voltage source of 8 V rms in series with an internal resistance of 2 kΩ is connected to a 50-Ω load resistance through an ideal transformer. Calculate the value of turns ratio for which maximum power is supplied to the load and the corresponding load power? Using MATLAB, plot the the power in milliwatts supplied to the load as a function of the transformer ratio, covering ratios from 1.0 to 10.0.

2.6 Repeat Problem 2.5 with the source resistance replaced by a 2-kΩ reactance.

2.7 A single-phase 60-Hz transformer has a nameplate voltage rating of 7.97 kV:266 V, which is based on its winding turns ratio. The manufacturer calculates that the primary (7.97-kV) leakage inductance is 165 mH and the primary magnetizing inductance is 135 H. For an applied primary voltage of 7970 V at 60 Hz, calculate the resultant open-circuit secondary voltage.

2.8 The manufacturer calculates that the transformer of Problem 2.7 has a secondary leakage inductance of 0.225 mH.

   a. Calculate the magnetizing inductance as referred to the secondary side.

   b. A voltage of 266 V, 60 Hz is applied to the secondary. Calculate (i) the resultant open-circuit primary voltage and (ii) the secondary current which would result if the primary were short-circuited.

2.9 A 120-V:2400-V, 60-Hz, 50-kVA transformer has a magnetizing reactance (as measured from the 120-V terminals) of 34.6 Ω. The 120-V winding has a leakage reactance of 27.4 mΩ and the 2400-V winding has a leakage reactance of 11.2 Ω.

   a. With the secondary open-circuited and 120 V applied to the primary (120-V) winding, calculate the primary current and the secondary voltage.

   b. With the secondary short-circuited, calculate the primary voltage which will result in rated current in the primary winding. Calculate the corresponding current in the secondary winding.

2.10 A 460-V:2400-V transformer has a series leakage reactance of 37.2 Ω as referred to the high-voltage side. A load connected to the low-voltage side is observed to be absorbing 25 kW, unity power factor, and the voltage is measured to be 450 V. Calculate the corresponding voltage and power factor as measured at the high-voltage terminals.

2.11 The resistances and leakage reactances of a 30-kVA, 60-Hz, 2400-V:240-V distribution transformer are

$$R_1 = 0.68 \ \Omega \quad R_2 = 0.0068 \ \Omega$$

$$X_{l_1} = 7.8 \ \Omega \quad X_{l_2} = 0.0780 \ \Omega$$

where subscript 1 denotes the 2400-V winding and subscript 2 denotes the 240-V winding. Each quantity is referred to its own side of the transformer.

a. Draw the equivalent circuit referred to (*i*) the high- and (*ii*) the low-voltage sides. Label the impedances numerically.

b. Consider the transformer to deliver its rated kVA to a load on the low-voltage side with 230 V across the load. (*i*) Find the high-side terminal voltage for a load power factor of 0.85 power factor lagging. (*ii*) Find the high-side terminal voltage for a load power factor of 0.85 power factor leading.

c. Consider a rated-kVA load connected at the low-voltage terminals operating at 240V. Use MATLAB to plot the high-side terminal voltage as a function of the power-factor angle as the load power factor varies from 0.6 leading through unity power factor to 0.6 pf lagging.

2.12 Repeat Problem 2.11 for a 75-kVA, 60-Hz, 4600-V:240-V distribution transformer whose resistances and leakage reactances are

$$R_1 = 0.846 \ \Omega \quad R_2 = 0.00261 \ \Omega$$

$$X_{l_1} = 26.8 \ \Omega \quad X_{l_2} = 0.0745 \ \Omega$$

where subscript 1 denotes the 4600-V winding and subscript 2 denotes the 240-V winding. Each quantity is referred to its own side of the transformer.

2.13 A single-phase load is supplied through a 35-kV feeder whose impedance is $95 + j360 \ \Omega$ and a 35-kV:2400-V transformer whose equivalent impedance is $0.23 + j1.27 \ \Omega$ referred to its low-voltage side. The load is 160 kW at 0.89 leading power factor and 2340 V.

a. Compute the voltage at the high-voltage terminals of the transformer.

b. Compute the voltage at the sending end of the feeder.

c. Compute the power and reactive power input at the sending end of the feeder.

2.14 Repeat Example 2.6 with the transformer operating at full load and unity power factor.

2.15 The nameplate on a 50-MVA, 60-Hz single-phase transformer indicates that it has a voltage rating of 8.0-kV:78-kV. An open-circuit test is conducted from the low-voltage side, and the corresponding instrument readings are 8.0 kV, 62.1 A, and 206 kW. Similarly, a short-circuit test from the low-voltage side gives readings of 674 V, 6.25 kA, and 187 kW.

a. Calculate the equivalent series impedance, resistance, and reactance of the transformer as referred to the low-voltage terminals.

b. Calculate the equivalent series impedance of the transformer as referred to the high-voltage terminals.

c. Making appropriate approximations, draw a T equivalent circuit for the transformer.

d. Determine the efficiency and voltage regulation if the transformer is operating at the rated voltage and load (unity power factor).

e. Repeat part (d), assuming the load to be at 0.9 power factor leading.

2.16 A 550-kVA, 60-Hz transformer with a 13.8-kV primary winding draws 4.93 A and 3420 W at no load, rated voltage and frequency. Another transformer has a core with all its linear dimensions $\sqrt{2}$ times as large as the corresponding dimensions of the first transformer. The core material and lamination thickness are the same in both transformers. If the primary windings of both transformers have the same number of turns, what no-load current and power will the second transformer draw with 27.6 kV at 60 Hz impressed on its primary?

2.17 The following data were obtained for a 20-kVA, 60-Hz, 2400:240-V distribution transformer tested at 60 Hz:

	Voltage, V	Current, A	Power, W
With high-voltage winding open-circuited	240	1.038	122
With low-voltage terminals short-circuited	61.3	8.33	257

a. Compute the efficiency at full-load current and the rated terminal voltage at 0.8 power factor.

b. Assume that the load power factor is varied while the load current and secondary terminal voltage are held constant. Use a phasor diagram to determine the load power factor for which the regulation is greatest. What is this regulation?

2.18 A 75-kVa, 240-V:7970-V, 60-Hz single-phase distribution transformer has the following parameters referred to the high-voltage side:

$$R_1 = 5.93\ \Omega \quad X_1 = 43.2\ \Omega$$
$$R_2 = 3.39\ \Omega \quad X_2 = 40.6\ \Omega$$
$$R_c = 244\ k\Omega \quad X_m = 114\ k\Omega$$

Assume that the transformer is supplying its rated kVA at its low-voltage terminals. Write a MATLAB script to determine the efficiency and regulation of the transformer for any specified load power factor (leading or lagging). You may use reasonable engineering approximations to simplify your analysis. Use your MATLAB script to determine the efficiency and regulation for a load power factor of 0.87 leading.

2.19 The transformer of Problem 2.11 is to be connected as an autotransformer. Determine (*a*) the voltage ratings of the high- and low-voltage windings for this connection and (*b*) the kVA rating of the autotransformer connection.

2.20 A 120:480-V, 10-kVA transformer is to be used as an autotransformer to supply a 480-V circuit from a 600-V source. When it is tested as a two-winding transformer at rated load, unity power factor, its efficiency is 0.979.

   a. Make a diagram of connections as an autotransformer.

   b. Determine its kVA rating as an autotransformer.

   c. Find its efficiency as an autotransformer at full load, with 0.85 power factor lagging.

2.21 Consider the 8-kV:78-kV, 50-MVA transformer of Problem 2.15 connected as an autotransformer.

   a. Determine the voltage ratings of the high- and low-voltage windings for this connection and the kVA rating of the autotransformer connection.

   b. Calculate the efficiency of the transformer in this connection when it is supplying its rated load at unity power factor.

 2.22 Write a MATLAB script whose inputs are the rating (voltage and kVA) and rated-load, unity-power-factor efficiency of a single-transformer and whose output is the transformer rating and rated-load, unity-power-factor efficiency when connected as an autotransformer.

2.23 The high-voltage terminals of a three-phase bank of three single-phase transformers are supplied from a three-wire, three-phase 13.8-kV (line-to-line) system. The low-voltage terminals are to be connected to a three-wire, three-phase substation load drawing up to 4500 kVA at 2300 V line-to-line. Specify the required voltage, current, and kVA ratings of each transformer (both high- and low-voltage windings) for the following connections:

	High-voltage Windings	Low-voltage Windings
a.	Y	Δ
b.	Δ	Y
c.	Y	Y
d.	Δ	Δ

2.24 Three 100-MVA single-phase transformers, rated at 13.8 kV:66.4 kV, are to be connected in a three-phase bank. Each transformer has a series impedance of $0.0045 + j0.19\ \Omega$ referred to its 13.8-kV winding.

   a. If the transformers are connected Y-Y, calculate (*i*) the voltage and power rating of the three-phase connection, (*ii*) the equivalent impedance as referred to its low-voltage terminals, and (*iii*) the equivalent impedance as referred to its high-voltage terminals.

   b. Repeat part (a) if the transformer is connected Y on its low-voltage side and Δ on its high-voltage side.

2.25 Repeat Example 2.8 for a load drawing rated current from the transformers at unity power factor.

2.26 A three-phase Y-Δ transformer is rated 225-kV:24-kV, 400 MVA and has a series reactance of 11.7 Ω as referred to its high-voltage terminals. The transformer is supplying a load of 325 MVA, with 0.93 power factor lagging at a voltage of 24 kV (line-to-line) on its low-voltage side. It is supplied from a feeder whose impedance is $0.11 + j2.2$ Ω connected to its high-voltage terminals. For these conditions, calculate (*a*) the line-to-line voltage at the high-voltage terminals of the transformer and (*b*) the line-to-line voltage at the sending end of the feeder.

2.27 Assume the total load in the system of Problem 2.26 to remain constant at 325 MVA. Write a MATLAB script to plot the line-to-line voltage which must be applied to the sending end of the feeder to maintain the load voltage at 24 kV line-to-line for load power factors in range from 0.75 lagging to unity to 0.75 leading. Plot the sending-end voltage as a function of power factor angle.

2.28 A Δ-Y-connected bank of three identical 100-kVA, 2400-V:120-V, 60-Hz transformers is supplied with power through a feeder whose impedance is $0.065 + j0.87$ Ω per phase. The voltage at the sending end of the feeder is held constant at 2400 V line-to-line. The results of a single-phase short-circuit test on one of the transformers with its low-voltage terminals short-circuited are

$$V_{\mathrm{H}} = 53.4\ \mathrm{V} \quad f = 60\ \mathrm{Hz} \quad I_{\mathrm{H}} = 41.7\ \mathrm{A} \quad P = 832\ \mathrm{W}$$

a. Determine the line-to-line voltage on the low-voltage side of the transformer when the bank delivers rated current to a balanced three-phase unity power factor load.

b. Compute the currents in the transformer's high- and low-voltage windings and in the feeder wires if a solid three-phase short circuit occurs at the secondary line terminals.

2.29 A 7970-V:120-V, 60-Hz potential transformer has the following parameters as seen from the high-voltage (primary) winding:

$$X_1 = 1721\ \Omega \quad X_2' = 1897\ \Omega \quad X_{\mathrm{m}} = 782\ \mathrm{k}\Omega$$

$$R_1 = 1378\ \Omega \quad R_2' = 1602\ \Omega$$

a. Assuming that the secondary is open-circuited and that the primary is connected to a 7.97-kV source, calculate the magnitude and phase angle (with respect to the high-voltage source) of the voltage at the secondary terminals.

b. Calculate the magnitude and phase angle of the secondary voltage if a 1-kΩ resistive load is connected to the secondary terminals.

c. Repeat part (b) if the burden is changed to a 1-kΩ reactance.

2.30 For the potential transformer of Problem 2.29, find the maximum reactive burden (mimimum reactance) which can be applied at the secondary terminals such that the voltage magnitude error does not exceed 0.5 percent.

2.31 Consider the potential transformer of Problem 2.29.

   a. Use MATLAB to plot the percentage error in voltage magnitude as a function of the magnitude of the burden impedance (i) for a resistive burden of 100 $\Omega \leq R_b \leq$ 3000 Ω and (ii) for a reactive burden of 100 $\Omega \leq X_b \leq$ 3000 Ω. Plot these curves on the same axis.

   b. Next plot the phase error in degrees as a function of the magnitude of the burden impedance (i) for a resistive burden of 100 $\Omega \leq R_b \leq$ 3000 Ω and (ii) for a reactive burden of 100 $\Omega \leq X_b \leq$ 3000 Ω. Again, plot these curves on the same axis.

2.32 A 200-A:5-A, 60-Hz current transformer has the following parameters as seen from the 200-A (primary) winding:

$$X_1 = 745 \ \mu\Omega \quad X_2' = 813 \ \mu\Omega \quad X_m = 307 \ m\Omega$$
$$R_1 = 136 \ \mu\Omega \quad R_2' = 128 \ \mu\Omega$$

   a. Assuming a current of 200 A in the primary and that the secondary is short-circuited, find the magnitude and phase angle of the secondary current.

   b. Repeat the calculation of part (a) if the CT is shorted through a 250 $\mu\Omega$ burden.

2.33 Consider the current transformer of Problem 2.32.

   a. Use MATLAB to plot the percentage error in current magnitude as a function of the magnitude of the burden impedance (i) for a resistive burden of 100 $\Omega \leq R_b \leq$ 1000 Ω and (ii) for a reactive burden of 100 $\Omega \leq X_b \leq$ 1000 Ω. Plot these curves on the same axis.

   b. Next plot the phase error in degrees as a function of the magnitude of the burden impedance (i) for a resistive burden of 100 $\Omega \leq R_b \leq$ 1000 Ω and (ii) for a reactive burden of 100 $\Omega \leq X_b \leq$ 1000 Ω. Again, plot these curves on the same axis.

2.34 A 15-kV:175-kV, 125-MVA, 60-Hz single-phase transformer has primary and secondary impedances of $0.0095 + j0.063$ per unit each. The magnetizing impedance is $j148$ per unit. All quantities are in per unit on the transformer base. Calculate the primary and secondary resistances and reactances and the magnetizing inductance (referred to the low-voltage side) in ohms and henrys.

2.35 The nameplate on a 7.97-kV:460-V, 75-kVA, single-phase transformer indicates that it has a series reactance of 12 percent (0.12 per unit).

   a. Calculate the series reactance in ohms as referred to (i) the low-voltage terminal and (ii) the high-voltage terminal.

   b. If three of these transformers are connected in a three-phase Y-Y connection, calculate (i) the three-phase voltage and power rating, (ii) the per unit impedance of the transformer bank, (iii) the series reactance in ohms as referred to the high-voltage terminal, and (*iv*) the series reactance in ohms as referred to the low-voltage terminal.

c.   Repeat part (b) if the three transformers are connected in Y on their HV side and Δ on their low-voltage side.

2.36 a.   Consider the Y-Y transformer connection of Problem 2.35, part (b). If the rated voltage is applied to the high-voltage terminals and the three low-voltage terminals are short-circuited, calculate the magnitude of the phase current in per unit and in amperes on (*i*) the high-voltage side and (*ii*) the low-voltage side.

b.   Repeat this calculation for the Y-Δ connection of Problem 2.35, part (c).

2.37 A three-phase generator step-up transformer is rated 26-kV:345-kV, 850 MVA and has a series impedance of $0.0035 + j0.087$ per unit on this base. It is connected to a 26-kV, 800-MVA generator, which can be represented as a voltage source in series with a reactance of $j1.57$ per unit on the generator base.

a.   Convert the per unit generator reactance to the step-up transformer base.

b.   The unit is supplying 700 MW at 345 kV and 0.95 power factor lagging to the system at the transformer high-voltage terminals. (*i*) Calculate the transformer low-side voltage and the generator internal voltage behind its reactance in kV. (*ii*) Find the generator output power in MW and the power factor.

3 CHAPTER

Electromechanical-Energy-Conversion Principles

We are concerned here with the electromechanical-energy-conversion process, which takes place through the medium of the electric or magnetic field of the conversion device. Although the various conversion devices operate on similar principles, their structures depend on their function. Devices for measurement and control are frequently referred to as *transducers;* they generally operate under linear input-output conditions and with relatively small signals. The many examples include microphones, pickups, sensors, and loudspeakers. A second category of devices encompasses *force-producing devices* and includes solenoids, relays, and electromagnets. A third category includes *continuous energy-conversion equipment* such as motors and generators.

This chapter is devoted to the principles of electromechanical energy conversion and the analysis of the devices which accomplish this function. Emphasis is placed on the analysis of systems which use magnetic fields as the conversion medium since the remaining chapters of the book deal with such devices. However, the analytical techniques for electric field systems are quite similar.

The purpose of such analysis is threefold: (1) to aid in understanding how energy conversion takes place, (2) to provide techniques for designing and optimizing the devices for specific requirements, and (3) to develop models of electromechanical-energy-conversion devices that can be used in analyzing their performance as components in engineering systems. Transducers and force-producing devices are treated in this chapter; continuous energy-conversion devices are treated in the rest of the book.

The concepts and techniques presented in this chapter are quite powerful and can be applied to a wide range of engineering situations involving electromechanical energy conversion. Sections 3.1 and 3.2 present a quantitative discussion of the forces in electromechanical systems and an overview of the energy method which forms the basis for the derivations presented here. Based upon the energy method, the remainder of the chapter develops expressions for forces and torques in magnetic-field-based electromechanical systems.

3.1 FORCES AND TORQUES IN MAGNETIC FIELD SYSTEMS

The *Lorentz Force Law*

$$\mathbf{F} = q(\mathbf{E} + \mathbf{v} \times \mathbf{B}) \tag{3.1}$$

gives the force $\mathbf{F}$ on a particle of charge q in the presence of electric and magnetic fields. In SI units, $\mathbf{F}$ is in *newtons, q* in *coulombs, E* in *volts per meter,* $\mathbf{B}$ in *teslas,* and v, which is the velocity of the particle relative to the magnetic field, in *meters per second.*

Thus, in a pure electric-field system, the force is determined simply by the charge on the particle and the electric field

$$\mathbf{F} = q\mathbf{E} \tag{3.2}$$

The force acts in the direction of the electric field and is independent of any particle motion.

In pure magnetic-field systems, the situation is somewhat more complex. Here the force

$$\mathbf{F} = q(\mathbf{v} \times \mathbf{B}) \tag{3.3}$$

is determined by the magnitude of the charge on the particle and the magnitude of the $\mathbf{B}$ field as well as the velocity of the particle. In fact, the direction of the force is always perpendicular to the direction of both the particle motion and that of the magnetic field. Mathematically, this is indicated by the vector cross product $\mathbf{v} \times \mathbf{B}$ in Eq. 3.3. The magnitude of this cross product is equal to the product of the magnitudes of $\mathbf{v}$ and $\mathbf{B}$ and the sine of the angle between them; its direction can be found from the right-hand rule, which states that when the thumb of the right hand points in the direction of $\mathbf{v}$ and the index finger points in the direction of $\mathbf{B}$, the force, which is perpendicular to the directions of both $\mathbf{B}$ and $\mathbf{v}$, points in the direction normal to the palm of the hand, as shown in Fig. 3.1.

For situations where large numbers of charged particles are in motion, it is convenient to rewrite Eq. 3.1 in terms of the *charge density* ρ (measured in units of *coulombs per cubic meter*) as

$$\mathbf{F}_{\mathrm{v}} = \rho(\mathbf{E} + \mathbf{v} \times \mathbf{B}) \tag{3.4}$$

where the subscript v indicates that $\mathbf{F}_{\mathrm{v}}$ is a *force density* (force per unit volume) which in SI units is measured in *newtons per cubic meter.*

The product $\rho\mathbf{v}$ is known as the *current density*

$$\mathbf{J} = \rho\mathbf{v} \tag{3.5}$$

which has the units of *amperes per square meter.* The magnetic-system force density

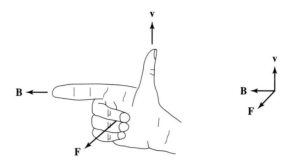

Figure 3.1 Right-hand rule for determining the direction magnetic-field component of the Lorentz force $\mathbf{F} = q(\mathbf{v} \times \mathbf{B})$.

corresponding to Eq. 3.3 can then be written as

$$\mathbf{F_v} = \mathbf{J} \times \mathbf{B} \tag{3.6}$$

For currents flowing in conducting media, Eq. 3.6 can be used to find the force density acting on the material itself. Note that a considerable amount of physics is hidden in this seemingly simple statement, since the mechanism by which the force is transferred from the moving charges to the conducting medium is a complex one.

EXAMPLE 3.1

A nonmagnetic rotor containing a single-turn coil is placed in a uniform magnetic field of magnitude B_0, as shown in Fig. 3.2. The coil sides are at radius R and the wire carries current I

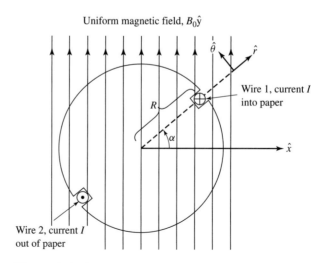

Figure 3.2 Single-coil rotor for Example 3.1.

as indicated. Find the θ-directed torque as a function of rotor position α when $I = 10$ A, $B_0 = 0.02$ T and $R = 0.05$ m. Assume that the rotor is of length $l = 0.3$ m.

■ **Solution**

The force per unit length on a wire carrying current I can be found by multiplying Eq. 3.6 by the cross-sectional area of the wire. When we recognize that the product of the cross-sectional area and the current density is simply the current **I**, the force per unit length acting on the wire is given by

$$\mathbf{F} = \mathbf{I} \times \mathbf{B}$$

Thus, for wire 1 carrying current I into the paper, the θ-directed force is given by

$$F_{1_\theta} = -I B_0 l \sin \alpha$$

and for wire 2 (which carries current in the opposite direction and is located 180° away from wire 1)

$$F_{2_\theta} = -I B_0 l \sin \alpha$$

where l is the length of the rotor. The torque T acting on the rotor is given by the sum of the force-moment-arm products for each wire

$$T = -2I B_0 R l \sin \alpha = -2(10)(0.02)(0.05)(0.3) \sin \alpha = -0.006 \sin \alpha \quad \text{N} \cdot \text{m}$$

Practice Problem 3.1

Repeat Example 3.1 for the situation in which the uniform magnetic field points to the right instead of vertically upward as in Fig. 3.2.

Solution

$$T = -0.006 \cos \alpha \quad \text{N} \cdot \text{m}$$

For situations in which the forces act only on current-carrying elements and which are of simple geometry (such as that of Example 3.1), Eq. 3.6 is generally the simplest and easiest way to calculate the forces acting on the system. Unfortunately, very few practical situations fall into this class. In fact, as discussed in Chapter 1, most electromechanical-energy-conversion devices contain magnetic material; in these systems, forces act directly on the magnetic material and clearly cannot be calculated from Eq. 3.6.

Techniques for calculating the detailed, localized forces acting on magnetic materials are extremely complex and require detailed knowledge of the field distribution throughout the structure. Fortunately, most electromechanical-energy-conversion devices are constructed of rigid, nondeforming structures. The performance of these devices is typically determined by the net force, or torque, acting on the moving component, and it is rarely necessary to calculate the details of the internal force distribution. For example, in a properly designed motor, the motor characteristics are determined by the net accelerating torque acting on the rotor; accompanying forces, which act to squash or deform the rotor, play no significant role in the performance of the motor and generally are not calculated.

To understand the behavior of rotating machinery, a simple physical picture is quite useful. Associated with the rotor structure is a magnetic field (produced in many machines by currents in windings on the rotor), and similarly with the stator; one can picture them as a set of north and south magnetic poles associated with each structure. Just as a compass needle tries to align with the earth's magnetic field, these two sets of fields attempt to align, and torque is associated with their displacement from alignment. Thus, in a motor, the stator magnetic field rotates ahead of that of the rotor, pulling on it and performing work. The opposite is true for a generator, in which the rotor does the work on the stator.

Various techniques have evolved to calculate the net forces of concern in the electromechanical-energy-conversion process. The technique developed in this chapter and used throughout the book is known as the *energy method* and is based on the principle of *conservation of energy*. The basis for this method can be understood with reference to Fig. 3.3a, where a magnetic-field-based electromechanical-energy-conversion device is indicated schematically as a lossless magnetic-energy-storage system with two terminals. The electric terminal has two terminal variables, a voltage e and a current i, and the mechanical terminal also has two terminal variables, a force f_{fld} and a position x.

This sort of representation is valid in situations where the loss mechanism can be separated (at least conceptually) from the energy-storage mechanism. In these cases the electrical losses, such as ohmic losses in windings, can be represented as external elements (i.e., resistors) connected to the electric terminals, and the mechanical losses, such as friction and windage, can be included external to the mechanical terminals. Figure 3.3b shows an example of such a system; a simple force-producing device with a single coil forming the electric terminal, and a movable plunger serving as the mechanical terminal.

The interaction between the electric and mechanical terminals, i.e., the electromechanical energy conversion, occurs through the medium of the magnetic stored

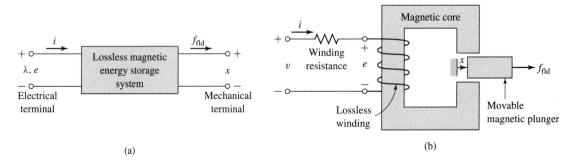

(a)   (b)

Figure 3.3 (a) Schematic magnetic-field electromechanical-energy-conversion device; (b) simple force-producing device.

energy. Since the energy-storage system is lossless, it is a simple matter to write that the time rate of change of W_{fld}, the stored energy in the magnetic field, is equal to the electric power input (given by the product of the terminal voltage and current) less the mechanical power output of the energy storage system (given by the product of the mechanical force and the mechanical velocity):

$$\frac{dW_{\text{fld}}}{dt} = ei - f_{\text{fld}}\frac{dx}{dt} \tag{3.7}$$

Recognizing that, from Eq. 1.27, the voltage at the terminals of our lossless winding is given by the time-derivative of the winding flux linkages

$$e = \frac{d\lambda}{dt} \tag{3.8}$$

and multiplying Eq. 3.7 by dt, we get

$$dW_{\text{fld}} = i\,d\lambda - f_{\text{fld}}\,dx \tag{3.9}$$

As shown in Section 3.4, Eq. 3.9 permits us to solve for the force simply as a function of the flux λ and the mechanical terminal position x. Note that this result comes about as a consequence of our assumption that it is possible to separate the losses out of the physical problem, resulting in a lossless energy-storage system, as in Fig. 3.3a.

Equations 3.7 and 3.9 form the basis for the energy method. This technique is quite powerful in its ability to calculate forces and torques in complex electromechanical-energy-conversion systems. The reader should recognize that this power comes at the expense of a detailed picture of the force-producing mechanism. The forces themselves are produced by such well-known physical phenomena as the Lorentz force on current carrying elements, described by Eq. 3.6, and the interaction of the magnetic fields with the dipoles in the magnetic material.

3.2 ENERGY BALANCE

The principle of conservation of energy states that energy is neither created nor destroyed; it is merely changed in form. For example, a golf ball leaves the tee with a certain amount of kinetic energy; this energy is eventually dissipated as heat due to air friction or rolling friction by the time the ball comes to rest on the fairway. Similarly, the kinetic energy of a hammer is eventually dissipated as heat as a nail is driven into a piece of wood. For isolated systems with clearly identifiable boundaries, this fact permits us to keep track of energy in a simple fashion: the net flow of energy into the system across its boundary is equal to the sum of the time rate of change of energy stored in the system.

This result, which is a statement of the first law of thermodynamics, is quite general. We apply it in this chapter to electromechanical systems whose predominant energy-storage mechanism is in magnetic fields. In such systems, one can account for energy transfer as

$$\begin{pmatrix}\text{Energy input}\\\text{from electric}\\\text{sources}\end{pmatrix} = \begin{pmatrix}\text{Mechanical}\\\text{energy}\\\text{output}\end{pmatrix} + \begin{pmatrix}\text{Increase in energy}\\\text{stored in magnetic}\\\text{field}\end{pmatrix} + \begin{pmatrix}\text{Energy}\\\text{converted}\\\text{into heat}\end{pmatrix}$$

$$(3.10)$$

Equation 3.10 is written so that the electric and mechanical energy terms have positive values for motor action. The equation applies equally well to generator action: these terms then simply have negative values. In either case, the sign of the heat generation term is such that heat generation within the system results in a flow of thermal energy out of the system.

In the systems which we consider here, the conversion of energy into heat occurs by such mechanisms as ohmic heating due to current flow in the windings of the electric terminals and mechanical friction due to the motion of the system components forming the mechanical terminals. As discussed in Section 3.1, it is generally possible to mathematically separate these loss mechanisms from the energy-storage mechanism. In such cases, the device can be represented as a lossless magnetic-energy-storage system with electric and mechanical terminals, as shown in Fig. 3.3a. The loss mechanisms can then be represented by external elements connected to these terminals, resistances to the electric terminals, and mechanical dampers to the mechanical terminals. Figure 3.3a can be readily generalized to situations with any number of electric or mechanical terminals. For this type of system, the magnetic field serves as the coupling medium between the electric and mechanical terminals.

The ability to identify a lossless-energy-storage system is the essence of the energy method. It is important to recognize that this is done mathematically as part of the modeling process. It is not possible, of course, to take the resistance out of windings or the friction out of bearings. Instead we are making use of the fact that a model in which this is done is a valid representation of the physical system.

For the lossless magnetic-energy-storage system of Fig. 3.3a, rearranging Eq. 3.9 in the form of Eq. 3.10 gives

$$dW_{\text{elec}} = dW_{\text{mech}} + dW_{\text{fld}} \tag{3.11}$$

where

$dW_{\text{elec}} = i\,d\lambda = $ differential electric energy input
$dW_{\text{mech}} = f_{\text{fld}}\,dx = $ differential mechanical energy output
$dW_{\text{fld}} = $ differential change in magnetic stored energy

From Eq. 3.8, we can write

$$dW_{\text{elec}} = ei\,dt \tag{3.12}$$

Here e is the voltage induced in the electric terminals by the changing magnetic stored energy. It is through this reaction voltage that the external electric circuit supplies power to the coupling magnetic field and hence to the mechanical output terminals. Thus the basic energy-conversion process is one involving the coupling field and its action and reaction on the electric and mechanical systems.

Combining Eqs. 3.11 and 3.12 results in

$$dW_{elec} = ei\,dt = dW_{mech} + dW_{fld} \qquad (3.13)$$

Equation 3.13, together with Faraday's law for induced voltage (Eq. 1.27), form the basis for the energy method; the following sections illustrate its use in the analysis of electromechanical-energy-conversion devices.

3.3 ENERGY IN SINGLY-EXCITED MAGNETIC FIELD SYSTEMS

In Chapters 1 and 2 we were concerned primarily with fixed-geometry magnetic circuits such as those used for transformers and inductors. Energy in those devices is stored in the leakage fields and to some extent in the core itself. However, the stored energy does not enter directly into the transformation process. In this chapter we are dealing with energy-conversion systems; the magnetic circuits have air gaps between the stationary and moving members in which considerable energy is stored in the magnetic field. This field acts as the energy-conversion medium, and its energy is the reservoir between the electric and mechanical systems.

Consider the electromagnetic relay shown schematically in Fig. 3.4. The resistance of the excitation coil is shown as an external resistance R, and the mechanical terminal variables are shown as a force f_{fld} produced by the magnetic field directed from the relay to the external mechanical system and a displacement x; mechanical losses can be included as external elements connected to the mechanical terminal. Similarly, the moving armature is shown as being massless; its mass represents mechanical energy storage and can be included as an external mass connected to the mechanical terminal. As a result, the magnetic core and armature constitute a lossless magnetic-energy-storage system, as is represented schematically in Fig. 3.3a.

This relay structure is essentially the same as the magnetic structures analyzed in Chapter 1. In Chapter 1 we saw that the magnetic circuit of Fig. 3.4 can be described by an inductance L which is a function of the geometry of the magnetic structure and the

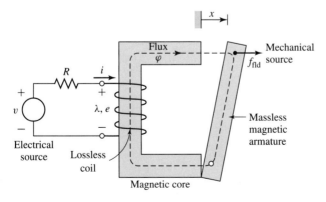

Figure 3.4 Schematic of an electromagnetic relay.

permeability of the magnetic material. Electromechanical-energy-conversion devices contain air gaps in their magnetic circuits to separate the moving parts. As discussed in Section 1.1, in most such cases the reluctance of the air gap is much larger than that of the magnetic material. Thus the predominant energy storage occurs in the air gap, and the properties of the magnetic circuit are determined by the dimensions of the air gap.

Because of the simplicity of the resulting relations, magnetic nonlinearity and core losses are often neglected in the analysis of practical devices. The final results of such approximate analyses can, if necessary, be corrected for the effects of these neglected factors by semi-empirical methods. Consequently, analyses are carried out under the assumption that the flux and mmf are directly proportional for the entire magnetic circuit. Thus the flux linkages λ and current i are considered to be linearly related by an inductance which depends solely on the geometry and hence on the armature position x.

$$\lambda = L(x)i \qquad (3.14)$$

where the explicit dependence of L on x has been indicated.

Since the magnetic force f_{fld} has been defined as acting from the relay upon the external mechanical system and dW_{mech} is defined as the mechanical energy output of the relay, we can write

$$dW_{\text{mech}} = f_{\text{fld}} \, dx \qquad (3.15)$$

Thus, using Eq. 3.15 and the substitution $dW_{\text{elec}} = i \, d\lambda$, we can write Eq. 3.11 as

$$dW_{\text{fld}} = i \, d\lambda - f_{\text{fld}} \, dx \qquad (3.16)$$

Since the magnetic energy storage system is lossless, it is a *conservative system* and the value of W_{fld} is uniquely specified by the values of λ and x; λ and x are thus referred to as *state variables* since their values uniquely determine the state of the system.

From this discussion we see that W_{fld}, being uniquely determined by the values of λ and x, *is the same regardless of how λ and x are brought to their final values.* Consider Fig. 3.5, in which two separate paths are shown over which Eq. 3.16 can be integrated to find W_{fld} at the point (λ_0, x_0). Path 1 is the general case and is difficult to

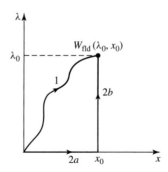

Figure 3.5 Integration paths for W_{fld}.

integrate unless both i and f_{fld} are known explicitly as a function of λ and x. However, because the integration of Eq. 3.16 is path independent, path 2 gives the same result and is much easier to integrate. From Eq. 3.16

$$W_{fld}(\lambda_0, x_0) = \int_{\text{path 2a}} dW_{fld} + \int_{\text{path 2b}} dW_{fld} \tag{3.17}$$

Notice that on path 2a, $d\lambda = 0$ and $f_{fld} = 0$ (since $\lambda = 0$ and there can be no magnetic force in the absence of magnetic fields). Thus from Eq. 3.16, $dW_{fld} = 0$ on path 2a. On path 2b, $dx = 0$, and, thus, from Eq. 3.16, Eq. 3.17 reduces to the integral of $i\,d\lambda$ over path 2b (for which $x = x_0$).

$$W_{fld}(\lambda_0, x_0) = \int_0^{\lambda_0} i(\lambda, x_0)\, d\lambda \tag{3.18}$$

For a linear system in which λ is proportional to i, as in Eq. 3.14, Eq. 3.18 gives

$$W_{fld}(\lambda, x) = \int_0^{\lambda} i(\lambda', x)\, d\lambda' = \int_0^{\lambda} \frac{\lambda'}{L(x)}\, d\lambda' = \frac{1}{2} \frac{\lambda^2}{L(x)} \tag{3.19}$$

It can be shown that the magnetic stored energy can also be expressed in terms of the energy density of the magnetic field integrated over the volume V of the magnetic field. In this case

$$W_{fld} = \int_V \left(\int_0^B \mathbf{H} \cdot d\mathbf{B}' \right) dV \tag{3.20}$$

For soft magnetic material of constant permeability ($\mathbf{B} = \mu\mathbf{H}$), this reduces to

$$W_{fld} = \int_V \left(\frac{B^2}{2\mu} \right) dV \tag{3.21}$$

EXAMPLE 3.2

The relay shown in Fig. 3.6a is made from infinitely-permeable magnetic material with a movable plunger, also of infinitely-permeable material. The height of the plunger is much greater than the air-gap length ($h \gg g$). Calculate the magnetic stored energy W_{fld} as a function of plunger position ($0 < x < d$) for $N = 1000$ turns, $g = 2.0$ mm, $d = 0.15$ m, $l = 0.1$ m, and $i = 10$ A.

■ **Solution**
Equation 3.19 can be used to solve for W_{fld} when λ is known. For this situation, i is held constant, and thus it would be useful to have an expression for W_{fld} as a function of i and x. This can be obtained quite simply by substituting Eq. 3.14 into Eq. 3.19, with the result

$$W_{fld} = \frac{1}{2} L(x) i^2$$

The inductance is given by

$$L(x) = \frac{\mu_0 N^2 A_{gap}}{2g}$$

where A_{gap} is the gap cross-sectional area. From Fig. 3.6b, A_{gap} can be seen to be

$$A_{gap} = l(d - x) = ld \left(1 - \frac{x}{d}\right)$$

Thus

$$L(x) = \frac{\mu_0 N^2 ld(1 - x/d)}{2g}$$

and

$$
\begin{aligned}
W_{fld} &= \frac{1}{2} \frac{N^2 \mu_0 ld(1 - x/d)}{2g} i^2 \\
&= \frac{1}{2} \frac{(1000^2)(4\pi \times 10^{-7})(0.1)(0.15)}{2(0.002)} \times 10^2 \left(1 - \frac{x}{d}\right) \\
&= 236 \left(1 - \frac{x}{d}\right) \text{J}
\end{aligned}
$$

Practice Problem 3.2

The relay of Fig. 3.6 is modified in such a fashion that the air gaps surrounding the plunger are no longer uniform. The top air gap length is increased to $g_{top} = 3.5$ mm and that of the bottom gap is increased to $g_{bot} = 2.5$ mm. The number of turns is increased to $N = 1500$. Calculate the stored energy as a function of plunger position ($0 < x < d$) for a current of $i = 5$ A.

Solution

$$W_{fld} = 88.5 \left(1 - \frac{x}{d}\right) \text{J}$$

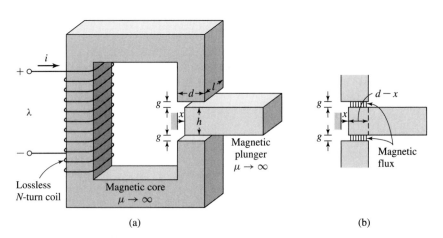

Figure 3.6 (a) Relay with movable plunger for Example 3.2. (b) Detail showing air-gap configuration with the plunger partially removed.

In this section we have seen the relationship between the magnetic stored energy and the electric and mechanical terminal variables for a system which can be represented in terms of a lossless-magnetic-energy-storage element. If we had chosen for our example a device with a rotating mechanical terminal instead of a linearly displacing one, the results would have been identical except that force would be replaced by torque and linear displacement by angular displacement. In Section 3.4 we see how knowledge of the magnetic stored energy permits us to solve for the mechanical force and torque.

3.4 DETERMINATION OF MAGNETIC FORCE AND TORQUE FROM ENERGY

As discussed in Section 3.3, for a lossless magnetic-energy-storage system, the magnetic stored energy W_{fld} is a *state function,* determined uniquely by the values of the independent state variables λ and x. This can be shown explicitly by rewriting Eq. 3.16 in the form

$$dW_{\text{fld}}(\lambda, x) = i \, d\lambda - f_{\text{fld}} \, dx \tag{3.22}$$

For any state function of two independent variables, e.g., $F(x_1, x_2)$, the total differential of F with respect to the two state variables x_1 and x_2 can be written

$$dF(x_1, x_2) = \left. \frac{\partial F}{\partial x_1} \right|_{x_2} dx_1 + \left. \frac{\partial F}{\partial x_2} \right|_{x_1} dx_2 \tag{3.23}$$

It is extremely important to recognize that the partial derivatives in Eq. 3.23 are each taken by holding the opposite state variable constant.

Equation 3.23 is valid for any state function F and hence it is certainly valid for W_{fld}; thus

$$dW_{\text{fld}}(\lambda, x) = \left. \frac{\partial W_{\text{fld}}}{\partial \lambda} \right|_{x} d\lambda + \left. \frac{\partial W_{\text{fld}}}{dx} \right|_{\lambda} dx \tag{3.24}$$

Since λ and x are independent variables, Eqs. 3.22 and 3.24 must be equal for all values of $d\lambda$ and dx, and so

$$i = \left. \frac{\partial W_{\text{fld}}(\lambda, x)}{\partial \lambda} \right|_{x} \tag{3.25}$$

where the partial derivative is taken while holding x constant and

$$f_{\text{fld}} = - \left. \frac{\partial W_{\text{fld}}(\lambda, x)}{\partial x} \right|_{\lambda} \tag{3.26}$$

in this case holding λ constant while taking the partial derivative.

This is the result we have been seeking. Once we know W_{fld} as a function of λ and x, Eq. 3.25 can be used to solve for $i(\lambda, x)$. More importantly, Eq. 3.26 can be used to solve for the mechanical force $f_{\text{fld}}(\lambda, x)$. It cannot be overemphasized that *the partial derivative of Eq. 3.26 is taken while holding the flux linkages λ constant. This* is easily done provided W_{fld} is a known function of λ and x. Note that this is purely a mathematical requirement and has nothing to do with whether λ is held constant when operating the actual device.

The force f_{fld} is determined from Eq. 3.26 directly in terms of the electrical state variable λ. If we then want to express the force as a function of i, we can do so by substituting the appropriate expression for λ as a function of i into the expression for f_{fld} that is obtained by using Eq. 3.26.

For linear magnetic systems for which $\lambda = L(x)i$, the energy is expressed by Eq. 3.19 and the force can be found by direct substitution in Eq. 3.26

$$f_{fld} = -\frac{\partial}{\partial x}\left(\frac{1}{2}\frac{\lambda^2}{L(x)}\right)\bigg|_{\lambda} = \frac{\lambda^2}{2L(x)^2}\frac{dL(x)}{dx} \tag{3.27}$$

If desired, the force can now be expressed in directly in terms of the current i simply by substitution of $\lambda = L(x)i$

$$f_{fld} = \frac{i^2}{2}\frac{dL(x)}{dx} \tag{3.28}$$

EXAMPLE 3.3

Table 3.1 contains data from an experiment in which the inductance of a solenoid was measured as a function of position x, where $x = 0$ corresponds to the solenoid being fully retracted.

Table 3.1   Data for Example 3.3.

x [cm]	0	0.2	0.4	0.6	0.8	1.0	1.2	1.4	1.6	1.8	2.0
L [mH]	2.8	2.26	1.78	1.52	1.34	1.26	1.20	1.16	1.13	1.11	1.10

Plot the solenoid force as a function of position for a current of 0.75 A over the range $0.2 \le x \le 1.8$ cm.

■ **Solution**

The solution is most easily obtained using MATLAB.[†] First, a fourth-order polynomial fit of the inductance as a function of x is obtained using the MATLAB function *polyfit*. The result is of the form

$$L(x) = a(1)x^4 + a(2)x^3 + a(3)x^2 + a(4)x + a(5)$$

Figure 3.7a shows a plot of the data points along with the results of the polynomial fit.

Once this fit has been obtained, it is a straight forward matter to calculate the force from Eq. 3.28.

$$f_{fld} = \frac{i^2}{2}\frac{dL(x)}{dx} = \frac{i^2}{2}(4a(1)x^3 + 3a(2)x^2 + 2a(3)x + a(4))$$

This force is plotted in Figure 3.7b. Note that the force is negative, which means that it is acting in such a direction as to pull the solenoid inwards towards $x = 0$.

[†] MATLAB is a registered trademark of The MathWorks, Inc.

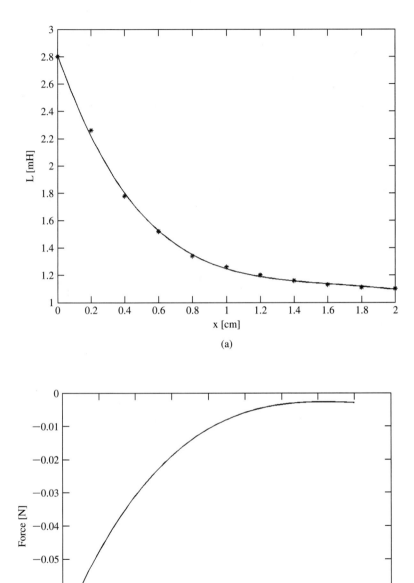

(a)

(b)

Figure 3.7 Example 3.3. (a) Polynomial curve fit of inductance. (b) Force as a function of position x for i = 0.75 A.

Here is the MATLAB script:

```
clc
clear

% Here is the data: x in cm, L in mH
xdata = [0 0.2 0.4 0.6 0.8 1.0 1.2 1.4 1.6 1.8 2.0];
Ldata = [2.8 2.26 1.78 1.52 1.34 1.26 1.20 1.16 1.13 1.11 1.10];

%Convert to SI units
x = xdata*1.e-2;
L = Ldata*1.e-3;

len = length(x);
xmax = x(len);

% Use polyfit to perform a 4'th order fit of L to x. Store
% the polynomial coefficients in vector a. The fit will be
% of the form:
%
%       Lfit = a(1)*x^4 + a(2)*x^3 + a(3)*x^2 + a(4)*x + a(5);
%

a = polyfit(x,L,4);

% Let's check the fit

for n = 1:101
  xfit(n) = xmax*(n-1)/100;
  Lfit(n) = a(1)*xfit(n)^4 + a(2)*xfit(n)^3 + a(3)*xfit(n)^2 ...
  + a(4)*xfit(n) + a(5);
end

% Plot the data and then the fit to compare (convert xfit to cm and
% Lfit to mH)

plot(xdata,Ldata,'*')
hold
plot(xfit*100,Lfit*1000)
hold
xlabel('x [cm]')
ylabel('L [mH]')

fprintf('\n Paused. Hit any key to plot the force.\n')
pause;

% Now plot the force. The force will be given by
%
%    i^2    dL     i^2
%    --- * ---- = --- ( 4*a(1)*x^3 + 3*a(2)*x^2 + 2*a(3)*x + a(4))
%     2     dx     2
```

```
%Set current to 0.75 A
I = 0.75;

for n = 1:101
  xfit(n)  = 0.002 + 0.016*(n-1)/100;
  F(n) = 4*a(1)*xfit(n)^3 + 3* a(2)*xfit(n)^2 + 2*a(3)*xfit(n) + a(4);
  F(n) = (I^2/2)*F(n);
end

plot(xfit*100,F)
xlabel('x [cm]')
ylabel('Force [N]')
```

An external controller is connected to the solenoid of Example 3.3 which maintains the coil flux linkages constant at $\lambda = 1.5$ mWb. Plot the resultant solenoid force over the range $0.2 \leq x \leq 1.8$ cm.

Solution

The resultant force is plotted in Fig. 3.8.

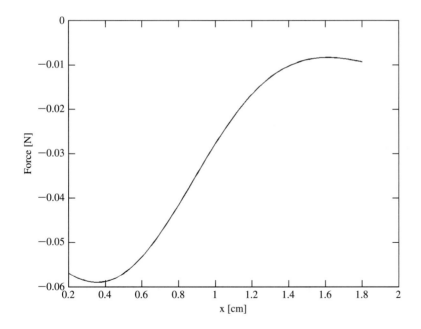

Figure 3.8 Practice problem 3.3. Plot of force vs. x for $\lambda = 1.5$ mWb.

For a system with a rotating mechanical terminal, the mechanical terminal variables become the angular displacement θ and the torque T_{fld}. In this case, Eq. 3.22 becomes

$$dW_{\text{fld}}(\lambda, \theta) = i\, d\lambda - T_{\text{fld}}\, d\theta \qquad (3.29)$$

where the explicit dependence of W_{fld} on state variables λ and θ has been indicated.

By analogy to the development that led to Eq. 3.26, the torque can be found from the negative of the partial derivative of the energy with respect to θ taken holding λ constant

$$T_{\text{fld}} = -\frac{\partial W_{\text{fld}}(\lambda, \theta)}{\partial \theta}\bigg|_{\lambda} \qquad (3.30)$$

For linear magnetic systems for which $\lambda = L(\theta)i$, by analogy to Eq. 3.19 the energy is given by

$$W_{\text{fld}}(\lambda, \theta) = \frac{1}{2}\frac{\lambda^2}{L(\theta)} \qquad (3.31)$$

The torque is therefore given by

$$T_{\text{fld}} = -\frac{\partial}{\partial \theta}\left(\frac{1}{2}\frac{\lambda^2}{L(\theta)}\right)\bigg|_{\lambda} = \frac{1}{2}\frac{\lambda^2}{L(\theta)^2}\frac{dL(\theta)}{d\theta} \qquad (3.32)$$

which can be expressed indirectly in terms of the current i as

$$T_{\text{fld}} = \frac{i^2}{2}\frac{dL(\theta)}{d\theta} \qquad (3.33)$$

EXAMPLE 3.4

The magnetic circuit of Fig. 3.9 consists of a single-coil stator and an oval rotor. Because the air-gap is nonuniform, the coil inductance varies with rotor angular position, measured between the magnetic axis of the stator coil and the major axis of the rotor, as

$$L(\theta) = L_0 + L_2\cos(2\theta)$$

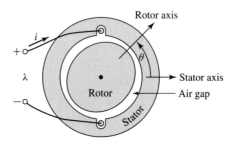

Figure 3.9 Magnetic circuit for Example 3.4.

where $L_0 = 10.6$ mH and $L_2 = 2.7$ mH. Note the second-harmonic variation of inductance with rotor angle θ. This is consistent with the fact that the inductance is unchanged if the rotor is rotated through an angle of 180°.

Find the torque as a function of θ for a coil current of 2 A.

■ Solution

From Eq. 3.33

$$T_{\text{fld}}(\theta) = \frac{i^2}{2}\frac{dL(\theta)}{d\theta} = \frac{i^2}{2}\left(-2L_2\sin(2\theta)\right)$$

Numerical substitution gives

$$T_{\text{fld}}(\theta) = -1.08 \times 10^{-2}\sin(2\theta) \quad \text{N}\cdot\text{m}$$

Note that in this case the torque acts in such a direction as to pull the rotor axis in alignment with the coil axis and hence to maximize the coil inductance.

Practice Problem 3.4

The inductance of a coil on a magnetic circuit similar to that of Fig. 3.9 is found to vary with rotor position as

$$L(\theta) = L_0 + L_2\cos(2\theta) + L_4\sin(4\theta)$$

where $L_0 = 25.4$ mH, $L_2 = 8.3$ mH and $L_4 = 1.8$ mH. (*a*) Find the torque as a function of θ for a winding current of 3.5 A. (*b*) Find a rotor position $\theta_{\max}$ that produces the largest negative torque.

Solution

a. $T_{\text{fld}}(\theta) = -0.1017\sin(2\theta) + 0.044\cos(4\theta) \quad \text{N}\cdot\text{m}$
b. The largest negative torque occurs when $\theta = 45°$ and $\theta = 225°$. This can be determined analytically, but it is helpful to plot the torque using MATLAB.

3.5 DETERMINATION OF MAGNETIC FORCE AND TORQUE FROM COENERGY

A mathematical manipulation of Eq. 3.22 can be used to define a new state function, known as the *coenergy*, from which the force can be obtained directly as a function of the current. The selection of energy or coenergy as the state function is purely a matter of convenience; they both give the same result, but one or the other may be simpler analytically, depending on the desired result and the characteristics of the system being analyzed.

The coenergy W'_{fld} is defined as a function of i and x such that

$$W'_{\text{fld}}(i, x) = i\lambda - W_{\text{fld}}(\lambda, x) \tag{3.34}$$

The desired derivation is carried out by using the differential of $i\lambda$

$$d(i\lambda) = i\,d\lambda + \lambda\,di \tag{3.35}$$

and the differential of $dW_{fld}(\lambda, x)$ from Eq. 3.22. From Eq. 3.34

$$dW'_{fld}(i, x) = d(i\lambda) - dW_{fld}(\lambda, x) \tag{3.36}$$

Substitution of Eqs. 3.22 and 3.35 into Eq. 3.36 results in

$$dW'_{fld}(i, x) = \lambda\,di + f_{fld}\,dx \tag{3.37}$$

From Eq. 3.37, the coenergy $W'_{fld}(i, x)$ can be seen to be a state function of the two independent variables i and x. Thus, its differential can be expressed as

$$dW'_{fld}(i, x) = \left.\frac{\partial W'_{fld}}{\partial i}\right|_{x} di + \left.\frac{\partial W'_{fld}}{\partial x}\right|_{i} dx \tag{3.38}$$

Equations 3.37 and 3.38 must be equal for all values of di and dx; thus

$$\lambda = \left.\frac{\partial W'_{fld}(i, x)}{\partial i}\right|_{x} \tag{3.39}$$

$$f_{fld} = \left.\frac{\partial W'_{fld}(i, x)}{\partial x}\right|_{i} \tag{3.40}$$

Equation 3.40 gives the mechanical force directly in terms of i and x. Note that *the partial derivative in Eq. 3.40 is taken while holding i constant;* thus W'_{fld} must be a known function of i and x. For any given system, Eqs. 3.26 and 3.40 will give the same result; the choice as to which to use to calculate the force is dictated by user preference and convenience.

By analogy to the derivation of Eq. 3.18, the coenergy can be found from the integral of $\lambda\,di$

$$W'_{fld}(i, x) = \int_{0}^{i} \lambda(i', x)\,di' \tag{3.41}$$

For linear magnetic systems for which $\lambda = L(x)i$, the coenergy is therefore given by

$$W'_{fld}(i, x) = \frac{1}{2}L(x)i^2 \tag{3.42}$$

and the force can be found from Eq. 3.40

$$f_{fld} = \frac{i^2}{2}\frac{dL(x)}{dx} \tag{3.43}$$

which, as expected, is identical to the expression given by Eq. 3.28.

Similarly, for a system with a rotating mechanical displacement, the coenergy can be expressed in terms of the current and the angular displacement θ

$$W'_{fld}(i, \theta) = \int_{0}^{i} \lambda(i', \theta)\,di' \tag{3.44}$$

and the torque is given by

$$T_{\text{fld}} = \left. \frac{\partial W'_{\text{fld}}(i, \theta)}{\partial \theta} \right|_i \tag{3.45}$$

If the system is magnetically linear,

$$W'_{\text{fld}}(i, \theta) = \frac{1}{2} L(\theta) i^2 \tag{3.46}$$

and

$$T_{\text{fld}} = \frac{i^2}{2} \frac{dL(\theta)}{d\theta} \tag{3.47}$$

which is identical to Eq. 3.33.

In field-theory terms, for soft magnetic materials (for which $\mathbf{B} = 0$ when $\mathbf{H} = 0$), it can be shown that

$$W'_{\text{fld}} = \int_V \left(\int_0^{H_0} \mathbf{B} \cdot \mathbf{dH} \right) dV \tag{3.48}$$

For soft magnetic material with constant permeability ($\mathbf{B} = \mu \mathbf{H}$), this reduces to

$$W'_{\text{fld}} = \int_V \frac{\mu H^2}{2} \, dV \tag{3.49}$$

For permanent-magnet (hard) materials such as those which are discussed in Chapter 1 and for which $B = 0$ when $H = H_c$, the energy and coenergy are equal to zero when $B = 0$ and hence when $H = H_c$. Thus, although Eq. 3.20 still applies for calculating the energy, Eq. 3.48 must be modified to the form

$$W'_{\text{fld}} = \int_V \left(\int_{H_c}^{H_0} \mathbf{B} \cdot \mathbf{dH} \right) dV \tag{3.50}$$

Note that Eq. 3.50 can be considered to apply in general since soft magnetic materials can be considered to be simply hard magnetic materials with $H_c = 0$, in which case Eq. 3.50 reduces to Eq. 3.48.

In some cases, magnetic circuit representations may be difficult to realize or may not yield solutions of the desired accuracy. Often such situations are characterized by complex geometries and/or magnetic materials driven deeply into saturation. In such situations, numerical techniques can be used to evaluate the system energy using Eq. 3.20 or the coenergy using either Eqs. 3.48 or 3.50.

One such technique, known as the *finite-element method*,[1] has become widely used. For example, such programs, which are available commercially from a number of vendors, can be used to calculate the system coenergy for various values of the displacement x of a linear-displacement actuator (making sure to hold the current constant as x is varied). The force can then be obtained from Eq. 3.40, with the derivative of coenergy with respect to x being calculated numerically from the results of the finite-element analysis.

---

[1] See, for example, P. P. Sylvester and R. L. Ferrari, *Finite Elements for Electrical Engineers*, Cambridge University Press, New York, 1983.

EXAMPLE 3.5

For the relay of Example 3.2, find the force on the plunger as a function of x when the coil is driven by a controller which produces a current as a function of x of the form

$$i(x) = I_0 \left(\frac{x}{d}\right) \text{ A}$$

■ **Solution**

From Example 3.2

$$L(x) = \frac{\mu_0 N^2 l d(1 - x/d)}{2g}$$

This is a magnetically-linear system for which the force can be calculated as

$$f_{\text{fld}} = \frac{i^2}{2}\frac{dL(x)}{dx} = -\frac{i^2}{2}\left(\frac{\mu_0 N^2 l}{2g}\right)$$

Substituting for $i(x)$, the expression for the force as a function of x can be determined as

$$f_{\text{fld}} = -\frac{I_0^2 \mu_0 N^2 l}{4g}\left(\frac{x}{d}\right)^2$$

Note that from Eq. 3.46, the coenergy for this system is equal to

$$W'_{\text{fld}}(i, x) = \frac{i^2}{2}L(x) = \frac{i^2}{2}\frac{N^2 \mu_0 l d(1 - x/d)}{2g}$$

Substituting for $i(x)$, this can be written as

$$W'_{\text{fld}}(i, x) = \frac{I_0^2 N^2 \mu_0 l d(1 - x/d)}{4g}\left(\frac{x}{d}\right)^2$$

Note that, although this is a perfectly correct expression for the coenergy as a function of x under the specified operating conditions, if one were to attempt to calculate the force from taking the partial derivative of this expression for W'_{fld} with respect to x, the resultant expression would not give the correct expression for the force. The reason for this is quite simple: As seen from Eq. 3.40, the partial derivative must be taken holding the current constant. Having substituted the expression for $i(x)$ to obtain the expression, the current is no longer a constant, and this requirement cannot be met. This illustrates the problems that can arise if the various force and torque expressions developed here are misapplied.

Practice Problem 3.5

Consider a plunger whose inductance varies as

$$L(x) = L_0(1 - (x/d)^2)$$

Find the force on the plunger as a function of x when the coil is driven by a controller which produces a current as a function of x of the form

$$i(x) = I_0 \left(\frac{x}{d}\right)^2 \text{ A}$$

Solution

$$f_{\text{fld}} = -\left(\frac{2L_0 I_0^2}{d}\right)\left(\frac{x}{d}\right)^3$$

For a magnetically-linear system, the energy and coenergy are numerically equal: $\frac{1}{2}\lambda^2/L = \frac{1}{2}Li^2$. The same is true for the energy and coenergy densities: $\frac{1}{2}B^2/\mu = \frac{1}{2}\mu H^2$. For a nonlinear system in which λ and i or B and H are not linearly proportional, the two functions are not even numerically equal. A graphical interpretation of the energy and coenergy for a nonlinear system is shown in Fig. 3.10. The area between the $\lambda - i$ curve and the vertical axis, equal to the integral of $i\,d\lambda$, is the energy. The area to the horizontal axis given by the integral of $\lambda\,di$ is the coenergy. For this singly-excited system, the sum of the energy and coenergy is, by definition (see Eq. 3.34),

$$W_{\text{fld}} + W'_{\text{fld}} = \lambda i \tag{3.51}$$

The force produced by the magnetic field in a device such as that of Fig. 3.4 for some particular value of x and i or λ cannot, of course, depend upon whether it is calculated from the energy or coenergy. A graphical illustration will demonstrate that both methods must give the same result.

Assume that the relay armature of Fig. 3.4 is at position x so that the device is operating at point a in Fig. 3.11a. The partial derivative of Eq. 3.26 can be interpreted as the limit of $-\Delta W_{\text{fld}}/\Delta x$ with λ constant as $\Delta x \to 0$. If we allow a change Δx, the change $-\Delta W_{\text{fld}}$ is shown by the shaded area in Fig. 3.11a. Hence, the force $f_{\text{fld}} =$ (shaded area)$/\Delta x$ as $\Delta x \to 0$. On the other hand, the partial derivative of Eq. 3.40 can be interpreted as the limit of $\Delta W'_{\text{fld}}/\Delta x$ with i constant as $\Delta x \to 0$. This perturbation of the device is shown in Fig. 3.11b; the force $f_{\text{fld}} =$ (shaded area)$/\Delta x$ as $\Delta x \to 0$. The shaded areas differ only by the small triangle abc of sides Δi and $\Delta\lambda$, so that in the limit the shaded areas resulting from Δx at constant λ or at constant i are

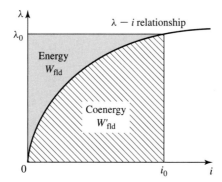

Figure 3.10 Graphical interpretation of energy and coenergy in a singly-excited system.

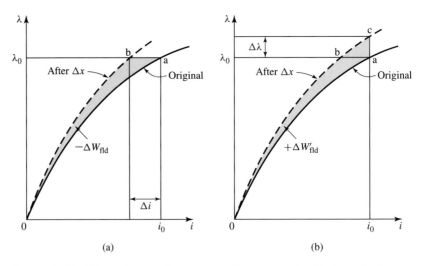

Figure 3.11   Effect of Δx on the energy and coenergy of a singly-excited device: (a) change of energy with λ held constant; (b) change of coenergy with i held constant.

equal. Thus the force produced by the magnetic field is independent of whether the determination is made with energy or coenergy.

Equations 3.26 and 3.40 express the mechanical force of electrical origin in terms of partial derivatives of the energy and coenergy functions $W_{\text{fld}}(\lambda, x)$ and $W'_{\text{fld}}(i, x)$. It is important to note two things about them: the variables in terms of which they must be expressed and their algebraic signs. Physically, of course, the force depends on the dimension x and the magnetic field. The field (and hence the energy or coenergy) can be specified in terms of flux linkage λ, or current i, or related variables. We again emphasize that the selection of the energy or coenergy function as a basis for analysis is a matter of convenience.

The algebraic signs in Eqs. 3.26 and 3.40 show that the force acts in a direction to decrease the magnetic field stored energy at constant flux or to increase the coenergy at constant current. In a singly-excited device, the force acts to increase the inductance by pulling on members so as to reduce the reluctance of the magnetic path linking the winding.

EXAMPLE 3.6

The magnetic circuit shown in Fig. 3.12 is made of high-permeability electrical steel. The rotor is free to turn about a vertical axis. The dimensions are shown in the figure.

a. Derive an expression for the torque acting on the rotor in terms of the dimensions and the magnetic field in the two air gaps. Assume the reluctance of the steel to be negligible (i.e., $\mu \to \infty$) and neglect the effects of fringing.

b. The maximum flux density in the overlapping portions of the air gaps is to be limited to approximately 1.65 T to avoid excessive saturation of the steel. Compute the maximum torque for $r_1 = 2.5$ cm, $h = 1.8$ cm, and $g = 3$ mm.

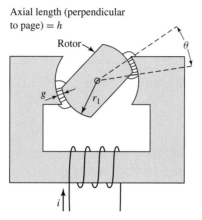

Axial length (perpendicular to page) = h

Rotor

θ

g

r_1

i

Figure 3.12   Magnetic system of Example 3.6.

■ Solution

a. There are two air gaps in series, each of length g, and hence the air-gap field intensity H_{ag} is equal to

$$H_{ag} = \frac{Ni}{2g}$$

Because the permeability of the steel is assumed infinite and B_{steel} must remain finite, $H_{steel} = B_{steel}/\mu$ is zero and the coenergy density (Eq. 3.49) in the steel is zero ($\mu H_{steel}^2/2 = B_{steel}^2/2\mu = 0$). Hence the system coenergy is equal to that of the air gaps, in which the coenergy density in the air gap is $\mu_0 H_{ag}^2/2$. The volume of the two overlapping air gaps is $2gh(r_1 + 0.5g)\theta$. Consequently, the coenergy is equal to the product of the air-gap coenergy density and the air-gap volume

$$W'_{ag} = \left(\frac{\mu_0 H_{ag}^2}{2}\right)(2gh(r_1 + 0.5g)\theta) = \frac{\mu_0(Ni)^2 h(r_1 + 0.5g)\theta}{4g}$$

and thus, from Eq. 3.40

$$T_{fld} = \left.\frac{\partial W'_{ag}(i, \theta)}{\partial \theta}\right|_i = \frac{\mu_0(Ni)^2 h(r_1 + 0.5g)}{4g}$$

The sign of the torque is positive, hence acting in the direction to increase the overlap angle θ and thus to align the rotor with the stator pole faces.

b. For $B_{ag} = 1.65$ T,

$$H_{ag} = \frac{B_{ag}}{\mu_0} = \frac{1.65}{4\pi \times 10^{-7}} = 1.31 \times 10^6 \text{ A/m}$$

and thus

$$Ni = 2g\,H_{ag} = 2(3 \times 10^{-3})1.31 \times 10^6 = 7860 \text{ A-turns}$$

T_{fld} can now be calculated as

$$T_{\text{fld}} = \frac{4\pi \times 10^{-7}(7860)^2(1.8 \times 10^{-2})(2.5 \times 10^{-2} + 0.5(3 \times 10^{-3}))}{4(3 \times 10^{-3})}$$

$$= 3.09 \text{ N} \cdot \text{m}$$

(a) Write an expression for the inductance of the magnetic circuit of Fig. 3.12 as a function of θ. (b) Using this expression, derive an expression for the torque acting on the rotor as a function of the winding current i and the rotor angle θ.

Solution

a.

$$L(\theta) = \frac{\mu_0 N^2 h(r_1 + 0.5g)\theta}{2g}$$

b.

$$T_{\text{fld}} = \frac{i^2}{2} \frac{dL(\theta)}{d\theta} = \frac{i^2}{2}\left(\frac{\mu_0 N^2 h(r_1 + 0.5g)}{2g}\right)$$

3.6 MULTIPLY-EXCITED MAGNETIC FIELD SYSTEMS

Many electromechanical devices have multiple electrical terminals. In measurement systems it is often desirable to obtain torques proportional to two electric signals; a meter which determines power as the product of voltage and current is one example. Similarly, most electromechanical-energy-conversion devices consist of multiply-excited magnetic field systems.

Analysis of these systems follows directly from the techniques discussed in previous sections. This section illustrates these techniques based on a system with two electric terminals. A schematic representation of a simple system with two electrical terminals and one mechanical terminal is shown in Fig. 3.13. In this case it represents a system with rotary motion, and the mechanical terminal variables are torque T_{fld} and angular displacement θ. Since there are three terminals, the system must be described in terms of three independent variables; these can be the mechanical angle θ along with the flux linkages λ_1 and λ_2, currents i_1 and i_2, or a hybrid set including one current and one flux.[2]

When the fluxes are used, the differential energy function $dW_{\text{fld}}(\lambda_1, \lambda_2, \theta)$ corresponding to Eq. 3.29 is

$$dW_{\text{fld}}(\lambda_1, \lambda_2, \theta) = i_1 \, d\lambda_1 + i_2 \, d\lambda_2 - T_{\text{fld}} \, d\theta \tag{3.52}$$

---

[2] See, for example, H. H. Woodson and J. R. Melcher, *Electromechanical Dynamics,* Wiley, New York, 1968, Pt. I, Chap. 3.

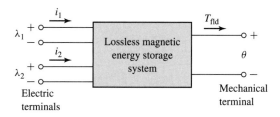

Figure 3.13 Multiply-excited magnetic energy storage system.

and in direct analogy to the previous development for a singly-excited system

$$i_1 = \left.\frac{\partial W_{fld}(\lambda_1, \lambda_2, \theta)}{\partial \lambda_1}\right|_{\lambda_2, \theta} \tag{3.53}$$

$$i_2 = \left.\frac{\partial W_{fld}(\lambda_1, \lambda_2, \theta)}{\partial \lambda_2}\right|_{\lambda_1, \theta} \tag{3.54}$$

and

$$T_{fld} = -\left.\frac{\partial W_{fld}(\lambda_1, \lambda_2, \theta)}{\partial \theta}\right|_{\lambda_1, \lambda_2} \tag{3.55}$$

Note that in each of these equations, *the partial derivative with respect to each independent variable must be taken holding the other two independent variables constant.*

The energy W_{fld} can be found by integrating Eq. 3.52. As in the singly-excited case, this is most conveniently done by holding λ_1 and λ_2 fixed at zero and integrating first over θ; under these conditions, T_{fld} is zero, and thus this integral is zero. One can then integrate over λ_2 (while holding λ_1 zero) and finally over λ_1. Thus

$$W_{fld}(\lambda_{1_0}, \lambda_{2_0}, \theta_0) = \int_0^{\lambda_{2_0}} i_2(\lambda_1 = 0, \lambda_2, \theta = \theta_0) \, d\lambda_2$$

$$+ \int_0^{\lambda_{1_0}} i_1(\lambda_1, \lambda_2 = \lambda_{2_0}, \theta = \theta_0) \, d\lambda_1 \tag{3.56}$$

This path of integration is illustrated in Fig. 3.14 and is directly analogous to that of Fig. 3.5. One could, of course, interchange the order of integration for λ_2 and λ_1. It is extremely important to recognize, however, that the state variables are integrated over a specific path over which only one state variable is varied at a time; for example, λ_1 is maintained at zero while integrating over λ_2 in Eq. 3.56. This is explicitly indicated in Eq. 3.56 and can also be seen from Fig. 3.14. Failure to observe this requirement is one of the most common errors made in analyzing these systems.

In a magnetically-linear system, the relationships between λ and i can be specified in terms of inductances as is discussed in Section 1.2

$$\lambda_1 = L_{11}i_1 + L_{12}i_2 \tag{3.57}$$

$$\lambda_2 = L_{21}i_1 + L_{22}i_2 \tag{3.58}$$

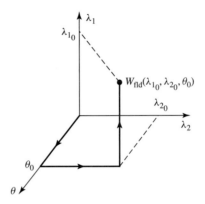

Figure 3.14 Integration path to obtain $W_{fld}(\lambda_{1_0}, \lambda_{2_0}, \theta_0)$.

where

$$L_{12} = L_{21} \tag{3.59}$$

Here the inductances are, in general, functions of angular position θ.

These equations can be inverted to obtain expressions for the i's as a function of the θ's

$$i_1 = \frac{L_{22}\lambda_1 - L_{12}\lambda_2}{D} \tag{3.60}$$

$$i_2 = \frac{-L_{21}\lambda_1 + L_{11}\lambda_2}{D} \tag{3.61}$$

where

$$D = L_{11}L_{22} - L_{12}L_{21} \tag{3.62}$$

The energy for this linear system can be found from Eq. 3.56

$$
\begin{aligned}
W_{fld}(\lambda_{1_0}, \lambda_{2_0}, \theta_0) &= \int_0^{\lambda_{2_0}} \frac{L_{11}(\theta_0)\lambda_2}{D(\theta_0)} \, d\lambda_2 \\
&+ \int_0^{\lambda_{1_0}} \frac{(L_{22}(\theta_0)\lambda_1 - L_{12}(\theta_0)\lambda_{2_0})}{D(\theta_0)} \, d\lambda_1 \\
&= \frac{1}{2D(\theta_0)} L_{11}(\theta_0)\lambda_{2_0}^2 + \frac{1}{2D(\theta_0)} L_{22}(\theta_0)\lambda_{1_0}^2 \\
&- \frac{L_{12}(\theta_0)}{D(\theta_0)} \lambda_{1_0}\lambda_{2_0}
\end{aligned}
\tag{3.63}
$$

where the dependence of the inductances and the determinant $D(\theta)$ on the angular displacement θ has been explicitly indicated.

In Section 3.5, the coenergy function was defined to permit determination of force and torque directly in terms of the current for a single-winding system. A similar coenergy function can be defined in the case of systems with two windings as

$$W'_{fld}(i_1, i_2, \theta) = \lambda_1 i_1 + \lambda_2 i_2 - W_{fld} \tag{3.64}$$

It is a state function of the two terminal currents and the mechanical displacement. Its differential, following substitution of Eq. 3.52, is given by

$$dW'_{fld}(i_1, i_2, \theta) = \lambda_1 \, di_1 + \lambda_2 \, di_2 + T_{fld} \, d\theta \tag{3.65}$$

From Eq. 3.65 we see that

$$\lambda_1 = \left. \frac{\partial W'_{fld}(i_1, i_2, \theta)}{\partial i_1} \right|_{i_2, \theta} \tag{3.66}$$

$$\lambda_2 = \left. \frac{\partial W'_{fld}(i_1, i_2, \theta)}{\partial i_2} \right|_{i_1, \theta} \tag{3.67}$$

Most significantly, the torque can now be determined directly in terms of the currents as

$$T_{fld} = \left. \frac{\partial W'_{fld}(i_1, i_2, \theta)}{\partial \theta} \right|_{i_1, i_2} \tag{3.68}$$

Analogous to Eq. 3.56, the coenergy can be found as

$$W'_{fld}(i_{1_0}, i_{2_0}, \theta_0) = \int_0^{i_{2_0}} \lambda_2(i_1 = 0, i_2, \theta = \theta_0) \, di_2$$

$$+ \int_0^{i_{1_0}} \lambda_1(i_1, i_2 = i_{2_0}, \theta = \theta_0) \, di_1 \tag{3.69}$$

For the linear system of Eqs. 3.57 to 3.59

$$W'_{fld}(i_1, i_2, \theta) = \frac{1}{2} L_{11}(\theta) i_1^2 + \frac{1}{2} L_{22}(\theta) i_2^2 + L_{12}(\theta) i_1 i_2 \tag{3.70}$$

For such a linear system, the torque can be found either from the energy of Eq. 3.63 using Eq. 3.55 or from the coenergy of Eq. 3.70 using Eq. 3.68. It is at this point that the utility of the coenergy function becomes apparent. The energy expression of Eq. 3.63 is a complex function of displacement, and its derivative is even more so. Alternatively, the coenergy function is a relatively simple function of displacement, and from its derivative a straightforward expression for torque can be determined as a function of the winding currents i_1 and i_2 as

$$T_{fld} = \left. \frac{\partial W'_{fld}(i_1, i_2, \theta)}{\partial \theta} \right|_{i_1, i_2}$$

$$= \frac{i_1^2}{2} \frac{dL_{11}(\theta)}{d\theta} + \frac{i_2^2}{2} \frac{dL_{22}(\theta)}{d\theta} + i_1 i_2 \frac{dL_{12}(\theta)}{d\theta} \tag{3.71}$$

Systems with more than two electrical terminals are handled in analogous fashion. As with the two-terminal-pair system above, the use of a coenergy function of the terminal currents greatly simplifies the determination of torque or force.

EXAMPLE 3.7

In the system shown in Fig. 3.15, the inductances in henrys are given as $L_{11} = (3 + \cos 2\theta) \times 10^{-3}$; $L_{12} = 0.3 \cos \theta$; $L_{22} = 30 + 10 \cos 2\theta$. Find and plot the torque $T_{fld}(\theta)$ for current $i_1 = 0.8$ A and $i_2 = 0.01$ A.

■ Solution

The torque can be determined from Eq. 3.71.

$$T_{fld} = \frac{i_1^2}{2} \frac{dL_{11}(\theta)}{d\theta} + \frac{i_2^2}{2} \frac{dL_{22}(\theta)}{d\theta} + i_1 i_2 \frac{dL_{12}(\theta)}{d\theta}$$

$$= \frac{i_1^2}{2}(-2 \times 10^{-3}) \sin 2\theta + \frac{i_2^2}{2}(-20 \sin 2\theta) - i_1 i_2(0.3) \sin \theta$$

For $i_1 = 0.8$ A and $i_2 = 0.01$ A, the torque is

$$T_{fld} = -1.64 \times 10^{-3} \sin 2\theta - 2.4 \times 10^{-3} \sin \theta$$

Notice that the torque expression consists of terms of two types. One term, proportional to $i_1 i_2 \sin \theta$, is due to the mutual interaction between the rotor and stator currents; it acts in a direction to align the rotor and stator so as to maximize their mutual inductance. Alternately, it can be thought of as being due to the tendency of two magnetic fields (in this case those of the rotor and stator) to align.

The torque expression also has two terms each proportional to $\sin 2\theta$ and to the square of one of the coil currents. These terms are due to the action of the individual winding currents alone and correspond to the torques one sees in singly-excited systems. Here the torque is due to the fact that the self inductances are a function of rotor position and the corresponding torque acts in a direction to maximize each inductance so as to maximize the coenergy. The 2θ variation is due to the corresponding variation in the self inductances (exactly as was seen previously in Example 3.4), which in turn is due to the variation of the air-gap reluctance;

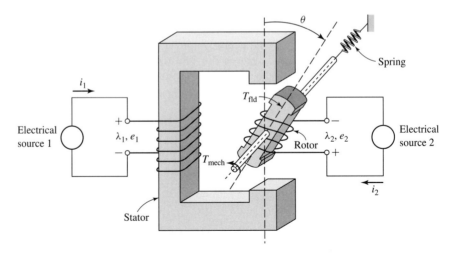

Figure 3.15   Multiply-excited magnetic system for Example 3.7.

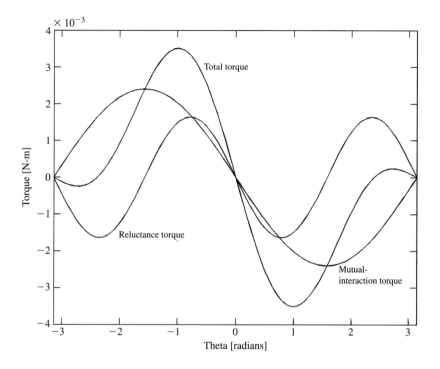

Figure 3.16 Plot of torque components for the multiply-excited system of Example 3.7.

notice that rotating the rotor by 180° from any given position gives the same air-gap reluctance (hence the twice-angle variation). This torque component is known as the *reluctance torque*. The two torque components (mutual and reluctance), along with the total torque, are plotted with MATLAB in Fig. 3.16.

Practice Problem 3.7

Find an expression for the torque of a symmetrical two-winding system whose inductances vary as

$$L_{11} = L_{22} = 0.8 + 0.27 \cos 4\theta$$

$$L_{12} = 0.65 \cos 2\theta$$

for the condition that $i_1 = -i_2 = 0.37$ A.

Solution

$$T_{fld} = -0.296 \sin (4\theta) + 0.178 \sin (2\theta)$$

The derivation presented above for angular displacement can be repeated in an analogous fashion for the systems with linear displacement. If this is done, the expressions for energy and coenergy will be found to be

$$W_{\text{fld}}(\lambda_{1_0}, \lambda_{2_0}, x_0) = \int_0^{\lambda_{2_0}} i_2(\lambda_1 = 0, \lambda_2, x = x_0)\, d\lambda_2$$

$$+ \int_0^{\lambda_{1_0}} i_1(\lambda_1, \lambda_2 = \lambda_{2_0}, x = x_0)\, d\lambda_1 \qquad (3.72)$$

$$W'_{\text{fld}}(i_{1_0}, i_{2_0}, x_0) = \int_0^{i_{2_0}} \lambda_2(i_1 = 0, i_2, x = x_0)\, di_2$$

$$+ \int_0^{i_{1_0}} \lambda_1(i_1, i_2 = i_{2_0}, x = x_0)\, di_1 \qquad (3.73)$$

Similarly the force can be found from

$$f_{\text{fld}} = -\left.\frac{\partial W_{\text{fld}}(\lambda_1, \lambda_2, x)}{\partial x}\right|_{\lambda_1, \lambda_2} \qquad (3.74)$$

or

$$f_{\text{fld}} = \left.\frac{\partial W'_{\text{fld}}(i_1, i_2, x)}{\partial x}\right|_{i_1, i_2} \qquad (3.75)$$

For a magnetically-linear system, the coenergy expression of Eq. 3.70 becomes

$$W'_{\text{fld}}(i_1, i_2, x) = \frac{1}{2}L_{11}(x)i_1^2 + \frac{1}{2}L_{22}(x)i_2^2 + L_{12}(x)i_1 i_2 \qquad (3.76)$$

and the force is thus given by

$$f_{\text{fld}} = \frac{i_1^2}{2}\frac{dL_{11}(x)}{dx} + \frac{i_2^2}{2}\frac{dL_{22}(x)}{dx} + i_1 i_2 \frac{dL_{12}(x)}{dx} \qquad (3.77)$$

3.7 FORCES AND TORQUES IN SYSTEMS WITH PERMANENT MAGNETS

The derivations of the force and torque expressions of Sections 3.4 through 3.6 focus on systems in which the magnetic fields are produced by the electrical excitation of specific windings in the system. However, in Section 3.5, it is seen that special care must be taken when considering systems which contain permanent magnets (also referred to as *hard* magnetic materials). Specifically, the discussion associated with the derivation of the coenergy expression of Eq. 3.50 points out that in such systems the magnetic flux density is zero when $H = H_c$, not when $H = 0$.

For this reason, the derivation of the expressions for force and torque in Sections 3.4 through 3.6 must be modified for systems which contain permanent magnets. Consider for example that the derivation of Eq. 3.18 depends on the fact that in Eq. 3.17 the force can be assumed zero when integrating over path 2a because there is no electrical excitation in the system. A similar argument applies in the derivation of the coenergy expressions of Eqs. 3.41 and 3.69.

In systems with permanent magnets, these derivations must be carefully revisited. In some cases, such systems have no windings at all, their magnetic fields are due solely to the presence of permanent-magnet material, and it is not possible to base a derivation purely upon winding fluxes and currents. In other cases, magnetic fields may be produced by a combination of permanent magnets and windings.

A modification of the techniques presented in the previous sections can be used in systems which contain permanent magnets. Although the derivation presented here applies specifically to systems in which the magnet appears as an element of a magnetic circuit with a uniform internal field, it can be generalized to more complex situations; in the most general case, the field theory expressions for energy (Eq. 3.20) and coenergy (Eq. 3.50) can be used.

The essence of this technique is to consider the system as having an additional *fictitious winding* acting upon the same portion of the magnetic circuit as does the permanent magnet. Under normal operating conditions, the fictitious winding carries zero current. Its function is simply that of a mathematical "crutch" which can be used to accomplish the required analysis. The current in this winding can be adjusted to zero-out the magnetic fields produced by the permanent magnet in order to achieve the "zero-force" starting point for the analyses such as that leading from Eq. 3.17 to Eq. 3.18.

For the purpose of calculating the energy and coenergy of the system, this winding is treated as any other winding, with its own set of current and flux linkages. As a result, energy and coenergy expressions can be obtained as a function of all the winding flux linkages or currents, including those of the fictitious winding. Since under normal operating conditions the current in this winding will be set equal to zero, it is useful to derive the expression for the force from the system coenergy since the winding currents are explicitly expressed in this representation.

Figure 3.17a shows a magnetic circuit with a permanent magnet and a movable plunger. To find the force on the plunger as a function of the plunger position, we assume that there is a fictitious winding of N_f turns carrying a current i_f wound so as to produce flux through the permanent magnet, as shown in Fig. 3.17b.

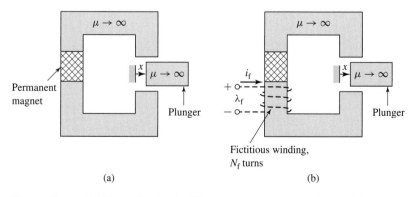

(a)   (b)

Figure 3.17 (a) Magnetic circuit with permanent magnet and movable plunger; (b) fictitious winding added.

For this single-winding system we can write the expression for the differential in coenergy from Eq. 3.37 as

$$dW'_{fld}(i_f, x) = \lambda_f \, di_f + f_{fld} \, dx \qquad (3.78)$$

where the subscript 'f' indicates the fictitious winding. Corresponding to Eq. 3.40, the force in this system can be written as

$$f_{fld} = \left. \frac{\partial W'_{fld}(i_f = 0, x)}{\partial x} \right|_{i_f} \qquad (3.79)$$

where the partial derivative is taken while holding i_f constant, and the resultant expression is then evaluated at $i_f = 0$, which is equivalent to setting $i_f = 0$ in the expression for W'_{fld} before taking the derivative. As we have seen, holding i_f constant for the derivative in Eq. 3.79 is a requirement of the energy method; it must be set to zero to properly calculate the force due to the magnet alone so as not to include a force component from current in the fictitious winding.

To calculate the coenergy $W'_{fld}(i_f, x)$ in this system, it is necessary to integrate Eq. 3.78. Since W'_{fld} is a state function of i_f and x, we are free to choose any integration path we wish. Figure 3.18 illustrates a path over which this integration is particularly simple. For this path we can write the expression for coenergy in this system as

$$W'_{fld}(i_f = 0, x) = \underbrace{\int dW'_{fld}}_{\text{path 1a}} + \underbrace{\int dW'_{fld}}_{\text{path 1b}}$$

$$= \int_0^x f_{fld}(i_f = I_{f0}, x') \, dx' + \int_{I_{f0}}^0 \lambda_f(i_f, x) \, di_f \qquad (3.80)$$

which corresponds directly to the analogous expression for energy found in Eq. 3.17.

Note that the integration is initially over x with the current i_f held fixed at $i_f = I_{f0}$. This is a very specific current, equal to that fictitious-winding current which reduces the magnetic flux in the system to zero. In other words, the current I_{f0} is that current in the fictitious winding which totally counteracts the magnetic field produced by the permanent magnet. Thus, the force f_{fld} is zero at point A in Fig. 3.18 and remains so for the integral over x of path 1a. Hence the integral over path 1a in Eq. 3.80 is zero,

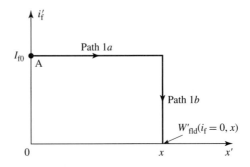

Figure 3.18 Integration path for calculating $W_{fld}(i_f = 0, x)$ in the permanent magnet system of Fig. 3.17.

and Eq. 3.80 reduces to

$$W'_{\text{fld}}(i_f = 0, x) = \int_{I_{f0}}^{0} \lambda_f(i'_f, x) \, di'_f \tag{3.81}$$

Note that Eq. 3.81 is perfectly general and does not require either the permanent magnet or the magnetic material in the magnetic circuit to be linear. Once Eq. 3.81 has been evaluated, the force at a given plunger position x can be readily found from Eq. 3.79. Note also that due to the presence of the permanent magnet, neither the coenergy nor the force is zero when i_f is zero, as we would expect.

EXAMPLE 3.8

The magnetic circuit of Fig. 3.19 is excited by a samarium-cobalt permanent magnet and includes a movable plunger. Also shown is the fictitious winding of N_f turns carrying a current i_f which is included here for the sake of the analysis. The dimensions are:

$$W_m = 2.0 \text{ cm} \quad W_g = 3.0 \text{ cm} \quad W_0 = 2.0 \text{ cm}$$
$$d = 2.0 \text{ cm} \quad g_0 = 0.2 \text{ cm} \quad D = 3.0 \text{ cm}$$

Find (*a*) an expression for the coenergy of the system as a function of plunger position x and (*b*) an expression for the force on the plunger as a function of x. Finally, (*c*) calculate the force at $x = 0$ and $x = 0.5$ cm. Neglect any effects of fringing fluxes in this calculation.

■ **Solution**

a. Because it is quite linear over most of its useful operating range, the dc magnetization curve for samarium-cobalt can be represented as a straight line of the form of Eq. 1.61

$$B_m = \mu_R(H_m - H'_c) = \mu_R H_m + B_r$$

where the subscript 'm' is used here to refer specifically to the fields within the samarium-cobalt magnet and

$\mu_R = 1.05\mu_0$
$H'_c = -712 \text{ kA/m}$
$B_r = 0.94 \text{ T}$

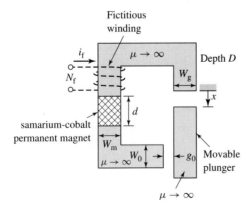

Figure 3.19 Magnetic circuit for Example 3.8.

Note from Fig. 1.19 that the DC magnetization curve for samarium-cobalt is not completely linear; it bends slightly downward for low flux densities. Hence, in the linearized B-H characteristic given above, the apparent coercivity H'_c is somewhat larger than the actual coercivity of samarium-cobalt.

From Eq. 1.5 we can write

$$N_f i_f = H_m d + H_g x + H_0 g_0$$

where the subscript 'g' refers to the variable gap and the subscript '0' refers to the fixed gap. Similarly from the continuity of flux condition, Eq. 1.3, we can write

$$B_m W_m D = B_g W_g D = B_0 W_0 D$$

Recognizing that in the air gaps $B_g = \mu_0 H_g$ and $B_0 = \mu_0 H_0$, we can solve the above equations for B_m:

$$B_m = \frac{\mu_R (N_f i_f - H'_c d)}{d + W_m \left(\frac{\mu_R}{\mu_0}\right) \left(\frac{x}{W_g} + \frac{g_0}{W_0}\right)}$$

Finally we can solve for the flux linkages λ_f of the fictitious winding as

$$\lambda_f = N_f W_m D B_m = \frac{N_f W_m D \mu_R (N_f i_f - H'_c d)}{d + W_m \left(\frac{\mu_R}{\mu_0}\right) \left(\frac{x}{W_g} + \frac{g_0}{W_0}\right)}$$

Thus we see that the flux linkages λ_f will be zero when $i_f = I_{f0} = H'_c d / N_f = -B_r d / (\mu_R N_f)$ and from Eq. 3.81 we can find the coenergy as

$$W'_{fld}(x) = \int_{H'_c d/N_f}^0 \left[\frac{N_f W_m D \mu_R (N_f i_f - H'_c d)}{d + W_m \left(\frac{\mu_R}{\mu_0}\right) \left(\frac{x}{W_g} + \frac{g_0}{W_0}\right)} \right] di_f$$

$$= \frac{W_m D (B_r d)^2}{2 \mu_R \left[d + W_m \left(\frac{\mu_R}{\mu_0}\right) \left(\frac{x}{W_g} + \frac{g_0}{W_0}\right) \right]}$$

Note that the answer does not depend upon N_f or i_f which is as we would expect since the fictitious winding does not actually exist in this system.

b. Once the coenergy has been found, the force can be found from Eq. 3.79 as

$$f_{fld} = -\frac{W_m^2 D (B_r d)^2}{2 \mu_0 W_g \left[d + W_m \left(\frac{\mu_R}{\mu_0}\right) \left(\frac{x}{W_g} + \frac{g_0}{W_0}\right) \right]^2}$$

Notice that the force is negative, indicating that the force is acting in the direction to decrease x, that is to pull the plunger in the direction which decreases the gap.

c. Finally, substitution into the force expression yields

$$f_{fld} = \begin{cases} -115 \text{ N} & \text{at } x = 0 \text{ cm} \\ -85.8 \text{ N} & \text{at } x = 0.5 \text{ cm} \end{cases}$$

(*a*) Derive an expression for the coenergy in the magnetic circuit of Fig. 3.20 as a function of the plunger position x. (*b*) Derive an expression for the x-directed force on the plunger and evaluate it at $x = W_g/2$. Neglect any effects of fringing fluxes. The dimensions are:

$$W_m = 2.0 \text{ cm} \quad W_g = 2.5 \text{ cm} \quad D = 3.0 \text{ cm}$$
$$d = 1.0 \text{ cm} \quad g_0 = 0.2 \text{ cm}$$

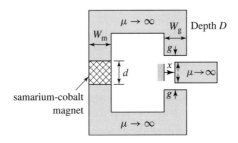

Figure 3.20 Magnetic circuit for Practice Problem 3.8.

Solution

a.

$$W'_{fld} = \frac{W_m D (B_r d)^2}{2\mu_R \left[d + \left(\frac{\mu_R}{\mu_0} \right) \left(\frac{2g W_m}{(W_g - x)} \right) \right]}$$

b.

$$f_{fld} = -\frac{g W_m^2 D B_r^2}{\mu_0 (W_g - x)^2 \left[1 + \left(\frac{\mu_R}{\mu_0} \right) \left(\frac{w g W_m}{(W_g - x)} \right) \right]^2}$$

At $x = W_g/2$, $f_{fld} = -107$ N.

Consider the schematic magnetic circuit of Fig. 3.21a. It consists of a section of linear, hard magnetic material ($B_m = \mu_R (H_m - H'_c)$) of area A and length d. It is connected in series with an external magnetic circuit of mmf $\mathcal{F}_e$.

From Eq. 1.21, since there are no ampere-turns acting on this magnetic circuit,

$$H_m d + \mathcal{F}_e = 0 \tag{3.82}$$

The flux produced in the external magnetic circuit by the permanent magnet is given by

$$\Phi = A B_m = \mu_R A (H_m - H'_c) \tag{3.83}$$

Substitution for H_m from Eq. 3.82 in Eq. 3.83 gives

$$\Phi = \mu_R A \left(-H'_c - \frac{\mathcal{F}_e}{d} \right) \tag{3.84}$$

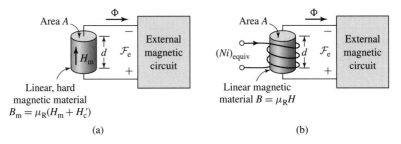

Figure 3.21   (a) Generic magnetic circuit containing a section of linear, permanent-magnet material. (b) Generic magnetic circuit in which the permanent-magnet material has been replaced by a section of linear magnetic material and a fictitious winding.

Now consider the schematic magnetic circuit of Fig. 3.21b in which the linear, hard magnetic material of Fig. 3.21a has been replaced by soft, linear magnetic material of the same permeability ($B = \mu_R H$) and of the same dimensions, length d and area A. In addition, a winding carrying $(Ni)_{\text{equiv}}$ ampere-turns has been included.

For this magnetic circuit, the flux can be shown to be given by

$$\Phi = \mu_R A \left(\frac{(Ni)_{\text{equiv}}}{d} - \frac{\mathcal{F}_e}{d} \right) \tag{3.85}$$

Comparing Eqs. 3.84 and 3.85, we see that the same flux is produced in the external magnetic circuit if the ampere-turns, $(Ni)_{\text{equiv}}$, in the winding of Fig. 3.21b is equal to $-H_c' d$.

This is a useful result for analyzing magnetic-circuit structures which contain linear, permanent-magnet materials whose B-H characteristic can be represented in the form of Eq. 1.61. In such cases, replacing the permanent-magnet section with a section of linear-magnetic material of the same permeability μ_R and geometry and an equivalent winding of ampere-turns

$$(Ni)_{\text{equiv}} = -H_c' d \tag{3.86}$$

results in the same flux in the external magnetic circuit. As a result, both the linear permanent magnet and the combination of the linear magnetic material and the winding are indistinguishable with regard to the production of magnetic fields in the external magnetic circuit, and hence they produce identical forces. Thus, the analysis of such systems may be simplified by this subsitution, as is shown in Example 3.9. This technique is especially useful in the analysis of magnetic circuits containing both a permanet magnetic and one or more windings.

EXAMPLE 3.9

Figure 3.22a shows an actuator consisting of an infinitely-permeable yoke and plunger, excited by a section of neodymium-iron-boron magnet and an excitation winding of $N_1 = 1500$ turns.

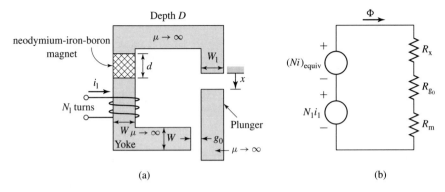

Figure 3.22 (a) Actuator for Example 3.9. (b) Equivalent circuit for the actuator with the permanent magnet replaced by linear material and an equivalent winding carrying $(Ni)_{\text{equiv}}$ ampere-turns.

The dimensions are:

$$W = 4.0 \text{ cm} \quad W_1 = 4.5 \text{ cm} \quad D = 3.5 \text{ cm}$$
$$d = 8 \text{ mm} \quad g_0 = 1 \text{ mm}$$

Find (a) the x-directed force on the plunger when the current in the excitation winding is zero and $x = 3$ mm. (b) Calculate the current in the excitation winding required to reduce the plunger force to zero.

■ Solution

a. As discussed in Section 1.6, the dc-magnetization characteristic of neodymium-iron-boron can be represented by a linear relationship of the form

$$B = \mu_{\text{R}}(H - H'_c) = B_{\text{r}} + \mu_{\text{R}} H$$

where $\mu_{\text{R}} = 1.06\mu_0$, $H'_c = -940$ kA/m and $B_{\text{r}} = 1.25$ T. As discussed in this section, we can replace the magnet by a section of linear material of permeability μ_{R} and an equivalent winding of ampere-turns

$$(Ni)_{\text{equiv}} = -H'_c d = -(-9.4 \times 10^5)(8 \times 10^{-3}) = 7520 \text{ ampere-turns}$$

Based upon this substitution, the equivalent circuit for the system becomes that of Fig. 3.22b. There are two sources of mmf in series with three reluctances: the variable gap $\mathcal{R}_x$, the fixed gap $\mathcal{R}_0$, and the magnet $\mathcal{R}_m$.

$$\mathcal{R}_x = \frac{x}{\mu_0 W_1 D}$$

$$\mathcal{R}_0 = \frac{g_0}{\mu_0 W D}$$

$$\mathcal{R}_m = \frac{d}{\mu_{\text{R}} W D}$$

With $i_1 = 0$, the actuator is equivalent to a single-winding system whose coenergy is given by

$$W'_{\text{fld}} = \frac{1}{2} L i_1^2 = \frac{1}{2} \left(\frac{(Ni)^2_{\text{equiv}}}{\mathcal{R}_x + \mathcal{R}_0 + \mathcal{R}_m} \right)$$

The force on the plunger can then be found from

$$f_{\text{fld}} = \left. \frac{\partial W'_{\text{fld}}}{\partial x} \right|_{i_{\text{equiv}}} = -\frac{(Ni)^2_{\text{equiv}}}{(\mathcal{R}_x + \mathcal{R}_0 + \mathcal{R}_m)^2} \left(\frac{d\mathcal{R}_x}{dx} \right)$$

$$= -\frac{(Ni)^2_{\text{equiv}}}{\mu_0 W_1 D (\mathcal{R}_x + \mathcal{R}_0 + \mathcal{R}_m)^2}$$

Substituting the given values gives $f_{\text{fld}} = -703$ N, where the minus sign indicates that the force acts in the direction to reduce x (i.e., to close the gap).

b. The flux in the actuator is proportional to the total effective ampere-turns $(Ni)_{\text{equiv}} + N_1 i_1$ acting on the magnetic circuit. Thus, the force will be zero when the net ampere-turns is equal to zero or when

$$i_1 = \frac{(Ni)_{\text{equiv}}}{N_1} = \frac{7520}{1500} = 5.01 \text{ A}$$

Note however that the sign of the current (i.e., in which direction it should be applied to the excitation winding) cannot be determined from the information given here since we do not know the direction of magnetization of the magnet. Since the force depends upon the square of the magnetic flux density, the magnet can be oriented to produce flux either upward or downward in the left-hand leg of the magnetic circuit, and the force calculated in part (a) will be the same. To reduce the force to zero, the excitation winding current of 5.01 amperes must be applied in such a direction as to reduce the flux to zero; if the opposite current is applied, the flux density will increase, as will the force.

Practice Problem 3.9

Practice Problem 3.8 is to be reworked replacing the samarium-cobalt magnet by a section of linear material and an equivalent winding. Write (a) expressions for $\mathcal{R}_m$, the reluctance of the section of linear material; $\mathcal{R}_g$, the reluctance of the air gap; and $(Ni)_{\text{equiv}}$, the ampere-turns of the equivalent winding; and (b) an expression for the inductance of the equivalent winding and the coenergy.

$$W_m = 2.0 \text{ cm} \quad W_g = 2.5 \text{ cm} \quad D = 3.0 \text{ cm}$$
$$d = 1.0 \text{ cm} \quad g_0 = 0.2 \text{ cm}$$

Solution

   a.

$$\mathcal{R}_m = \frac{d}{\mu_R W_m D}$$

$$\mathcal{R}_g = \frac{2g}{\mu_0 (W_g - x) D}$$

$$(Ni)_{\text{equiv}} = -H'_c d = \frac{(B_r d)}{\mu_R}$$

b.

$$L = \frac{N_{\text{equiv}}^2}{(\mathcal{R}_{\text{m}} + \mathcal{R}_{\text{g}})}$$

$$W'_{\text{fld}} = \frac{Li_{\text{equiv}}^2}{2} = \frac{(B_{\text{r}}d)^2}{2\mu_{\text{R}}^2(\mathcal{R}_{\text{m}} + \mathcal{R}_{\text{g}})} = \frac{W_{\text{m}}D(B_{\text{r}}d)^2}{2\mu_{\text{R}}\left[d + \left(\frac{\mu_{\text{R}}}{\mu_0}\right)\left(\frac{2gW_{\text{m}}}{(W_{\text{g}}-x)}\right)\right]}$$

Clearly the methods described in this chapter can be extended to handle situations in which there are permanent magnets and multiple current-carrying windings. In many devices of practical interest, the geometry is sufficiently complex, independent of the number of windings and/or permanent magnets, that magnetic-circuit analysis is not necessarily applicable, and analytical solutions can be expected to be inaccurate, if they can be found at all. In these cases, numerical techniques, such as the finite-element method discussed previously, can be employed. Using this method, the coenergy of Eq. 3.48, or Eq. 3.50 if permanet magnets are involved, can be evaluated numerically at constant winding currents and for varying values of displacement.

3.8 DYNAMIC EQUATIONS

We have derived expressions for the forces and torques produced in electromechanical-energy-conversion devices as functions of electrical terminal variables and mechanical displacement. These expressions were derived for conservative energy-conversion systems for which it can be assumed that losses can be assigned to external electrical and mechanical elements connected to the terminals of the energy-conversion system. Such energy-conversion devices are intended to operate as a coupling means between electric and mechanical systems. Hence, we are ultimately interested in the operation of the complete electromechanical system and not just of the electromechanical-energy-conversion system around which it is built.

The model of a simple electromechancal system shown in Fig. 3.23 shows the basic system components, the details of which may vary from system to system. The system shown consists of three parts: an external electric system, the electromechanical-energy-conversion system, and an external mechanical system. The electric system

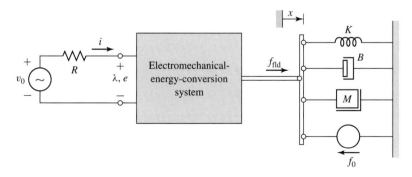

Figure 3.23 Model of a singly-excited electromechanical system.

is represented by a voltage source v_0 and resistance R; the source could alternatively be represented by a current source and a parallel conductance G.

Note that all the electrical losses in the system, including those which are inherent to the electromechanical-energy-conversion system are assigned to the resistance R in this model. For example, if the voltage source has an equivalent resistance R_s and the winding resistance of the electromechanical-energy-conversion system is R_w, the resistance R would equal the sum of these two resistances; $R = R_s + R_w$.

The electric equation for this model is

$$v_0 = iR + \frac{d\lambda}{dt} \tag{3.87}$$

If the flux linkage λ can be expressed as $\lambda = L(x)i$, the external equation becomes

$$v_0 = iR + L(x)\frac{di}{dt} + i\frac{dL(x)}{dx}\frac{dx}{dt} \tag{3.88}$$

The second term on the right, $L(di/dt)$, is the self-inductance voltage term. The third term $i(dL/dx)(dx/dt)$ includes the multiplier dx/dt. This is the speed of the mechanical terminal, and the third term is often called simply the *speed voltage*. The speed-voltage term is common to all electromechanical-energy-conversion systems and is responsible for energy transfer to and from the mechanical system by the electrical system.

For a multiply-excited system, electric equations corresponding to Eq. 3.87 are written for each input pair. If the expressions for the λ's are to be expanded in terms of inductances, as in Eq. 3.88, both self- and mutual-inductance terms will be required.

The mechanical system of Fig. 3.23 includes the representation for a spring (spring constant K), a damper (damping constant B), a mass M, and an external mechanical excitation force f_0. Here, as for the electrical system, the damper represents the losses both of the external mechanical system as well as any mechanical losses of the electromechanical-energy-conversion system.

The x-directed forces and displacement x are related as follows:

Spring:

$$f_K = -K(x - x_0) \tag{3.89}$$

Damper:

$$f_D = -B\frac{dx}{dt} \tag{3.90}$$

Mass:

$$f_M = -M\frac{d^2x}{dt^2} \tag{3.91}$$

where x_0 is the value of x with the spring normally unstretched. Force equilibrium thus requires that

$$f_{fld} + f_K + f_D + f_M - f_0 = f_{fld} - K(x - x_0) - B\frac{dx}{dt} - M\frac{d^2x}{dt^2} - f_0 = 0 \tag{3.92}$$

Combining Eqs. 3.88 and 3.92, the differential equations for the overall system of Fig. 3.23 for arbitrary inputs $v_0(t)$ and $f_0(t)$ are thus

$$v_0(t) = iR + L(x)\frac{di}{dt} + i\frac{dL(x)}{dx} \tag{3.93}$$

$$f_0(t) = -M\frac{d^2x}{dt^2} - B\frac{dx}{dt} - K(x - x_0) + f_{\text{fld}}(x, i) \tag{3.94}$$

The functions $L(x)$ and $f_{\text{fld}}(x, i)$ depend on the properties of the electromechanical-energy-conversion system and are calculated as previously discussed.

EXAMPLE 3.10

Figure 3.24 shows in cross section a cylindrical solenoid magnet in which the cylindrical plunger of mass M moves vertically in brass guide rings of thickness g and mean diameter d.

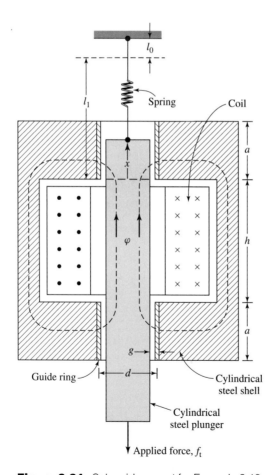

Figure 3.24 Solenoid magnet for Example 3.10.

The permeability of brass is the same as that of free space and is $\mu_0 = 4\pi \times 10^{-7}$ H/m in SI units. The plunger is supported by a spring whose spring constant is K. Its unstretched length is l_0. A mechanical load force f_t is applied to the plunger from the mechanical system connected to it, as shown in Fig. 3.24. Assume that frictional force is linearly proportional to the velocity and that the coefficient of friction is B. The coil has N turns and resistance R. Its terminal voltage is v_t and its current is i. The effects of magnetic leakage and reluctance of the steel are negligible.

Derive the dynamic equations of motion of the electromechanical system, i.e., the differential equations expressing the dependent variables i and x in terms of v_t, f_t, and the given constants and dimensions.

■ Solution

We begin by expressing the inductance as functions of x. The coupling terms, i.e., the magnetic force f_{fld} and induced emf e, can then be expressed in terms of x and i and these relations substituted in the equations for the mechanical and electric systems.

The reluctance of the magnetic circuit is that of the two guide rings in series, with the flux directed radially through them, as shown by the dashed flux lines φ in Fig. 3.24. Because $g \ll d$, the flux density in the guide rings is very nearly constant with respect to the radial distance. In a region where the flux density is constant, the reluctance is

$$\frac{\text{Length of flux path in direction of field}}{\mu \ (\text{area of flux path perpendicular to field})}$$

The reluctance of the upper gap is

$$\mathcal{R}_1 = \frac{g}{\mu_0 \pi x d}$$

in which it is assumed that the field is concentrated in the area between the upper end of the plunger and the lower end of the upper guide ring. Similarly, the reluctance of the lower gap is

$$\mathcal{R}_2 = \frac{g}{\mu_0 \pi a d}$$

The total reluctance is

$$\mathcal{R} = \mathcal{R}_1 + \mathcal{R}_2 = \frac{g}{\mu_0 \pi d} \left(\frac{1}{x} + \frac{1}{a} \right) = \frac{g}{\mu_0 \pi a d} \left(\frac{a+x}{x} \right)$$

Hence, the inductance is

$$L(x) = \frac{N^2}{\mathcal{R}} = \frac{\mu_0 \pi a d N^2}{g} \left(\frac{x}{a+x} \right) = L' \left(\frac{x}{a+x} \right)$$

where

$$L' = \frac{\mu_0 \pi a d N^2}{g}$$

The magnetic force acting upward on the plunger in the positive x direction is

$$f_{fld} = \left. \frac{\partial W'_{fld}(i, x)}{\partial x} \right|_i = \frac{i^2}{2} \frac{dL}{dx} = \frac{i^2}{2} \frac{a L'}{(a+x)^2}$$

The induced emf in the coil is

$$e = \frac{d}{dt}(Li) = L\frac{di}{dt} + i\frac{dL}{dx}\frac{dx}{dt}$$

or

$$e = L'\left(\frac{x}{a+x}\right)\frac{di}{dt} + L'\left(\frac{ai}{(a+x)^2}\right)\frac{dx}{dt}$$

Substitution of the magnetic force in the differential equation of motion of the mechanical system (Eq. 3.94) gives

$$f_t = -M\frac{d^2x}{dt^2} - B\frac{dx}{dt} - K(x-l_0) + \frac{1}{2}L'\frac{ai^2}{(a+x)^2}$$

The voltage equation for the electric system is (from Eq. 3.93)

$$v_t = iR + L'\left(\frac{x}{a+x}\right)\frac{di}{dt} + iL'\left(\frac{a}{(a+x)^2}\right)\frac{dx}{dt}$$

These last two equations are the desired results. They are valid only as long as the upper end of the plunger is well within the upper guide ring, say, between the limits $0.1a < x < 0.9a$. This is the normal working range of the solenoid.

3.9 ANALYTICAL TECHNIQUES

We have described relatively simple devices in this chapter. The devices have one or two electrical terminals and one mechanical terminal, which is usually constrained to incremental motion. More complicated devices capable of continuous energy conversion are treated in the following chapters. The analytical techniques discussed here apply to the simple devices, but the principles are applicable to the more complicated devices as well.

Some of the devices described in this chapter are used to produce gross motion, such as in relays and solenoids, where the devices operate under essentially "on" and "off" conditions. Analysis of these devices is carried out to determine force as a function of displacement and reaction on the electric source. Such calculations have already been made in this chapter. If the details of the motion are required, such as the displacement as a function of time after the device is energized, nonlinear differential equations of the form of Eqs. 3.93 and 3.94 must be solved.

In contrast to gross-motion devices, other devices such as loudspeakers, pickups, and transducers of various kinds are intended to operate with relatively small displacements and to produce a linear relationship between electrical signals and mechanical motion, and vice versa. The relationship between the electrical and mechanical variables is made linear either by the design of the device or by restricting the excursion of the signals to a linear range. In either case the differential equations are linear and can be solved using standard techniques for transient response, frequency response, and so forth, as required.

3.9.1 Gross Motion

The differential equations for a singly-excited device as derived in Example 3.10 are
of the form

$$\frac{1}{2}L'\left(\frac{ai^2}{(a+x)^2}\right) = M\frac{d^2x}{dt^2} + B\frac{dx}{dt} + K(x - l_0) + f_t \tag{3.95}$$

$$v_t = iR + L'\left(\frac{x}{a+x}\right)\frac{di}{dt} + L'\left(\frac{ai}{(a+x)^2}\right)\frac{dx}{dt} \tag{3.96}$$

A typical problem using these differential equations is to find the excursion $x(t)$
when a prescribed voltage $v_t = V_0$ is applied at $t = 0$. An even simpler problem is
to find the time required for the armature to move from its position $x(0)$ at $t = 0$
to a given displacement $x = X$ when a voltage $v_t = V$ is applied at $t = 0$. There
is no general analytical solution for these differential equations; they are nonlinear,
involving products and powers of the variables x and i and their derivatives. They
can be solved using computer-based numerical integration techniques.

In many cases the gross-motion problem can be simplified and a solution found by
relatively simple methods. For example, when the winding of the device is connected
to the voltage source with a relatively large resistance, the iR term dominates on the
right-hand side of Eq. 3.96 compared with the di/dt self-inductance voltage term and
the dx/dt speed-voltage term. The current i can then be assumed equal to V/R and
inserted directly into Eq. 3.95. The same assumption can be made when the winding
is driven from power electronic circuitry which directly controls the current to the
winding. With the assumption that $i = V/R$, two cases can be solved easily.

Case 1   The first case includes those devices in which the dynamic motion is dominated by damping rather than inertia, e.g., devices purposely having low inertia or
relays having dashpots or dampers to slow down the motion. For example, under such
conditions, with $f_t = 0$, the differential equation of Eq. 3.95 reduces to

$$B\frac{dx}{dt} = f(x) = \frac{1}{2}L'\left(\frac{a}{(a+x)^2}\right)\left(\frac{V}{R}\right)^2 - K(x - l_0) \tag{3.97}$$

where $f(x)$ is the difference between the force of electrical origin and the spring force
in the device of Fig. 3.24. The velocity at any value of x is merely $dx/dt = f(x)/B$;
the time t to reach $x = X$ is given by

$$t = \int_0^X \frac{B}{f(x)}\,dx \tag{3.98}$$

The integration of Eq. 3.98 can be carried out analytically or numerically.

Case 2   In this case, the dynamic motion is governed by the inertia rather than the
damping. Again with $f_t = 0$, the differential equation of Eq. 3.95 reduces to

$$M\frac{d^2x}{dt^2} = f(x) = \frac{1}{2}L'\left(\frac{a}{(a+x)^2}\right)\left(\frac{V}{R}\right)^2 - K(x - l_0) \tag{3.99}$$

Equation 3.99 can be written in the form

$$\frac{M}{2}\frac{d}{dx}\left(\frac{dx}{dt}\right)^2 = f(x) \tag{3.100}$$

and the velocity $v(x)$ at any value x is then given by

$$v(x) = \frac{dx}{dt} = \sqrt{\frac{2}{M}\int_0^x f(x')\,dx'} \tag{3.101}$$

The integration of Eq. 3.101 can be carried out analytically or numerically to find $v(x)$ and to find the time t to reach any value of x.

3.9.2 Linearization

Devices characterized by nonlinear differential equations such as Eqs. 3.95 and 3.96 will yield nonlinear responses to input signals when used as transducers. To obtain linear behavior, such devices must be restricted to small excursions of displacement and electrical signals about their equilibrium values. The equilibrium displacement is determined either by a bias mmf produced by a dc winding current, or a permanent magnet acting against a spring, or by a pair of windings producing mmf's whose forces cancel at the equilibrium point. The equilibrium point must be stable; the transducer following a small disturbance should return to the equilibrium position.

With the current and applied force set equal to their equilibrium values, I_0 and f_{t_0} respectively, the equilibrium displacement X_0 and voltage V_0 can be determined for the system described by Eqs. 3.95 and 3.96 by setting the time derivatives equal to zero. Thus

$$\frac{1}{2}L'\left(\frac{aI_0^2}{(a+l_0)^2}\right) = K(X_0 - l_0) + f_{t_0} \tag{3.102}$$

$$V_0 = I_0 R \tag{3.103}$$

The incremental operation can be described by expressing each variable as the sum of its equilibrium and incremental values; thus $i = I_0 + i'$, $f_t = f_{t_0} + f'$, $v_t = V_0 + v'$, and $x = X_0 + x'$. The equations are then linearized by canceling any products of increments as being of second order. Equations 3.95 and 3.96 thus become

$$\frac{1}{2}\frac{L'a(I_0+i')^2}{(a+X_0+x')^2} = M\frac{d^2x'}{dt^2} + B\frac{dx'}{dt} + K(X_0+x'-l_0) + f_{t_0} + f' \tag{3.104}$$

and

$$V_0 + v' = (I_0+i')R + \frac{L'(X_0+x')}{a+X_0+x'}\frac{di'}{dt} + \frac{L'a(I_0+i')}{(a+X_0+x')^2}\frac{dx'}{dt} \tag{3.105}$$

The equilibrium terms cancel, and retaining only first-order incremental terms yields a set of linear differential equations in terms of just the incremental variables of first order as

$$\frac{L'aI_0}{(a + X_0)^2} i' = M \frac{d^2 x'}{dt^2} + B \frac{dx'}{dt} + \left[K + \frac{L'aI_0^2}{(a + X_0)^3} \right] x' + f' \quad (3.106)$$

$$v' = i'R + \frac{L'X_0}{a + X_0} \frac{di'}{dt} + \frac{L'aI_0}{(a + X_0)^2} \frac{dx'}{dt} \quad (3.107)$$

Standard techniques can be used to solve for the time response of this set of linear differential equations. Alternatively, sinusoidal-steady-state operation can be assumed, and Eqs. 3.106 and 3.107 can be converted to a set of linear, complex algebraic equations and solved in the frequency domain.

3.10 SUMMARY

In electromechanical systems, energy is stored in magnetic and electric fields. When the energy in the field is influenced by the configuration of the mechanical parts constituting the boundaries of the field, mechanical forces are created which tend to move the mechanical elements so that energy is transmitted from the field to the mechanical system.

Singly excited magnetic systems are considered first in Section 3.3. By removing electric and mechanical loss elements from the electromechanical-energy-conversion system (and incorporating them as loss elements in the external electrical and mechanical systems), the energy conversion device can be modeled as a conservative system. Its energy then becomes a state function, determined by its state variables λ and x. Section 3.4 derives expressions for determining the force and torque as the negative of partial derivative of the energy with respect to the displacement, taken while holding the flux-linkage λ constant.

In Section 3.5 the state function coenergy, with state variables i and x or θ, is introduced. The force and torque are then shown to be given by the partial derivative of the coenergy with respect to displacement, taken while holding the current i constant.

These concepts are extended in Section 3.6 to include systems with multiple windings. Section 3.7 further extends the development to include systems in which permanent magnets are included among the sources of the magnetic energy storage.

Energy conversion devices operate between electric and mechanical systems. Their behavior is described by differential equations which include the coupling terms between the systems, as discussed in Section 3.8. These equations are usually nonlinear and can be solved by numerical methods if necessary. As discussed in Section 3.9, in some cases approximations can be made to simplify the equations. For example, in many cases, linearized analyses can provide useful insight, both with respect to device design and performance.

This chapter has been concerned with basic principles applying broadly to the electromechanical-energy-conversion process, with emphasis on magnetic-field systems. Basically, rotating machines and linear-motion transducers work in the same way. The remainder of this text is devoted almost entirely to rotating machines. Rotating machines typically include multiple windings and may include permanent magnets. Their performance can be analyzed by using the techniques and principles developed in this chapter.

3.11 PROBLEMS

3.1 The rotor of Fig. 3.25 is similar to that of Fig. 3.2 (Example 3.1) except that it has two coils instead of one. The rotor is nonmagnetic and is placed in a uniform magnetic field of magnitude B_0. The coil sides are of radius R and are uniformly spaced around the rotor surface. The first coil is carrying a current I_1 and the second coil is carrying a current I_2.

Assuming that the rotor is 0.30 m long, $R = 0.13$ m, and $B_0 = 0.85$ T, find the θ-directed torque as a function of rotor position α for (a) $I_1 = 0$ A and $I_2 = 5$ A, (b) $I_1 = 5$ A and $I_2 = 0$ A, and (c) $I_1 = 8$ A and $I_2 = 8$ A.

3.2 The winding currents of the rotor of Problem 3.1 are controlled as a function of rotor angle α such that

$$I_1 = 8 \sin \alpha \quad \text{A} \qquad \text{and} \qquad I_2 = 8 \cos \alpha \quad \text{A}$$

Write an expression for the rotor torque as a function of the rotor position α.

3.3 Calculate the magnetic stored energy in the magnetic circuit of Example 1.2.

3.4 An inductor has an inductance which is found experimentally to be of the form

$$L = \frac{2L_0}{1 + x/x_0}$$

where $L_0 = 30$ mH, $x_0 = 0.87$ mm, and x is the displacement of a movable element. Its winding resistance is measured and found to equal 110 mΩ.

a. The displacement x is held constant at 0.90 mm, and the current is increased from 0 to 6.0 A. Find the resultant magnetic stored energy in the inductor.

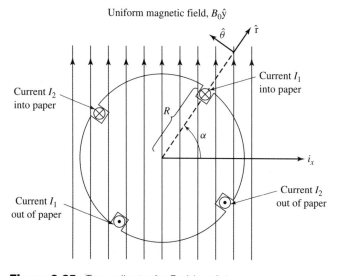

Figure 3.25 Two-coil rotor for Problem 3.1.

b.   The current is then held constant at 6.0 A, and the displacement is increased to 1.80 mm. Find the corresponding change in magnetic stored energy.

3.5   Repeat Problem 3.4, assuming that the inductor is connected to a voltage source which increases from 0 to 0.4 V (part [a]) and then is held constant at 0.4 V (part [b]). For both calculations, assume that all electric transients can be ignored.

3.6   The inductor of Problem 3.4 is driven by a sinusoidal current source of the form

$$i(t) = I_0 \sin \omega t$$

where $I_0 = 5.5$ A and $\omega = 100\pi$ (50 Hz). With the displacement held fixed at $x = x_0$, calculate (a) the time-averaged magnetic stored energy (W_{fld}) in the inductor and (b) the time-averaged power dissipated in the winding resistance.

3.7   An actuator with a rotating vane is shown in Fig. 3.26. You may assume that the permeability of both the core and the vane are infinite ($\mu \rightarrow \infty$). The total air-gap length is $2g$ and shape of the vane is such that the effective area of the air gap can be assumed to be of the form

$$A_g = A_0 \left(1 - \left(\frac{4\theta}{\pi} \right)^2 \right)$$

(valid only in the range $|\theta| \leq \pi/6$). The actuator dimensions are $g = 0.8$ mm, $A_0 = 6.0$ mm^2, and $N = 650$ turns.

a.   Assuming the coil to be carrying current i, write an expression for the magnetic stored energy in the actuator as a function of angle θ for $|\theta| \leq \pi/6$.

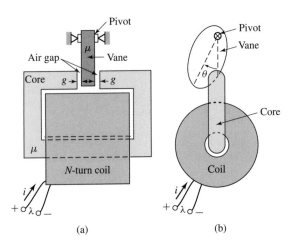

Figure 3.26   Actuator with rotating vane for Problem 3.7. (a) Side view. (b) End view.

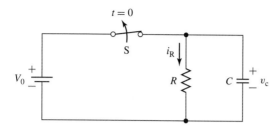

Figure 3.27 An RC circuit for Problem 3.8.

b. Find the corresponding inductance $L(\theta)$. Use MATLAB to plot this inductance as a function of θ.

3.8 An RC circuit is connected to a battery, as shown in Fig. 3.27. Switch S is initially closed and is opened at time $t = 0$.

a. Find the capacitor voltage $v_C(t)$ for $t \geq 0$

b. What are the initial and final ($t = \infty$) values of the stored energy in the capacitor? (Hint: $W_{\text{fld}} = \frac{1}{2}q^2/C$, where $q = CV_0$.) What is the energy stored in the capacitor as a function of time?

c. What is the power dissipated in the resistor as a function of time? What is the total energy dissipated in the resistor?

3.9 An RL circuit is connected to a battery, as shown in Fig. 3.28. Switch S is initially closed and is opened at time $t = 0$.

a. Find the inductor current $i_L(t)$ for $t \geq 0$. (Hint: Note that while the switch is closed, the diode is reverse-biased and can be assumed to be an open circuit. Immediately after the switch is opened, the diode becomes forward-biased and can be assumed to be a short circuit.)

b. What are the initial and final ($t = \infty$) values of the stored energy in the inductor? What is the energy stored in the inductor as a function of time?

c. What is the power dissipated in the resistor as a function of time? What is the total energy dissipated in the resistor?

3.10 The L/R time constant of the field winding of an 500-MVA synchronous generator is 4.8 s. At normal operating conditions, the field winding is known to be dissipating 1.3 MW. Calculate the corresponding magnetic stored energy.

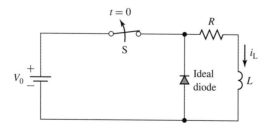

Figure 3.28 An RL circuit for Problem 3.9.

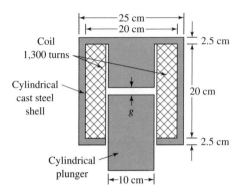

Figure 3.29 Plunger actuator for Problem 3.12.

3.11 The inductance of a phase winding of a three-phase salient-pole motor is measured to be of the form

$$L(\theta_m) = L_0 + L_2 \cos 2\theta_m$$

where θ_m is the angular position of the rotor.

a. How many poles are on the rotor of this motor?

b. Assuming that all other winding currents are zero and that this phase is excited by a constant current I_0, find the torque $T_{fld}(\theta)$ acting on the rotor.

3.12 Cylindrical iron-clad solenoid actuators of the form shown in Fig. 3.29 are used for tripping circuit breakers, for operating valves, and in other applications in which a relatively large force is applied to a member which moves a relatively short distance. When the coil current is zero, the plunger drops against a stop such that the gap g is 2.25 cm. When the coil is energized by a direct current of sufficient magnitude, the plunger is raised until it hits another stop set so that g is 0.2 cm. The plunger is supported so that it can move freely in the vertical direction. The radial air gap between the shell and the plunger can be assumed to be uniform and 0.05 cm in length.

For this problem neglect the magnetic leakage and fringing in the air gaps. The exciting coil has 1300 turns and carries a constant current of 2.3 A. Assume that the mmf in the iron can be neglected and use MATLAB to

a. plot the flux density in the variable gap between the yoke and the plunger for the range of travel of the plunger,

b. plot the corresponding values of the total energy stored in the magnetic field in μJ, and

c. plot the corresponding values of the coil inductance in μH.

3.13 Consider the plunger actuator of Fig. 3.29. Assume that the plunger is initially fully opened ($g = 2.25$ cm) and that a battery is used to supply a current of 2.5 A to the winding.

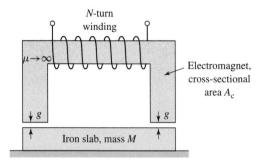

Figure 3.30 Electromagnet lifting an iron slab
(Problem 3.14).

a. If the plunger is constrained to move very slowly (i.e., slowly compared to the electrical time constant of the actuator), reducing the gap g from 2.25 to 0.20 cm, how much mechanical work in joules will be supplied to the plunger?

b. For the conditions of part (a), how much energy will be supplied by the battery (in excess of the power dissipated in the coil)?

3.14 As shown in Fig. 3.30, an N-turn electromagnet is to be used to lift a slab of iron of mass M. The surface roughness of the iron is such that when the iron and the electromagnet are in contact, there is a minimum air gap of $g_{min} = 0.18$ mm in each leg. The electromagnet cross-sectional area $A_c = 32$ cm and coil resistance is 2.8 Ω. Calculate the minimum coil voltage which must be used to lift a slab of mass 95 kg against the force of gravity. Neglect the reluctance of the iron.

3.15 Data for the magnetization curve of the iron portion of the magnetic circuit of the plunger actuator of Problem 3.12 are given below:

Flux (mWb)	5.12	8.42	9.95	10.6	10.9	11.1	11.3	11.4	11.5	11.6
mmf (A · turns)	68	135	203	271	338	406	474	542	609	677

a. Use the MATLAB polyfit function to obtain a 3'rd-order fit of reluctance and total flux versus mmf for the iron portions of the magnetic circuit. Your fits will be of the form:

$$\mathcal{R}_{iron} = a_1 \mathcal{F}_{iron}^3 + a_2 \mathcal{F}_{iron}^2 + a_3 \mathcal{F}_{iron} + a_4$$

$$\phi_{iron} = b_1 \mathcal{F}_{iron}^3 + b_2 \mathcal{F}_{iron}^2 + b_3 \mathcal{F}_{iron} + b_4$$

List the coefficients.

b. (*i*) Using MATLAB and the functional forms found in part (a), plot the magnetization curve for the complete magnetic circuit (flux linkages λ versus winding current i) for a variable-gap length of $g = 0.2$ cm. On the

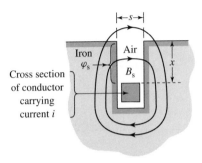

Figure 3.31   Conductor in a slot
(Problem 3.17).

same axes, plot the magnetization curve corresponding to the assumption that the iron is of infinite permeability. The maximum current in your plot should correspond to a flux in the magnetic circuit of 600 mWb.
(*ii*) Calculate the magnetic field energy and coenergy for each of these cases corresponding to a winding current of 2.0 A.

c. Repeat part (b) for a variable-gap length of $g = 2.25$ cm. In part (ii), calculate the magnetic field energy and coenergy corresponding to a winding current of 20 A.

3.16 An inductor is made up of a 525-turn coil on a core of 14-cm² cross-sectional area and gap length 0.16 mm. The coil is connected directly to a 120-V 60-Hz voltage source. Neglect the coil resistance and leakage inductance. Assuming the coil reluctance to be negligible, calculate the time-averaged force acting on the core tending to close the air gap. How would this force vary if the air-gap length were doubled?

3.17 Figure 3.31 shows the general nature of the slot-leakage flux produced by current i in a rectangular conductor embedded in a rectangular slot in iron. Assume that the iron reluctance is negligible and that the slot leakage flux goes straight across the slot in the region between the top of the conductor and the top of the slot.

a. Derive an expression for the flux density B_s in the region between the top of the conductor and the top of the slot.

b. Derive an expression for the slot-leakage φ_s sits crossing the slot above the conductor, in terms of the height x of the slot above the conductor, the slot width s, and the embedded length l perpendicular to the paper.

c. Derive an expression for the force f created by this magnetic field on a conductor of length l. In what direction does this force act on the conductor?

d. When the conductor current is 850 A, compute the force per meter on a conductor in a slot 2.5 cm wide.

3.18 A long, thin solenoid of radius r_0 and height h is shown in Fig. 3.32. The magnetic field inside such a solenoid is axially directed, essentially uniform and equal to $H = Ni/h$. The magnetic field outside the solenoid can be

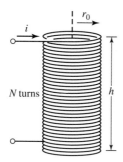

Figure 3.32
Solenoid coil
(Problem 3.18).

shown to be negligible. Calculate the radial pressure in newtons per square meter acting on the sides of the solenoid for constant coil current $i = I_0$.

3.19 An electromechanical system in which electric energy storage is in electric fields can be analyzed by techniques directly analogous to those derived in this chapter for magnetic field systems. Consider such a system in which it is possible to separate the loss mechanism mathematically from those of energy storage in electric fields. Then the system can be represented as in Fig. 3.33. For a single electric terminal, Eq. 3.11 applies, where

$$dW_{elec} = vi\,dt = v\,dq$$

where v is the electric terminal voltage and q is the net charge associated with electric energy storage. Thus, by analogy to Eq. 3.16,

$$dW_{fld} = v\,dq - f_{fld}\,dx$$

a. Derive an expression for the electric stored energy $W_{fld}(q, x)$ analogous to that for the magnetic stored energy in Eq. 3.18.

b. Derive an expression for the force of electric origin f_{fld} analogous to that of Eq. 3.26. State clearly which variable must be held constant when the derivative is taken.

c. By analogy to the derivation of Eqs. 3.34 to 3.41, derive an expression for the coenergy $W'_{fld}(v, x)$ and the corresponding force of electric origin.

Figure 3.33 Lossless electric energy storage system.

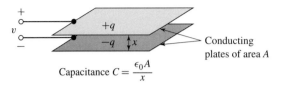

$$\text{Capacitance } C = \frac{\epsilon_0 A}{x}$$

Figure 3.34   Capacitor plates (Problem 3.20).

3.20   A capacitor (Fig. 3.34) is made of two conducting plates of area A separated in air by a spacing x. The terminal voltage is v, and the charge on the plates is q. The capacitance C, defined as the ratio of charge to voltage, is

$$C = \frac{q}{v} = \frac{\epsilon_0 A}{x}$$

where ϵ_0 is the dielectric constant of free space (in SI units $\epsilon_0 = 8.85 \times 10^{-12}$ F/m).

a.   Using the results of Problem 3.19, derive expressions for the energy $W_{\text{fld}}(q, x)$ and the coenergy $W'_{\text{fld}}(v, x)$.

b.   The terminals of the capacitor are connected to a source of constant voltage V_0. Derive an expression for the force required to maintain the plates separated by a constant spacing $x = \delta$.

3.21   Figure 3.35 shows in schematic form an *electrostatic voltmeter*, a capacitive system consisting of a fixed electrode and a moveable electrode. The moveable electrode is connected to a vane which rotates on a pivot such that the air gap between the two electrodes remains fixed as the vane rotates. The capacitance of this system is given by

$$C(\theta) = \frac{\epsilon_0 R d(\alpha - |\theta|)}{g} \qquad (|\theta| \leq \alpha)$$

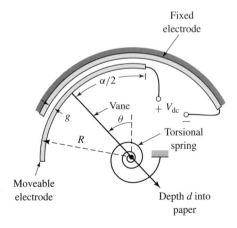

Figure 3.35   Schematic electrostatic voltmeter (Problem 3.21).

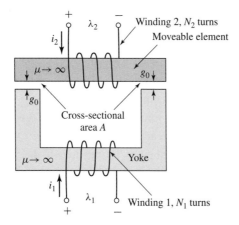

Figure 3.36 Two-winding magnetic circuit for Problem 3.22.

A torsional spring is connected to the moveable vane, producing a torque

$$T_{\text{spring}} = -K(\theta - \theta_0)$$

a. For $0 \le \theta \le \alpha$, using the results of Problem 3.19, derive an expression for the electromagnetic torque T_{fld} in terms of the applied voltage V_{dc}.

b. Find an expression for the angular position of the moveable vane as a function of the applied voltage V_{dc}.

c. For a system with

$$R = 12\text{ cm}, \quad d = 4\text{ cm}, \quad g = 0.2\text{ mm}$$
$$\alpha = \pi/3\text{ rad}, \quad \theta_0 = 0\text{ rad}, \quad K = 3.65\text{ N} \cdot \text{m/rad}$$

Plot the vane position in degrees as a function of applied voltage for $0 \le V_{\text{dc}} \le 1500$ V.

3.22 The two-winding magnetic circuit of Fig. 3.36 has a winding on a fixed yoke and a second winding on a moveable element. The moveable element is constrained to motion such that the lengths of both air gaps remain equal.

a. Find the self-inductances of windings 1 and 2 in terms of the core dimensions and the number of turns.

b. Find the mutual inductance between the two windings.

c. Calculate the coenergy $W'_{\text{fld}}(i_1, i_2)$.

d. Find an expression for the force acting on the moveable element as a function of the winding currents.

3.23 Two coils, one mounted on a stator and the other on a rotor, have self- and mutual inductances of

$$L_{11} = 3.5\text{ mH} \quad L_{22} = 1.8\text{ mH} \quad L_{12} = 2.1\cos\theta\text{ mH}$$

where θ is the angle between the axes of the coils. The coils are connected in series and carry a current

$$i = \sqrt{2}I \sin \omega t$$

a. Derive an expression for the instantaneous torque T on the rotor as a function of the angular position θ.

b. Find an expression for the time-averaged torque T_{avg} as a function of θ.

c. Compute the numerical value of T_{avg} for $I = 10$ A and $\theta = 90°$.

d. Sketch curves of T_{avg} versus θ for currents $I = 5, 7.07$, and 10 A.

e. A helical restraining spring which tends to hold the rotor at $\theta = 90°$ is now attached to the rotor. The restraining torque of the spring is proportional to the angular deflection from $\theta = 90°$ and is -0.1 N·m when the rotor is turned to $\theta = 0°$. Show on the curves of part (d) how you could find the angular position of the rotor-plus-spring combination for coil currents $I = 5, 7.07$, and 10 A. From your curves, estimate the rotor angle for each of these currents.

f. Write a MATLAB script to plot the angular position of the rotor as a function of rms current for $0 \leq I \leq 10$ A.

(Note that this problem illustrates the principles of the dynamometer-type ac ammeter.)

3.24 Two windings, one mounted on a stator and the other on a rotor, have self- and mutual inductances of

$$L_{11} = 4.5\,\text{H} \quad L_{22} = 2.5\,\text{H} \quad L_{12} = 2.8 \cos \theta \,\text{H}$$

where θ is the angle between the axes of the windings. The resistances of the windings may be neglected. Winding 2 is short-circuited, and the current in winding 1 as a function of time is $i_1 = 10 \sin \omega t$ A.

a. Derive an expression for the numerical value in newton-meters of the instantaneous torque on the rotor in terms of the angle θ.

b. Compute the time-averaged torque in newton-meters when $\theta = 45°$.

c. If the rotor is allowed to move, will it rotate continuously or will it tend to come to rest? If the latter, at what value of θ_0?

3.25 A loudspeaker is made of a magnetic core of infinite permeability and circular symmetry, as shown in Figs. 3.37a and b. The air-gap length g is much less than the radius r_0 of the central core. The voice coil is constrained to move only in the x direction and is attached to the speaker cone, which is not shown in the figure. A constant radial magnetic field is produced in the air gap by a direct current in coil 1, $i_1 = I_1$. An audio-frequency signal $i_2 = I_2 \cos \omega t$ is then applied to the voice coil. Assume the voice coil to be of negligible thickness and composed of N_2 turns uniformly distributed over its height h. Also assume that its displacement is such that it remains in the air gap $(0 \leq x \leq l - h)$.

a. Calculate the force on the voice coil, using the Lorentz Force Law (Eq. 3.1).

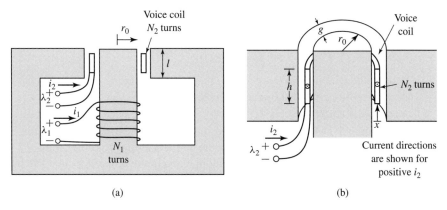

(a)   (b)

Figure 3.37 Loudspeaker for Problem 3.25.

    b.   Calculate the self-inductance of each coil.

    c.   Calculate the mutual inductance between the coils. (Hint: Assume that current is applied to the voice coil, and calculate the flux linkages of coil 1. Note that these flux linkages vary with the displacement x.)

    d.   Calculate the force on the voice coil from the coenergy W'_{fld}.

3.26 Repeat Example 3.8 with the samarium-cobalt magnet replaced by a neodymium-iron-boron magnet.

3.27 The magnetic structure of Fig. 3.38 is a schematic view of a system designed to support a block of magnetic material ($\mu \to \infty$) of mass M against the force of gravity. The system includes a permanent magnet and a winding. Under normal conditions, the force is supplied by the permanent magnet alone. The function of the winding is to counteract the field produced by the magnet so that the mass can be removed from the device. The system is designed such that the air gaps at each side of the mass remain constant at length $g_0/2$.

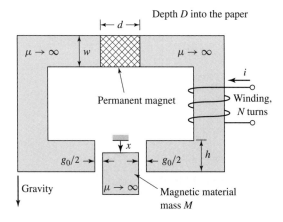

Figure 3.38 Magnetic support system for Problem 3.27.

Assume that the permanent magnet can be represented by a linear characteristic of the form

$$B_m = \mu_R(H_m - H_c)$$

and that the winding direction is such that positive winding current reduces the air-gap flux produced by the permanent magnet. Neglect the effects of magnetic fringing.

a. Assume the winding current to be zero. (*i*) Find the force f_{fld} acting on the mass in the x direction due to the permanent magnet alone as a function of x ($0 \le x \le h$). (*ii*) Find the maximum mass M_{max} that can be supported against gravity for $0 \le x \le h$.

b. For $M = M_{max}/2$, find the minimum current required to ensure that the mass will fall out of the system when the current is applied.

3.28 Winding 1 in the loudspeaker of Problem 3.25 (Fig. 3.37) is replaced by a permanent magnet as shown in Fig. 3.39. The magnet can be represented by the linear characteristic $B_m = \mu_R(H_m - H_c)$.

a. Assuming the voice coil current to be zero, ($i_2 = 0$), calculate the magnetic flux density in the air gap.

b. Calculate the flux linkage of the voice coil due to the permanent magnet as a function of the displacement x.

c. Calculate the coenergy $W'_{fld}(i_2, x)$ assuming that the voice coil current is sufficiently small so that the component of W'_{fld} due to the voice coil self inductance can be ignored.

d. Calculate the force on the voice coil.

3.29 Figure 3.40 shows a circularly symmetric system in which a moveable plunger (constrained to move only in the vertical direction) is supported by a spring of spring constant $K = 5.28$ N/m. The system is excited by a

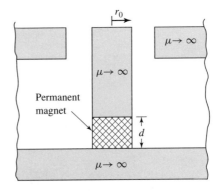

Figure 3.39 Central core of loudspeaker of Fig. 3.37 with winding 1 replaced by a permanent magnet (Problem 3.28).

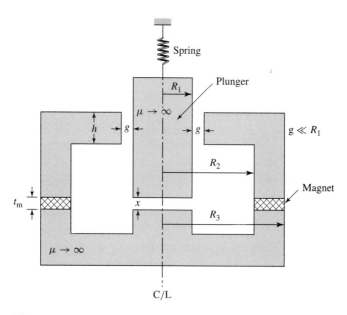

Figure 3.40 Permanent-magnet system for Problem 3.29.

samarium-cobalt permanent-magnet in the shape of a washer of outer radius R_3, inner radius R_2, and thickness t_m. The system dimensions are:

$$R_1 = 2.1 \text{ cm}, \quad R_2 = 4 \text{ cm}, \quad R_3 = 4.5 \text{ cm}$$
$$h = 1 \text{ cm}, \quad g = 1 \text{ mm}, \quad t_m = 3 \text{ mm}$$

The equilibrium position of the plunger is observed to be $x = 1.0$ mm.

a. Find the magnetic flux density B_g in the fixed gap and B_x in the variable gap.

b. Calculate the x-directed magnetic force pulling down on the plunger.

c. The spring force is of the form $f_{spring} = K(X_0 - x)$. Find X_0.

3.30 The plunger of a solenoid is connected to a spring. The spring force is given by $f = K_0(0.9a - x)$, where x is the air-gap length. The inductance of the solenoid is of the form $L = L_0(1 - x/a)$, and its winding resistance is R.

The plunger is initially stationary at position $x = 0.9a$ when a dc voltage of magnitude V_0 is applied to the solenoid.

a. Find an expression for the force as a function of time required to hold the plunger at position $a/2$.

b. If the plunger is then released and allowed to come to equilibrium, find the equilibrium position X_0. You may assume that this position falls in the range $0 \le X_0 \le a$.

3.31 Consider the solenoid system of Problem 3.30. Assume the following parameter values:

$$L_0 = 4.0 \text{ mH} \quad a = 2.2 \text{ cm} \quad R = 1.5 \, \Omega \quad K_0 = 3.5 \text{ N/cm}$$

The plunger has mass $M = 0.2$ kg. Assume the coil to be connected to a dc source of magnitude 4 A. Neglect any effects of gravity.

a. Find the equilibrium displacement X_0.

b. Write the dynamic equations of motion for the system.

c. Linearize these dynamic equations for incremental motion of the system around its equilibrium position.

d. If the plunger is displaced by an incremental distance ϵ from its equilibrium position X_0 and released with zero velocity at time $t = 0$, find (*i*) The resultant motion of the plunger as a function of time, and (*ii*) The corresponding time-varying component of current induced across the coil terminals.

3.32 The solenoid of Problem 3.31 is now connected to a dc voltage source of magnitude 6 V.

a. Find the equilibrium displacement X_0.

b. Write the dynamic equations of motion for the system.

c. Linearize these dynamic equations for incremental motion of the system around its equilibrium position.

3.33 Consider the single-coil rotor of Example 3.1. Assume the rotor winding to be carrying a constant current of $I = 8$ A and the rotor to have a moment of inertia $J = 0.0125$ kg $\cdot$ m^2.

a. Find the equilibrium position of the rotor. Is it stable?

b. Write the dynamic equations for the system.

c. Find the natural frequency in hertz for incremental rotor motion around this equilibrium position.

3.34 Consider a solenoid magnet similar to that of Example 3.10 (Fig. 3.24) except that the length of the cylindrical plunger is reduced to $a + h$. Derive the dynamic equations of motion of the system.

CHAPTER 4

Introduction to Rotating Machines

T he object of this chapter is to introduce and discuss some of the principles underlying the performance of electric machinery. As will be seen, these principles are common to both ac and dc machines. Various techniques and approximations involved in reducing a physical machine to simple mathematical models, sufficient to illustrate the basic principles, will be developed.

4.1 ELEMENTARY CONCEPTS

Equation 1.27, $e = d\lambda/dt$, can be used to determine the voltages induced by time-varying magnetic fields. Electromagnetic energy conversion occurs when changes in the flux linkage λ result from mechanical motion. In rotating machines, voltages are generated in windings or groups of coils by rotating these windings mechanically through a magnetic field, by mechanically rotating a magnetic field past the winding, or by designing the magnetic circuit so that the reluctance varies with rotation of the rotor. By any of these methods, the flux linking a specific coil is changed cyclically, and a time-varying voltage is generated.

A set of such coils connected together is typically referred to as an *armature winding*. In general, the term armature winding is used to refer to a winding or a set of windings on a rotating machine which carry ac currents. In *ac machines* such as synchronous or induction machines, the armature winding is typically on the stationary portion of the motor referred to as the *stator,* in which case these windings may also be referred to as *stator windings*. Figure 4.1 shows the stator winding of a large, multipole, three-phase synchronous motor under construction.

In a *dc machine,* the armature winding is found on the rotating member, referred to as the *rotor*. Figure 4.2 shows a dc-machine rotor. As we will see, the armature winding of a dc machine consists of many coils connected together to form a closed loop. A rotating mechanical contact is used to supply current to the armature winding as the rotor rotates.

Figure 4.1   Stator of a 190-MVA three-phase 12-kV 37-r/min hydroelectric generator. The conductors have hollow passages through which cooling water is circulated. (*Brown Boveri Corporation.*)

Synchronous and dc machines typically include a second winding (or set of windings) which carry dc current and which are used to produce the main operating flux in the machine. Such a winding is typically referred to as *field winding*. The field winding on a dc machine is found on the stator, while that on a synchronous machine is found on the rotor, in which case current must be supplied to the field winding via a rotating mechanical contact. As we have seen, permanent magnets also produce dc magnetic flux and are used in the place of field windings in some machines.

In most rotating machines, the stator and rotor are made of electrical steel, and the windings are installed in slots on these structures. As is discussed in Chapter 1, the use of such high-permeability material maximizes the coupling between the coils and increases the magnetic energy density associated with the electromechanical interaction. It also enables the machine designer to shape and distribute the magnetic fields according to the requirements of each particular machine design. The time-varying flux present in the armature structures of these machines tends to induce currents, known as *eddy currents,* in the electrical steel. Eddy currents can be a large source of loss in such machines and can significantly reduce machine performance. In order to minimize the effects of eddy currents, the armature structure is typically built from thin laminations of electrical steel which are insulated from each other. This is illustrated in Fig. 4.3, which shows the stator core of an ac motor being constructed as a stack of individual laminations.

In some machines, such as *variable reluctance machines* and *stepper motors,* there are no windings on the rotor. Operation of these machines depends on the

Figure 4.2 Armature of a dc motor. (*General Electric Company.*)

Figure 4.3 Partially completed stator core for an ac motor. (*Westinghouse Electric Corporation.*)

nonuniformity of air-gap reluctance associated with variations in rotor position in conjunction with time-varying currents applied to their stator windings. In such machines, both the stator and rotor structures are subjected to time-varying magnetic flux and, as a result, both may require lamination to reduce eddy-current losses.

Rotating electric machines take many forms and are known by many names: dc, synchronous, permanent-magnet, induction, variable reluctance, hysteresis, brushless, and so on. Although these machines appear to be quite dissimilar, the physical principles governing their behavior are quite similar, and it is often helpful to think of them in terms of the same physical picture. For example, analysis of a dc machine shows that associated with both the rotor and the stator are magnetic flux distributions which are fixed in space and that the torque-producing characteristic of the dc machine stems from the tendency of these flux distributions to align. An induction machine, in spite of many fundamental differences, works on exactly the same principle; one can identify flux distributions associated with the rotor and stator. Although they are not stationary but rather rotate in synchronism, just as in a dc motor they are displaced by a constant angular separation, and torque is produced by the tendency of these flux distribution to align.

Certainly, analytically based models are essential to the analysis and design of electric machines, and such models will be derived thoughout this book. However, it is also important to recognize that physical insight into the performance of these devices is equally useful. One objective of this and subsequent chapters is to guide the reader in the development of such insight.

4.2 INTRODUCTION TO AC AND DC MACHINES

4.2.1 AC Machines

Traditional ac machines fall into one of two categories: *synchronous* and *induction*. In synchronous machines, rotor-winding currents are supplied directly from the stationary frame through a rotating contact. In induction machines, rotor currents are induced in the rotor windings by a combination of the time-variation of the stator currents and the motion of the rotor relative to the stator.

Synchronous Machines   A preliminary picture of synchronous-machine performance can be gained by discussing the voltage induced in the armature of the very much simplified *salient-pole* ac synchronous generator shown schematically in Fig. 4.4. The field-winding of this machine produces a single pair of magnetic poles (similar to that of a bar magnet), and hence this machine is referred to as a *two-pole* machine.

With rare exceptions, the armature winding of a synchronous machine is on the stator, and the field winding is on the rotor, as is true for the simplified machine of Fig. 4.4. The field winding is excited by direct current conducted to it by means of stationary carbon *brushes* which contact rotatating *slip rings* or *collector rings*. Practical factors usually dictate this orientation of the two windings: It is advantageous

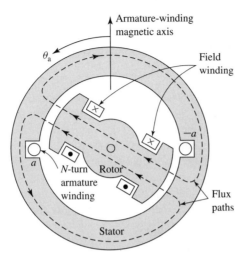

Figure 4.4 Schematic view of a simple, two-pole, single-phase synchronous generator.

to have the single, low-power field winding on the rotor while having the high-power, typically multiple-phase, armature winding on the stator.

The armature winding, consisting here of only a single coil of N turns, is indicated in cross section by the two coil sides a and $-a$ placed in diametrically opposite narrow slots on the inner periphery of the stator of Fig. 4.4. The conductors forming these coil sides are parallel to the shaft of the machine and are connected in series by end connections (not shown in the figure). The rotor is turned at a constant speed by a source of mechanical power connected to its shaft. The armature winding is assumed to be open-circuited and hence the flux in this machine is produced by the field winding alone. Flux paths are shown schematically by dashed lines in Fig. 4.4.

A highly idealized analysis of this machine would assume a sinusoidal distribution of magnetic flux in the air gap. The resultant radial distribution of air-gap flux density B is shown in Fig. 4.5a as a function of the spatial angle θ_a (measured with respect to the magnetic axis of the armature winding) around the rotor periphery. In

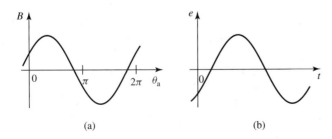

Figure 4.5 (a) Space distribution of flux density and (b) corresponding waveform of the generated voltage for the single-phase generator of Fig. 4.4.

practice, the air-gap flux-density of practical salient-pole machines can be made to approximate a sinusoidal distribution by properly shaping the pole faces.

As the rotor rotates, the flux-linkages of the armature winding change with time. Under the assumption of a sinusoidal flux distribution and constant rotor speed, the resulting coil voltage will be sinusoidal in time as shown in Fig. 4.5b. The coil voltage passes through a complete cycle for each revolution of the two-pole machine of Fig. 4.4. Its frequency in cycles per second (Hz) is the same as the speed of the rotor in revolutions per second: the electric frequency of the generated voltage is synchronized with the mechanical speed, and this is the reason for the designation "synchronous" machine. Thus a two-pole synchronous machine must revolve at 3600 revolutions per minute to produce a 60-Hz voltage.

A great many synchronous machines have more than two poles. As a specific example, Fig. 4.6 shows in schematic form a *four-pole* single-phase generator. The field coils are connected so that the poles are of alternate polarity. There are two complete wavelengths, or cycles, in the flux distribution around the periphery, as shown in Fig. 4.7. The armature winding now consists of two coils a_1, $-a_1$ and a_2, $-a_2$ connected in series by their end connections. The span of each coil is one wavelength of flux. The generated voltage now goes through two complete cycles per revolution of the rotor. The frequency in hertz will thus be twice the speed in revolutions per second.

When a machine has more than two poles, it is convenient to concentrate on a single pair of poles and to recognize that the electric, magnetic, and mechanical conditions associated with every other pole pair are repetitions of those for the pair under consideration. For this reason it is convenient to express angles in *electrical degrees* or *electrical radians* rather than in physical units. One pair of poles in a multipole machine or one cycle of flux distribution equals 360 electrical degrees or 2π electrical radians. Since there are poles/2 complete wavelengths, or cycles, in one

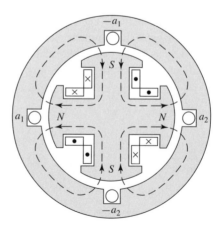

Figure 4.6 Schematic view of a simple, four-pole, single-phase synchronous generator.

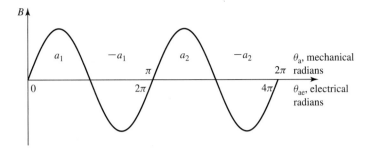

Figure 4.7 Space distribution of the air-gap flux density in a idealized, four-pole synchronous generator.

complete revolution, it follows, for example, that

$$\theta_{ae} = \left(\frac{\text{poles}}{2}\right)\theta_a \tag{4.1}$$

where θ_{ae} is the angle in electrical units and θ_a is the spatial angle. This same relationship applies to all angular measurements in a multipole machine; their values in electrical units will be equal to (poles/2) times their actual spatial values.

The coil voltage of a multipole machine passes through a complete cycle every time a pair of poles sweeps by, or (poles/2) times each revolution. The electrical frequency f_e of the voltage generated in a synchronous machine is therefore

$$f_e = \left(\frac{\text{poles}}{2}\right)\frac{n}{60} \quad \text{Hz} \tag{4.2}$$

where n is the mechanical speed in revolutions per minute, and hence $n/60$ is the speed in revolutions per second. The electrical frequency of the generated voltage in radians per second is $\omega_e = (\text{poles}/2)\,\omega_m$ where ω_m is the mechanical speed in radians per second.

The rotors shown in Figs. 4.4 and 4.6 have *salient*, or *projecting,* poles with *concentrated windings.* Figure 4.8 shows diagrammatically a *nonsalient-pole,* or *cylindrical,* rotor. The field winding is a two-pole *distributed winding;* the coil sides are distributed in multiple slots around the rotor periphery and arranged to produce an approximately sinusoidal distribution of radial air-gap flux.

The relationship between electrical frequency and rotor speed of Eq. 4.2 can serve as a basis for understanding why some synchronous generators have salient-pole rotor structures while others have cylindrical rotors. Most power systems in the world operate at frequencies of either 50 or 60 Hz. A salient-pole construction is characteristic of hydroelectric generators because hydraulic turbines operate at relatively low speeds, and hence a relatively large number of poles is required to produce the desired frequency; the salient-pole construction is better adapted mechanically to this situation. The rotor of a large hydroelectric generator is shown in Fig. 4.9. Steam turbines and gas turbines, however, operate best at relatively high speeds, and turbine-driven alternators or turbine generators are commonly two- or four-pole cylindrical-rotor

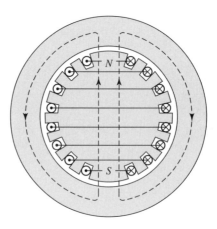

Figure 4.8 Elementary two-pole cylindrical-rotor field winding.

Figure 4.9 Water-cooled rotor of the 190-MVA hydroelectric generator whose stator is shown in Fig. 4.1. (*Brown Boveri Corporation.*)

Figure 4.10 Rotor of a two-pole 3600 r/min turbine generator. (*Westinghouse Electric Corporation.*)

machines. The rotors are made from a single steel forging or from several forgings, as shown in Figs. 4.10 and 4.11.

Most of the world's power systems are three-phase systems and, as a result, with very few exceptions, synchronous generators are three-phase machines. For the production of a set of three voltages phase-displaced by 120 electrical degrees in time, a minimum of three coils phase-displaced 120 electrical degrees in space must be used. A simplified schematic view of a three-phase, two-pole machine with one coil per phase is shown in Fig. 4.12a. The three phases are designated by the letters a, b, and c. In an elementary four-pole machine, a minimum of two such sets of coils must be used, as illustrated in Fig. 4.12b; in an elementary multipole machine, the minimum number of coils sets is given by one half the number of poles.

The two coils in each phase of Fig. 4.12b are connected in series so that their voltages add, and the three phases may then be either Y- or Δ-connected. Figure 4.12c shows how the coils may be interconnected to form a Y connection. Note however, since the voltages in the coils of each phase are indentical, a parallel connection is also possible, e.g., coil $(a, -a)$ in parallel with coil $(a', -a')$, and so on.

When a synchronous generator supplies electric power to a load, the armature current creates a magnetic flux wave in the air gap which rotates at synchronous speed, as shown in Section 4.5. This flux reacts with the flux created by the field current, and electromechanical torque results from the tendency of these two magnetic fields to align. In a generator this torque opposes rotation, and mechanical torque must be applied from the prime mover to sustain rotation. This electromechanical torque is the mechanism through which the synchronous generator converts mechanical to electric energy.

The counterpart of the synchronous generator is the synchronous motor. A cut-away view of a three-phase, 60-Hz synchronous motor is shown in Fig. 4.13. Alternating current is supplied to the armature winding on the stator, and dc excitation is supplied to the field winding on the rotor. The magnetic field produced by the

Figure 4.11 Parts of multipiece rotor for a 1333-MVA three-phase 1800 r/min turbine generator. The separate forgings will be shrunk on the shaft before final machining and milling slots for the windings. The total weight of the rotor is 435,000 lb. (*Brown Boveri Corporation.*)

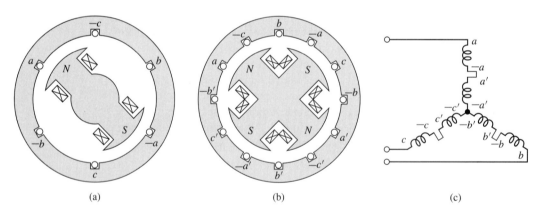

Figure 4.12 Schematic views of three-phase generators: (a) two-pole, (b) four-pole, and (c) *Y* connection of the windings.

armature currents rotates at synchronous speed. To produce a steady electromechanical torque, the magnetic fields of the stator and rotor must be constant in amplitude and stationary with respect to each other. In a synchronous motor, the steady-state speed is determined by the number of poles and the frequency of the armature current. Thus a synchronous motor operated from a constant-frequency ac source will operate at a constant steady-state speed.

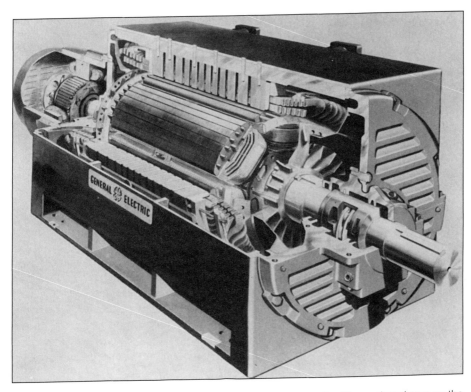

Figure 4.13 Cutaway view of a high-speed synchronous motor. The exciter shown on the left end of the rotor is a small ac generator with a rotating semiconductor rectifier assembly. (*General Electric Company.*)

In a motor the electromechanical torque is in the direction of rotation and balances the opposing torque required to drive the mechanical load. The flux produced by currents in the armature of a synchronous motor rotates ahead of that produced by the field, thus pulling on the field (and hence on the rotor) and doing work. This is the opposite of the situation in a synchronous generator, where the field does work as its flux pulls on that of the armature, which is lagging behind. In both generators and motors, an electromechanical torque and a rotational voltage are produced. These are the essential phenomena for electromechanical energy conversion.

Induction Machines   A second type of ac machine is the *induction machine*. Like the synchronous machine, the stator winding of an induction machine is excited with alternating currents. In contrast to a synchronous machine in which a field winding on the rotor is excited with dc current, alternating currents flow in the rotor windings of an induction machine. In induction machines, alternating currents are applied directly to the stator windings. Rotor currents are then produced by induction, i.e., transformer action. The induction machine may be regarded as a generalized transformer in which electric power is transformed between rotor and stator together with a change of frequency and a flow of mechanical power. Although the induction motor is the most common of all motors, it is seldom used as a generator; its performance characteristics as a generator are unsatisfactory for most applications, although in recent years it has been found to be well suited for wind-power applications. The induction machine may also be used as a frequency changer.

In the induction motor, the stator windings are essentially the same as those of a synchronous machine. However, the rotor windings are electrically short-circuited and frequently have no external connections; currents are induced by transformer action from the stator winding. A cutaway view of a squirrel-cage induction motor is shown in Fig. 4.14. Here the rotor "windings" are actually solid aluminum bars which are cast into the slots in the rotor and which are shorted together by cast aluminum rings at each end of the rotor. This type of rotor construction results in induction motors which are relatively inexpensive and highly reliable, factors contributing to their immense popularity and widespread application.

As in a synchronous motor, the armature flux in the induction motor leads that of the rotor and produces an electromechanical torque. In fact, we will see that, just as in a synchronous machine, the rotor and stator fluxes rotate in synchronism with each other and that torque is related to the relative displacement between them. However, unlike a synchronous machine, the rotor of an induction machine does not itself rotate synchronously; it is the "slipping" of the rotor with respect to the synchronous armature flux that gives rise to the induced rotor currents and hence the torque. Induction motors operate at speeds less than the synchronous mechanical speed. A typical speed-torque characteristic for an induction motor is shown in Fig. 4.15.

Figure 4.14   Cutaway view of a squirrel-cage induction motor. (*Westinghouse Electric Corporation.*)

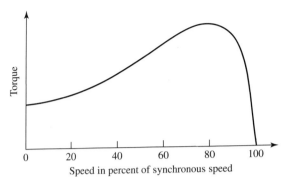

Figure 4.15 Typical induction-motor speed-torque characteristic.

Figure 4.16 Cutaway view of a typical integral-horsepower dc motor. (*ASEA Brown Boveri.*)

4.2.2 DC Machines

As has been discussed, the armature winding of a dc generator is on the rotor with current conducted from it by means of carbon brushes. The field winding is on the stator and is excited by direct current. A cutaway view of a dc motor is shown in Fig. 4.16.

A very elementary two-pole dc generator is shown in Fig. 4.17. The armature winding, consisting of a single coil of N turns, is indicated by the two coil sides

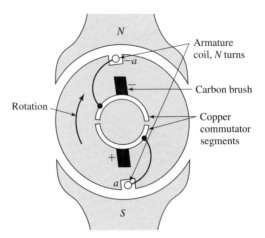

Figure 4.17   Elementary dc machine with commutator.

a and $-a$ placed at diametrically opposite points on the rotor with the conductors parallel to the shaft. The rotor is normally turned at a constant speed by a source of mechanical power connected to the shaft. The air-gap flux distribution usually approximates a flat-topped wave, rather than the sine wave found in ac machines, and is shown in Fig. 4.18a. Rotation of the coil generates a coil voltage which is a time function having the same waveform as the spatial flux-density distribution.

Although the ultimate purpose is the generation of a direct voltage, the voltage induced in an individual armature coil is an alternating voltage, which must therefore be rectified. The output voltage of an ac machine can be rectified using external semiconductor rectifiers. This is in contrast to the conventional dc machine in which rectification is produced mechanically by means of a *commutator,* which is a cylinder formed of copper segments insulated from each other by mica or some other highly insulating material and mounted on, but insulated from, the rotor shaft. Stationary carbon brushes held against the commutator surface connect the winding to the external armature terminals. The commutator and brushes can readily be seen in Fig. 4.16. The need for commutation is the reason why the armature windings of dc machines are placed on the rotor.

For the elementary dc generator, the commutator takes the form shown in Fig. 4.17. For the direction of rotation shown, the commutator at all times connects the coil side, which is under the south pole, to the positive brush and that under the north pole to the negative brush. The commutator provides full-wave rectification, transforming the voltage waveform between brushes to that of Fig. 4.18b and making available a unidirectional voltage to the external circuit. The dc machine of Fig. 4.17 is, of course, simplified to the point of being unrealistic in the practical sense, and later it will be essential to examine the action of more realistic commutators.

The effect of direct current in the field winding of a dc machine is to create a magnetic flux distribution which is stationary with respect to the stator. Similarly, the

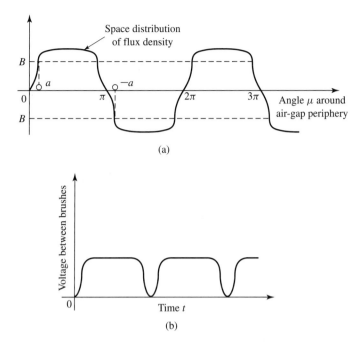

(a)

(b)

Figure 4.18 (a) Space distribution of air-gap flux density in an elementary dc machine; (b) waveform of voltage between brushes.

effect of the commutator is such that when direct current flows through the brushes, the armature creates a magnetic flux distribution which is also fixed in space and whose axis, determined by the design of the machine and the position of the brushes, is typically perpendicular to the axis of the field flux.

Thus, just as in the ac machines discussed previously, it is the interaction of these two flux distributions that creates the torque of the dc machine. If the machine is acting as a generator, this torque opposes rotation. If it is acting as a motor, the electromechanical torque acts in the direction of the rotation. Remarks similar to those already made concerning the roles played by the generated voltage and electromechanical torque in the energy conversion process in synchronous machines apply equally well to dc machines.

4.3 MMF OF DISTRIBUTED WINDINGS

Most armatures have distributed windings, i.e., windings which are spread over a number of slots around the air-gap periphery, as in Figs. 4.2 and 4.1. The individual coils are interconnected so that the result is a magnetic field having the same number of poles as the field winding.

The study of the magnetic fields of distributed windings can be approached by examining the magnetic field produced by a winding consisting of a single N-turn coil which spans 180 electrical degrees, as shown in Fig. 4.19a. A coil which spans

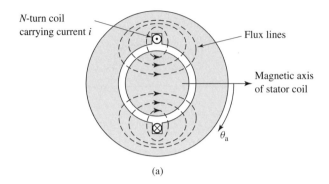

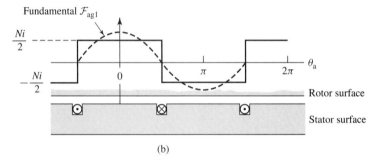

Figure 4.19 (a) Schematic view of flux produced by a concentrated, full-pitch winding in a machine with a uniform air gap. (b) The air-gap mmf produced by current in this winding.

180 electrical degrees is known as a *full-pitch coil*. The dots and crosses indicate current flow towards and away from the reader, respectively. For simplicity, a concentric cylindrical rotor is shown. The general nature of the magnetic field produced by the current in the coil is shown by the dashed lines in Fig. 4.19a. Since the permeability of the armature and field iron is much greater than that of air, it is sufficiently accurate for our present purposes to assume that all the reluctance of the magnetic circuit is in the air gap. From symmetry of the structure it is evident that the magnetic field intensity H_{ag} in the air gap at angle θ_a under one pole is the same in magnitude as that at angle $\theta_a + \pi$ under the opposite pole, but the fields are in the opposite direction.

Around any of the closed paths shown by the flux lines in Fig. 4.19a the mmf is Ni. The assumption that all the reluctance of this magnetic circuit is in the air gap leads to the result that the line integral of **H** inside the iron is negligibly small, and thus it is reasonable to neglect the mmf drops associated with portions of the magnetic circuit inside the iron. By symmetry we argued that the air-gap fields H_{ag} on opposite sides of the rotor are equal in magnitude but opposite in direction. It follows that the air-gap mmf should be similarly distributed; since each flux line crosses the air gap twice, the mmf drop across the air gap must be equal to half of the total or $Ni/2$.

Figure 4.19b shows the air gap and winding in developed form, i.e., laid out flat. The air-gap mmf distribution is shown by the steplike distribution of amplitude

When they are connected in series to form the phase winding, their phasor sum is then less than their numerical sum. (See Appendix B for details.) For most three-phase windings, k_w typically falls in the range of 0.85 to 0.95.

The factor $k_w N_{ph}$ is the effective series turns per phase for the fundamental mmf. The peak amplitude of this mmf wave is

$$(F_{ag1})_{peak} = \frac{4}{\pi} \left(\frac{k_w N_{ph}}{poles} \right) i_a \tag{4.6}$$

EXAMPLE 4.1

The phase-a two-pole armature winding of Fig. 4.20a can be considered to consist of 8 N_c-turn, full-pitch coils connected in series, with each slot containing two coils. There are a total of 24 armature slots, and thus each slot is separated by $360°/24 = 15°$. Assume angle θ_a is measured from the magnetic axis of phase a such that the four slots containing the coil sides labeled a are at $\theta_a = 67.5°, 82.5°, 97.5°$, and $112.5°$. The opposite sides of each coil are thus found in the slots found at $-112.5°, -97.5°, -82.5°$ and $-67.5°$, respectively. Assume this winding to be carrying current i_a.

(a) Write an expression for the space-fundamental mmf produced by the two coils whose sides are in the slots at $\theta_a = 112.5°$ and $-67.5°$. (b) Write an expression for the space-fundamental mmf produced by the two coils whose sides are in the slots at $\theta_a = 67.5°$ and $-112.5°$. (c) Write an expression for the space-fundamental mmf of the complete armature winding. (d) Determine the winding factor k_w for this distributed winding.

■ Solution

a. Noting that the magnetic axis of this pair of coils is at $\theta_a = (112.5° - 67.5°)/2 = 22.5°$ and that the total ampere-turns in the slot is equal to $2N_c i_a$, the mmf produced by this pair of coils can be found from analogy with Eq. 4.3 to be

$$(\mathcal{F}_{ag1})_{22.5°} = \frac{4}{\pi} \left(\frac{2N_c i_a}{2} \right) \cos(\theta_a - 22.5°)$$

b. This pair of coils produces the same space-fundamental mmf as the pair of part (a) with the exception that this mmf is centered at $\theta_a = -22.5°$. Thus

$$(\mathcal{F}_{ag1})_{-22.5°} = \frac{4}{\pi} \left(\frac{2N_c i_a}{2} \right) \cos(\theta_a + 22.5°)$$

c. By analogy with parts (a) and (b), the total space-fundamental mmf can be written as

$$(\mathcal{F}_{ag1})_{total} = (\mathcal{F}_{ag1})_{-22.5°} + (\mathcal{F}_{ag1})_{-7.5°} + (\mathcal{F}_{ag1})_{7.5°} + (\mathcal{F}_{ag1})_{22.5°}$$

$$= \frac{4}{\pi} \left(\frac{2N_c}{2} \right) i_a [\cos(\theta_a + 22.5°) + \cos(\theta_a + 7.5°)$$

$$+ \cos(\theta_a - 7.5°) + \cos(\theta_a - 22.5°)]$$

$$= \frac{4}{\pi} \left(\frac{7.66N_c}{2} \right) i_a \cos\theta_a$$

$$= 4.88 N_c i_a \cos\theta_a$$

d. Recognizing that, for this winding $N_{ph} = 8N_c$, the total mmf of part (c) can be rewritten as

$$(\mathcal{F}_{ag1})_{total} = \frac{4}{\pi}\left(\frac{0.958N_{ph}}{2}\right) i_a \cos\theta_a$$

Comparison with Eq. 4.5 shows that for this winding, the winding factor is $k_w = 0.958$.

Practice Problem 4.1

Calculate the winding factor of the phase-a winding of Fig. 4.20 if the number of turns in the four coils in the two outer pairs of slots is reduced to six while the number of turns in the four coils in the inner slots remains at eight.

Solution

$$k_w = 0.962$$

Equation 4.5 describes the space-fundamental component of the mmf wave produced by current in phase a of a distributed winding. If the phase-a current is sinusoidal in time, e.g., $i_a = I_m \cos\omega t$, the result will be an mmf wave which is stationary in space and varies sinusoidally both with respect to θ_a and in time. In Section 4.5 we will study the effect of currents in all three phases and will see that the application of three-phase currents will produce a rotating mmf wave.

In a directly analogous fashion, rotor windings are often distributed in slots to reduce the effects of space harmonics. Figure 4.21a shows the rotor of a typical two-pole round-rotor generator. Although the winding is symmetric with respect to the rotor axis, the number of turns per slot can be varied to control the various harmonics. In Fig. 4.21b it can be seen that there are fewer turns in the slots nearest the pole face. In addition, the designer can vary the spacing of the slots. As for distributed armature windings, the fundamental air-gap mmf wave of a multipole rotor winding can be found from Eq. 4.5 in terms of the total number of series turns N_r, the winding current I_r and a winding factor k_r as

$$\mathcal{F}_{ag1} = \frac{4}{\pi}\left(\frac{k_r N_r}{\text{poles}}\right) I_r \cos\left(\frac{\text{poles}}{2}\theta_r\right) \tag{4.7}$$

where θ_r is the spatial angle measured with respect to the rotor magnetic axis, as shown in Fig. 4.21b. Its peak amplitude is

$$(F_{ag1})_{peak} = \frac{4}{\pi}\left(\frac{k_r N_r}{\text{poles}}\right) I_r \tag{4.8}$$

4.3.2   DC Machines

Because of the restrictions imposed on the winding arrangement by the commutator, the mmf wave of a dc machine armature approximates a sawtooth waveform

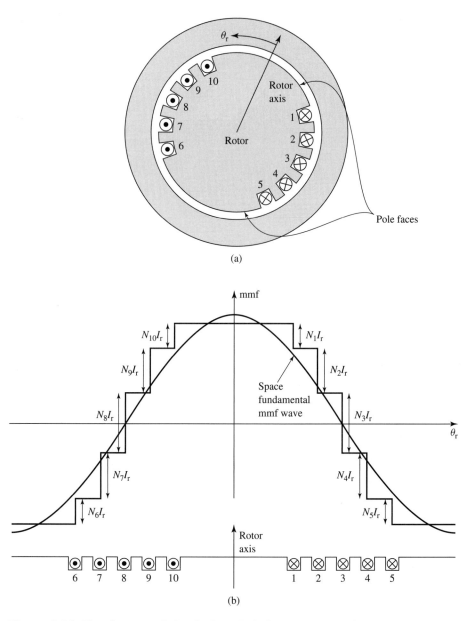

Figure 4.21 The air-gap mmf of a distributed winding on the rotor of a round-rotor generator.

more nearly than the sine wave of ac machines. For example, Fig. 4.22 shows diagrammatically in cross section the armature of a two-pole dc machine. (In practice, in all but the smallest of dc machines, a larger number of coils and slots would probably be used.) The current directions are shown by dots and crosses. The armature winding coil connections are such that the armature winding produces a magnetic

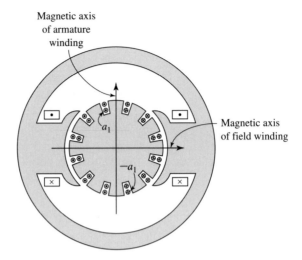

Figure 4.22   Cross section of a two-pole dc machine.

field whose axis is vertical and thus is perpendicular to the axis of the field winding. As the armature rotates, the coil connections to the external circuit are changed by the commutator such that the magnetic field of the armature remains vertical. Thus, the armature flux is always perpendicular to that produced by the field winding and a continuous unidirectional torque results. Commutator action is discussed in some detail in Section 7.2.

Figure 4.23a shows this winding laid out flat. The mmf wave is shown in Fig. 4.23b. On the assumption of narrow slots, it consists of a series of steps. The height of each step equals the number of ampere-turns $2N_c i_c$ in a slot, where N_c is the number of turns in each coil and i_c is the coil current, with a two-layer winding and full-pitch coils being assumed. The peak value of the mmf wave is along the magnetic axis of the armature, midway between the field poles. This winding is equivalent to a coil of $12N_c i_c$ A·turns distributed around the armature. On the assumption of symmetry at each pole, the peak value of the mmf wave at each armature pole is $6N_c i_c$ A·turns.

This mmf wave can be represented approximately by the sawtooth wave drawn in Fig. 4.23b and repeated in Fig. 4.23c. For a more realistic winding with a larger number of armature slots per pole, the triangular distribution becomes a close approximation. This mmf wave would be produced by a rectangular distribution of current density at the armature surface, as shown in Fig. 4.23c.

For our preliminary study, it is convenient to resolve the mmf waves of distributed windings into their Fourier series components. The fundamental component of the sawtooth mmf wave of Fig. 4.23c is shown by the sine wave. Its peak value is $8/\pi^2 = 0.81$ times the height of the sawtooth wave. This fundamental mmf wave is that which would be produced by the fundamental space-harmonic component of the rectangular current-density distribution of Fig. 4.23c. This sinusoidally-distributed current sheet is shown dashed in Fig. 4.23c.

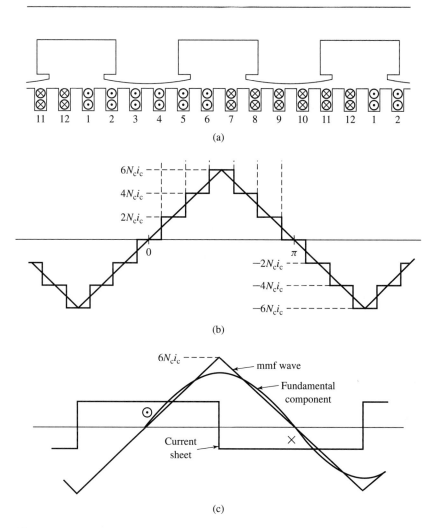

Figure 4.23   (a) Developed sketch of the dc machine of Fig. 4.22; (b) mmf wave; (c) equivalent sawtooth mmf wave, its fundamental component, and equivalent rectangular current sheet.

Note that the air-gap mmf distribution depends on only the winding arrangement and symmetry of the magnetic structure at each pole. The air-gap flux density, however, depends not only on the mmf but also on the magnetic boundary conditions, primarily the length of the air gap, the effect of the slot openings, and the shape of the pole face. The designer takes these effects into account by means of detailed analyses, but these details need not concern us here.

DC machines often have a magnetic structure with more than two poles. For example, Fig. 4.24a shows schematically a four-pole dc machine. The field winding

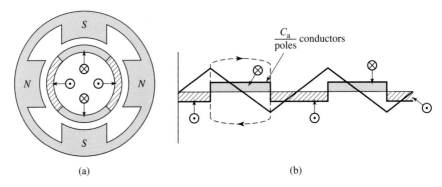

(a)   (b)

Figure 4.24 (a) Cross section of a four-pole dc machine; (b) development of current sheet and mmf wave.

produces alternate north-south-north-south polarity, and the armature conductors are distributed in four belts of slots carrying currents alternately toward and away from the reader, as symbolized by the cross-hatched areas. This machine is shown in laid-out form in Fig. 4.24b. The corresponding sawtooth armature-mmf wave is also shown. On the assumption of symmetry of the winding and field poles, each successive pair of poles is like every other pair of poles. Magnetic conditions in the air gap can then be determined by examining any pair of adjacent poles, that is, 360 electrical degrees.

The peak value of the sawtooth armature mmf wave can be written in terms of the total number of conductors in the armature slots as

$$(F_{ag})_{peak} = \left(\frac{C_a}{2m \cdot poles}\right) i_a \quad \text{A} \cdot \text{turns/pole} \qquad (4.9)$$

where

C_a = total number of conductors in armature winding

m = number of parallel paths through armature winding

i_a = armature current, A

This equation takes into account the fact that in some cases the armature may be wound with multiple current paths in parallel. It is for this reason that it is often more convenient to think of the armature in terms of the number of conductors (each conductor corresponding to a single current-carrying path within a slot). Thus i_a/m is the current in each conductor. This equation comes directly from the line integral around the dotted closed path in Fig. 4.24b which crosses the air gap twice and encloses $C_a/poles$ conductors, each carrying current i_a/m in the same direction. In more compact form,

$$(F_{ag})_{peak} = \left(\frac{N_a}{poles}\right) i_a \qquad (4.10)$$

where $N_a = C_a/(2m)$ is the number of series armature turns. From the Fourier series for the sawtooth mmf wave of Fig. 4.24b, the peak value of the space fundamental is given by

$$(F_{ag1})_{peak} = \frac{8}{\pi^2}\left(\frac{N_a}{poles}\right) i_a \tag{4.11}$$

4.4 MAGNETIC FIELDS IN ROTATING MACHINERY

We base our preliminary investigations of both ac and dc machines on the assumption of sinusoidal spatial distributions of mmf. This assumption will be found to give very satisfactory results for most problems involving ac machines because their windings are commonly distributed so as to minimize the effects of space harmonics. Because of the restrictions placed on the winding arrangement by the commutator, the mmf waves of dc machines inherently approach more nearly a sawtooth waveform. Nevertheless, the theory based on a sinusoidal model brings out the essential features of dc machine theory. The results can readily be modified whenever necessary to account for any significant discrepancies.

It is often easiest to begin by examination of a two-pole machine, in which the electrical and mechanical angles and velocities are equal. The results can immediately be extrapolated to a multipole machine when it is recalled that electrical angles and angular velocities are related to mechanical angles and angular velocities by a factor of poles/2 (see, for example, Eq. 4.1).

The behavior of electric machinery is determined by the magnetic fields created by currents in the various windings of the machine. This section discusses how these magnetic fields and currents are related.

4.4.1 Machines with Uniform Air Gaps

Figure 4.25a shows a single full-pitch, N-turn coil in a high-permeability magnetic structure ($\mu \to \infty$), with a concentric, cylindrical rotor. The air-gap mmf $\mathcal{F}_{ag}$ of this configuration is shown plotted versus angle θ_a in Fig. 4.25b. For such a structure, with a uniform air gap of length g at radius r_r (very much larger than g), it is quite accurate to assume that the magnetic field **H** in the air gap is directed only radially and has constant magnitude across the air gap.

The air-gap mmf distribution of Fig. 4.25b is equal to the line integral of H_{ag} across the air gap. For this case of constant radial H_{ag}, this integral is simply equal to the product of the air-gap radial magnetic field H_{ag} times the air-gap length g, and thus H_{ag} can be found simply by dividing the air-gap mmf by the air-gap length:

$$H_{ag} = \frac{\mathcal{F}_{ag}}{g} \tag{4.12}$$

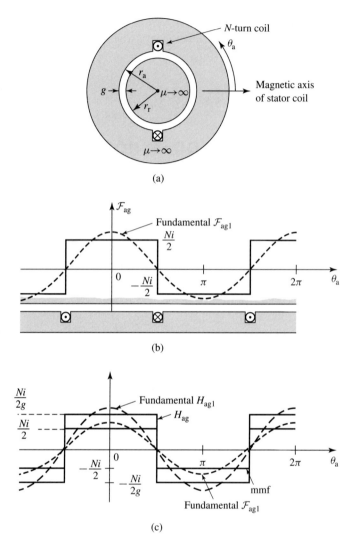

Figure 4.25 The air-gap mmf and radial component of H_{ag} for a concentrated full-pitch winding.

Thus, in Fig. 4.25c, the radial H_{ag} field and mmf can be seen to be identical in form, simply related by a factor of $1/g$.

The fundamental space-harmonic component of H_{ag} can be found directly from the fundamental component $\mathcal{F}_{ag1}$, given by Eq. 4.3.

$$H_{ag1} = \frac{\mathcal{F}_{ag1}}{g} = \frac{4}{\pi} \left(\frac{Ni}{2g} \right) \cos \theta_a \tag{4.13}$$

It is a sinusoidal space wave of amplitude

$$(H_{ag1})_{peak} = \frac{4}{\pi} \left(\frac{Ni}{2g} \right) \qquad (4.14)$$

For a distributed winding such as that of Fig. 4.20, the air-gap magnetic field intensity is easily found once the air-gap mmf is known. Thus the fundamental component of H_{ag} can be found by dividing the fundamental component of the air-gap mmf (Eq. 4.5) by the air-gap length g

$$H_{ag1} = \frac{4}{\pi} \left(\frac{k_w N_{ph}}{g \cdot poles} \right) i_a \cos \left(\frac{poles}{2} \theta_a \right) \qquad (4.15)$$

This equation has been written for the general case of a multipole machine, and N_{ph} is the total number of series turns per phase.

Note that the space-fundamental air-gap mmf $\mathcal{F}_{ag1}$ and air-gap magnetic field H_{ag1} produced by a distributed winding of winding factor k_w and $N_{ph}/poles$ series turns per pole is equal to that produced by a concentrated, full pitch winding of $(k_w N_{ph})/poles$ turns per pole. In the analysis of machines with distributed windings, this result is useful since in considering space-fundamental quantities it permits the distributed solution to be obtained from the single N-turn, full-pitch coil solution simply by replacing N by the effective number of turns, $k_w N_{ph}$, of the distributed winding.

EXAMPLE 4.2

A four-pole synchronous ac generator with a smooth air gap has a distributed rotor winding with 263 series turns, a winding factor of 0.935, and an air gap of length 0.7 mm. Assuming the mmf drop in the electrical steel to be negligible, find the rotor-winding current required to produce a peak, space-fundamental magnetic flux density of 1.6 T in the machine air gap.

■ Solution

The space-fundamental air-gap magnetic flux density can be found by multiplying the air-gap magnetic field by the permeability of free space μ_0, which in turn can be found from the space-fundamental component of the air-gap mmf by dividing by the air-gap length g. Thus, from Eq. 4.8

$$(B_{ag1})_{peak} = \frac{\mu_0 (\mathcal{F}_{ag1})_{peak}}{g} = \frac{4\mu_0}{\pi g} \left(\frac{k_r N_r}{poles} \right) I_r$$

and I_r can be found from

$$I_r = \left(\frac{\pi g \cdot poles}{4\mu_0 k_r N_r} \right) (B_{ag1})_{peak}$$

$$= \left(\frac{\pi \times 0.0007 \times 4}{4 \times 4\pi \times 10^{-7} \times 0.935 \times 263} \right) 1.6$$

$$= 11.4 \text{ A}$$

A 2-pole synchronous machine has an air-gap length of 2.2 cm and a field winding with a total of 830 series turns. When excited by a field current of 47 A, the peak, space-fundamental magnetic flux density in the machine air-gap is measured to be 1.35 T.

Based upon the measured flux density, calculate the field-winding winding factor k_r.

Solution

$$k_r = 0.952$$

4.4.2 Machines with Nonuniform Air Gaps

Figure 4.26a shows the structure of a typical dc machine, and Fig. 4.26b shows the structure of a typical salient-pole synchronous machine. Both machines consist of magnetic structures with extremely nonuniform air gaps. In such cases the air-gap magnetic-field distribution is more complex than that of uniform-air-gap machines.

Detailed analysis of the magnetic field distributions in such machines requires complete solutions of the field problem. For example, Fig. 4.27 shows the magnetic field distribution in a salient-pole dc generator (obtained by a finite-element solution). However, experience has shown that through various simplifying assumptions, analytical techniques which yield reasonably accurate results can be developed. These techniques are illustrated in later chapters, where the effects of saliency on both dc and ac machines are discussed.

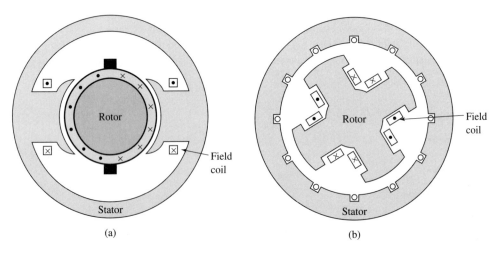

Figure 4.26 Structure of typical salient-pole machines: (a) dc machine and (b) salient-pole synchronous machine.

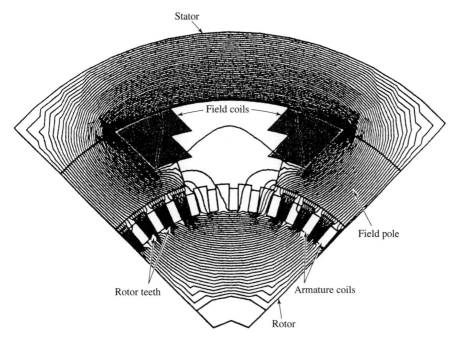

Stator

Field coils

Field pole

Rotor teeth

Armature coils

Rotor

Figure 4.27 Finite-element solution of the magnetic field distribution in a salient-pole dc generator. Field coils excited; no current in armature coils. (*General Electric Company.*)

4.5 ROTATING MMF WAVES IN AC MACHINES

To understand the theory and operation of polyphase ac machines, it is necessary to study the nature of the mmf wave produced by a polyphase winding. Attention will be focused on a two-pole machine or one pair of a multipole winding. To develop insight into the polyphase situation, it is helpful to begin with an analysis of a single-phase winding.

4.5.1 MMF Wave of a Single-Phase Winding

Figure 4.28a shows the space-fundamental mmf distribution of a single-phase winding, where, from Eq. 4.5,

$$\mathcal{F}_{ag1} = \frac{4}{\pi} \left(\frac{k_w N_{ph}}{poles} \right) i_a \cos \left(\frac{poles}{2} \theta_a \right) \tag{4.16}$$

When this winding is excited by a sinusoidally varying current in time at electrical frequency ω_e

$$i_a = I_a \cos \omega_e t \tag{4.17}$$

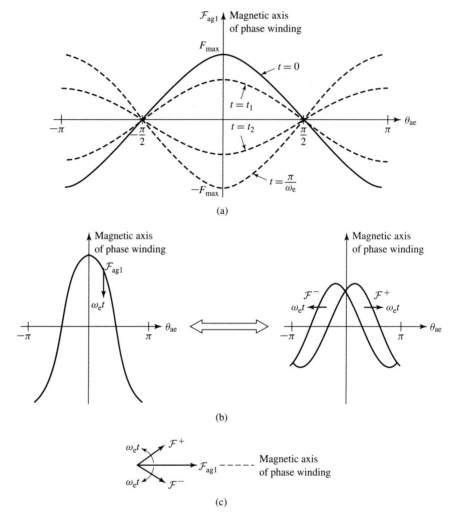

Figure 4.28 Single-phase-winding space-fundamental air-gap mmf: (a) mmf distribution of a single-phase winding at various times; (b) total mmf $\mathcal{F}_{ag1}$ decomposed into two traveling waves $\mathcal{F}^-$ and $\mathcal{F}^+$; (c) phasor decomposition of $\mathcal{F}_{ag1}$.

the mmf distribution is given by

$$\mathcal{F}_{ag1} = F_{max} \cos \left(\frac{\text{poles}}{2} \theta_a \right) \cos \omega_e t$$

$$= F_{max} \cos (\theta_{ae}) \cos \omega_e t \tag{4.18}$$

Equation 4.18 has been written in a form to emphasize the fact that the result is an mmf distribution of maximum amplitude.

$$F_{max} = \frac{4}{\pi} \left(\frac{k_w N_{ph}}{\text{poles}} \right) I_a \tag{4.19}$$

This mmf distribution remains fixed in space with an amplitude that varies sinusoidally in time at frequency ω_e, as shown in Fig. 4.28a. Note that, to simplify the notation, Eq. 4.1 has been used to express the mmf distribution of Eq. 4.18 in terms of the electrical angle θ_{ae}.

Use of a common trigonometric identity[1] permits Eq. 4.18 to be rewritten in the form

$$\mathcal{F}_{ag1} = F_{max} \left[\frac{1}{2} \cos (\theta_{ae} - \omega_e t) + \frac{1}{2} \cos (\theta_{ae} + \omega_e t) \right] \qquad (4.20)$$

which shows that the mmf of a single-phase winding can be resolved into two rotating mmf waves each of amplitude one-half the maximum amplitude of $\mathcal{F}_{ag1}$ with one, $\mathcal{F}_{ag1}^+$, traveling in the $+\theta_a$ direction and the other, $\mathcal{F}_{ag1}^-$, traveling in the $-\theta_a$ direction, both with electrical angular velocity ω_e (equal to a mechanical angular velocity of $2\omega_e/poles$):

$$\mathcal{F}_{ag1}^+ = \frac{1}{2} F_{max} \cos (\theta_{ae} - \omega_e t) \qquad (4.21)$$

$$\mathcal{F}_{ag1}^- = \frac{1}{2} F_{max} \cos (\theta_{ae} + \omega_e t) \qquad (4.22)$$

This decomposition is shown graphically in Fig. 4.28b and in a phasor representation in Fig. 4.28c.

The fact that the air-gap mmf of a single-phase winding excited by a source of alternating current can be resolved into rotating traveling waves is an important conceptual step in understanding ac machinery. As shown in Section 4.5.2, in polyphase ac machinery the windings are equally displaced in space phase, and the winding currents are similarly displaced in time phase, with the result that the negative-traveling flux waves of the various windings sum to zero while the positive-traveling flux waves reinforce, giving a single positive-traveling flux wave.

In single-phase ac machinery, the positive-traveling flux wave produces useful torque while the negative-traveling flux wave produces both negative and pulsating torque as well as losses. These machines are designed so as to minimize the effects of the negative-traveling flux wave, although, unlike in the case of polyphase machinery, these effects cannot be totally eliminated.

4.5.2 MMF Wave of a Polyphase Winding

In this section we study the mmf distributions of three-phase windings such as those found on the stator of three-phase induction and synchronous machines. The analyses presented can be readily extended to a polyphase winding with any number of phases. Once again attention is focused on a two-pole machine or one pair of poles of a multipole winding.

In a three-phase machine, the windings of the individual phases are displaced from each other by 120 electrical degrees in space around the airgap circumference, as shown by coils a, $-a$, b, $-b$, and c, $-c$ in Fig. 4.29. The concentrated full-pitch

---

[1] $\cos \alpha \cos \beta = \frac{1}{2} \cos (\alpha - \beta) + \frac{1}{2} \cos (\alpha + \beta)$

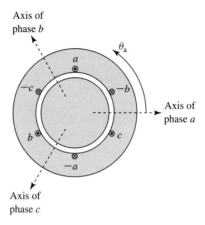

Figure 4.29 Simplified two-pole three-phase stator winding.

coils shown here may be considered to represent distributed windings producing sinusoidal mmf waves centered on the magnetic axes of the respective phases. The space-fundamental sinusoidal mmf waves of the three phases are accordingly displaced 120 electrical degrees in space. Each phase is excited by an alternating current which varies in magnitude sinusoidally with time. Under balanced three-phase conditions, the instantaneous currents are

$$i_a = I_m \cos \omega_e t \tag{4.23}$$

$$i_b = I_m \cos (\omega_e t - 120°) \tag{4.24}$$

$$i_c = I_m \cos (\omega_e t + 120°) \tag{4.25}$$

where I_m is the maximum value of the current and the time origin is arbitrarily taken as the instant when the phase-a current is a positive maximum. The phase sequence is assumed to be abc. The instantaneous currents are shown in Fig. 4.30. The dots and crosses in the coil sides (Fig. 4.29) indicate the reference directions for positive phase currents.

The mmf of phase a has been shown to be

$$\mathcal{F}_{a1} = \mathcal{F}_{a1}^+ + \mathcal{F}_{a1}^- \tag{4.26}$$

where

$$\mathcal{F}_{a1}^+ = \frac{1}{2} F_{max} \cos (\theta_{ae} - \omega_e t) \tag{4.27}$$

$$\mathcal{F}_{a1}^- = \frac{1}{2} F_{max} \cos (\theta_{ae} + \omega_e t) \tag{4.28}$$

and

$$F_{max} = \frac{4}{\pi} \left(\frac{k_w N_{ph}}{poles} \right) I_m \tag{4.29}$$

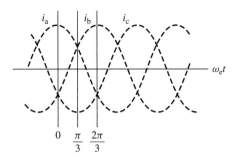

Figure 4.30 Instantaneous phase currents under balanced three-phase conditions.

Note that to avoid excessive notational complexity, the subscript ag has been dropped; here the subscript a1 indicates the space-fundamental component of the phase-a air-gap mmf.

Similarly, for phases b and c, whose axes are at $\theta_a = 120°$ and $\theta_a = -120°$, respectively,

$$\mathcal{F}_{b1} = \mathcal{F}_{b1}^+ + \mathcal{F}_{b1}^- \tag{4.30}$$

$$\mathcal{F}_{b1}^+ = \frac{1}{2} F_{max} \cos{(\theta_{ae} - \omega_e t)} \tag{4.31}$$

$$\mathcal{F}_{b1}^- = \frac{1}{2} F_{max} \cos{(\theta_{ae} + \omega_e t + 120°)} \tag{4.32}$$

and

$$\mathcal{F}_{c1} = \mathcal{F}_{c1}^+ + \mathcal{F}_{c1}^- \tag{4.33}$$

$$\mathcal{F}_{c1}^+ = \frac{1}{2} F_{max} \cos{(\theta_{ae} - \omega_e t)} \tag{4.34}$$

$$\mathcal{F}_{c1}^- = \frac{1}{2} F_{max} \cos{(\theta_{ae} + \omega_e t - 120°)} \tag{4.35}$$

The total mmf is the sum of the contributions from each of the three phases

$$\mathcal{F}(\theta_{ae}, t) = \mathcal{F}_{a1} + \mathcal{F}_{b1} + \mathcal{F}_{c1} \tag{4.36}$$

This summation can be performed quite easily in terms of the positive- and negative-traveling waves. The negative-traveling waves sum to zero

$$\mathcal{F}^-(\theta_{ae}, t) = \mathcal{F}_{a1}^- + \mathcal{F}_{b1}^- + \mathcal{F}_{c1}^-$$

$$= \frac{1}{2} F_{max} [\cos{(\theta_{ae} + \omega_e t)} + \cos{(\theta_{ae} + \omega_e t - 120°)}$$

$$+ \cos{(\theta_{ae} + \omega_e t + 120°)}]$$

$$= 0 \tag{4.37}$$

while the positive-traveling waves reinforce

$$\mathcal{F}^+(\theta_{ae}, t) = \mathcal{F}_{a1}^+ + \mathcal{F}_{b1}^+ + \mathcal{F}_{c1}^+$$
$$= \frac{3}{2} F_{max} \cos (\theta_{ae} - \omega_e t) \tag{4.38}$$

Thus, the result of displacing the three windings by 120° in space phase and displacing the winding currents by 120° in time phase is a single positive-traveling mmf wave

$$\mathcal{F}(\theta_{ae}, t) = \frac{3}{2} F_{max} \cos (\theta_{ae} - \omega_e t)$$
$$= \frac{3}{2} F_{max} \cos \left(\left(\frac{\text{poles}}{2} \right) \theta_a - \omega_e t \right) \tag{4.39}$$

The air-gap mmf wave described by Eq. 4.39 is a space-fundamental sinusoidal function of the electrical space angle θ_{ae} (and hence of the space angle $\theta_a = (2/\text{poles})\theta_{ae}$). It has a constant amplitude of $(3/2)F_{max}$, i.e., 1.5 times the amplitude of the air-gap mmf wave produced by the individual phases alone. It has a positive peak at angle $\theta_a = (2/\text{poles})\omega_e t$. Thus, under balanced three-phase conditions, the three-phase winding produces an air-gap mmf wave which rotates at *synchronous angular velocity* ω_s

$$\omega_s = \left(\frac{2}{\text{poles}} \right) \omega_e \tag{4.40}$$

where

ω_e = angular frequency of the applied electrical excitation [rad/sec]

ω_s = synchronous spatial angular velocity of the air-gap mmf wave [rad/sec]

The corresponding *synchronous speed* n_s in r/min can be expressed in terms of the applied electrical frequency $f_e = \omega_e/(2\pi)$ in Hz as

$$n_s = \left(\frac{120}{\text{poles}} \right) f_e \quad \text{r/min} \tag{4.41}$$

In general, a rotating field of constant amplitude will be produced by a q-phase winding excited by balanced q-phase currents of frequency f_e when the respective phase axes are located $2\pi/q$ electrical radians apart in space. The amplitude of this flux wave will be $q/2$ times the maximum contribution of any one phase, and the synchronous angular velocity will remain $\omega_s = (\frac{2}{\text{poles}})\omega_e$ radians per second.

In this section, we have seen that a polyphase winding excited by balanced polyphase currents produces a rotating mmf wave. Production of a rotating mmf wave and the corresponding rotating magnetic flux is key to the operation of polyphase rotating electrical machinery. It is the interaction of this magnetic flux wave with that of the rotor which produces torque. Constant torque is produced when rotor-produced magnetic flux rotates in sychronism with that of the stator.

4.5.3 Graphical Analysis of Polyphase MMF

For balanced three-phase currents as given by Eqs. 4.23 to 4.25, the production of a rotating mmf can also be shown graphically. Consider the state of affairs at $t = 0$ (Fig. 4.30), the moment when the phase-a current is at its maximum value I_m. The mmf of phase a then has its maximum value F_{max}, as shown by the vector $F_a = F_{max}$ drawn along the magnetic axis of phase a in the two-pole machine shown schematically in Fig. 4.31a. At this moment, currents i_b and i_c are both $I_m/2$ in the negative direction, as shown by the dots and crosses in Fig. 4.31a indicating the actual instantaneous directions. The corresponding mmf's of phases b and c are shown by the vectors F_b and F_c, both of magnitude $F_{max}/2$ drawn in the negative direction along the magnetic axes of phases b and c, respectively. The resultant, obtained by adding the individual contributions of the three phases, is a vector of magnitude $F = \frac{3}{2}F_{max}$ centered on the axis of phase a. It represents a sinusoidal space wave with its positive peak centered on the axis of phase a and having an amplitude $\frac{3}{2}$ times that of the phase-a contribution alone.

At a later time $\omega_e t = \pi/3$ (Fig. 4.30), the currents in phases a and b are a positive half maximum, and that in phase c is a negative maximum. The individual mmf components and their resultant are now shown in Fig. 4.31b. The resultant has the same amplitude as at $t = 0$, but it has now rotated counterclockwise 60 electrical degrees in space. Similarly, at $\omega_e t = 2\pi/3$ (when the phase-b current is a positive maximum and the phase-a and phase-c currents are a negative half maximum) the same resultant mmf distribution is again obtained, but it has rotated counterclockwise 60 electrical degrees still farther and is now aligned with the magnetic axis of phase b (see Fig. 4.31c). As time passes, then, the resultant mmf wave retains its sinusoidal form and amplitude but rotates progressively around the air gap; the net result can be seen to be an mmf wave of constant amplitude rotating at a uniform angular velocity.

In one cycle the resultant mmf must be back in the position of Fig. 4.31a. The mmf wave therefore makes one revolution per electrical cycle in a two-pole machine. In a multipole machine the mmf wave travels one pole-pair per electrical cycle and hence one revolution in poles/2 electrical cycles.

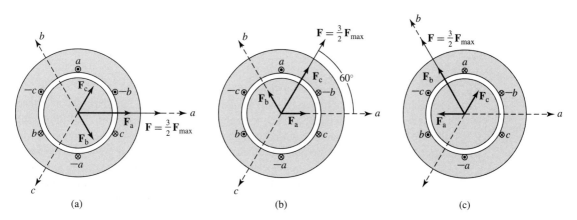

Figure 4.31 The production of a rotating magnetic field by means of three-phase currents.

EXAMPLE 4.3

Consider a three-phase stator excited with balanced, 60-Hz currents. Find the synchronous angular velocity in rad/sec and speed in r/min for stators with two, four, and six poles.

■ **Solution**

For a frequency of $f_e = 60$ Hz, the electrical angular frequency is equal to

$$\omega_e = 2\pi f_e = 120\pi \approx 377 \text{ rad/sec}$$

Using Eqs. 4.40 and 4.41, the following table can be constructed:

Poles	n_s (r/min)	ω_s (rad/sec)
2	3600	$120\pi \approx 377$
4	1800	60π
6	1200	40π

Repeat Example 4.3 for a three-phase stator excited by balanced 50-Hz currents.

Solution

Poles	n_s (r/min)	ω_s (rad/sec)
2	3000	100π
4	1500	50π
6	1000	$100\pi/3$

4.6 GENERATED VOLTAGE

The general nature of the induced voltage has already been discussed in Section 4.2. Quantitative expressions for the induced voltage will now be determined.

4.6.1 AC Machines

An elementary ac machine is shown in cross section in Fig. 4.32. The coils on both the rotor and the stator have been shown as concentrated, multiple-turn, full-pitch coils. As we have seen, a machine with distributed windings can be represented in this form simply by multiplying the number of series turns in the winding by a winding factor. Under the assumption of a small air gap, the field winding can be assumed to produce radial space-fundamental air-gap flux of peak flux density B_{peak}. Although Fig. 4.32 shows a two-pole machine, the analysis presented here is for the general case of a multipole machine. As is derived in Example 4.2, if the air gap is uniform, B_{peak} can

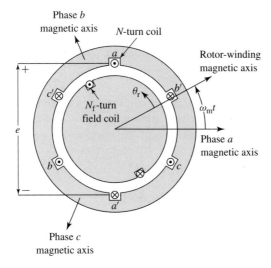

Phase *b* magnetic axis   *N*-turn coil

Rotor-winding magnetic axis

θ_r

N_f-turn field coil

$\omega_m t$

Phase *a* magnetic axis

Phase *c* magnetic axis

Figure 4.32 Cross-sectional view of an elementary three-phase ac machine.

be found from

$$B_{\text{peak}} = \frac{4\mu_0}{\pi g} \left(\frac{k_f N_f}{\text{poles}} \right) I_f \qquad (4.42)$$

where

g = air-gap length
N_f = total series turns in the field winding
k_f = field-winding winding factor
I_f = field current

When the rotor poles are in line with the magnetic axis of a stator phase, the flux linkage with a stator phase winding is $k_w N_{ph} \Phi_p$, where Φ_p is the air-gap flux per pole [Wb]. For the assumed sinusoidal air-gap flux-density

$$B = B_{\text{peak}} \cos \left(\frac{\text{poles}}{2} \theta_r \right) \qquad (4.43)$$

Φ_p can be found as the integral of the flux density over the pole area

$$\Phi_p = l \int_{-\pi/\text{poles}}^{+\pi/\text{poles}} B_{\text{peak}} \cos \left(\frac{\text{poles}}{2} \theta_r \right) r \, d\theta_r$$

$$= \left(\frac{2}{\text{poles}} \right) 2 B_{\text{peak}} l r \qquad (4.44)$$

Here,

θ_r = angle measured from the rotor magnetic axis

r = radius to air gap

l = axial length of the stator/rotor iron

As the rotor turns, the flux linkage varies cosinusoidally with the angle between the magnetic axes of the stator coil and rotor. With the rotor spinning at constant angular velocity ω_m, the flux linkage with the phase-a stator coil is

$$\lambda_a = k_w N_{ph} \Phi_p \cos\left(\left(\frac{poles}{2}\right)\omega_m t\right)$$

$$= k_w N_{ph} \Phi_p \cos \omega_{me} t \qquad (4.45)$$

where time t is arbitrarily chosen as zero when the peak of the flux-density wave coincides with the magnetic axis of phase a. Here,

$$\omega_{me} = \left(\frac{poles}{2}\right)\omega_m \qquad (4.46)$$

is the mechanical rotor velocity expressed in electrical rad/sec.

By Faraday's law, the voltage induced in phase a is

$$e_a = \frac{d\lambda_a}{dt} = k_w N_{ph} \frac{d\Phi_p}{dt} \cos \omega_{me} t$$

$$-\omega_{me} k_w N_{ph} \Phi_p \sin \omega_{me} t \qquad (4.47)$$

The polarity of this induced voltage is such that if the stator coil were short-circuited, the induced voltage would cause a current to flow in the direction that would oppose any change in the flux linkage of the stator coil. Although Eq. 4.47 is derived on the assumption that only the field winding is producing air-gap flux, the equation applies equally well to the general situation where Φ_p is the net air-gap flux per pole produced by currents on both the rotor and the stator.

The first term on the right-hand side of Eq. 4.47 is a transformer voltage and is present only when the amplitude of the air-gap flux wave changes with time. The second term is the *speed voltage* generated by the relative motion of the air-gap flux wave and the stator coil. In the normal steady-state operation of most rotating machines, the amplitude of the air-gap flux wave is constant; under these conditions the first term is zero and the generated voltage is simply the speed voltage. The term *electromotive force* (abbreviated *emf*) is often used for the speed voltage. Thus, for constant air-gap flux,

$$e_a = -\omega_{me} k_w N_{ph} \Phi_p \sin \omega_{me} t \qquad (4.48)$$

EXAMPLE 4.4

The so-called *cutting-of-flux* equation states that the voltage v induced in a wire of length l (in the frame of the wire) moving with respect to a constant magnetic field with flux density of magnitude B is given by

$$v = lv_\perp B$$

where $v_\perp$ is the component of the wire velocity perpendicular to the direction of the magnetic flux density.

Consider the two-pole elementary three-phase machine of Fig. 4.32. Assume the rotor-produced air-gap flux density to be of the form

$$B_{ag}(\theta_r) = B_{peak} \sin \theta_r$$

and the rotor to rotate at constant angular velocity ω_e. (Note that since this is a two-pole machine, $\omega_m = \omega_e$). Show that if one assumes that the armature-winding coil sides are in the air gap and not in the slots, the voltage induced in a full-pitch, N-turn concentrated armature phase coil can be calculated from the cutting-of-flux equation and that it is identical to that calculated using Eq. 4.48. Let the average air-gap radius be r and the air-gap length be g ($g \ll r$).

■ Solution

We begin by noting that the cutting-of-flux equation requires that the conductor be moving and the magnetic field to be nontime varying. Thus in order to apply it to calculating the stator magnetic field, we must translate our reference frame to the rotor.

In the rotor frame, the magnetic field is constant and the stator coil sides, when moved to the center of the air gap at radius r, appear to be moving with velocity $\omega_{me}r$ which is perpendicular to the radially-directed air-gap flux. If the rotor and phase-coil magnetic axes are assumed to be aligned at time $t = 0$, the location of a coil side as a function of time will be given by $\theta_r = -\omega_{me}t$. The voltage induced in one side of one turn can therefore be calculated as

$$e_1 = lv_\perp B_{ag}(\theta_r) = l\omega_{me}r B_{peak} \sin(-\omega_{me}t)$$

There are N turns per coil and two sides per turn. Thus the total coil voltage is given by

$$e = 2Ne_1 = -2Nl\omega_{me}r B_{peak} \sin \omega_{me}t$$

From Eq. 4.48, the voltage induced in the full-pitched, 2-pole stator coil is given by

$$e = -\omega_{me}N\Phi_p \sin \omega_{me}t$$

Substituting $\Phi_p = 2B_{peak}lr$ from Eq. 4.44 gives

$$e = -\omega_{me}N(2B_{peak}lr) \sin \omega_{me}t$$

which is identical to the voltage determined using the cutting-of-flux equation.

In the normal steady-state operation of ac machines, we are usually interested in the rms values of voltages and currents rather than their instantaneous values. From Eq. 4.48 the maximum value of the induced voltage is

$$E_{max} = \omega_{me}k_w N_{ph}\Phi_p = 2\pi f_{me}k_w N_{ph}\Phi_p \qquad (4.49)$$

Its rms value is

$$E_{rms} = \frac{2\pi}{\sqrt{2}} f_{me} k_w N_{ph} \Phi_p = \sqrt{2}\, \pi f_{me} k_w N_{ph} \Phi_p \qquad (4.50)$$

where f_{me} is the electrical speed of the rotor measured in Hz, which is also equal to the electrical frequency of the generated voltage. Note that these equations are identical in form to the corresponding emf equations for a transformer. Relative motion of a coil and a constant-amplitude spatial flux-density wave in a rotating machine produces voltage in the same fashion as does a time-varying flux in association with stationary coils in a transformer. Rotation introduces the element of time variation and transforms a space distribution of flux density into a time variation of voltage.

The voltage induced in a single winding is a single-phase voltage. For the production of a set of balanced, three-phase voltages, it follows that three windings displaced 120 electrical degrees in space must be used, as shown in elementary form in Fig. 4.12. The machine of Fig. 4.12 is shown to be Y-connected and hence each winding voltage is a phase-neutral voltage. Thus, Eq. 4.50 gives the rms line-neutral voltage produced in this machine when N_{ph} is the total series turns per phase. For a Δ-connected machine, the voltage winding voltage calculated from Eq. 4.50 would be the machine line-line voltage.

EXAMPLE 4.5

A two-pole, three-phase, Y-connected 60-Hz round-rotor synchronous generator has a field winding with N_f distributed turns and winding factor k_f. The armature winding has N_a turns per phase and winding factor k_a. The air-gap length is g, and the mean air-gap radius is r. The armature-winding active length is l. The dimensions and winding data are

$$N_f = 68 \text{ series turns} \qquad k_f = 0.945$$
$$N_a = 18 \text{ series turns/phase} \qquad k_a = 0.933$$
$$r = 0.53 \text{ m} \qquad g = 4.5 \text{ cm}$$
$$l = 3.8 \text{ m}$$

The rotor is driven by a steam turbine at a speed of 3600 r/min. For a field current of $I_f = 720$ A dc, compute (a) the peak fundamental mmf $(F_{ag1})_{peak}$ produced by the field winding, (b) the peak fundamental flux density $(B_{ag1})_{peak}$ in the air gap, (c) the fundamental flux per pole Φ_p, and (d) the rms value of the open-circuit voltage generated in the armature.

■ Solution
a. From Eq. 4.8

$$(F_{ag1})_{peak} = \frac{4}{\pi}\left(\frac{k_f N_f}{\text{poles}}\right) I_f = \frac{4}{\pi}\left(\frac{0.945 \times 68}{2}\right) 720$$

$$= \frac{4}{\pi}(32.1)720 = 2.94 \times 10^4 \text{ A} \cdot \text{turns/pole}$$

b. Using Eq. 4.12, we get

$$(B_{ag1})_{peak} = \frac{\mu_0 (F_{ag1})_{peak}}{g} = \frac{4\pi \times 10^{-7} \times 2.94 \times 10^4}{4.5 \times 10^{-2}} = 0.821 \text{ T}$$

Because of the effect of the slots containing the armature winding, most of the air-gap flux is confined to the stator teeth. The flux density in the teeth at a pole center is higher than the value calculated in part (b), probably by a factor of about 2. In a detailed design this flux density must be calculated to determine whether the teeth are excessively saturated.

c. From Eq. 4.44

$$\Phi_p = 2(B_{ag1})_{peak} lr = 2(0.821)(3.8)(0.53) = 3.31 \text{ Wb}$$

d. From Eq. 4.50 with $f_{me} = 60$ Hz

$$E_{rms, line-neutral} = \sqrt{2}\, \pi f_{me} k_a N_a \Phi_p = \sqrt{2}\, \pi (60)(0.933)(18)(3.31)$$

$$= 14.8 \text{ kV rms}$$

The line-line voltage is thus

$$E_{rms, line-line} = \sqrt{3}\,(14.8 \text{ kV}) = 25.7 \text{ kV rms}$$

Practice Problem 4.4

The rotor of the machine of Example 4.5 is to be rewound. The new field winding will have a total of 76 series turns and a winding factor of 0.925. (a) Calculate the field current which will result in a peak air-gap flux density of 0.83 T. (b) Calculate the corresponding rms line-line open-circuit voltage which will result if this modified machine is operated at this value of field current and 3600 rpm.

Solution

a. $I_f = 696$ A
b. $E_{rms, line-line} = 26.0$ kV rms

4.6.2 DC Machines

In a dc machine, although the ultimate objective is the generation of dc voltage, ac voltages are produced in the armature-winding coils as these coils rotate through the dc flux distribution of the stationary field winding. The armature-winding alternating voltage must therefore be rectified. Mechanical rectification is provided by the commutator as has been discussed in Section 4.2.2.

Consider the single N-turn armature coil of the elementary, two-pole dc machine of Fig. 4.17. The simple two-segment commutator provides full-wave rectification of the coil voltage. Although the spatial distribution of the air-gap flux in dc machines is typically far from sinusoidal, we can approximate the magnitude of the generated voltage by assuming a sinusoidal distribution. As we have seen, such a flux distribution will produce a sinusoidal ac voltage in the armature coil. The rectification action of

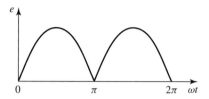

Figure 4.33 Voltage between the brushes in the elementary dc machine of Fig. 4.17.

the commutator will produce a dc voltage across the brushes as in Fig. 4.33. The average, or dc, value of this voltage can be found from taking the average of Eq. 4.48,

$$E_a = \frac{1}{\pi} \int_0^\pi \omega_{me} N \Phi_p \sin(\omega_{me}t) \, d(\omega_{me}t) = \frac{2}{\pi} \omega_{me} N \Phi_p \qquad (4.51)$$

For dc machines it is usually more convenient to express the voltage E_a in terms of the mechanical speed ω_m (rad/sec) or n (r/min). Substitution of Eq. 4.46 in Eq. 4.51 for a multipole machine then yields

$$E_a = \left(\frac{\text{poles}}{\pi}\right) N \Phi_p \omega_m = \text{poles } N \Phi_p \left(\frac{n}{30}\right) \qquad (4.52)$$

The single-coil dc winding implied here is, of course, unrealistic in the practical sense, and it will be essential later to examine the action of commutators more carefully. Actually, Eq. 4.52 gives correct results for the more practical distributed ac armature windings as well, provided N is taken as the total number of turns in series between armature terminals. Usually the voltage is expressed in terms of the total number of active conductors C_a and the number m of parallel paths through the armature winding. Because it takes two coil sides to make a turn and $1/m$ of these are connected in series, the number of series turns is $N_a = C_a/(2m)$. Substitution in Eq. 4.52 then gives

$$E_a = \left(\frac{\text{poles}}{2\pi}\right) \left(\frac{C_a}{m}\right) \Phi_p \omega_m = \left(\frac{\text{poles}}{60}\right) \left(\frac{C_a}{m}\right) \Phi_p n \qquad (4.53)$$

4.7 TORQUE IN NONSALIENT-POLE MACHINES

The behavior of any electromagnetic device as a component in an electromechanical system can be described in terms of its electrical-terminal equations and its displacement and electromechanical torque. The purpose of this section is to derive the voltage and torque equations for an idealized elementary machine, results which can be readily extended later to more complex machines. We derive these equations from two viewpoints and show that basically they stem from the same ideas.

The first viewpoint is essentially the same as that of Section 3.6. The machine will be regarded as a circuit element whose inductances depend on the angular position of the rotor. The flux linkages λ and magnetic field coenergy will be expressed in terms of the currents and inductances. The torque can then be found from the partial derivative of the energy or coenergy with respect to the rotor position and the terminal voltages from the sum of the resistance drops Ri and the Faraday-law voltages $d\lambda/dt$. The result will be a set of nonlinear differential equations describing the dynamic performance of the machine.

The second viewpoint regards the machine as two groups of windings producing magnetic flux in the air gap, one group on the stator, and the other on the rotor. By making suitable assumptions regarding these fields (similar to those used to derive analytic expressions for the inductances), simple expressions can be derived for the flux linkages and the coenergy in the air gap in terms of the field quantities. The torque and generated voltage can then be found from these expressions. In this fashion, torque can be expressed explicitly as the tendency for two magnetic fields to align, in the same way that permanent magnets tend to align, and generated voltage can be expressed in terms of the relative motion between a field and a winding. These expressions lead to a simple physical picture of the normal steady-state behavior of rotating machines.

4.7.1 Coupled-Circuit Viewpoint

Consider the elementary smooth-air-gap machine of Fig. 4.34 with one winding on the stator and one on the rotor and with θ_m being the mechanical angle between the axes of the two windings. These windings are distributed over a number of slots so that their mmf waves can be approximated by space sinusoids. In Fig. 4.34a the coil sides s, $-s$ and r, $-r$ mark the positions of the centers of the belts of conductors comprising the distributed windings. An alternative way of drawing these windings is shown in Fig. 4.34b, which also shows reference directions for voltages and currents. Here it is assumed that current in the arrow direction produces a magnetic field in the air gap in the arrow direction, so that a single arrow defines reference directions for both current and flux.

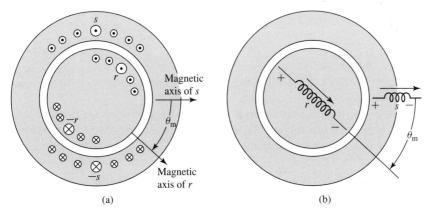

(a)   (b)

Figure 4.34 Elementary two-pole machine with smooth air gap: (a) winding distribution and (b) schematic representation.

The stator and rotor are concentric cylinders, and slot openings are neglected. Consequently, our elementary model does not include the effects of salient poles, which are investigated in later chapters. We also assume that the reluctances of the stator and rotor iron are negligible. Finally, although Fig. 4.34 shows a two-pole machine, we will write the derivations that follow for the general case of a multipole machine, replacing θ_m by the electrical rotor angle

$$\theta_{me} = \left(\frac{poles}{2}\right)\theta_m \tag{4.54}$$

Based upon these assumptions, the stator and rotor self-inductances L_{ss} and L_{rr} can be seen to be constant, but the stator-to-rotor mutual inductance depends on the electrical angle θ_{me} between the magnetic axes of the stator and rotor windings. The mutual inductance is at its positive maximum when $\theta_{me} = 0$ or 2π, is zero when $\theta_{me} = \pm\pi/2$, and is at its negative maximum when $\theta_{me} = \pm\pi$. On the assumption of sinusoidal mmf waves and a uniform air gap, the space distribution of the air-gap flux wave is sinusoidal, and the mutual inductance will be of the form

$$\mathcal{L}_{sr}(\theta_{me}) = L_{sr}\cos(\theta_{me}) \tag{4.55}$$

where the script letter $\mathcal{L}$ denotes an inductance which is a function of the electrical angle θ_{me}. The italic capital letter L denotes a constant value. Thus L_{sr} is the magnitude of the mutual inductance; its value when the magnetic axes of the stator and rotor are aligned ($\theta_{me} = 0$). In terms of the inductances, the stator and rotor flux linkages λ_s and λ_r are

$$\lambda_s = L_{ss}i_s + \mathcal{L}_{sr}(\theta_{me})i_r = L_{ss}i_s + L_{sr}\cos(\theta_{me})i_r \tag{4.56}$$

$$\lambda_r = \mathcal{L}_{sr}(\theta_{me})i_s + L_{rr}i_r = L_{sr}\cos(\theta_{me})i_s + L_{rr}i_r \tag{4.57}$$

where the inductances can be calculated as in Appendix B. In matrix notation

$$\begin{bmatrix} \lambda_s \\ \lambda_r \end{bmatrix} = \begin{bmatrix} L_{ss} & \mathcal{L}_{sr}(\theta_{me}) \\ \mathcal{L}_{sr}(\theta_{me}) & L_{rr} \end{bmatrix}\begin{bmatrix} i_s \\ i_r \end{bmatrix} \tag{4.58}$$

The terminal voltages v_s and v_r are

$$v_s = R_s i_s + \frac{d\lambda_s}{dt} \tag{4.59}$$

$$v_r = R_r i_r + \frac{d\lambda_r}{dt} \tag{4.60}$$

where R_s and R_r are the resistances of the stator and rotor windings respectively.

When the rotor is revolving, θ_{me} must be treated as a variable. Differentiation of Eqs. 4.56 and 4.57 and substitution of the results in Eqs. 4.59 and 4.60 then give

$$v_s = R_s i_s + L_{ss}\frac{di_s}{dt} + L_{sr}\cos(\theta_{me})\frac{di_r}{dt} - L_{sr}i_r\sin(\theta_{me})\frac{d\theta_{me}}{dt} \tag{4.61}$$

$$v_r = R_r i_r + L_{rr}\frac{di_r}{dt} + L_{sr}\cos(\theta_{me})\frac{di_r}{dt} - L_{sr}i_s\sin(\theta_{me})\frac{d\theta_{me}}{dt} \tag{4.62}$$

where

$$\frac{d\theta_{me}}{dt} = \omega_{me} = \left(\frac{\text{poles}}{2}\right)\omega_m \tag{4.63}$$

is the instantaneous speed in electrical radians per second. In a two-pole machine (such as that of Fig. 4.34), θ_{me} and ω_{me} are equal to the instantaneous shaft angle θ_m and the shaft speed ω_m respectively. In a multipole machine, they are related by Eqs. 4.54 and 4.46. The second and third terms on the right-hand sides of Eqs. 4.61 and 4.62 are $L(di/dt)$ induced voltages like those induced in stationary coupled circuits such as the windings of transformers. The fourth terms are caused by mechanical motion and are proportional to the instantaneous speed. These are the speed voltage terms which correspond to the interchange of power between the electric and mechanical systems.

The electromechanical torque can be found from the coenergy. From Eq. 3.70

$$
\begin{aligned}
W'_{fld} &= \frac{1}{2}L_{ss}i_s^2 + \frac{1}{2}L_{rr}i_r^2 + L_{sr}i_s i_r \cos\theta_{me} \\
&= \frac{1}{2}L_{ss}i_s^2 + \frac{1}{2}L_{rr}i_r^2 + L_{sr}i_s i_r \cos\left(\left(\frac{\text{poles}}{2}\right)\theta_m\right)
\end{aligned}
\tag{4.64}
$$

Note that the coenergy of Eq. 4.64 has been expressed specifically in terms of the shaft angle θ_m because the torque expression of Eq. 3.68 requires that the torque be obtained from the derivative of the coenergy with respect to the spatial angle θ_m and not with respect to the electrical angle θ_{me}. Thus, from Eq. 3.68

$$
\begin{aligned}
T &= \left.\frac{\partial W'_{fld}(i_s, i_r, \theta_m)}{\partial\theta_m}\right|_{i_s, i_r} = -\left(\frac{\text{poles}}{2}\right)L_{sr}i_s i_r \sin\left(\frac{\text{poles}}{2}\theta_m\right) \\
&= -\left(\frac{\text{poles}}{2}\right)L_{sr}i_s i_r \sin\theta_{me}
\end{aligned}
\tag{4.65}
$$

where T is the electromechanical torque acting to accelerate the rotor (i.e., a positive torque acts to increase θ_m). The negative sign in Eq. 4.65 means that the electromechanical torque acts in the direction to bring the magnetic fields of the stator and rotor into alignment.

Equations 4.61, 4.62, and 4.65 are a set of three equations relating the electrical variables v_s, i_s, v_r, i_r and the mechanical variables T and θ_m. These equations, together with the constraints imposed on the electrical variables by the networks connected to the terminals (sources or loads and external impedances) and the constraints imposed on the rotor (applied torques and inertial, frictional, and spring torques), determine the performance of the device and its characteristics as a conversion device between the external electrical and mechanical systems. These are nonlinear differential equations and are difficult to solve except under special circumstances. We are not specifically concerned with their solution here; rather we are using them merely as steps in the development of the theory of rotating machines.

EXAMPLE 4.6

Consider the elementary two-pole, two-winding machine of Fig. 4.34. Its shaft is coupled to a mechanical device which can be made to absorb or deliver mechanical torque over a wide range of speeds. This machine can be connected and operated in several ways. For this example, let us consider the situation in which the rotor winding is excited with direct current I_r and the stator winding is connected to an ac source which can either absorb or deliver electric power. Let the stator current be

$$i_s = I_s \cos \omega_e t$$

where $t = 0$ is arbitrarily chosen as the moment when the stator current has its peak value.

a. Derive an expression for the magnetic torque developed by the machine as the speed is varied by control of the mechanical device connected to its shaft.
b. Find the speed at which average torque will be produced if the stator frequency is 60 Hz.
c. With the assumed current-source excitations, what voltages are induced in the stator and rotor windings at synchronous speed ($\omega_m = \omega_e$)?

■ **Solution**

a. From Eq. 4.65 for a two-pole machine

$$T = -L_{sr} i_s i_r \sin \theta_m$$

For the conditions of this problem, with $\theta_m = \omega_m t + \delta$

$$T = -L_{sr} I_s I_r \cos \omega_e t \sin (\omega_m t + \delta)$$

where ω_m is the clockwise angular velocity impressed on the rotor by the mechanical drive and δ is the angular position of the rotor at $t = 0$. Using a trigonometric identity,[2] we have

$$T = -\frac{1}{2} L_{sr} I_s I_r \{\sin [(\omega_m + \omega_e)t + \delta] + \sin [(\omega_m - \omega_e)t + \delta]\}$$

The torque consists of two sinusoidally time-varying terms of frequencies $\omega_m + \omega_e$ and $\omega_m - \omega_e$. As shown in Section 4.5, ac current applied to the two-pole, single-phase stator winding in the machine of Fig. 4.34 creates two flux waves, one traveling in the positive θ_m direction with angular velocity ω_e and the second traveling in the negative θ_m direction also with angular velocity ω_e. It is the interaction of the rotor with these two flux waves which results in the two components of the torque expression.

b. Except when $\omega_m = \pm\omega_e$, the torque averaged over a sufficiently long time is zero. But if $\omega_m = \omega_e$, the rotor is traveling in synchronism with the positive-traveling stator flux wave, and the torque becomes

$$T = -\frac{1}{2} L_{sr} I_s I_r [\sin (2\omega_e t + \delta) + \sin \delta]$$

---

[2] $\sin \alpha \cos \beta = \frac{1}{2}[\sin (\alpha + \beta) + \sin (\alpha - \beta)]$

The first sine term is a double-frequency component whose average value is zero. The second term is the average torque

$$T_{avg} = -\frac{1}{2}L_{sr}I_sI_r\sin\delta$$

A nonzero average torque will also be produced when $\omega_m = -\omega_e$ which merely means rotation in the counterclockwise direction; the rotor is now traveling in synchronism with the negative-traveling stator flux wave. The negative sign in the expression for T_{avg} means that a positive value of T_{avg} acts to reduce δ.

This machine is an idealized single-phase synchronous machine. With a stator frequency of 60 Hz, it will produce nonzero average torque for speeds of $\pm\omega_m = \omega_e = 2\pi60$ rad/sec, corresponding to speeds of ±3600 r/min as can be seen from Eq. 4.41.

c. From the second and fourth terms of Eq. 4.61 (with $\theta_e = \theta_m = \omega_m t + \delta$), the voltage induced in the stator when $\omega_m = \omega_e$ is

$$e_s = -\omega_e L_{ss}I_s\sin\omega_e t - \omega_e L_{sr}I_r\sin(\omega_e t + \delta)$$

From the third and fourth terms of Eq. 4.62, the voltage induced in the rotor is

$$e_r = -\omega_0 L_{sr}I_s[\sin\omega_e t\cos(\omega_e t + \delta) + \cos\omega_s t\sin(\omega_e t + \delta)]$$

$$= -\omega_e L_{sr}I_s\sin(2\omega_e t + \delta)$$

The backwards-rotating component of the stator flux induces a double-frequency voltage in the rotor, while the forward-rotating component, which is rotating in sychronism with the rotor, appears as a dc flux to the rotor, and hence induces no voltage in the rotor winding.

Now consider a uniform-air-gap machine with several stator and rotor windings. The same general principles that apply to the elementary model of Fig. 4.34 also apply to the multiwinding machine. Each winding has its own self-inductance as well as mutual inductances with other windings. The self-inductances and mutual inductances between pairs of windings on the same side of the air gap are constant on the assumption of a uniform gap and negligible magnetic saturation. However, the mutual inductances between pairs of stator and rotor windings vary as the cosine of the angle between their magnetic axes. Torque results from the tendency of the magnetic field of the rotor windings to line up with that of the stator windings. It can be expressed as the sum of terms like that of Eq. 4.65.

EXAMPLE 4.7

Consider a 4-pole, three-phase synchronous machine with a uniform air gap. Assume the armature-winding self- and mutual inductances to be constant

$$L_{aa} = L_{bb} = L_{cc}$$

$$L_{ab} = L_{bc} = L_{ca}$$

Similarly, assume the field-winding self-inductance L_f to be constant while the mutual inductances between the field winding and the three armature phase windings will vary with the angle θ_m between the magnetic axis of the field winding and that of phase a

$$\mathcal{L}_{af} = L_{af} \cos 2\theta_m$$

$$\mathcal{L}_{bf} = L_{af} \cos (2\theta_m - 120°)$$

$$\mathcal{L}_{cf} = L_{af} \cos (2\theta_m + 120°)$$

Show that when the field is excited with constant current I_f and the armature is excited by balanced-three-phase currents of the form

$$i_a = I_a \cos (\omega_e t + \delta)$$

$$i_b = I_a \cos (\omega_e t - 120° + \delta)$$

$$i_c = I_a \cos (\omega_e t + 120° + \delta)$$

the torque will be constant when the rotor travels at synchronous speed ω_s as given by Eq. 4.40.

■ Solution

The torque can be calculated from the coenergy as described in Section 3.6. This particular machine is a four-winding system and the coenergy will consist of four terms involving 1/2 the self-inductance multiplied by the square of the corresponding winding current as well as product-terms consisting of the mutual inductances between pairs of windings multiplied by the corresponding winding currents. Noting that only the terms involving the mutual inductances between the field winding and the three armature phase windings will contain terms that vary with θ_m, we can write the coenergy in the form

$$W'_{fld}(i_a, i_b, i_c, i_f, \theta_m) = \text{(constant terms)} + \mathcal{L}_{af}i_a i_f + \mathcal{L}_{bf}i_b i_f + \mathcal{L}_{cf}i_c i_f$$

$$= \text{(constant terms)} + L_{af}I_a I_f [\cos 2\theta_m \cos (\omega_e t + \delta)$$

$$+ \cos (2\theta_m - 120°) \cos (\omega_e t - 120° + \delta)$$

$$+ \cos (2\theta_m + 120°) \cos (\omega_e t + 120° + \delta)]$$

$$= \text{(constant terms)} + \frac{3}{2} L_{af}I_a I_f \cos (2\theta_m - \omega_e t - \delta)$$

The torque can now be found from the partial derivative of W'_{fld} with respect to θ_m

$$T = \frac{\partial W'_{fld}}{\partial \theta_m}\bigg|_{i_a, i_b, i_c, i_f}$$

$$= -3L_{af}I_a I_f \sin (2\theta_m - \omega_e t - \delta)$$

From this expression, we see that the torque will be constant when the rotor rotates at synchronous velocity ω_s such that

$$\theta_m = \omega_s t = \left(\frac{\omega_e}{2}\right) t$$

in which case the torque will be equal to

$$T = 3L_{af}I_a I_f \sin \delta$$

Note that unlike the case of the single-phase machine of Example 4.6, the torque for this three-phase machine operating at synchronous velocity under balanced-three-phase conditions is constant. As we have seen, this is due to the fact that the stator mmf wave consists of a single rotating flux wave, as opposed to the single-phase case in which the stator phase current produces both a forward- and a backward-rotating flux wave. This backwards flux wave is not in synchronism with the rotor and hence is responsible for the double-frequency time-varying torque component seen in Example 4.6.

For the four-pole machine of Example 4.7, find the synchronous speed at which a constant torque will be produced if the rotor currents are of the form

$$i_a = I_a \cos(\omega_e t + \delta)$$

$$i_b = I_a \cos(\omega_e t + 120° + \delta)$$

$$i_c = I_a \cos(\omega_e t - 120° + \delta)$$

Solution

$$\omega_s = -(\omega_e/2)$$

In Example 4.7 we found that, under balanced conditions, a four-pole synchronous machine will produce constant torque at a rotational angular velocity equal to half of the electrical excitation frequency. This result can be generalized to show that, under balanced operating conditions, a multiphase, multipole synchronous machine will produce constant torque at a rotor speed, at which the rotor rotates in synchronism with the rotating flux wave produced by the stator currents. Hence, this is known as the *synchronous speed* of the machine. From Eqs. 4.40 and 4.41, the synchronous speed is equal to $\omega_s = (2/\text{poles})\omega_e$ in rad/sec or $n_s = (120/\text{poles})f_e$ in r/min.

4.7.2 Magnetic Field Viewpoint

In the discussion of Section 4.7.1 the characteristics of a rotating machine as viewed from its electric and mechanical terminals have been expressed in terms of its winding inductances. This viewpoint gives little insight into the physical phenomena which occur within the machine. In this section, we will explore an alternative formulation in terms of the interacting magnetic fields.

As we have seen, currents in the machine windings create magnetic flux in the air gap between the stator and rotor, the flux paths being completed through the stator and rotor iron. This condition corresponds to the appearance of magnetic poles on both the stator and the rotor, centered on their respective magnetic axes, as shown in Fig. 4.35a for a two-pole machine with a smooth air gap. Torque is produced by the tendency of the two component magnetic fields to line up their magnetic axes. A useful physical picture is that this is quite similar to the situation of two bar magnets pivoted at their centers on the same shaft; there will be a torque, proportional to the angular displacement of the bar magnets, which will act to align them. In the machine

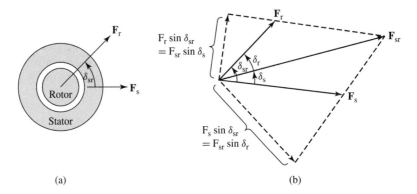

(a)   (b)

Figure 4.35 Simplified two-pole machine: (a) elementary model and
(b) vector diagram of mmf waves. Torque is produced by the tendency of the
rotor and stator magnetic fields to align. Note that these figures are drawn with
δ_{sr} positive, i.e., with the rotor mmf wave **F**$_r$ leading that of the stator **F**$_s$.

of Fig. 4.35a, the resulting torque is proportional to the product of the amplitudes of
the stator and rotor mmf waves and is also a function of the angle δ_{sr} measured from
the axis of the stator mmf wave to that of the rotor. In fact, we will show that, for a
smooth-air-gap machine, the torque is proportional to $\sin \delta_{sr}$.

In a typical machine, most of the flux produced by the stator and rotor windings
crosses the air gap and links both windings; this is termed the *mutual flux,* directly
analogous to the mutual or magnetizing flux in a transformer. However, some of the
flux produced by the rotor and stator windings does not cross the air gap; this is
analogous to the leakage flux in a transformer. These flux components are referred
to as the *rotor leakage flux* and the *stator leakage flux*. Components of this leakage
flux include slot and toothtip leakage, end-turn leakage, and space harmonics in the
air-gap field.

Only the mutual flux is of direct concern in torque production. The leakage
fluxes do affect machine performance however, by virtue of the voltages they induce
in their respective windings. Their effect on the electrical characteristics is accounted
for by means of leakage inductances, analogous to the use of inclusion of leakage
inductances in the transformer models of Chapter 2.

When expressing torque in terms of the winding currents or their resultant mmf's,
the resulting expressions do not include terms containing the leakage inductances.
Our analysis here, then, will be in terms of the resultant mutual flux. We shall derive
an expression for the magnetic coenergy stored in the air gap in terms of the stator
and rotor mmfs and the angle δ_{sr} between their magnetic axes. The torque can then
be found from the partial derivative of the coenergy with respect to angle δ_{sr}.

For analytical simplicity, we will assume that the radial length g of the air gap
(the radial clearance between the rotor and stator) is small compared with the radius
of the rotor or stator. For a smooth-air-gap machine constructed from electrical steel
with high magnetic permeability, it is possible to show that this will result in air-gap
flux which is primarily radially directed and that there is relatively little difference

between the flux density at the rotor surface, at the stator surface, or at any intermediate radial distance in the air gap. The air-gap field then can be represented as a radial field H_{ag} or B_{ag} whose intensity varies with the angle around the periphery. The line integral of H_{ag} across the gap then is simply $H_{ag}g$ and equals the *resultant air-gap mmf* $\mathcal{F}_{sr}$ produced by the stator and rotor windings; thus

$$H_{ag}g = \mathcal{F}_{sr} \tag{4.66}$$

where the script $\mathcal{F}$ denotes the mmf wave as a function of the angle around the periphery.

The mmf waves of the stator and rotor are spatial sine waves with δ_{sr} being the phase angle between their magnetic axes in electrical degrees. They can be represented by the space vectors $\mathbf{F}_s$ and $\mathbf{F}_r$ drawn along the magnetic axes of the stator- and rotor-mmf waves respectively, as in Fig. 4.35b. The resultant mmf $\mathbf{F}_{sr}$ acting across the air gap, also a sine wave, is their vector sum. From the trigonometric formula for the diagonal of a parallelogram, its peak value is found from

$$F_{sr}^2 = F_s^2 + F_r^2 + 2F_s F_r \cos \delta_{sr} \tag{4.67}$$

in which the F's are the peak values of the mmf waves. The resultant radial H_{ag} field is a sinusoidal space wave whose peak value $H_{ag,peak}$ is, from Eq. 4.66,

$$(H_{ag})_{peak} = \frac{F_{sr}}{g} \tag{4.68}$$

Now consider the magnetic field coenergy stored in the air gap. From Eq. 3.49, the coenergy density at a point where the magnetic field intensity is H is $(\mu_0/2)H^2$ in SI units. Thus, the coenergy density averaged over the volume of the air gap is $\mu_0/2$ times the average value of H_{ag}^2. The average value of the square of a sine wave is half its peak value. Hence,

$$\text{Average coenergy density} = \frac{\mu_0}{2} \left(\frac{(H_{ag})_{peak}^2}{2} \right) = \frac{\mu_0}{4} \left(\frac{F_{sr}}{g} \right)^2 \tag{4.69}$$

The total coenergy is then found as

$$W'_{fld} = (\text{average coenergy density})(\text{volume of air gap})$$

$$= \frac{\mu_0}{4} \left(\frac{F_{sr}}{g} \right)^2 \pi D l g = \frac{\mu_0 \pi D l}{4g} F_{sr}^2 \tag{4.70}$$

where l is the axial length of the air gap and D is its average diameter.

From Eq. 4.67 the coenergy stored in the air gap can now be expressed in terms of the peak amplitudes of the stator- and rotor-mmf waves and the space-phase angle between them; thus

$$W'_{fld} = \frac{\mu_0 \pi D l}{4g} \left(F_s^2 + F_r^2 + 2F_s F_r \cos \delta_{sr} \right) \tag{4.71}$$

Recognizing that holding mmf constant is equivalent to holding current constant, an expression for the electromechanical torque T can now be obtained in terms of the interacting magnetic fields by taking the partial derivative of the field coenergy with respect to angle. For a two-pole machine

$$T = \left.\frac{\partial W'_{\text{fld}}}{\partial \delta_{\text{sr}}}\right|_{F_s, F_r} = -\left(\frac{\mu_0 \pi Dl}{2g}\right) F_s F_r \sin \delta_{\text{sr}} \tag{4.72}$$

The general expression for the torque for a multipole machine is

$$T = -\left(\frac{\text{poles}}{2}\right)\left(\frac{\mu_0 \pi Dl}{2g}\right) F_s F_r \sin \delta_{\text{sr}} \tag{4.73}$$

In this equation, δ_{sr} is the electrical space-phase angle between the rotor and stator mmf waves and the torque T acts in the direction to accelerate the rotor. Thus when δ_{sr} is positive, the torque is negative and the machine is operating as a generator. Similarly, a negative value of δ_{sr} corresponds to positive torque and, correspondingly, motor action.

This important equation states that the torque is proportional to the peak values of the stator- and rotor-mmf waves F_s and F_r and to the sine of the electrical space-phase angle δ_{sr} between them. The minus sign means that the fields tend to align themselves. Equal and opposite torques are exerted on the stator and rotor. The torque on the stator is transmitted through the frame of the machine to the foundation.

One can now compare the results of Eq. 4.73 with that of Eq. 4.65. Recognizing that F_s is proportional to i_s and F_r is proportional to i_r, one sees that they are similar in form. In fact, they must be equal, as can be verified by substitution of the appropriate expressions for F_s, F_r (Section 4.3.1), and L_{sr} (Appendix B). Note that these results have been derived with the assumption that the iron reluctance is negligible. However, the two techniques are equally valid for finite iron permeability.

On referring to Fig. 4.35b, it can be seen that $F_r \sin \delta_{\text{sr}}$ is the component of the F_r wave in electrical space quadrature with the F_s wave. Similarly $F_s \sin \delta_{\text{sr}}$ is the component of the F_s wave in quadrature with the F_r wave. Thus, the torque is proportional to the product of one magnetic field and the component of the other in quadrature with it, much like the cross product of vector analysis. Also note that in Fig. 4.35b

$$F_s \sin \delta_{\text{sr}} = F_{\text{sr}} \sin \delta_r \tag{4.74}$$

and

$$F_r \sin \delta_{\text{sr}} = F_{\text{sr}} \sin \delta_s \tag{4.75}$$

where, as seen in Fig. 4.35, δ_r is the angle measured from the axis of the resultant mmf wave to the axis of the rotor mmf wave. Similarly, δ_s is the angle measured from the axis of the stator mmf wave to the axis of the resultant mmf wave.

The torque, acting to accelerate the rotor, can then be expressed in terms of the resultant mmf wave F_{sr} by substitution of either Eq. 4.74 or Eq. 4.75 in Eq. 4.73; thus

$$T = -\left(\frac{\text{poles}}{2}\right)\left(\frac{\mu_0 \pi Dl}{2g}\right) F_s F_{\text{sr}} \sin \delta_s \tag{4.76}$$

$$T = -\left(\frac{\text{poles}}{2}\right)\left(\frac{\mu_0 \pi Dl}{2g}\right) F_r F_{\text{sr}} \sin \delta_r \tag{4.77}$$

Comparison of Eqs. 4.73, 4.76, and 4.77 shows that the torque can be expressed in terms of the component magnetic fields due to *each* current acting alone, as in Eq. 4.73, or in terms of the *resultant* field and *either* of the components, as in Eqs. 4.76 and 4.77, *provided that we use the corresponding angle between the axes of the fields.* Ability to reason in any of these terms is a convenience in machine analysis.

In Eqs. 4.73, 4.76, and 4.77, the fields have been expressed in terms of the peak values of their mmf waves. When magnetic saturation is neglected, the fields can, of course, be expressed in terms of the peak values of their flux-density waves or in terms of total flux per pole. Thus the peak value B_{ag} of the field due to a sinusoidally distributed mmf wave in a uniform-air-gap machine is $\mu_0 F_{ag,peak}/g$, where $F_{ag,peak}$ is the peak value of the mmf wave. For example, the resultant mmf F_{sr} produces a resultant flux-density wave whose peak value is $B_{sr} = \mu_0 F_{sr}/g$. Thus, $F_{sr} = g B_{sr}/\mu_0$ and substitution in Eq. 4.77 gives

$$T = -\left(\frac{poles}{2}\right)\left(\frac{\pi Dl}{2}\right) B_{sr} F_r \sin \delta_r \qquad (4.78)$$

One of the inherent limitations in the design of electromagnetic apparatus is the saturation flux density of magnetic materials. Because of saturation in the armature teeth the peak value B_{sr} of the resultant flux-density wave in the air gap is limited to about 1.5 to 2.0 T. The maximum permissible value of the winding current, and hence the corresponding mmf wave, is limited by the temperature rise of the winding and other design requirements. Because the resultant flux density and mmf appear explicitly in Eq. 4.78, this equation is in a convenient form for design purposes and can be used to estimate the maximum torque which can be obtained from a machine of a given size.

EXAMPLE 4.8

An 1800-r/min, four-pole, 60-Hz synchronous motor has an air-gap length of 1.2 mm. The average diameter of the air-gap is 27 cm, and its axial length is 32 cm. The rotor winding has 786 turns and a winding factor of 0.976. Assuming that thermal considerations limit the rotor current to 18 A, estimate the maximum torque and power output one can expect to obtain from this machine.

■ Solution

First, we can determine the maximum rotor mmf from Eq. 4.8

$$(F_r)_{max} = \frac{4}{\pi}\left(\frac{k_r N_r}{poles}\right)(I_r)_{max} = \frac{4}{\pi}\left(\frac{0.976 \times 786}{4}\right)18 = 4395 \text{ A}$$

Assuming that the peak value of the resultant air-gap flux is limited to 1.5 T, we can estimate the maximum torque from Eq. 4.78 by setting δ_r equal to $-\pi/2$ (recognizing that negative values of δ_r, with the rotor mmf lagging the resultant mmf, correspond to positive, motoring torque)

$$T_{max} = \left(\frac{poles}{2}\right)\left(\frac{\pi Dl}{2}\right) B_{sr}(F_r)_{max}$$

$$= \left(\frac{4}{2}\right)\left(\frac{\pi \times 0.27 \times 0.32}{2}\right) 1.5 \times 4400 = 1790 \text{ N} \cdot \text{m}$$

For a synchronous speed of 1800 r/min, $\omega_m = n_s (\pi/30) = 1800 (\pi/30) = 60\pi$ rad/sec, and thus the corresponding power can be calculated as $P_{max} = \omega_m T_{max} = 337$ kW.

Practice Problem 4.6

Repeat Example 4.8 for a two-pole, 60-Hz synchronous motor with an air-gap length of 1.3 mm, an average air-gap diameter of 22 cm and an axial length of 41 cm. The rotor winding has a 900 turns and a winding factor of 0.965. The maximum rotor current is 22 A.

Solution

$T_{max} = 2585$ N $\cdot$ m and $P_{max} = 975$ kW

Alternative forms of the torque equation arise when it is recognized that the resultant flux per pole is

$$\Phi_p = (\text{average value of B over a pole})(\text{pole area}) \tag{4.79}$$

and that the average value of a sinusoid over one-half wavelength is $2/\pi$ times its peak value. Thus

$$\Phi_p = \frac{2}{\pi} B_{peak} \left(\frac{\pi Dl}{poles} \right) = \left(\frac{2Dl}{poles} \right) B_{peak} \tag{4.80}$$

where B_{peak} is the peak value of the corresponding flux-density wave. For example, using the peak value of the resultant flux B_{sr} and substitution of Eq. 4.80 into Eq. 4.78 gives

$$T = -\frac{\pi}{2} \left(\frac{poles}{2} \right)^2 \Phi_{sr} F_r \sin \delta_r \tag{4.81}$$

where Φ_{sr} is the resultant flux per pole produced by the combined effect of the stator and rotor mmf's.

To recapitulate, we now have several forms in which the torque of a uniform-air-gap machine can be expressed in terms of its magnetic fields. *All are merely statements that the torque is proportional to the product of the magnitudes of the interacting fields and to the sine of the electrical space angle between their magnetic axes.* The negative sign indicates that the electromechanical torque acts in a direction to decrease the displacement angle between the fields. In our preliminary discussion of machine types, Eq. 4.81 will be the preferred form.

One further remark can be made concerning the torque equations and the thought process leading to them. There was no restriction in the derivation that the mmf wave or flux-density wave remain stationary in space. They may remain stationary, or they may be traveling waves, as discussed in Section 4.5. As we have seen, if the magnetic fields of the stator and rotor are constant in amplitude and travel around the air gap at the same speed, a steady torque will be produced by the tendency of the stator and rotor fields to align themselves in accordance with the torque equations.

4.8 LINEAR MACHINES

In general, each of the machine types discussed in this book can be produced in linear versions in addition to the rotary versions which are commonly found and which are discussed extensively in the following chapters. In fact, for clarity of discussion, many of the machine types discussed in this book are drawn in their developed (Cartesian coordinate) form, such as in Fig. 4.19b.

Perhaps the most widely known use of linear motors is in the transportation field. In these applications, linear induction motors are used, typically with the ac "stator" on the moving vehicle and with a conducting stationary "rotor" constituting the rails. In these systems, in addition to providing propulsion, the induced currents in the rail may be used to provide levitation, thus offering a mechanism for high-speed transportation without the difficulties associated with wheel-rail interactions on more conventional rail transport.

Linear motors have also found application in the machine tool industry and in robotics where linear motion (required for positioning and in the operation of manipulators) is a common requirement. In addition, reciprocating linear machines are being constructed for driving reciprocating compressors and alternators.

The analysis of linear machines is quite similar to that of rotary machines. In general, linear dimensions and displacements replace angular ones, and forces replace torques. With these exceptions, the expressions for machine parameters are derived in an analogous fashion to those presented here for rotary machines, and the results are similar in form.

Consider the linear winding shown in Fig. 4.36. This winding, consisting of N turns per slot and carrying a current i, is directly analogous to the rotary winding shown in developed form in Fig. 4.25. In fact, the only difference is that the angular position θ_a is replaced by the linear position z.

The fundamental component of the mmf wave of Fig. 4.36 can be found directly from Eq. 4.13 simply by recognizing that this winding has a wavelength equal to β and that the fundamental component of this mmf wave varies as $\cos(2\pi z/\beta)$. Thus replacing the angle θ_a in Eq. 4.13 by $2\pi z/\beta$, we can find the fundamental component of the mmf wave directly as

$$H_{\text{ag}1} = \frac{4}{\pi}\left(\frac{Ni}{2g}\right)\cos\left(\frac{2\pi z}{\beta}\right) \qquad (4.82)$$

If an actual machine has a distributed winding (similar to its rotary counterpart, shown in Fig. 4.20) consisting of a total of N_{ph} turns distributed over p periods in z (i.e., over a length of $p\beta$), the fundamental component of H_{ag} can be found by analogy with Eq. 4.15

$$H_{\text{ag}1} = \frac{4}{\pi}\left(\frac{k_{\text{w}}N_{\text{ph}}i}{2pg}\right)\cos\left(\frac{2\pi z}{\beta}\right) \qquad (4.83)$$

where k_{w} is the winding factor.

In a fashion analogous to the discussion of Section 4.5.2, a three-phase linear winding can be made from three windings such as those of Fig. 4.31, with each phase

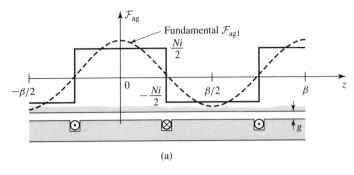

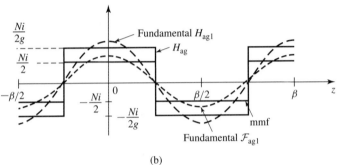

Figure 4.36 The mmf and H field of a concentrated full-pitch linear winding.

displaced in position by a distance $\beta/3$ and with each phase excited by balanced three-phase currents of angular frequency ω_e

$$i_a = I_m \cos \omega_e t \tag{4.84}$$

$$i_b = I_m \cos (\omega_e t - 120°) \tag{4.85}$$

$$i_c = I_m \cos (\omega_e t + 120°) \tag{4.86}$$

Following the development of Eqs. 4.26 through 4.38, we can see that there will be a single positive-traveling mmf which can be written directly from Eq. 4.38 simply by replacing θ_a by $2\pi z/\beta$ as

$$\mathcal{F}^+(z, t) = \frac{3}{2} F_{max} \cos \left(\frac{2\pi z}{\beta} - \omega_e t \right) \tag{4.87}$$

where F_{max} is given by

$$F_{max} = \frac{4}{\pi} \left(\frac{k_w N_{ph}}{2p} \right) I_m \tag{4.88}$$

From Eq. 4.87 we see that the result is an mmf which travels in the z direction with a linear velocity

$$v = \frac{\omega_e \beta}{2\pi} = f_e \beta \qquad (4.89)$$

where f_e is the exciting frequency in hertz.

EXAMPLE 4.9

A three-phase linear ac motor has a winding with a wavelength of $\beta = 0.5$ m and an air gap of 1.0 cm in length. A total of 45 turns, with a winding factor $k_w = 0.92$, are distributed over a total winding length of $3\beta = 1.5$ m. Assume the windings to be excited with balanced three-phase currents of peak amplitude 700 A and frequency 25 Hz. Calculate (a) the amplitude of the resultant mmf wave, (b) the corresponding peak air-gap flux density, and (c) the velocity of this traveling mmf wave.

■ **Solution**

a. From Eqs. 4.87 and 4.88, the amplitude of the resultant mmf wave is

$$F_{peak} = \frac{3}{2} \frac{4}{\pi} \left(\frac{k_w N_{ph}}{2p} \right) I_m$$

$$= \frac{3}{2} \frac{4}{\pi} \left(\frac{0.92 \times 45}{2 \times 3} \right) 700$$

$$= 8.81 \times 10^3 \text{ A/m}$$

b. The peak air-gap flux density can be found from the result of part (a) by dividing by the air-gap length and multiplying by μ_0:

$$B_{peak} = \frac{\mu_0 F_{peak}}{g}$$

$$= \frac{(4\pi \times 10^{-7})(8.81 \times 10^3)}{0.01}$$

$$= 1.11 \text{ T}$$

c. Finally, the velocity of the traveling wave can be determined from Eq. 4.89:

$$v = f_e \beta = 25 \times 0.5 = 12.5 \text{ m/s}$$

A three-phase linear synchronous motor has a wavelength of 0.93 m. It is observed to be traveling at speed of 83 km/hr. Calculate the frequency of the electrical excitation required under this operating condition.

Solution

$$f = 24.8 \text{ Hz}$$

Linear machines are not discussed specifically in this book. Rather, the reader is urged to recognize that the fundamentals of their performance and analysis correspond directly to those of their rotary counterparts. One major difference between these two machine types is that linear machines have *end effects,* corresponding to the magnetic fields which "leak" out of the air gap ahead of and behind the machine. These effects are beyond the scope of this book and have been treated in detail in the published literature.[3]

4.9   MAGNETIC SATURATION

The characteristics of electric machines depend heavily upon the use of magnetic materials. These materials are required to form the magnetic circuit and are used by the machine designer to obtain specific machine characteristics. As we have seen in Chapter 1, magnetic materials are less than ideal. As their magnetic flux is increased, they begin to saturate, with the result that their magnetic permeabilities begin to decrease, along with their effectiveness in contributing to the overall flux density in the machine.

Both electromechanical torque and generated voltage in all machines depend on the winding flux linkages. For specific mmf's in the windings, the fluxes depend on the reluctances of the iron portions of the magnetic circuits and on those of the air gaps. Saturation may therefore appreciably influence the characteristics of the machines.

Another aspect of saturation, more subtle and more difficult to evaluate without experimental and theoretical comparisons, concerns its influence on the basic premises from which the analytic approach to machinery is developed. Specifically, relations for the air-gap mmf are typically based on the assumption of negligible reluctance in the iron. When these relations are applied to practical machines with varying degrees of saturation in the iron, significant errors in the analytical results can be expected. To improve these analytical relationships, the actual machine can be replaced for these considerations by an equivalent machine, one whose iron has negligible reluctance but whose air-gap length is increased by an amount sufficient to absorb the magnetic-potential drop in the iron of the actual machine.

Similarly, the effects of air-gap nonuniformities such as slots and ventilating ducts are also incorporated by increasing the effective air-gap length. Ultimately, these various approximate techniques must be verified and confirmed experimentally. In cases where such simple techniques are found to be inadequate, detailed analyses, such as those employing finite-element or other numerical techniques, can be used. However, it must be recognized that the use of these techniques represents a significant increase in modeling complexity.

Saturation characteristics of rotating machines are typically presented in the form of an *open-circuit characteristic,* also called a *magnetization curve* or *saturation*

---

[3] See, for example, S. Yamamura, *Theory of Linear Induction Motors,* 2d ed., Halsted Press, 1978. Also, S. Nasar and I. Boldea, *Linear Electric Motors: Theory, Design and Practical Applications,* Prentice-Hall, 1987.

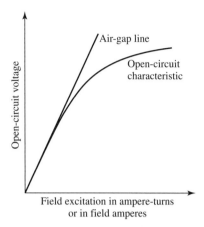

Figure 4.37 Typical open-circuit characteristic and air-gap line.

curve. An example is shown in Fig. 4.37. This characteristic represents the magnetization curve for the particular iron and air geometry of the machine under consideration. For a synchronous machine, the open-circuit saturation curve is obtained by operating the machine at constant speed and measuring the open-circuit armature voltage as a function of the field current. The straight line tangent to the lower portion of the curve is the *air-gap line,* corresponding to low levels of flux within the machine. Under these conditions the reluctance of the machine iron is typically negligible, and the mmf required to excite the machine is simply that required to overcome the reluctance of the air gap. If it were not for the effects of saturation, the air-gap line and open-circuit characteristic would coincide. Thus, the departure of the curve from the air-gap line is an indication of the degree of saturation present. In typical machines the ratio at rated voltage of the total mmf to that required by the air gap alone usually is between 1.1 and 1.25.

At the design stage, the open-circuit characteristic can be calculated from design data techniques such as finite-element analyses. A typical finite-element solution for the flux distribution around the pole of a salient-pole machine is shown in Fig. 4.38. The distribution of the air-gap flux found from this solution, together with the fundamental and third-harmonic components, is shown in Fig. 4.39.

In addition to saturation effects, Fig. 4.39 clearly illustrates the effect of a nonuniform air gap. As expected, the flux density over the pole face, where the air gap is small, is much higher than that away from the pole. This type of detailed analysis is of great use to a designer in obtaining specific machine properties.

As we have seen, the magnetization curve for an existing synchronous machine can be determined by operating the machine as an unloaded generator and measuring the values of terminal voltage corresponding to a series of values of field current. For an induction motor, the machine is operated at or close to synchronous speed (in which case very little current will be induced in the rotor windings), and values of the magnetizing current are obtained for a series of values of impressed stator voltage.

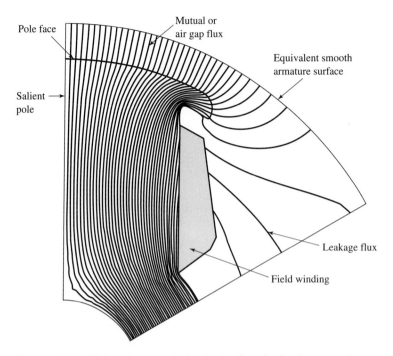

Figure 4.38 Finite-element solution for the flux distribution around a salient pole. (*General Electric Company.*)

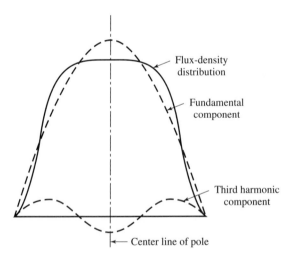

Figure 4.39 Flux-density wave corresponding to Fig. 4.38 with its fundamental and third-harmonic components.

It should be emphasized, however, that saturation in a fully loaded machine occurs as a result of the total mmf acting on the magnetic circuit. Since the flux distribution under load generally differs from that of no-load conditions, the details of the machine saturation characteristics may vary from the open-circuit curve of Fig. 4.37.

4.10 LEAKAGE FLUXES

In Section 2.4 we showed that in a two-winding transformer the flux created by each winding can be separated into two components. One component consists of flux which links both windings, and the other consists of flux which links only the winding creating the flux. The first component, called *mutual flux,* is responsible for coupling between the two coils. The second, known as *leakage flux,* contributes only to the self-inductance of each coil.

Note that the concept of mutual and leakage flux is meaningful only in the context of a multiwinding system. For systems of three or more windings, the bookkeeping must be done very carefully. Consider, for example, the three-winding system of Fig. 4.40. Shown schematically are the various components of flux created by a current in winding 1. Here φ_{123} is clearly mutual flux that links all three windings, and φ_{1l} is clearly leakage flux since it links only winding 1. However, φ_{12} is mutual flux with respect to winding 2 yet is leakage flux with respect to winding 3, while φ_{13} mutual flux with respect to winding 3 and leakage flux with respect to winding 2.

Electric machinery often contains systems of multiple windings, requiring careful bookkeeping to account for the flux contributions of the various windings. Although the details of such analysis are beyond the scope of this book, it is useful to discuss these effects in a qualitative fashion and to describe how they affect the basic machine inductances.

Air-Gap Space-Harmonic Fluxes   In this chapter we have seen that although single distributed coils create air-gap flux with a significant amount of space-harmonic

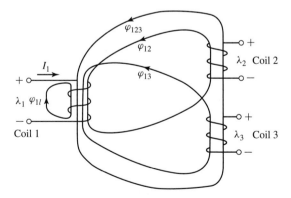

Figure 4.40 Three-coil system showing components of mutual and leakage flux produced by current in coil 1.

content, it is possible to distribute these windings so that the space-fundamental component is emphasized while the harmonic effects are greatly reduced. As a result, we can neglect harmonic effects and consider only space-fundamental fluxes in calculating the self and mutual-inductance expressions of Eqs. B.26 and B.27.

Though often small, the space-harmonic components of air-gap flux do exist. In dc machines they are useful torque-producing fluxes and therefore can be counted as mutual flux between the rotor and stator windings. In ac machines, however, they may generate time-harmonic voltages or asynchronously rotating flux waves. These effects generally cannot be rigorously accounted for in most standard analyses. Nevertheless, it is consistent with the assumptions basic to these analyses to recognize that these fluxes form a part of the leakage flux of the individual windings which produce them.

Slot-Leakage Flux   Figure 4.41 shows the flux created by a single coil side in a slot. Notice that in addition to flux which crosses the air gap, contributing to the air-gap flux, there are flux components which cross the slot. Since this flux links only the coil that is producing it, it also forms a component of the leakage inductance of the winding producing it.

End-Turn Fluxes   Figure 4.42 shows the stator end windings on an ac machine. The magnetic field distribution created by end turns is extremely complex. In general these fluxes do not contribute to useful rotor-to-stator mutual flux, and thus they, too, contribute to leakage inductance.

From this discussion we see that the self-inductance expression of Eq. B.26 must, in general, be modified by an additional term L_l, which represents the winding leakage inductance. This leakage inductance corresponds directly to the leakage inductance of a transformer winding as discussed in Chapter 1. Although the leakage inductance is usually difficult to calculate analytically and must be determined by approximate or empirical techniques, it plays an important role in machine performance.

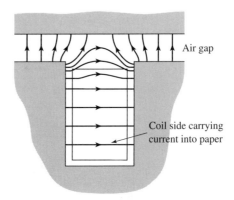

Figure 4.41 Flux created by a single coil side in a slot.

Figure 4.42 End view of the stator of a 26-kV 908-MVA 3600 r/min turbine generator with water-cooled windings. Hydraulic connections for coolant flow are provided for each winding end turn. (*General Electric Company.*)

4.11 SUMMARY

This chapter presents a brief and elementary description of three basic types of rotating machines: synchronous, induction, and dc machines. In all of them the basic principles are essentially the same. Voltages are generated by the relative motion of a magnetic field with respect to a winding, and torques are produced by the interaction of the magnetic fields of the stator and rotor windings. The characteristics of the various machine types are determined by the methods of connection and excitation of the windings, but the basic principles are essentially similar.

The basic analytical tools for studying rotating machines are expressions for the generated voltages and for the electromechanical torque. Taken together, they express the coupling between the electric and mechanical systems. To develop a reasonably quantitative theory without the confusion arising from too much detail, we have made several simplifying approximations. In the study of ac machines we have assumed sinusoidal time variations of voltages and currents and sinusoidal space waves of air-gap flux density and mmf. On examination of the mmf produced by distributed ac windings we found that the space-fundamental component is the most important. On the other hand, in dc machines the armature-winding mmf is more nearly a sawtooth wave. For our preliminary study in this chapter, however, we have assumed sinusoidal mmf distributions for both ac and dc machines. We examine this assumption more thoroughly for dc machines in Chapter 7. Faraday's law results in Eq. 4.50 for the rms voltage generated in an ac machine winding or Eq. 4.53 for the average voltage generated between brushes in a dc machine.

On examination of the mmf wave of a three-phase winding, we found that balanced three-phase currents produce a constant-amplitude air-gap magnetic field rotating at synchronous speed, as shown in Fig. 4.31 and Eq. 4.39. The importance of this fact cannot be overstated, for it means that it is possible to operate such machines, either as motors or generators, under conditions of constant torque (and hence constant electrical power as is discussed in Appendix A), eliminating the double-frequency, time-varying torque inherently associated with single-phase machines. For example, imagine a multimegawatt single-phase 60-Hz generator subjected to multimegawatt instantaneous power pulsation at 120 Hz! The discovery of rotating fields led to the invention of the simple, rugged, reliable, self-starting polyphase induction motor, which is analyzed in Chapter 6. (A single-phase induction motor will not start; it needs an auxiliary starting winding, as shown in Chapter 9.)

In single-phase machines, or in polyphase machines operating under unbalanced conditions, the backward-rotating component of the armature mmf wave induces currents and losses in the rotor structure. Thus, the operation of polyphase machines under balanced conditions not only eliminates the second-harmonic component of generated torque, it also eliminates a significant source of rotor loss and rotor heating. It was the invention of polyphase machines operating under balanced conditions that made possible the design and construction of large synchronous generators with ratings as large as 1000 MW.

Having assumed sinusoidally-distributed magnetic fields in the air gap, we then derived expressions for the magnetic torque. The simple physical picture for torque production is that of two magnets, one on the stator and one on the rotor, as shown schematically in Fig. 4.35a. The torque acts in the direction to align the magnets. To get a reasonably close quantitative analysis without being hindered by details, we assumed a smooth air gap and neglected the reluctance of the magnetic paths in the iron parts, with a mental note that this assumption may not be valid in all situations and a more detailed model may be required.

In Section 4.7 we derived expressions for the magnetic torque from two viewpoints, both based on the fundamental principles of Chapter 3. The first viewpoint regards the machine as a set of magnetically-coupled circuits with inductances which depend on the angular position of the rotor, as in Section 4.7.1. The second regards the machine from the viewpoint of the magnetic fields in the air gap, as in Section 4.7.2. It is shown that the torque can be expressed as the product of the stator field, the rotor field, and the sine of the angle between their magnetic axes, as in Eq. 4.73 or any of the forms derived from Eq. 4.73. The two viewpoints are complementary, and ability to reason in terms of both is helpful in reaching an understanding of how machines work.

This chapter has been concerned with basic principles underlying rotating-machine theory. By itself it is obviously incomplete. Many questions remain unanswered. How do we apply these principles to the determination of the characteristics of synchronous, induction, and dc machines? What are some of the practical problems that arise from the use of iron, copper, and insulation in physical machines? What are some of the economic and engineering considerations affecting rotating-machine applications? What are the physical factors limiting the conditions under which a machine can operate successfully? Appendix D discusses some of these problems. Taken together, Chapter 4 along with Appendix D serve as an introduction to the more detailed treatments of rotating machines in the following chapters.

4.12 PROBLEMS

4.1 The rotor of a six-pole synchronous generator is rotating at a mechanical speed of 1200 r/min.

a. Express this mechanical speed in radians per second.

b. What is the frequency of the generated voltage in hertz and in radians per second?

c. What mechanical speed in revolutions per minute would be required to generate voltage at a frequency of 50 Hz?

4.2 The voltage generated in one phase of an unloaded three-phase synchronous generator is of the form $v(t) = V_0 \cos \omega t$. Write expressions for the voltage in the remaining two phases.

4.3 A three-phase motor is used to drive a pump. It is observed (by the use of a stroboscope) that the motor speed decreases from 898 r/min when the pump is unloaded to 830 r/min as the pump is loaded.

a. Is this a synchronous or an induction motor?

b. Estimate the frequency of the applied armature voltage in hertz.

c. How many poles does this motor have?

4.4 The object of this problem is to illustrate how the armature windings of certain machines, i.e., dc machines, can be approximately represented by uniform current sheets, the degree of correspondence growing better as the winding is distributed in a greater number of slots around the armature periphery. For this purpose, consider an armature with eight slots uniformly distributed over 360 electrical degrees (corresponding to a span of one pole pair). The air gap is of uniform length, the slot openings are very small, and the reluctance of the iron is negligible.

Lay out 360 electrical degrees of the armature with its slots in developed form in the manner of Fig. 4.23a and number the slots 1 to 8 from left to right. The winding consists of eight single-turn coils, each carrying a direct current of 10 A. Coil sides placed in any of the slots 1 to 4 carry current directed into the paper; those placed in any of the slots 5 to 8 carry current out of the paper.

a. Consider that all eight slots are placed with one side in slot 1 and the other in slot 5. The remaining slots are empty. Draw the rectangular mmf wave produced by these slots.

b. Next consider that four coils have one side in slot 1 and the other in slot 5, while the remaining four coils have one side in slot 3 and the other in slot 7. Draw the component rectangular mmf waves produced by each group of coils, and superimpose the components to give the resultant mmf wave.

   c. Now consider that two coils are placed in slots 1 and 5, two in slots 2 and 6, two in 3 and 7, and two in 4 and 8. Again superimpose the component rectangular waves to produce the resultant wave. Note that the task can be systematized and simplified by recognizing that the mmf wave is symmetric about its axis and takes a step at each slot which is directly proportional to the number of ampere-conductors in that slot.

   d. Let the armature now consist of 16 slots per 360 electrical degrees with one coil side per slot. Draw the resultant mmf wave.

4.5 A three-phase Y-connected ac machine is initially operating under balanced three-phase conditions when one of the phase windings becomes open-circuited. Because there is no neutral connection on the winding, this requires that the currents in the remaining two windings become equal and opposite. Under this condition, calculate the relative magnitudes of the resultant positive- and negative-traveling mmf waves.

4.6 What is the effect on the rotating mmf and flux waves of a three-phase winding produced by balanced-three-phase currents if two of the phase connections are interchanges?

4.7 In a balanced two-phase machine, the two windings are displaced 90 electrical degrees in space, and the currents in the two windings are phase-displaced 90 electrical degrees in time. For such a machine, carry out the process leading to an equation for the rotating mmf wave corresponding to Eq. 4.39 (which is derived for a three-phase machine).

4.8 This problem investigates the advantages of short-pitching the stator coils of an ac machine. Figure 4.43a shows a single full-pitch coil in a two-pole machine. Figure 4.43b shows a fractional-pitch coil for which the coil sides are β radians apart, rather than π radians (180°) as is the case for the full-pitch coil.

For an air-gap radial flux distribution of the form

$$B_r = \sum_{n \text{ odd}} B_n \cos n\theta$$

where $n = 1$ corresponds to the fundamental space harmonic, $n = 3$ to the third space harmonic, and so on, the flux linkage of each coil is the integral of B_r over the surface spanned by that coil. Thus for the nth space harmonic, the ratio of the maximum fractional-pitch coil flux linkage to that of the full-pitch coil is

$$\frac{\int_{-\beta/2}^{\beta/2} B_n \cos n\theta \, d\theta}{\int_{-\pi/2}^{\pi/2} B_n \cos n\theta \, d\theta} = \frac{\int_{-\beta/2}^{\beta/2} \cos n\theta \, d\theta}{\int_{-\pi/2}^{\pi/2} \cos n\theta \, d\theta} = |\sin (n\beta/2)|$$

It is common, for example, to fractional-pitch the coils of an ac machine by 30 electrical degrees ($\beta = 5\pi/6 = 150°$). For $n = 1, 3, 5$ calculate the fractional reduction in flux linkage due to short pitching.

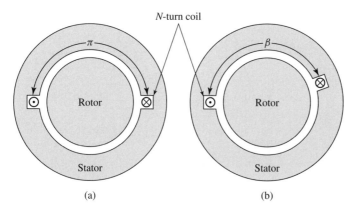

N-turn coil

Rotor

Stator

(a)

Rotor

Stator

(b)

Figure 4.43 Problem 4.8: (a) full-pitch coil and (b) fractional-pitch coil.

4.9 A six-pole, 60-Hz synchronous machine has a rotor winding with a total of 138 series turns and a winding factor $k_r = 0.935$. The rotor length is 1.97 m, the rotor radius is 58 cm, and the air-gap length $= 3.15$ cm.

   a. What is the rated operating speed in r/min?

   b. Calculate the rotor-winding current required to achieve a peak fundamental air-gap flux density of 1.23 T.

   c. Calculate the corresponding flux per pole.

4.10 Assume that a phase winding of the synchronous machine of Problem 4.9 consists of one full-pitch, 11-turn coil per pole pair, with the coils connected in series to form the phase winding. If the machine is operating at rated speed and under the operating conditions of Problem 4.9, calculate the rms generated voltage per phase.

4.11 The synchronous machine of Problem 4.9 has a three-phase winding with 45 series turns per phase and a winding factor $k_w = 0.928$. For the flux condition and rated speed of Problem 4.9, calculate the rms-generated voltage per phase.

4.12 The three-phase synchronous machine of Problem 4.9 is to be moved to an application which requires that its operating frequency be reduced from 60 to 50 Hz. This application requires that, for the operating condition considered in Problem 4.9, the rms generated voltage equal 13.0 kV line-to-line. As a result, the machine armature must be rewound with a different number of turns. Assuming a winding factor of $k_w = 0.928$, calculate the required number of series turns per phase.

4.13 Figure 4.44 shows a two-pole rotor revolving inside a smooth stator which carries a coil of 110 turns. The rotor produces a sinusoidal space distribution of flux at the stator surface; the peak value of the flux-density wave being 0.85 T when the current in the rotor is 15 A. The magnetic circuit is linear. The inside diameter of the stator is 11 cm, and its axial length is 0.17 m. The rotor is driven at a speed of 50 r/sec.

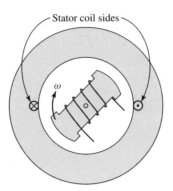

Figure 4.44 Elementary
generator for Problem 4.13.

a. The rotor is excited by a current of 15 A. Taking zero time as the instant
   when the axis of the rotor is vertical, find the expression for the
   instantaneous voltage generated in the open-circuited stator coil.

b. The rotor is now excited by a 50-Hz sinusoidal alternating currrent whose
   peak value is 15 A. Consequently, the rotor current reverses every half
   revolution; it is timed to be at its maximum just as the axis of the rotor is
   vertical (i.e., just as it becomes aligned with that of the stator coil).
   Taking zero time as the instant when the axis of the rotor is vertical, find
   the expression for the instantaneous voltage generated in the
   open-circuited stator coil. This scheme is sometimes suggested as a dc
   generator without a commutator; the thought being that if alternative half
   cycles of the alternating voltage generated in part (a) are reversed by
   reversal of the polarity of the field (rotor) winding, then a pulsating direct
   voltage will be generated in the stator. Discuss whether or not this scheme
   will work.

4.14 A three-phase two-pole winding is excited by balanced three-phase 60-Hz
currents as described by Eqs. 4.23 to 4.25. Although the winding distribution
has been designed to minimize harmonics, there remains some third and fifth
spatial harmonics. Thus the phase-*a* mmf can be written as

$$\mathcal{F}_a = i_a(A_1 \cos\theta_a + A_3 \cos 3\theta_a + A_5 \cos 5\theta_a)$$

Similar expressions can be written for phases *b* (replace θ_a by $\theta_a - 120°$) and
c (replace θ_a by $\theta_a + 120°$). Calculate the total three-phase mmf. What is the
angular velocity and rotational direction of each component of the mmf?

4.15 The nameplate of a dc generator indicates that it will produce an output
voltage of 24 V dc when operated at a speed of 1200 r/min. By what factor
must the number of armature turns be changed such that, for the same
field-flux per pole, the generator will produce an output voltage of 18 V dc at
a speed of 1400 r/min?

4.16 The armature of a two-pole dc generator has a total of 320 series turns. When operated at a speed of 1800 r/min, the open-circuit generated voltage is 240 V. Calculate Φ_p, the air-gap flux per pole.

4.17 The design of a four-pole, three-phase, 230-V, 60-Hz induction motor is to be based on a stator core of length 21 cm and inner diameter 9.52 cm. The stator winding distribution which has been selected has a winding factor $k_w = 0.925$. The armature is to be Y-connected, and thus the rated phase voltage will be $230/\sqrt{3}$ V.

    a. The designer must pick the number of armature turns so that the flux density in the machine is large enough to make efficient use of the magnetic material without being so large as to result in excessive saturation. To achieve this objective, the machine is to be designed with a peak fundamental air-gap flux density of 1.25 T. Calculate the required number of series turns per phase.

    b. For an air-gap length of 0.3 mm, calculate the self-inductance of an armature phase based upon the result of part (a) and using the inductance formulas of Appendix B. Neglect the reluctance of the rotor and stator iron and the armature leakage inductance.

4.18 A two-pole, 60-Hz, three-phase, laboratory-size synchronous generator has a rotor radius of 5.71 cm, a rotor length of 18.0 cm, and an air-gap length of 0.25 mm. The rotor field winding consists of 264 turns with a winding factor of $k_r = 0.95$. The Y-connected armature winding consists of 45 turns per phase with a winding factor $k_w = 0.93$.

    a. Calculate the flux per pole and peak fundamental air-gap flux density which will result in an open-circuit, 60-Hz armature voltage of 120 V rms/phase (line-to-neutral).

    b. Calculate the dc field current required to achieve the operating condition of part (a).

    c. Calculate the peak value of the field-winding to armature-phase-winding mutual inductance.

4.19 Write a MATLAB script which calculates the required total series field- and armature-winding turns for a three-phase, Y-connected synchronous motor given the following information:

    Rotor radius, R (meters)   Rotor length, l (meters)
    Air-gap length, g (meters)   Number of poles, poles
    Electrical frequency, f_e Peak fundamental air-gap flux density, B_{peak}
    Field-winding factor, k_f Armature-winding factor, k_w
    Rated rms open-circuit line-to-line terminal voltage, V_{rated}
    Field-current at rated-open-circuit terminal voltage, I_f

4.20 A four-pole, 60-Hz synchronous generator has a rotor length of 5.2 m, diameter of 1.24 m, and air-gap length of 5.9 cm. The rotor winding consists of a series connection of 63 turns per pole with a winding factor of $k_r = 0.91$. The peak value of the fundamental air-gap flux density is limited to 1.1 T and the rotor winding current to 2700 A. Calculate the maximum torque (N·m) and power output (MW) which can be supplied by this machine.

4.21 Thermal considerations limit the field-current of the laboratory-size synchronous generator of Problem 4.18 to a maximum value of 2.4 A. If the peak fundamental air-gap flux density is limited to a maximum of 1.3 T, calculate the maximum torque (N·m) and power (kW) which can be produced by this generator.

4.22 Figure 4.45 shows in cross section a machine having a rotor winding f and two identical stator windings a and b whose axes are in quadrature. The self-inductance of each stator winding is L_{aa} and of the rotor is L_{ff}. The air gap is uniform. The mutual inductance between a stator winding depends on the angular position of the rotor and may be assumed to be of the form

$$M_{af} = M \cos \theta_0 \quad M_{bf} = M \sin \theta_0$$

where M is the maximum value of the mutual inductance. The resistance of each stator winding is R_a.

a. Derive a general expression for the torque T in terms of the angle θ_0, the inductance parameters, and the instantaneous currents i_a, I_b, and i_f. Does this expression apply at standstill? When the rotor is revolving?

b. Suppose the rotor is stationary and constant direct currents $I_a = I_0$, $I_b = I_0$, and $I_f = 2I_0$ are supplied to the windings in the directions indicated by the dots and crosses in Fig. 4.45. If the rotor is allowed to move, will it rotate continuously or will it tend to come to rest? If the latter, at what value of θ_0?

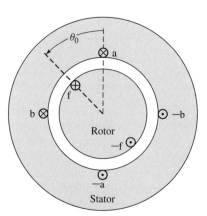

Figure 4.45 Elementary cylindrical-rotor, two-phase synchronous machine for Problem 4.22.

c. The rotor winding is now excited by a constant direct current I_f while the stator windings carry balanced two-phase currents

$$i_a = \sqrt{2}I_a \cos \omega t \quad i_b = \sqrt{2}I_a \sin \omega t$$

The rotor is revolving at synchronous speed so that its instantaneous angular position is given by $\theta_0 = \omega t - \delta$, where δ is a phase angle describing the position of the rotor at $t = 0$. The machine is an elementary two-phase synchronous machine. Derive an expression for the torque under these conditions.

d. Under the conditions of part (c), derive an expression for the instantaneous terminal voltages of stator phases a and b.

4.23 Consider the two-phase synchronous machine of Problem 4.22. Derive an expression for the torque acting on the rotor if the rotor is rotating at constant angular velocity, such that $\theta_0 = \omega t + \delta$, and the phase currents become unbalanced such that

$$i_a = \sqrt{2}I_a \cos \omega t \quad i_b = \sqrt{2}(I_a + I') \sin \omega t$$

What are the instantaneous and time-averaged torque under this condition?

4.24 Figure 4.46 shows in schematic cross section a salient-pole synchronous machine having two identical stator windings a and b on a laminated steel core. The salient-pole rotor is made of steel and carries a field winding f connected to slip rings.

Because of the nonuniform air gap, the self- and mutual inductances are functions of the angular position θ_0 of the rotor. Their variation with θ_0 can be approximated as:

$$L_{aa} = L_0 + L_2 \cos 2\theta_0 \quad L_{bb} = L_0 - L_2 \cos 2\theta_0 \quad M_{ab} = L_2 \sin 2\theta_0$$

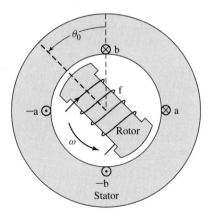

Figure 4.46 Schematic two-phase, salient-pole synchronous machine for Problem 4.24.

where L_0 and L_2 are positive constants. The mutual inductance between the rotor and the stator windings are functions of θ_0

$$M_{af} = M \cos \theta_0 \quad M_{bf} = M \sin \theta_0$$

where M is also a positive constant. The self-inductance of the field winding, L_{ff}, is constant, independent of θ_0.

Consider the operating condition in which the field winding is excited by direct current I_f and the stator windings are connected to a balanced two-phase voltage source of frequency ω. With the rotor revolving at synchronous speed, its angular position will be given by $\theta_0 = \omega t$.

Under this operating condition, the stator currents will be of the form

$$i_a = \sqrt{2} I_a \cos (\omega t + \delta) \quad i_b = \sqrt{2} I_a \sin (\omega t + \delta)$$

a. Derive an expression for the electromagnetic torque acting on the rotor.

b. Can the machine be operated as a motor and/or a generator? Explain.

c. Will the machine continue to run if the field current I_f is reduced to zero? Support you answer with an expression for the torque and an explanation as to why such operation is or is not possible.

4.25 A three-phase linear ac motor has an armature winding of wavelength 25 cm. A three-phase balanced set of currents at a frequency of 100 Hz is applied to the armature.

a. Calculate the linear velocity of the armature mmf wave.

b. For the case of a synchronous rotor, calculate the linear velocity of the rotor.

c. For the case of an induction motor operating at a slip of 0.045, calculate the linear velocity of the rotor.

4.26 The linear-motor armature of Problem 4.25 has a total active length of 7 wavelengths, with a total of 280 turns per phase with a winding factor $k_w = 0.91$. For an air-gap length of 0.93 cm, calculate the rms magnitude of the balanced three-phase currents which must be supplied to the armature to achieve a peak space-fundamental air-gap flux density of 1.45 T.

4.27 A two-phase linear permanent-magnet synchronous motor has an air-gap of length 1.0 mm, a wavelength of 12 cm, and a pole width of 4 cm. The rotor is 5 wavelengths in length. The permanent magnets on the rotor are arranged to produce an air-gap magnetic flux distribution that is uniform over the width of a pole but which varies sinusoidally in space in the direction of rotor travel. The peak density of this air-gap flux is 0.97 T.

a. Calculate the net flux per pole.

b. Each armature phase consists of 10 turns per pole, with all the poles connected in series. Assuming that the armature winding extends many wavelengths past either end of the rotor, calculate the peak flux linkages of the armature winding.

c. If the rotor is traveling at a speed of 6.3 m/sec, calculate the rms voltage induced in the armature winding.

CHAPTER 5

Synchronous Machines

A s we have seen in Section 4.2.1, a synchronous machine is an ac machine whose speed under steady-state conditions is proportional to the frequency of the current in its armature. The rotor, along with the magnetic field created by the dc field current on the rotor, rotates at the same speed as, or in synchronism with, the rotating magnetic field produced by the armature currents, and a steady torque results. An elementary picture of how a synchronous machine works is given in Section 4.2.1 with emphasis on torque production in terms of the interactions between the machine's magnetic fields.

Analytical methods of examining the steady-state performance of polyphase synchronous machines will be developed in this chapter. Initial consideration will be given to cylindrical-rotor machines; the effects of salient poles are taken up in Sections 5.6 and 5.7.

5.1 INTRODUCTION TO POLYPHASE SYNCHRONOUS MACHINES

As indicated in Section 4.2.1, a synchronous machine is one in which alternating current flows in the armature winding, and dc excitation is supplied to the field winding. The armature winding is almost invariably on the stator and is usually a three-phase winding, as discussed in Chapter 4. The field winding is on the rotor. The cylindrical-rotor construction shown in Figs. 4.10 and 4.11 is used for two- and four-pole turbine generators. The salient-pole construction shown in Fig. 4.9 is best adapted to multipolar, slow-speed, hydroelectric generators and to most synchronous motors. The dc power required for excitation—approximately one to a few percent of the rating of the synchronous machine—is supplied by the *excitation system*.

In older machines, the excitation current was typically supplied through *slip rings* from a dc machine, referred to as the *exciter,* which was often mounted on the same shaft as the synchronous machine. In more modern systems, the excitation is supplied from ac exciters and solid-state rectifiers (either simple diode bridges or phase-controlled rectifiers). In some cases, the rectification occurs in the stationary frame, and the rectified excitation current is fed to the rotor via slip rings. In other systems, referred to as *brushless excitation systems,* the alternator of the ac exciter is on the rotor, as is the rectification system, and the current is supplied directly to the field-winding without the need for slip rings. One such system is described in Appendix D.

As is discussed in Chapter 4, a single synchronous generator supplying power to an impedance load acts as a voltage source whose frequency is determined by the speed of its mechanical drive (or *prime mover*), as can be seen from Eq. 4.2. From Eqs. 4.42, 4.44, and 4.50, the amplitude of the generated voltage is proportional to the frequency and the field current. The current and power factor are then determined by the generator field excitation and the impedance of the generator and load.

Synchronous generators can be readily operated in parallel, and, in fact, the electricity supply systems of industrialized countries typically have scores or even hundreds of them operating in parallel, interconnected by thousands of miles of transmission lines, and supplying electric energy to loads scattered over areas of many thousands of square miles. These huge systems have grown in spite of the necessity for designing the system so that synchronism is maintained following disturbances and the problems, both technical and administrative, which must be solved to coordinate the operation of such a complex system of machines and personnel. The principal reasons for these interconnected systems are reliability of service and economies in plant investment and operating costs.

When a synchronous generator is connected to a large interconnected system containing many other synchronous generators, the voltage and frequency at its armature terminals are substantially fixed by the system. As a result, armature currents will produce a component of the air-gap magnetic field which rotates at synchronous speed (Eq. 4.41) as determined by the system electrical frequency f_e. As is discussed in Chapter 4, for the production of a steady, unidirectional electromechanical torque, the fields of the stator and rotor must rotate at the same speed, and therefore the rotor must turn at precisely synchronous speed. Because any individual generator is a small fraction of the total system generation, it cannot significantly affect the system voltage or frequency. It is thus often useful, when studying the behavior of an individual generator or group of generators, to represent the remainder of the system as a constant-frequency, constant-voltage source, commonly referred to as an *infinite bus.*

Many important features of synchronous-machine behavior can be understood from the analysis of a single machine connected to an infinite bus. The steady-state behavior of a synchronous machine can be visualized in terms of the torque equation. From Eq. 4.81, with changes in notation appropriate to synchronous-machine theory,

$$T = \frac{\pi}{2}\left(\frac{\text{poles}}{2}\right)^2 \Phi_R F_f \sin\delta_{RF} \tag{5.1}$$

where

$$\Phi_R = \text{resultant air-gap flux per pole}$$
$$F_f = \text{mmf of the dc field winding}$$
$$\delta_{RF} = \text{electrical phase angle between magnetic axes of } \Phi_R \text{ and } F_f$$

The minus sign of Eq. 4.81 has been omitted with the understanding that the electromechanical torque acts in the direction to bring the interacting fields into alignment. In normal steady-state operation, the electromechanical torque balances the mechanical torque applied to the shaft. In a generator, the prime-mover torque acts in the direction of rotation of the rotor, pushing the rotor mmf wave ahead of the resultant air-gap flux. The electromechanical torque then opposes rotation. The opposite situation exists in a synchronous motor, where the electromechanical torque is in the direction of rotation, in opposition to the retarding torque of the mechanical load on the shaft.

Variations in the electromechanical torque result in corresponding variations in the *torque angle, δ_{RF}*, as seen from Eq. 5.1. The relationship is shown in the form of a *torque-angle curve* in Fig. 5.1, where the field current (rotor mmf) and resultant air-gap flux are assumed constant. Positive values of torque represent generator action, corresponding to positive values of δ_{RF} for which the rotor mmf wave leads the resultant air-gap flux.

As the prime-mover torque is increased, the magnitude of δ_{RF} must increase until the electromechanical torque balances the shaft torque. The readjustment process is actually a dynamic one, requiring a change in the mechanical speed of the rotor, typically accompanied by a damped mechanical oscillation of the rotor about its new steady-state torque angle. This oscillation is referred to as a *hunting transient*. In a practical machine undergoing such a transient, some changes in the amplitudes of the resultant flux-density and field-winding mmf wave may also occur because of various factors such as saturation effects, the effect of the machine leakage impedance, the response of the machine's excitation system, and so on. To emphasize the fundamental

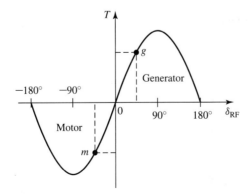

Figure 5.1 Torque-angle characteristic.

principles of synchronous-machine operation, such effects will be neglected in the present discussion.

The adjustment of the rotor to its new angular position following a load change can be observed experimentally in the laboratory by viewing the machine rotor with stroboscopic light triggered from the applied armature voltage (thus having a flashing frequency which causes the rotor to appear stationary when it is turning at its normal synchronous speed). Alternatively, electronic sensors can be used to determine the shaft position relative to the synchronous reference frame associated with the stator voltage. The resultant signal can be displayed on an oscilloscope or recorded with a data-acquisition system.

As can be seen from Fig. 5.1, an increase in prime-mover torque will result in a corresponding increase in the torque angle. When δ_{RF} becomes $90°$, the electromechanical torque reaches its maximum value, known as the *pull-out torque*. Any further increase in prime-mover torque cannot be balanced by a corresponding increase in synchronous electromechanical torque, with the result that synchronism will no longer be maintained and the rotor will speed up. This phenomenon is known as *loss of synchronism* or *pulling out of step*. Under these conditions, the generator is usually disconnected from the external electrical system by the automatic operation of circuit breakers, and the prime mover is quickly shut down to prevent dangerous overspeed. Note from Eq. 5.1 that the value of the pull-out torque can be increased by increasing either the field current or the resultant air-gap flux. However, this cannot be done without limit; the field current is limited by the ability to cool the field winding, and the air-gap flux is limited by saturation of the machine iron.

As seen from Fig. 5.1, a similar situation occurs in a synchronous motor for which an increase in the shaft load torque beyond the pull-out torque will cause the rotor to lose synchronism and thus to slow down. Since a synchronous motor develops torque only at synchronous speed, it cannot be started simply by the application of armature voltages of rated frequency. In some cases, a squirrel-cage structure is included in the rotor, and the motor can be started as an induction motor and then synchronized when it is close to synchronous speed.

5.2 SYNCHRONOUS-MACHINE INDUCTANCES; EQUIVALENT CIRCUITS

In Section 5.1, synchronous-machine torque-angle characteristics are described in terms of the interacting air-gap flux and mmf waves. Our purpose now is to derive an equivalent circuit which represents the steady-state terminal volt-ampere characteristics.

A cross-sectional sketch of a three-phase cylindrical-rotor synchronous machine is shown schematically in Fig. 5.2. The figure shows a two-pole machine; alternatively, this can be considered as two poles of a multipole machine. The three-phase armature winding on the stator is of the same type used in the discussion of rotating magnetic fields in Section 4.5. Coils aa', bb', and cc' represent distributed windings producing sinusoidal mmf and flux-density waves in the air gap. The reference directions for

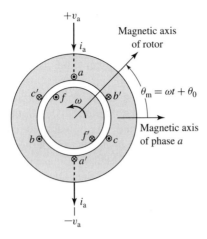

Figure 5.2 Schematic diagram of a two-pole, three-phase cylindrical-rotor synchronous machine.

the currents are shown by dots and crosses. The field winding ff' on the rotor also represents a distributed winding which produces a sinusoidal mmf and flux-density wave centered on its magnetic axis and rotating with the rotor.

When the flux linkages with armature phases a, b, c and field winding f are expressed in terms of the inductances and currents as follows,

$$\lambda_a = \mathcal{L}_{aa}i_a + \mathcal{L}_{ab}i_b + \mathcal{L}_{ac}i_c + \mathcal{L}_{af}i_f \tag{5.2}$$

$$\lambda_b = \mathcal{L}_{ba}i_a + \mathcal{L}_{bb}i_b + \mathcal{L}_{bc}i_c + \mathcal{L}_{bf}i_f \tag{5.3}$$

$$\lambda_c = \mathcal{L}_{ca}i_a + \mathcal{L}_{cb}i_b + \mathcal{L}_{cc}i_c + \mathcal{L}_{cf}i_f \tag{5.4}$$

$$\lambda_f = \mathcal{L}_{fa}i_a + \mathcal{L}_{fb}i_b + \mathcal{L}_{fc}i_c + \mathcal{L}_{ff}i_f \tag{5.5}$$

the induced voltages can be found from Faraday's law. Here, two like subscripts denote a self-inductance, and two unlike subscripts denote a mutual inductance between the two windings. The script $\mathcal{L}$ is used to indicate that, in general, both the self- and mutual inductances of a three-phase machine may vary with rotor angle, as is seen, for example, in Section C.2, where the effects of salient poles are analyzed.

Before we proceed, it is useful to investigate the nature of the various inductances. Each of these inductances can be expressed in terms of constants which can be computed from design data or measured by tests on an existing machine.

5.2.1 Rotor Self-Inductance

With a cylindrical stator, the self-inductance of the field winding is independent of the rotor position θ_m when the harmonic effects of stator slot openings are neglected. Hence

$$\mathcal{L}_{ff} = L_{ff} = L_{ff0} + L_{fl} \tag{5.6}$$

where the italic L is used for an inductance which is independent of θ_m. The component L_{ff0} corresponds to that portion of $\mathcal{L}_{\mathrm{ff}}$ due to the space-fundamental component of air-gap flux. This component can be computed from air-gap dimensions and winding data, as shown in Appendix B. The additional component L_{fl} accounts for the field-winding leakage flux.

Under transient or unbalanced conditions, the flux linkages with the field winding, Eq. 5.5, vary with time, and the voltages induced in the rotor circuits have an important effect on machine performance. With balanced three-phase armature currents, however, the constant-amplitude magnetic field of the armature currents rotates in synchronism with the rotor. Thus the field-winding flux linkages produced by the armature currents do not vary with time, and the voltage induced in the field winding is therefore zero. With constant dc voltage V_f applied to the field-winding terminals, the field direct current I_f is given by Ohm's law, $I_f = V_f/R_f$.

5.2.2 Stator-to-Rotor Mutual Inductances

The stator-to-rotor mutual inductances vary periodically with θ_{me}, the electrical angle between the magnetic axes of the field winding and the armature phase a as shown in Fig. 5.2 and as defined by Eq. 4.54. With the space-mmf and air-gap flux distribution assumed sinusoidal, the mutual inductance between the field winding f and phase a varies as $\cos \theta_{\mathrm{me}}$; thus

$$\mathcal{L}_{\mathrm{af}} = \mathcal{L}_{\mathrm{fa}} = L_{\mathrm{af}} \cos \theta_{\mathrm{me}} \tag{5.7}$$

Similar expressions apply to phases b and c, with θ_{me} replaced by $\theta_{\mathrm{me}} - 120°$ and $\theta_{\mathrm{me}} + 120°$, respectively. Attention will be focused on phase a. The inductance L_{af} can be calculated as discussed in Appendix B.

With the rotor rotating at synchronous speed ω_s (Eq. 4.40), the rotor angle will vary as

$$\theta_m = \omega_s t + \delta_0 \tag{5.8}$$

where δ_0 is the angle of the rotor at time $t = 0$. From Eq. 4.54

$$\theta_{\mathrm{me}} = \left(\frac{\text{poles}}{2}\right)\theta_m = \omega_e t + \delta_{e0} \tag{5.9}$$

Here, $\omega_e = (\text{poles}/2)\omega_s$ is the electrical frequency and δ_{e0} is the electrical angle of the rotor at time $t = 0$.

Thus, substituting into Eq. 5.7 gives

$$\mathcal{L}_{\mathrm{af}} = \mathcal{L}_{\mathrm{fa}} = L_{\mathrm{af}} \cos (\omega_e t + \delta_{e0}) \tag{5.10}$$

5.2.3 Stator Inductances; Synchronous Inductance

With a cylindrical rotor, the air-gap geometry is independent of θ_m if the effects of rotor slots are neglected. The stator self-inductances then are constant; thus

$$\mathcal{L}_{\mathrm{aa}} = \mathcal{L}_{\mathrm{bb}} = \mathcal{L}_{\mathrm{cc}} = L_{\mathrm{aa}} = L_{\mathrm{aa0}} + L_{\mathrm{al}} \tag{5.11}$$

where L_{aa0} is the component of self-inductance due to space-fundamental air-gap flux (Appendix B) and L_{al} is the additional component due to armature leakage flux (see Section 4.10).

The armature phase-to-phase mutual inductances can be found on the assumption that the mutual inductance is due solely to space-fundamental air-gap flux.[1] From Eq. B.28 of Appendix B, we see that the air-gap mutual inductance of two identical windings displaced by α electrical degrees is equal to the air-gap component of their self-inductance multiplied by $\cos \alpha$. Thus, because the armature phases are displaced by $120°$ electrical degrees and $\cos (\pm 120°) = -\frac{1}{2}$, the mutual inductances between the armature phases are equal and given by

$$\mathcal{L}_{ab} = \mathcal{L}_{ba} = \mathcal{L}_{ac} = \mathcal{L}_{ca} = \mathcal{L}_{bc} = \mathcal{L}_{cb} = -\frac{1}{2} L_{aa0} \tag{5.12}$$

Substituting Eqs. 5.11 and 5.12 for the self- and mutual inductances into the expression for the phase-a flux linkages (Eq. 5.2) gives

$$\lambda_a = (L_{aa0} + L_{al})i_a - \frac{1}{2} L_{aa0}(i_b + i_c) + \mathcal{L}_{af}i_f \tag{5.13}$$

Under balanced three-phase armature currents (see Fig. 4.30 and Eqs. 4.23 to 4.25)

$$i_a + i_b + i_c = 0 \tag{5.14}$$

$$i_b + i_c = -i_a \tag{5.15}$$

Substitution of Eq. 5.15 into Eq. 5.13 gives

$$\lambda_a = (L_{aa0} + L_{al})i_a + \frac{1}{2} L_{aa0}i_a + \mathcal{L}_{af}i_f$$

$$= \left(\frac{3}{2} L_{aa0} + L_{al}\right) i_a + \mathcal{L}_{af}i_f \tag{5.16}$$

It is useful to define the *synchronous inductance* L_s as

$$L_s = \frac{3}{2} L_{aa0} + L_{al} \tag{5.17}$$

and thus

$$\lambda_a = L_s i_a + \mathcal{L}_{af}i_f \tag{5.18}$$

Note that the synchronous inductance L_s is the *effective inductance seen by phase a under steady-state, balanced three-phase machine operating conditions*. It is made up of three components. The first, L_{aa0}, is due to the space-fundamental

---

[1] Since the armature windings in practical machines are generally wound with overlapping phase windings (i.e., portions of adjacent windings share the same slots), there is an additional component of the phase-to-phase mutual inductance which is due to slot leakage flux. This component is relatively small and is neglected in most analyses.

air-gap component of the phase-*a* self-flux linkages. The second, L_{al}, known as the armature-winding *leakage inductance,* is due to the leakage component of phase-*a* flux linkages. The third component, $\frac{1}{2}L_{aa0}$, is due to the phase-*a* flux linkages from the space-fundamental component of air-gap flux produced by currents in phases *b* and *c*. Under balanced three-phase conditions, the phase-*b* and -*c* currents are related to the current in phase *a* by Eq. 5.15. Thus the synchronous inductance is an apparent inductance in that it accounts for the flux linkages of phase *a* in terms of the current in phase *a*, even though some of this flux linkage is due to currents in phases *a* and *b*. Hence, although synchronous inductance appears in that role in Eq. 5.18, it is not the self-inductance of phase *a* alone.

The significance of the synchronous inductance can be further appreciated with reference to the discussion of rotating magnetic fields in Section 4.5.2, where it was shown that under balanced three-phase conditions, the armature currents create a rotating magnetic flux wave in the air gap of magnitude equal to $\frac{3}{2}$ times the magnitude of that due to phase *a* alone, the additional component being due to the phase-*b* and -*c* currents. This corresponds directly to the $\frac{3}{2}L_{aa0}$ component of the synchronous inductance in Eq. 5.17; this component of the synchronous inductance accounts for the total space-fundamental air-gap component of phase-*a* flux linkages produced by the three armature currents under balanced three-phase conditions.

5.2.4 Equivalent Circuit

The phase-*a* terminal voltage is the sum of the armature-resistance voltage drop $R_a i_a$ and the induced voltage. The voltage e_{af} induced by the field winding flux (often referred to as the *generated voltage* or *internal voltage*) can be found from the time derivative of Eq. 5.18 with the armature current i_a set equal to zero. With dc excitation I_f in the field winding, substitution of Eq. 5.10 gives

$$e_{af} = \frac{d}{dt}(\mathcal{L}_{af}i_f) = -\omega_e L_{af} I_f \sin(\omega_e t + \delta_{e0}) \tag{5.19}$$

Using Eq. 5.18, the terminal voltage can then be expressed as

$$v_a = R_a i_a + \frac{d\lambda_a}{dt}$$

$$= R_a i_a + L_s \frac{di_a}{dt} + e_{af} \tag{5.20}$$

The generated voltage e_{af} of Eq. 5.19 is at frequency ω_e, equal to the electrical frequency of the generator terminal voltage. Its rms amplitude is given by

$$E_{af} = \frac{\omega_e L_{af} I_f}{\sqrt{2}} \tag{5.21}$$

Under this synchronous operating condition, all machine armature quantities (current and flux linkage) will also vary sinusoidally in time at frequency ω_e. Thus, we can write the above equations in term of their rms, complex amplitudes. From Eq. 5.19 we can write the rms complex amplitude of the generated voltage as

$$\hat{E}_{af} = j\left(\frac{\omega_e L_{af} I_f}{\sqrt{2}}\right)e^{j\delta_{e0}} \tag{5.22}$$

Similarly, the terminal-voltage equation, Eq. 5.20, can be written in terms of rms complex amplitudes as

$$\hat{V}_a = R_a \hat{I}_a + j X_s \hat{I}_a + \hat{E}_{af} \tag{5.23}$$

where $X_s = \omega_e L_s$ is known as the *synchronous reactance*.

An equivalent circuit in complex form is shown in Fig. 5.3a. The reader should note that Eq. 5.23 and Fig. 5.3a are written with the reference direction for $\hat{I}_a$ defined as positive into the machine terminals. This is known as the *motor reference direction* for the current.

Alternatively, the *generator reference direction* is defined with the reference direction for $\hat{I}_a$ chosen as positive out of the machine terminals, as shown in Fig. 5.3b. Under this choice of current reference direction, Eq. 5.23 becomes

$$\hat{V}_a = -R_a \hat{I}_a - j X_s \hat{I}_a + \hat{E}_{af} \tag{5.24}$$

Note that these two representations are equivalent; when analyzing a particular synchronous-machine operating condition the actual current will be the same. The sign of I_a will simply be determined by the choice of reference direction. Either choice is acceptable, independent of whether the synchronous machine under investigation is operating as a motor or a generator. However, since power tends to flow into a motor, it is perhaps intuitively more satisfying to choose a reference direction with current flowing into the machine for the analysis of motor operation. The opposite is true for generator operation, for which power tends to flow out of the machine. Most of the synchronous-machine analysis techniques presented here were first developed to analyze the performance of synchronous generators in electric power systems. As a result, the generator reference direction is more common and is generally used from this point on in the text.

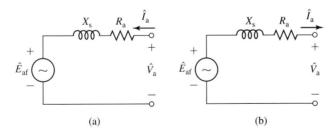

(a)   (b)

Figure 5.3 Synchronous-machine equivalent circuits: (a) motor reference direction and (b) generator reference direction.

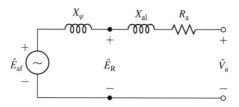

Figure 5.4 Synchronous-machine equivalent circuit showing air-gap and leakage components of synchronous reactance and air-gap voltage.

Figure 5.4 shows an alternative form of the equivalent circuit in which the synchronous reactance is shown in terms of its components. From Eq. 5.17

$$X_s = \omega_e L_s = \omega_e L_{al} + \omega_e \left(\frac{3}{2} L_{aa0}\right)$$

$$= X_{al} + X_\varphi \tag{5.25}$$

where $X_{al} = \omega L_{al}$ is the armature *leakage reactance* and $X_\varphi = \omega(\frac{3}{2}L_{aa0})$ is the reactance corresponding to the rotating space-fundamental air-gap flux produced by the three armature currents. The reactance X_φ is the effective *magnetizing reactance* of the armature winding under balanced three-phase conditions. The rms voltage $\hat{E}_R$ is the internal voltage generated by the resultant air-gap flux and is usually referred to as the *air-gap voltage* or the *voltage "behind" leakage reactance*.

It is important to recognize that the equivalent circuits of Figs. 5.3 and 5.4 are *single-phase, line-to-neutral equivalent circuits for a three-phase machine operating under balanced, three-phase conditions*. Thus, once the phase-a voltages and currents are found, either from the equivalent circuit or directly from the voltage equations (Eqs. 5.23 and 5.24), the currents and voltages for phases b and c can be found simply by phase-shifting those of phase a by $-120°$ and $120°$ respectively. Similarly, the total three-phase power of the machine can be found simply by multiplying that of phase a by three, unless the analysis is being performed in per unit (see Section 2.9), in which case the three-phase, per-unit power is equal to that found from solving for phase a alone and the factor of three is not needed.

EXAMPLE 5.1

A 60-Hz, three-phase synchronous motor is observed to have a terminal voltage of 460 V (line-line) and a terminal current of 120 A at a power factor of 0.95 lagging. The field-current under this operating condition is 47 A. The machine synchronous reactance is equal to 1.68 Ω (0.794 per unit on a 460-V, 100-kVA, 3-phase base). Assume the armature resistance to be negligible.

Calculate (*a*) the generated voltage E_{af} in volts, (*b*) the magnitude of the field-to-armature mutual inductance L_{af}, and (*c*) the electrical power input to the motor in kW and in horsepower.

■ **Solution**

a. Using the motor reference direction for the current and neglecting the armature resistance, the generated voltage can be found from the equivalent circuit of Fig. 5.3a or Eq. 5.23 as

$$\hat{E}_{af} = \hat{V}_a - jX_s\hat{I}_a$$

We will choose the terminal voltage as our phase reference. Because this is a line-to-neutral equivalent, the terminal voltage V_a must be expressed as a line-to-neutral voltage

$$\hat{V}_a = \frac{460}{\sqrt{3}} = 265.6 \text{ V, line-to-neutral}$$

A lagging power factor of 0.95 corresponds to a power factor angle $\theta = -\cos^{-1}(0.95) = -18.2°$. Thus, the phase-$a$ current is

$$\hat{I}_a = 120 \, e^{-j18.2°} \text{ A}$$

Thus

$$\hat{E}_{af} = 265.6 - j1.68(120 \, e^{-j18.2°})$$

$$= 278.8 \, e^{-j43.4°} \text{ V, line-to-neutral}$$

and hence, the generated voltage E_{af} is equal to 278.8 V rms, line-to-neutral.

b. The field-to-armature mutual inductance can be found from Eq. 5.21. With $\omega_e = 120\pi$,

$$L_{af} = \frac{\sqrt{2} \, E_{af}}{\omega_e I_f} = \frac{\sqrt{2} \times 279}{120\pi \times 47} = 22.3 \text{ mH}$$

c. The three-phase power input to the motor P_{in} can be found as three times the power input to phase a. Hence,

$$P_{in} = 3V_aI_a(\text{power factor}) = 3 \times 265.6 \times 120 \times 0.95$$

$$= 90.8 \text{ kW} = 122 \text{ hp}$$

EXAMPLE 5.2

Assuming the input power and terminal voltage for the motor of Example 5.1 remain constant, calculate (*a*) the phase angle δ of the generated voltage and (*b*) the field current required to achieve unity power factor at the motor terminals.

■ **Solution**

a. For unity power factor at the motor terminals, the phase-a terminal current will be in phase with the phase-a line-to-neutral voltage $\hat{V}_a$. Thus

$$\hat{I}_a = \frac{P_{in}}{3V_a} = \frac{90.6 \text{ kW}}{3 \times 265.6 \text{ V}} = 114 \text{ A}$$

From Eq. 5.23,

$$\hat{E}_{af} = \hat{V}_a - jX_s\hat{I}_a$$

$$= 265.6 - j1.68 \times 114 = 328 \, e^{-j35.8°} \text{ V, line-to-neutral}$$

Thus, $E_{af} = 328$ V line-to-neutral and $\delta = -35.8°$.

b. Having found L_{af} in Example 5.1, we can find the required field current from Eq. 5.21.

$$I_f = \frac{\sqrt{2}\,E_{af}}{\omega_e L_{af}} = \frac{\sqrt{2} \times 328}{377 \times 0.0223} = 55.2\ \text{A}$$

Practice Problem 5.1

The synchronous machine of Examples 5.1 and 5.2 is to be operated as a synchronous generator. For operation at 60 Hz with a terminal voltage of 460 V line-to-line, calculate the field current required to supply a load of 85 kW, 0.95 power-factor leading.

Solution

46.3 A

It is helpful to have a rough idea of the order of magnitude of the impedance components. For machines with ratings above a few hundred kVA, the armature-resistance voltage drop at rated current usually is less than 0.01 times rated voltage; i.e., the armature resistance usually is less than 0.01 per unit on the machine rating as a base. (The per-unit system is described in Section 2.9.) The armature leakage reactance usually is in the range of 0.1 to 0.2 per unit, and the synchronous reactance is typically in the range of 1.0 to 2.0 per unit. In general, the per-unit armature resistance increases and the per-unit synchronous reactance decreases with decreasing size of the machine. In small machines, such as those in educational laboratories, the armature resistance may be in the vicinity of 0.05 per unit and the synchronous reactance in the vicinity of 0.5 per unit. In all but small machines, the armature resistance can usually be neglected in most analyses, except insofar as its effect on losses and heating is concerned.

5.3 OPEN- AND SHORT-CIRCUIT CHARACTERISTICS

The fundamental characteristics of a synchronous machine can be determined by a pair of tests, one made with the armature terminals open-circuited and the second with the armature terminals short-circuited. These tests are discussed here. Except for a few remarks on the degree of validity of certain assumptions, the discussions apply to both cylindrical-rotor and salient-pole machines.

5.3.1 Open-Circuit Saturation Characteristic and No-Load Rotational Losses

Like the magnetization curve for a dc machine, the *open-circuit characteristic* (also referred to as the *open-circuit saturation curve*) of a synchronous machine is a curve of the open-circuit armature terminal voltage (either in volts or in per unit) as a function of the field excitation when the machine is running at synchronous speed, as shown by curve *occ* in Fig. 5.5. Typically, the base voltage is chosen equal to the rated voltage of the machine.

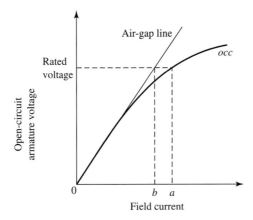

Figure 5.5 Open-circuit characteristic of a synchronous machine.

The open-circuit characteristic represents the relation between the space-fundamental component of the air-gap flux and the mmf acting on the magnetic circuit when the field winding constitutes the only mmf source. Note that the effects of magnetic saturation can be clearly seen; the characteristic bends downward with increasing field current as saturation of the magnetic material increases the reluctance of the flux paths in the machine and reduces the effectiveness of the field current in producing magnetic flux. As can be seen from Fig. 5.5, the open-circuit characteristic is initially linear as the field current is increased from zero. This portion of the curve (and its linear extension for higher values of field current) is known as the *air-gap line*. It represents the machine open-circuit voltage characteristic corresponding to unsaturated operation. Deviations of the actual open-circuit characteristic from this curve are a measure of the degree of saturation in the machine.

Note that with the machine armature winding open-circuited, the terminal voltage is equal to the generated voltage E_{af}. Thus the open-circuit characteristic is a measurement of the relationship between the field current I_f and E_{af}. It can therefore provide a direct measurement of the field-to-armature mutual inductance L_{af}.

EXAMPLE 5.3

An open-circuit test performed on a three-phase, 60-Hz synchronous generator shows that the rated open-circuit voltage of 13.8 kV is produced by a field current of 318 A. Extrapolation of the air-gap line from a complete set of measurements on the machine shows that the field-current corresponding to 13.8 kV on the air-gap line is 263 A. Calculate the saturated and unsaturated values of L_{af}.

■ **Solution**
From Eq. 5.21, L_{af} is found from

$$L_{af} = \frac{\sqrt{2}\,E_{af}}{\omega_e I_f}$$

Here, $E_{af} = 13.8 \text{ kV}/\sqrt{3} = 7.97 \text{ kV}$. Hence the saturated value of L_{af} is given by

$$(L_{af})_{sat} = \frac{\sqrt{2}(7.97 \times 10^3)}{377 \times 318} = 94 \text{ mH}$$

and the unsaturated value is

$$(L_{af})_{unsat} = \frac{\sqrt{2}(7.97 \times 10^3)}{377 \times 263} = 114 \text{ mH}$$

In this case, we see that saturation reduces the magnetic coupling between the field and armature windings by approximately 18 percent.

Practice Problem 5.2

If the synchronous generator of Example 5.3 is operated at a speed corresponding to a generated voltage of 50 Hz, calculate (*a*) the open-circuit line-to-line terminal voltage corresponding to a field current of 318 A and (*b*) the field-current corresponding to that same voltage on the 50-Hz air-gap line.

Solution

   a. 11.5 kV
   b. 263 A

When the machine is an existing one, the open-circuit characteristic is usually determined experimentally by driving the machine mechanically at synchronous speed with its armature terminals on open circuit and by reading the terminal voltage corresponding to a series of values of field current. If the mechanical power required to drive the synchronous machine during the open-circuit test is measured, the *no-load rotational losses* can be obtained. These losses consist of friction and windage losses associated with rotation as well as the core loss corresponding to the flux in the machine at no load. The friction and windage losses at synchronous speed are constant, while the open-circuit core loss is a function of the flux, which in turn is proportional to the open-circuit voltage.

The mechanical power required to drive the machine at synchronous speed and unexcited is its friction and windage loss. When the field is excited, the mechanical power equals the sum of the friction, windage, and open-circuit core loss. The open-circuit core loss therefore can be found from the difference between these two values of mechanical power. A typical curve of open-circuit core loss as a function of open-circuit voltage takes the form of that found in Fig. 5.6.

5.3.2   Short-Circuit Characteristic and Load Loss

A short-circuit characteristic can be obtained by applying a three-phase short circuit through suitable current sensors to the armature terminals of a synchronous machine. With the machine driven at synchronous speed, the field current can be increased

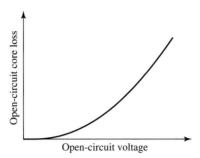

Figure 5.6 Typical form of an open-circuit core-loss curve.

and a plot of armature current versus field current can be obtained. This relation is known as the *short-circuit characteristic*. An open-circuit characteristic *occ* and a short-circuit characteristic *scc* are shown in Fig. 5.7.

With the armature short-circuited, $V_a = 0$ and, from Eq. 5.24 (using the generator reference direction for current)

$$\hat{E}_{af} = \hat{I}_a(R_a + jX_s) \qquad (5.26)$$

The corresponding phasor diagram is shown in Fig. 5.8. Because the resistance is much smaller than the synchronous reactance, the armature current lags the excitation voltage by very nearly 90°. Consequently the armature-reaction-mmf wave is very nearly in line with the axis of the field poles and in opposition to the field mmf, as shown by phasors $\hat{A}$ and $\hat{F}$ representing the space waves of armature reaction and field mmf, respectively.

The resultant mmf creates the resultant air-gap flux wave which generates the air-gap voltage $\hat{E}_R$ (see Fig. 5.4) equal to the voltage consumed in armature resistance

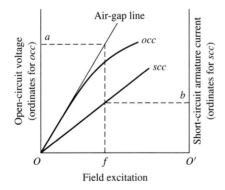

Figure 5.7 Open- and short-circuit characteristics of a synchronous machine.

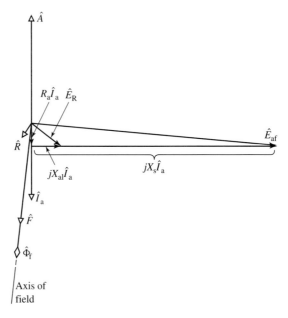

Figure 5.8 Phasor diagram for short-circuit conditions.

R_a and leakage reactance X_{al}; as an equation,

$$\hat{E}_R = \hat{I}_a(R_a + jX_{al}) \tag{5.27}$$

In many synchronous machines the armature resistance is negligible, and the leakage reactance is between 0.10 and 0.20 per unit; a representative value is about 0.15 per unit. That is, at rated armature current the leakage reactance voltage drop is about 0.15 per unit. From Eq. 5.27, therefore, the air-gap voltage at rated armature current on short circuit is about 0.15 per unit; i.e., the resultant air-gap flux is only about 0.15 times its normal voltage value. Consequently, the machine is operating in an unsaturated condition. The short-circuit armature current, therefore, is directly proportional to the field current over the range from zero to well above rated armature current; it is thus a straight line as can be seen in Fig. 5.7.

The *unsaturated synchronous reactance* (corresponding to unsaturated operating conditions within the machine) can be found from the open- and short-circuit characteristics. At any convenient field excitation, such as Of in Fig. 5.7, the armature current on short circuit is $O'b$, and the unsaturated generated voltage for the same field current corresponds to Oa, as read from the air-gap line. Note that the voltage on the air-gap line should be used because the machine is assumed to be operating in an unsaturated condition. If the line-to-neutral voltage corresponding to Oa is $V_{a,ag}$ and the armature current per phase corresponding to $O'b$ is $I_{a,sc}$, then from Eq. 5.26, with armature resistance neglected, the unsaturated synchronous reactance $X_{s,u}$ is

$$X_{s,u} = \frac{V_{a,ag}}{I_{a,sc}} \tag{5.28}$$

where the subscripts "ag" and "sc" indicate air-gap line conditions and short-circuit conditions, respectively. If $V_{a,ag}$ and $I_{a,sc}$ are expressed in per unit, the synchronous reactance will be in per unit. If $V_{a,ag}$ and $I_{a,sc}$ are expressed in rms line-to-neutral volts and rms amperes per phase, respectively, the synchronous reactance will be in ohms per phase.

Note that the synchronous reactance in ohms is calculated by using the phase or line-to-neutral voltage. Often the open-circuit saturation curve is given in terms of the line-to-line voltage, in which case the voltage must be converted to the line-to-neutral value by dividing by $\sqrt{3}$.

For operation at or near rated terminal voltage, it is sometimes assumed that the machine is equivalent to an unsaturated one whose magnetization line is a straight line through the origin and the rated-voltage point on the open-circuit characteristic, as shown by the dashed line Op in Fig. 5.9. According to this approximation, the *saturated value of the synchronous reactance* at rated voltage $V_{a,rated}$ is

$$X_s = \frac{V_{a,rated}}{I'_a} \tag{5.29}$$

where I'_a is the armature current $O'c$ read from the short-circuit characteristic at the field current Of' corresponding to $V_{a,rated}$ on the open-circuit characteristic, as shown in Fig. 5.9. As with the unsaturated synchronous reactance, if $V_{a,rated}$ and I'_a are expressed in per unit, the synchronous reactance will be in per unit. If $V_{a,rated}$ and I'_a are expressed in rms line-to-neutral volts and rms amperes per phase, respectively, the synchronous reactance will be in ohms per phase. This method of handling the

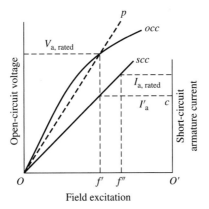

Figure 5.9 Open- and short-circuit characteristics showing equivalent magnetization line for saturated operating conditions.

effects of saturation, which assumes that the effects of saturation can be described by a single value of saturated reactance, usually gives satisfactory results when great accuracy is not required.

The *short-circuit ratio* (SCR) is defined as the ratio of the field current required for rated voltage on open circuit to the field current required for rated armature current on short circuit. That is, in Fig. 5.9

$$SCR = \frac{Of'}{Of''} \tag{5.30}$$

It can be shown that the SCR is the reciprocal of the per-unit value of the saturated synchronous reactance found from Eq. 5.29. It is common to refer to the field current Of' required to achieve rated-open-circuit voltage as AFNL (*Amperes Field No Load*) and the field current Of'' required to achieve rated-short-circuit current as AFSC (*Amperes Field Short Circuit*). Thus, the short-circuit ratio can also be written as

$$SCR = \frac{AFNL}{AFSC} \tag{5.31}$$

EXAMPLE 5.4

The following data are taken from the open- and short-circuit characteristics of a 45-kVA, three-phase, Y-connected, 220-V (line-to-line), six-pole, 60-Hz synchronous machine. From the open-circuit characteristic:

Line-to-line voltage = 220 V   Field current = 2.84 A

From the short-circuit characteristic:

Armature current, A	118	152
Field current, A	2.20	2.84

From the air-gap line:

Field current = 2.20 A   Line-to-line voltage = 202 V

Compute the unsaturated value of the synchronous reactance, its saturated value at rated voltage in accordance with Eq. 5.29, and the short-circuit ratio. Express the synchronous reactance in ohms per phase and in per unit on the machine rating as a base.

■ Solution

At a field current of 2.20 A the line-to-neutral voltage on the air-gap line is

$$V_{a,ag} = \frac{202}{\sqrt{3}} = 116.7 \text{ V}$$

and for the same field current the armature current on short circuit is

$$I_{a,sc} = 118 \text{ A}$$

From Eq. 5.28

$$X_{s,u} = \frac{116.7}{118} = 0.987 \; \Omega/\text{phase}$$

Note that rated armature current is

$$I_{a,\text{rated}} = \frac{45,000}{\sqrt{3} \times 220} = 118 \; \text{A}$$

Therefore, $I_{a,sc} = 1.00$ per unit. The corresponding air-gap-line voltage is

$$V_{a,ag} = \frac{202}{220} = 0.92 \; \text{per unit}$$

From Eq. 5.28 in per unit

$$X_{s,u} = \frac{0.92}{1.00} = 0.92 \; \text{per unit}$$

The saturated synchronous reactance can be found from the open- and short-circuit characteristics and Eq. 5.29

$$X_s = \frac{V_{a,\text{rated}}}{I'_a} = \frac{(220/\sqrt{3})}{152} = 0.836 \; \Omega/\text{phase}$$

In per unit $I'_a = \frac{152}{118} = 1.29$, and from Eq. 5.29

$$X_s = \frac{1.00}{1.29} = 0.775 \; \text{per unit}$$

Finally, from the open- and short-circuit characteristics and Eq. 5.30, the short-circuit ratio is given by

$$\text{SCR} = \frac{2.84}{2.20} = 1.29$$

Note that as was indicated following Eq. 5.30, the inverse of the short-circuit ratio is equal to the per-unit saturated synchronous reactance

$$X_s = \frac{1}{\text{SCR}} = \frac{1}{1.29} = 0.775 \; \text{per unit}$$

Practice Problem 5.3

Calculate the saturated synchronous reactance (in Ω/phase and per unit) of a 85-kVA synchronous machine which achieves its rated open-circuit voltage of 460 V at a field current 8.7 A and which achieves rated short-circuit current at a field current of 11.2 A.

Solution

$$X_s = 3.21 \; \Omega/\text{phase} = 1.29 \; \text{per unit}$$

If the mechanical power required to drive the machine is measured while the short-circuit test is being made, information can be obtained regarding the losses

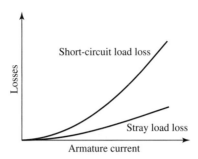

Figure 5.10 Typical form of short-circuit load loss and stray load loss curves.

caused by the armature current. Because the machine flux level is low under short-circuit conditions, the core loss under this condition is typically considered to be negligible. The mechanical power required to drive the synchronous machine during the short-circuit test then equals the sum of friction and windage loss (determined from the open-circuit test at zero field current) plus losses caused by the armature current. The losses caused by the armature current can then be found by subtracting friction and windage from the driving power. The losses caused by the short-circuit armature current are known collectively as the *short-circuit load loss*. A curve showing the typical form of short-circuit load loss plotted against armature current is shown in Fig. 5.10. Typically, it is approximately parabolic with armature current.

The short-circuit load loss consists of I^2R loss in the armature winding, local core losses caused by the armature leakage flux, and the very small core loss caused by the resultant flux. The dc resistance loss can be computed if the dc resistance is measured and corrected, when necessary, for the temperature of the windings during the short-circuit test. For copper conductors

$$\frac{R_T}{R_t} = \frac{234.5 + T}{234.5 + t} \tag{5.32}$$

where R_T and R_t are the resistances at Celsius temperatures T and t, respectively. If this dc resistance loss is subtracted from the short-circuit load loss, the difference will be the loss due to skin effect and eddy currents in the armature conductors plus the local core losses caused by the armature leakage flux. This difference between the short-circuit load loss and the dc resistance loss is the additional loss caused by the alternating current in the armature. It is the *stray-load loss* described in Appendix D, commonly considered to have the same value under normal load conditions as on short circuit. It is a function of the armature current, as shown by the curve in Fig. 5.10.

As with any ac device, the *effective resistance of the armature* $R_{a,\text{eff}}$ can be computed as the power loss attributable to the armature current divided by the square of the current. On the assumption that the stray load loss is a function of only the armature current, the effective resistance of the armature can be determined from the short-circuit load loss:

$$R_{a,eff} = \frac{\text{short-circuit load loss}}{(\text{short-circuit armature current})^2} \qquad (5.33)$$

If the short-circuit load loss and armature current are in per unit, the effective resistance will be in per unit. If they are in watts per phase and amperes per phase, respectively, the effective resistance will be in ohms per phase. Usually it is sufficiently accurate to find the value of $R_{a,eff}$ at rated current and then to assume it to be constant.

EXAMPLE 5.5

For the 45-kVA, three-phase, Y-connected synchronous machine of Example 5.4, at rated armature current (118 A) the short-circuit load loss (total for three phases) is 1.80 kW at a temperature of 25°C. The dc resistance of the armature at this temperature is 0.0335 Ω/phase. Compute the effective armature resistance in per unit and in ohms per phase at 25°C.

■ **Solution**

The short-circuit load loss is $1.80/45 = 0.040$ per unit at $I_a = 1.00$ per unit. Therefore,

$$R_{a,eff} = \frac{0.040}{(1.00)^2} = 0.040 \text{ per unit}$$

On a per-phase basis the short-circuit load loss is $1800/3 = 600$ W/phase and consequently the effective resistance is

$$R_{a,eff} = \frac{600}{(118)^2} = 0.043 \ \Omega/\text{phase}$$

The ratio of ac-to-dc resistance is

$$\frac{R_{a,eff}}{R_{a,dc}} = \frac{0.043}{0.0335} = 1.28$$

Because this is a small machine, its per-unit resistance is relatively high. The effective armature resistance of machines with ratings above a few hundred kilovoltamperes usually is less than 0.01 per unit.

Practice Problem 5.4

Consider a three-phase 13.8 kV 25-MVA synchronous generator whose three-phase short-circuit loss is 52.8 kW at rated armature current. Calculate (*a*) its rated armature current and (*b*) its effective armature resistance in Ω/phase and in per unit.

Solution

    a. 1046 A

    b. $R_{a,eff} = 0.0161 \ \Omega/\text{phase} = 0.0021$ per unit

5.4 STEADY-STATE POWER-ANGLE CHARACTERISTICS

The maximum power a synchronous machine can deliver is determined by the maximum torque which can be applied without loss of synchronism with the external system to which it is connected. The purpose of this section is to derive expressions for the steady-state power limits of synchronous machines in simple situations for which the external system can be represented as an impedance in series with a voltage source.

Since both the external system and the machine itself can be represented as an impedance in series with a voltage source, the study of power limits becomes merely a special case of the more general problem of the limitations on power flow through a series impedance. The impedance will include the synchronous impedance of the synchronous machine as well as an equivalent impedance of the external system (which may consist of transmission lines and transformer banks as well as additional synchronous machines).

Consider the simple circuit of Fig. 5.11a, consisting of two ac voltages $\hat{E}_1$ and $\hat{E}_2$ connected by an impedance $Z = R + jX$ through which the current is $\hat{I}$. The phasor diagram is shown in Fig. 5.11b. Note that in this phasor diagram, the reference direction for positive angles is counter-clockwise. Thus, in Fig. 5.11b, the angle δ is positive while the angle ϕ can be seen to be negative.

The power P_2 delivered through the impedance to the load-end voltage source $\hat{E}_2$ is

$$P_2 = E_2 I \cos \phi \tag{5.34}$$

where ϕ is the phase angle of $\hat{I}$ with respect to $\hat{E}_2$. The phasor current is

$$\hat{I} = \frac{\hat{E}_1 - \hat{E}_2}{Z} \tag{5.35}$$

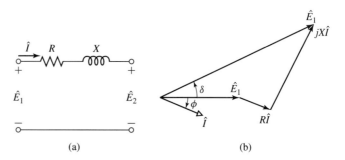

(a)   (b)

Figure 5.11 (a) Impedance interconnecting two voltages; (b) phasor diagram.

If the phasor voltages and the impedance are expressed in polar form,

$$\hat{E}_1 = E_1\, e^{j\delta} \tag{5.36}$$

$$\hat{E}_2 = E_2 \tag{5.37}$$

$$Z = R + jX = |Z|\, e^{j\phi_Z} \tag{5.38}$$

where δ is the phase angle by which $\hat{E}_1$ leads $\hat{E}_2$ and $\phi_Z = \tan^{-1}(X/R)$ is the phase angle of the impedance Z, then

$$\hat{I} = I\, e^{j\phi} = \frac{E_1\, e^{j\delta} - E_2}{|Z|\, e^{j\phi_Z}} = \frac{E_1}{|Z|}\, e^{j(\delta - \phi_Z)} - \frac{E_2}{|Z|}\, e^{-j\phi_Z} \tag{5.39}$$

Taking the real part of Eq. 5.39 gives

$$I \cos \phi = \frac{E_1}{|Z|} \cos(\delta - \phi_Z) - \frac{E_2}{|Z|} \cos(-\phi_Z) \tag{5.40}$$

Noting that $\cos(-\phi_Z) = \cos \phi_Z = R/|Z|$ we see that substitution of Eq. 5.40 in Eq. 5.34 gives

$$P_2 = \frac{E_1 E_2}{|Z|} \cos(\delta - \phi_Z) - \frac{E_2^2 R}{|Z|^2} \tag{5.41}$$

or

$$P_2 = \frac{E_1 E_2}{|Z|} \sin(\delta + \alpha_Z) - \frac{E_2^2 R}{|Z|^2} \tag{5.42}$$

where

$$\alpha_Z = 90° - \phi_Z = \tan^{-1}\left(\frac{R}{X}\right) \tag{5.43}$$

Similarly power P_1 at source end $\hat{E}_1$ of the impedance can be expressed as

$$P_1 = \frac{E_1 E_2}{|Z|} \sin(\delta - \alpha_Z) + \frac{E_1^2 R}{|Z|^2} \tag{5.44}$$

If, as is frequently the case, the resistance is negligible, then $R \ll |Z|$, $|Z| \approx X$ and $\alpha_Z \approx 0$ and hence

$$P_1 = P_2 = \frac{E_1 E_2}{X} \sin \delta \tag{5.45}$$

Equation 5.45 is a very important equation in the study of synchronous machines and indeed in the study of ac power systems in general. When applied to the situation of a synchronous machine connected to an ac system, Eq. 5.45 is commonly referred to as the *power-angle characteristic* for a synchronous machine, and the angle δ is known as the *power angle*. If the resistance is negligible and the voltages are constant, then from Eq. 5.45 the maximum power transfer

$$P_{1,\text{max}} = P_{2,\text{max}} = \frac{E_1 E_2}{X} \tag{5.46}$$

occurs when $\delta = \pm 90°$. Note that if δ is positive, $\hat{E}_1$ leads $\hat{E}_2$ and, from Eq. 5.45, power flows from source $\hat{E}_1$ to $\hat{E}_2$. Similarly, when δ is negative, $\hat{E}_1$ lags $\hat{E}_2$ and power flows from source $\hat{E}_2$ to $\hat{E}_1$.

Equation 5.45 is valid for any voltage sources $\hat{E}_1$ and $\hat{E}_2$ separated by a reactive impedance jX. Thus for a synchronous machine with generated voltage $\hat{E}_{\text{af}}$ and synchronous reactance X_s connected to a system whose Thevenin equivalent is a voltage source $\hat{V}_{\text{EQ}}$ in series with a reactive impedance jX_{EQ}, as shown in Fig. 5.12, the power-angle characteristic can be written

$$P = \frac{E_{\text{af}} V_{\text{EQ}}}{X_s + X_{\text{EQ}}} \sin \delta \tag{5.47}$$

where P is the power transferred from the synchronous machine to the system and δ is the phase angle of $\hat{E}_{\text{af}}$ with respect to $\hat{V}_{\text{EQ}}$.

In a similar fashion, it is possible to write a power-angle characteristic in terms of X_s, E_{af}, the terminal voltage V_a, and the relative angle between them, or alternatively X_{EQ}, V_a and V_{EQ} and their relative angle. Although these various expressions are equally valid, they are not equally useful. For example, while E_{af} and V_{EQ} remain constant as P is varied, V_a will not. Thus, while Eq. 5.47 gives an easily solved relation between P and δ, a power-angle characteristic based upon V_a cannot be solved without an additional expression relating V_a to P.

It should be emphasized that the derivation of Eqs. 5.34 to 5.47 is based on a single-phase ac circuit. For a balanced three-phase system, if E_1 and E_2 are the line-neutral voltages, the results must be multiplied by three to get the total three-phase power; alternatively E_1 and E_2 can be expressed in terms of the line-to-line voltage (equal to $\sqrt{3}$ times the line-neutral voltage), in which case the equations give three-phase power directly.

When the power expression of Eq. 5.45 is compared with the expression of Eq. 5.1 for torque in terms of interacting flux and mmf waves, they are seen to

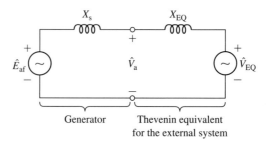

Generator   Thevenin equivalent
                   for the external system

Figure 5.12 Equivalent-circuit representation of a synchronous machine connected to an external system.

be of the same form. This is no coincidence. Remember that torque and power are proportional when, as here, speed is constant. What we are really saying is that Eq. 5.1, applied specifically to an idealized cylindrical-rotor machine and translated to circuit terms, becomes Eq. 5.45. A quick mental review of the background of each relation should show that they stem from the same fundamental considerations.

From Eq. 5.47 we see that the maximum power transfer associated with synchronous-machine operation is proportional to the magnitude of the system voltage, corresponding to V_{EQ}, as well as to that of the generator internal voltage E_{af}. Thus, for constant system voltage, the maximum power transfer can be increased by increasing the synchronous-machine field current and thus the internal voltage. Of course, this cannot be done without limit; neither the field current nor the machine fluxes can be raised past the point where cooling requirements fail to be met.

In general, stability considerations dictate that a synchronous machine achieve steady-state operation for a power angle considerably less than 90°. Thus, for a given system configuration, it is necessary to ensure that the machine will be able to achieve its rated operation and that this operating condition will be within acceptable operating limits for both the machine and the system.

EXAMPLE 5.6

A three-phase, 75-MVA, 13.8-kV synchronous generator with saturated synchronous reactance $X_s = 1.35$ per unit and unsaturated synchronous reactance $X_{s,u} = 1.56$ per unit is connected to an external system with equivalent reactance $X_{EQ} = 0.23$ per unit and voltage $V_{EQ} = 1.0$ per unit, both on the generator base. It achieves rated open-circuit voltage at a field current of 297 amperes.

a. Find the maximum power P_{max} (in MW and per unit) that can be supplied to the external system if the internal voltage of the generator is held equal to 1.0 per unit.

b. Using MATLAB,[†] plot the terminal voltage of the generator as the generator output is varied from zero to P_{max} under the conditions of part (a).

c. Now assume that the generator is equipped with an *automatic voltage regulator* which controls the field current to maintain constant terminal voltage. If the generator is loaded to its rated value, calculate the corresponding power angle, per-unit internal voltage, and field current. Using MATLAB, plot per-unit E_{af} as a function of per-unit power.

■ Solution

a. From Eq. 5.47

$$P_{max} = \frac{E_{af} V_{EQ}}{X_s + X_{EQ}}$$

Note that although this is a three-phase generator, no factor of 3 is required because we are working in per unit.

---

† MATLAB is a registered trademark of The MathWorks, Inc.

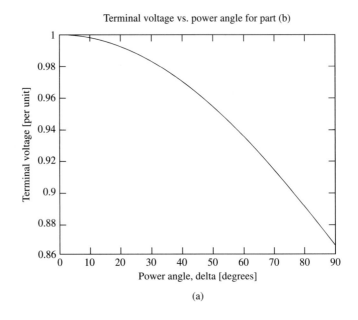

(a)

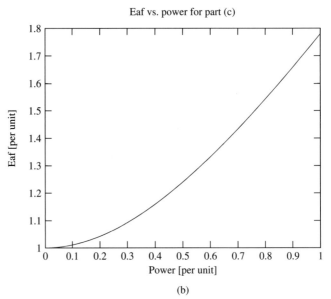

(b)

Figure 5.13 Example 5.6. (a) MATLAB plot of terminal voltage vs. δ for part (b). (b) MATLAB plot of E_{af} vs. power for part (c).

Because the machine is operating with a terminal voltage near its rated value, we should express P_{max} in terms of the saturated synchronous reactance. Thus

$$P_{max} = \frac{1}{1.35 + 0.23} = 0.633 \text{ per unit} = 47.5 \text{ MW}$$

b. From Fig 5.12, the generator terminal current is given by

$$\hat{I}_a = \frac{\hat{E}_{af} - \hat{V}_{EQ}}{j(X_s + X_{EQ})} = \frac{E_{af}\, e^{j\delta} - V_{EQ}}{j(X_s + X_{EQ})} = \frac{e^{j\delta} - 1.0}{j1.58}$$

The generator terminal voltage is then given by

$$\hat{V}_a = \hat{V}_{EQ} + jX_{EQ}\hat{I}_a = 1.0 + \frac{.23}{1.58}(e^{j\delta} - 1.0)$$

Figure 5.13a is the desired MATLAB plot. The terminal voltage can be seen to vary from 1.0 at $\delta = 0°$ to approximately 0.87 at $\delta = 90°$.

c. With the terminal voltage held constant at $V_a = 1.0$ per unit, the power can be expressed as

$$P = \frac{V_a V_{EQ}}{X_{EQ}} \sin \delta_t = \frac{1}{0.23} \sin \delta_t = 4.35 \sin \delta_t$$

where δ_t is the angle of the terminal voltage with respect to $\hat{V}_{EQ}$.

For $P = 1.0$ per unit, $\delta_t = 13.3°$ and hence $\hat{I}$ is equal to

$$\hat{I}_a = \frac{V_a\, e^{j\delta_t} - V_{EQ}}{jX_{EQ}} = 1.007\, e^{j6.65°}$$

and

$$\hat{E}_{af} = \hat{V}_{EQ} + j(X_{EQ} + X_s)\hat{I}_a = 1.78\, e^{j62.7°}$$

or $E_{af} = 1.78$ per unit, corresponding to a field current of $I_f = 1.78 \times 297 = 529$ amperes. The corresponding power angle is $62.7°$.

Figure 5.13b is the desired MATLAB plot. E_{af} can be seen to vary from 1.0 at $P = 0$ to 1.78 at $P = 1.0$.

Here is the MATLAB script:

```
clc
clear

% Solution for part (b)

%System parameters

Veq = 1.0;
Eaf = 1.0;
Xeq = .23;
Xs = 1.35;

% Solve for Va as delta varies from 0 to 90 degrees
for n = 1:101
delta(n) = (pi/2.)*(n-1)/100;
```

```
Ia(n) = (Eaf *exp(j*delta(n)) - Veq)/(j*(Xs + Xeq));
Va(n) = abs(Veq + j*Xeq*Ia(n));
degrees(n) = 180*delta(n)/pi;
end

%Now plot the results
plot(degrees,Va)
xlabel('Power angle, delta [degrees]')
ylabel('Terminal voltage [per unit]')
title('Terminal voltage vs. power angle for part (b)')

fprintf('\n\nHit any key to continue\n')
pause

% Solution for part (c)

%Set terminal voltage to unity

Vterm = 1.0;

for n = 1:101
P(n) = (n-1)/100;
deltat(n) = asin(P(n)*Xeq/(Vterm*Veq));
Ia(n) = (Vterm *exp(j*deltat(n)) - Veq)/(j*Xeq);
Eaf(n) = abs(Vterm + j*(Xs+Xeq)*Ia(n));
end

%Now plot the results
plot(P,Eaf)
xlabel('Power [per unit]')
ylabel('Eaf [per unit]')
title('Eaf vs. power for part (c)')
```

Practice Problem 5.5

Consider the 75-MVA, 13.8 kV machine of Example 5.6. It is observed to be operating at terminal voltage of 13.7 kV and an output power of 53 MW at 0.87 pf lagging. Find (a) the phase current in kA, (b) the internal voltage in per unit, and (c) the corresponding field current in amperes.

Solution

    a. $I_a = 2.57$ kA
    b. $E_{af} = 1.81$ per unit
    c. $I_f = 538$ amperes

EXAMPLE 5.7

A 2000-hp, 2300-V, unity-power-factor, three-phase, Y-connected, 30-pole, 60-Hz synchronous motor has a synchronous reactance of 1.95 Ω/phase. For this problem all losses may be neglected.

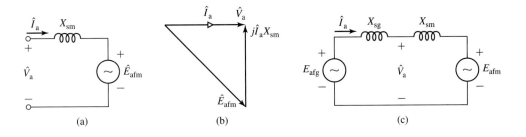

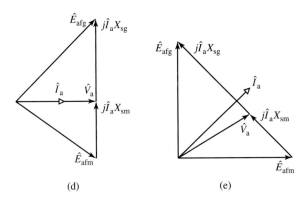

Figure 5.14 Equivalent circuits and phasor diagrams for Example 5.7.

a. Compute the maximum power and torque which this motor can deliver if it is supplied
   with power directly from a 60-Hz, 2300-V infinite bus. Assume its field excitation is
   maintained constant at the value which would result in unity power factor at rated load.

b. Instead of the infinite bus of part (a), suppose that the motor is supplied with power from
   a three-phase, Y-connected, 2300-V, 1500-kVA, two-pole, 3600 r/min turbine generator
   whose synchronous reactance is 2.65 Ω/phase. The generator is driven at rated speed, and
   the field excitations of generator and motor are adjusted so that the motor runs at unity
   power factor and rated terminal voltage at full load. Calculate the maximum power and
   torque which could be supplied corresponding to these values of field excitation.

■ Solution

Although this machine is undoubtedly of the salient-pole type, we will solve the problem
by simple cylindrical-rotor theory. The solution accordingly neglects reluctance torque. The
machine actually would develop a maximum torque somewhat greater than our computed value,
as discussed in Section 5.7.

a. The equivalent circuit is shown in Fig. 5.14a and the phasor diagram at full load in
   Fig. 5.14b, where $\hat{E}_{afm}$ is the generated voltage of the motor and X_{sm} is its synchronous
   reactance. From the motor rating with losses neglected,

$$\text{Rated kVA} = 2000 \times 0.746 = 1492 \text{ kVA, three-phase}$$

$$= 497 \text{ kVA/phase}$$

$$\text{Rated voltage} = \frac{2300}{\sqrt{3}} = 1328 \text{ V line-to-neutral}$$

$$\text{Rated current} = \frac{497{,}000}{1328} = 374 \text{ A/phase-Y}$$

From the phasor diagram at full load

$$E_{afm} = \sqrt{V_a^2 + (I_a X_{sm})^2} = 1515 \text{ V}$$

When the power source is an infinite bus and the field excitation is constant, V_a and E_{afm} are constant. Substitution of V_a for E_1, E_{afm} for E_2, and X_{sm} for X in Eq. 5.46 then gives

$$P_{max} = \frac{V_a E_{afm}}{X_{sm}} = \frac{1328 \times 1515}{1.95} = 1032 \text{ kW/phase}$$

$$= 3096 \text{ kW, three-phase}$$

In per unit, $P_{max} = 3096/1492 = 2.07$ per unit. Because this power exceeds the motor rating, the motor cannot deliver this power for any extended period of time.

With 30 poles at 60 Hz, the synchronous angular velocity is found from Eq. 4.40

$$\omega_s = \left(\frac{2}{\text{poles}}\right)\omega_e = \left(\frac{2}{30}\right)(2\pi 60) = 8\pi \text{ rad/sec}$$

and hence

$$T_{max} = \frac{P_{max}}{\omega_s} = \frac{3096 \times 10^3}{8\pi} = 123.2 \text{ kN} \cdot \text{m}$$

b. When the power source is the turbine generator, the equivalent circuit becomes that shown in Fig. 5.14c, where $\hat{E}_{afg}$ is the generated voltage of the generator and X_{sg} is its synchronous reactance. Here the synchronous generator is equivalent to an external voltage $\hat{V}_{EQ}$ and reactance X_{EQ} as in Fig. 5.12. The phasor diagram at full motor load, unity power factor, is shown in Fig. 5.14d. As before, $V_a = 1330$ V/phase at full load and $E_{afm} = 1515$ V/phase.

From the phasor diagram

$$E_{afg} = \sqrt{V_{ta}^2 + (I_a X_{sg})^2} = 1657 \text{ V}$$

Since the field excitations and speeds of both machines are constant, E_{afg} and E_{afm} are constant. Substitution of E_{afg} for E_1, E_{afm} for E_2, and $X_{sg} + X_{sm}$ for X in Eq. 5.46 then gives

$$P_{max} = \frac{E_{afg} E_{afm}}{X_{sg} + X_{sm}} = \frac{1657 \times 1515}{4.60} = 546 \text{ kW/phase}$$

$$= 1638 \text{ kW, three-phase}$$

In per unit, $P_{max} = 1638/1492 = 1.10$ per unit.

$$T_{max} = \frac{P_{max}}{\omega_s} = \frac{1635 \times 10^3}{8\pi} = 65.2 \text{ kN} \cdot \text{m}$$

Synchronism would be lost if a load torque greater than this value were applied to the motor shaft. Of course, as in part (a), this loading exceeds the rating of the motor and could not be sustained under steady-state operating conditions.

If the excitation system of the generator of Example 5.7 becomes damaged and must be limited to supplying only one half of the field excitation of part (b) of the example, calculate the maximum power which can be supplied to the motor.

Solution

819 kW

5.5 STEADY-STATE OPERATING CHARACTERISTICS

The principal steady-state operating characteristics of a synchronous machine are described by the interrelations between terminal voltage, field current, armature current, power factor, and efficiency. A selection of performance curves of importance in practical application of synchronous machines are presented here.

Consider a synchronous generator delivering power at constant frequency and rated terminal voltage to a load whose power factor is constant. The curve showing the field current required to maintain rated terminal voltage as the constant-power-factor load is varied is known as a *compounding curve*. The characteristic form of three compounding curves at various constant power factors are shown in Fig. 5.15.

Synchronous generators are usually rated in terms of the maximum apparent power (kVA or MVA) load at a specific voltage and power factor (often 80, 85, or 90 percent lagging) which they can carry continuously without overheating. The real power output of the generator is usually limited to a value within the apparent

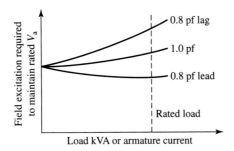

Figure 5.15 Characteristic form of synchronous-generator compounding curves.

power rating by the capability of its prime mover. By virtue of its *voltage-regulating system* (which controls the field current in response to the measured value of terminal voltage), the machine normally operates at a constant terminal voltage whose value is within ±5 percent of rated voltage. When the real-power loading and voltage are fixed, the allowable reactive-power loading is limited by either armature- or field-winding heating. A typical set of *capability curves* for a large, hydrogen-cooled turbine generator is shown in Fig. 5.16. They give the maximum reactive-power loadings corresponding to various real power loadings with operation at rated terminal voltage. Note that the three-curves seen in the figure correspond to differing pressure of the hydrogen cooling gas. As can be seen, increasing the hydrogen pressure improves cooling and permits a larger overall loading of the machine.

Armature-winding heating is the limiting factor in the region from unity to rated power factor (0.85 lagging power factor in Fig. 5.16). For example, for a given real-power loading, increasing the reactive power past a point on the armature-heating limited portion of the capability curve will result in an armature current in excess of that which can be succesfully cooled, resulting in armature-winding temperatures which will damage the armature-winding insulation and degrade its life. Similarly, for lower power factors, field-winding heating is the limiting factor.

Capability curves provide a valuable guide both to power system planners and to operators. As system planners consider modifications and additions to power systems,

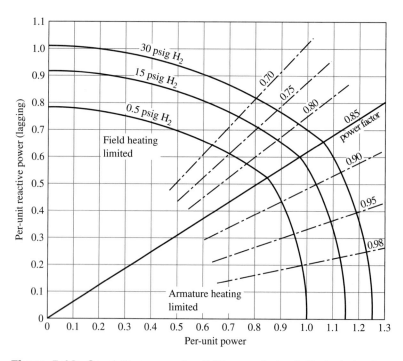

Figure 5.16 Capability curves of an 0.85 power factor, 0.80 short-circuit ratio, hydrogen-cooled turbine generator. Base MVA is rated MVA at 0.5 psig hydrogen.

they can readily see whether the various existing or proposed generators can safely supply their required loadings. Similarly, power system operators can quickly see whether individual generators can safely respond to changes in system loadings which occur during the normal course of system operation.

The derivation of capability curves such as those in Fig. 5.16 can be seen as follows. Operation under conditions of constant terminal voltage and armature current (at the maximum value permitted by heating limitations) corresponds to a constant value of apparent output power determined by the product of terminal voltage and current. Since the per-unit apparent power is given by

$$\text{Apparent power} = \sqrt{P^2 + Q^2} = V_a I_a \qquad (5.48)$$

where P represents the per-unit real power and Q represents the per-unit reactive power, we see that a constant apparent power corresponds to a circle centered on the origin on a plot of reactive power versus real power. Note also from Eq. 5.48, that, for constant terminal voltage, constant apparent power corresponds to constant armature-winding current and hence constant armature-winding heating. Such a circle, corresponding to the maximum acceptable level of armature heating, is shown in Fig. 5.17.

Similarly, consider operation when terminal voltage is constant and field current (and hence E_{af}) is limited to a maximum value, also determined by heating limitations.

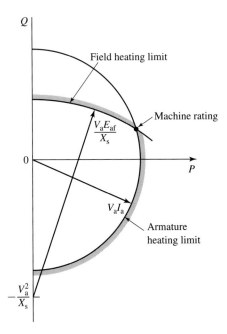

Figure 5.17 Construction used for the derivation of a synchronous generator capability curve.

In per unit,

$$P - jQ = \hat{V}_a \hat{I}_a \qquad (5.49)$$

From Eq. 5.24 (with $R_a = 0$)

$$\hat{E}_{af} = \hat{V}_a + jX_s\hat{I}_a \qquad (5.50)$$

Equations 5.49 and 5.50 can be solved to yield

$$P^2 + \left(Q + \frac{V_a^2}{X_s} \right)^2 = \left(\frac{V_a E_{af}}{X_s} \right)^2 \qquad (5.51)$$

This equation corresponds to a circle centered at $Q = -(V_a^2/X_s)$ in Fig. 5.17 and determines the field-heating limitation on machine operation shown in Fig. 5.16. It is common to specify the rating (apparent power and power factor) of the machine as the point of intersection of the armature- and field-heating limitation curves.

For a given real-power loading, the power factor at which a synchronous machine operates, and hence its armature current, can be controlled by adjusting its field excitation. The curve showing the relation between armature current and field current at a constant terminal voltage and with a constant real power is known as a *V curve* because of its characteristic shape. A family of V curves for a synchronous generator takes the form of those shown in Fig. 5.18.

For constant power output, the armature current is minimum at unity power factor and increases as the power factor decreases. The dashed lines are loci of constant power factor; they are the synchronous-generator compounding curves (see Fig. 5.15) showing how the field current must be varied as the load is changed to maintain constant power factor. Points to the right of the unity-power-factor compounding curve correspond to overexcitation and lagging power factor; points to the left correspond

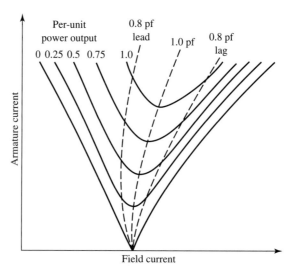

Figure 5.18 Typical form of synchronous-generator V curves.

to underexcitation and leading power factor. Synchronous-motor V curves and compounding curves are very similar to those of synchronous generators. In fact, if it were not for the small effects of armature resistance, motor and generator compounding curves would be identical except that the lagging- and leading-power-factor curves would be interchanged.

As in all electromechanical machines, the efficiency of a synchronous machine at any particular operating point is determined by the losses which consist of I^2R losses in the windings, core losses, stray-load losses, and mechanical losses. Because these losses change with operating condition and are somewhat difficult to measure accurately, various standard procedures have been developed to calculate the efficiency of synchronous machines.[2] The general principles for these calculations are described in Appendix D.

EXAMPLE 5.8

Data are given in Fig. 5.19 with respect to the losses of the 45-kVA synchronous machine of Examples 5.4 and 5.5. Compute its efficiency when it is running as a synchronous motor at a terminal voltage of 220 V and with a power input to its armature of 45 kVA at 0.80 lagging power factor. The field current measured in a load test taken under these conditions is I_f (test) = 5.50 A. Assume the armature and field windings to be at a temperature of 75°C.

■ Solution

For the specified operating conditions, the armature current is

$$I_a = \frac{45 \times 10^3}{\sqrt{3} \times 230} = 113 \text{ A}$$

The I^2R losses must be computed on the basis of the dc resistances of the windings at 75°C. Correcting the winding resistances by means of Eq. 5.32 gives

$$\text{Field-winding resistance} \, R_f \text{ at } 75°C = 35.5 \, \Omega$$

$$\text{Armature dc resistance} \, R_a \text{ at } 75°C = 0.0399 \, \Omega/\text{phase}$$

The field I^2R loss is therefore

$$I_f^2 R_f = 5.50^2 \times 35.5 = 1.07 \text{ kW}$$

According to ANSI standards, losses in the excitation system, including those in any field-rheostat, are not charged against the machine.

The armature I^2R loss is

$$3I_a^2 R_a = 3 \times 113^2 \times 0.0399 = 1.53 \text{ kW}$$

and from Fig. 5.19 at $I_a = 113$ A the stray-load loss = 0.37 kW. The stray-load loss is considered to account for the losses caused by the armature leakage flux. According to ANSI standards, no temperature correction is to be applied to the stray load loss.

---

[2] See, for example, IEEE Std. 115-1995, "IEEE Guide: Test Procedures for Synchronous Machines," Institute of Electrical and Electronic Engineers, Inc., 345 East 47th Street, New York, New York, 10017 and NEMA Standards Publication No. MG-1-1998, "Motors and Generators," National Electrical Manufacturers Association, 1300 North 17th Street, Suite 1847, Rosslyn, Virginia, 22209.

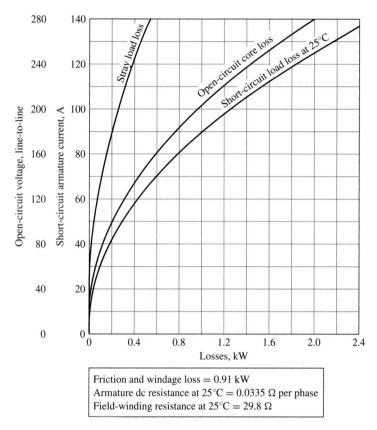

Friction and windage loss = 0.91 kW
Armature dc resistance at 25°C = 0.0335 Ω per phase
Field-winding resistance at 25°C = 29.8 Ω

Figure 5.19 Losses in a three-phase, 45-kVA, Y-connected, 220-V, 60-Hz, six-pole synchronous machine (Example 5.8).

Core loss under load is primarily a function of the main core flux in the motor. As is discussed in Chapter 2, the voltage across the magnetizing branch in a transformer (corresponding to the transformer core flux) is calculated by subtracting the leakage impedance drop from the terminal voltage. In a directly analogous fashion, the main core flux in a synchronous machine (i.e., the air-gap flux) can be calculated as the voltage behind the leakage impedance of the machine. Typically the armature resistance is small, and hence it is common to ignore the resistance and to calculate the voltage behind the leakage reactance. The core loss can then be estimated from the open-circuit core-loss curve at the voltage behind leakage reactance.

In this case, we do not know the machine leakage reactance. Thus, one approach would be simply to assume that the air-gap voltage is equal to the terminal voltage and to determine the core-loss under load from the core-loss curve at the value equal to terminal voltage.[3] In this case, the motor terminal voltage is 230 V line-to-line and thus from Fig. 5.19, the open-circuit core loss is 1.30 kW.

---

[3] Although not rigorously correct, it has become common practice to ignore the leakage impedance drop when determining the under-load core loss.

To estimate the effect of ignoring the leakage reactance drop, let us assume that the leakage reactance of this motor is 0.20 per unit or

$$X_{al} = 0.2 \left(\frac{220^2}{45 \times 10^3} \right) = 0.215 \ \Omega$$

Under this assumption, the air-gap voltage is equal to

$$\hat{V}_a - jX_{al}\hat{I}_a = \frac{230}{\sqrt{3}} - j0.215 \times 141(0.8 + j0.6)$$

$$= 151 - j24.2 = 153 \ e^{-j9.1°} \ \text{V, line-to-neutral}$$

which corresponds to a line-to-line voltage of $\sqrt{3}$ (153) = 265 V. From Fig. 5.19, the corresponding core-loss is 1.8 kW, 500 W higher than the value determined using the terminal voltage. We will use this value for the purposes of this example.

Including the friction and windage loss of 0.91 kW, all losses have now been found:

$$\text{Total losses} = 1.07 + 1.53 + 0.37 + 1.80 + 0.91 = 5.68 \ \text{kW}$$

The total motor input power is the input power to the armature plus that to the field.

$$\text{Input power} = 0.8 \times 45 + 1.07 = 37.1 \ \text{kW}$$

and the output power is equal to the total input power minus the total losses

$$\text{Output power} = 37.1 - 5.68 = 31.4 \ \text{kW}$$

Therefore

$$\text{Efficiency} = \frac{\text{Output power}}{\text{Input power}} = 1 - \frac{31.4}{37.1} = 0.846 = 84.6\%$$

Practice Problem 5.7

Calculate the efficiency of the motor of Example 5.8 if the motor is operating at a power input of 45 kW, unity power factor. You may assume that the motor stray-load losses remain unchanged and that the motor field current is 4.40 A.

Solution

$$\text{Efficiency} = 88.4\%$$

5.6 EFFECTS OF SALIENT POLES; INTRODUCTION TO DIRECT- AND QUADRATURE-AXIS THEORY

The essential features of salient-pole machines are developed in this section based on physical reasoning. A mathematical treatment, based on an inductance formulation like that presented in Section 5.2, is given in Appendix C, where the dq0 transformation is developed.

5.6.1   Flux and MMF Waves

The flux produced by an mmf wave in a uniform-air-gap machine is independent of the spatial alignment of the wave with respect to the field poles. In a salient-pole machine, such as that shown schematically in Fig. 5.20, however, the preferred direction of magnetization is determined by the protruding field poles. The permeance along the polar axis, commonly referred to as the rotor *direct axis,* is appreciably greater than that along the interpolar axis, commonly referred to as the rotor *quadrature axis*.

Note that, by definition, the field winding produces flux which is oriented along the rotor direct axis. Thus, when phasor diagrams are drawn, the field-winding mmf and its corresponding flux $\hat{\Phi}_f$ are found along the rotor direct axis. The generated internal voltage is proportional to the time derivative of the field-winding flux, and thus its phasor $\hat{E}_{af}$ leads the flux $\hat{\Phi}_f$ by 90°. Since by convention the quadrature axis leads the direct axis by 90°, we see that *the generated-voltage phasor $\hat{E}_{af}$ lies along the quadrature axis*. Thus a key point in the analysis of synchronous-machine phasor diagrams is that, by locating the phasor $\hat{E}_{af}$, the location of both the quadrature axis and the direct axis is immediately determined. This forms the basis of the direct- and quadrature-axis formulation for the analysis of salient-pole machines in which all machine voltages and currents can be resolved into their *direct-* and *quadrature-axis components*.

The armature-reaction flux wave $\hat{\Phi}_{ar}$ lags the field flux wave by a space angle of $90° + \phi_{lag}$, where ϕ_{lag} is the time-phase angle by which the armature current lags the generated voltage. If the armature current $\hat{I}_a$ lags the generated voltage $\hat{E}_{af}$ by 90°, the armature-reaction flux wave is directly opposite the field poles and in the opposite direction to the field flux $\hat{\Phi}_f$, as shown in the phasor diagram of Fig. 5.20a.

The corresponding component flux-density waves at the armature surface produced by the field current and by the synchronously-rotating space-fundamental component of armature-reaction mmf are shown in Fig. 5.20b, in which the effects of slots

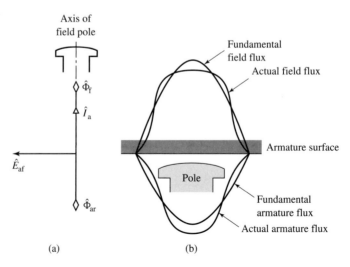

(a)   (b)

Figure 5.20   Direct-axis air-gap fluxes in a salient-pole synchronous machine.

are neglected. The waves consist of a space fundamental and a family of odd-harmonic components. In a well-designed machine the harmonic effects are usually small. Accordingly, only the space-fundamental components will be considered. It is the fundamental components which are represented by the flux-per-pole phasors $\hat{\Phi}_f$ and $\hat{\Phi}_{ar}$ in Fig. 5.20a.

Conditions are quite different when the armature current is in phase with the generated voltage, as illustrated in the phasor diagram of Fig. 5.21a. The axis of the armature-reaction wave then is opposite an interpolar space, as shown in Fig. 5.21b. The armature-reaction flux wave is badly distorted, comprising principally a fundamental and a prominent third space harmonic. The third-harmonic flux wave generates third-harmonic emf's in the armature phase (line-to-neutral) voltages. They will be of the form

$$E_{3,a} = \sqrt{2}V_3 \cos{(3\omega_e t + \phi_3)} \tag{5.52}$$

$$E_{3,b} = \sqrt{2}V_3 \cos{(3(\omega_e t - 120°) + \phi_3)} = \sqrt{2}V_3 \cos{(3\omega_e t + \phi_3)} \tag{5.53}$$

$$E_{3,c} = \sqrt{2}V_3 \cos{(3(\omega_e t - 120°) + \phi_3)} = \sqrt{2}V_3 \cos{(3\omega_e t + \phi_3)} \tag{5.54}$$

Note that these third-harmonic phase voltages are equal in magnitude and in phase. Hence they do not appear as components of the line-to-line voltages, which are equal to the differences between the various phase voltages.

Because of the longer air gap between the poles and the correspondingly larger reluctance, the space-fundamental armature-reaction flux when the armature reaction is along the quadrature axis (Fig. 5.21) is less than the space fundamental armature-reaction flux which would be created by the same armature current if the armature flux wave were directed along the direct axis (Fig. 5.20). Hence, the quadrature-axis magnetizing reactance is less than that of the direct axis.

Focusing our attention on the space-fundamental components of the air-gap flux and mmf, the effects of salient poles can be taken into account by resolving the

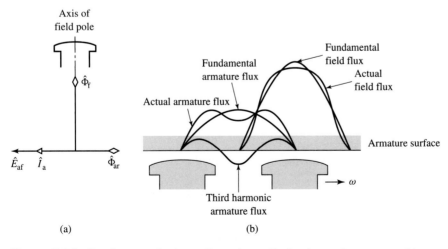

Figure 5.21 Quadrature-axis air-gap fluxes in a salient-pole synchronous machine.

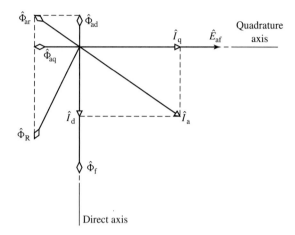

Figure 5.22 Phasor diagram of a salient-pole synchronous generator.

armature current $\hat{I}_a$ into two components, one along the direct axis and the other along the quadrature axis as shown in the phasor diagram of Fig. 5.22. This diagram is drawn for a unsaturated salient-pole generator operating at a lagging power factor. The direct-axis component $\hat{I}_d$ of the armature current, in time-quadrature with the generated voltage $\hat{E}_{af}$, produces a component of the space-fundamental armature-reaction flux $\hat{\Phi}_{ad}$ along the axis of the field poles (the direct axis), as in Fig. 5.20. The quadrature-axis component $\hat{I}_q$, in phase with the generated voltage, produces a component of the space-fundamental armature-reaction flux $\hat{\Phi}_{aq}$ in space-quadrature with the field poles, as in Fig. 5.21. Note that the subscripts d ("direct") and q ("quadrature") on the armature-reaction fluxes refer to their space phase and not to the time phase of the component currents producing them.

Thus a *direct-axis quantity* is one whose magnetic effect is aligned with the axes of the field poles; direct-axis mmf's produce flux along these axes. A *quadrature-axis quantity* is one whose magnetic effect is centered on the interpolar space. For an unsaturated machine, the armature-reaction flux $\hat{\Phi}_{ar}$ is the sum of the components $\hat{\Phi}_{ad}$ and $\hat{\Phi}_{aq}$. The resultant flux $\hat{\Phi}_R$ is the sum of $\hat{\Phi}_{ar}$ and the field flux $\hat{\Phi}_f$.

5.6.2 Phasor Diagrams for Salient-Pole Machines

With each of the component currents $\hat{I}_d$ and $\hat{I}_q$ there is associated a component synchronous-reactance voltage drop, $j\hat{I}_d X_d$ and $j\hat{I}_q X_q$ respectively. The reactances X_d and X_q are, respectively, the *direct-* and *quadrature-axis synchronous reactances;* they account for the inductive effects of all the space-fundamental fluxes created by the armature currents along the direct and quadrature axes, including both armature-leakage and armature-reaction fluxes. Thus, the inductive effects of the direct- and quadrature-axis armature-reaction flux waves can be accounted for by *direct-* and *quadrature-axis magnetizing reactances, $X_{\varphi d}$* and *$X_{\varphi q}$* respectively, similar to the

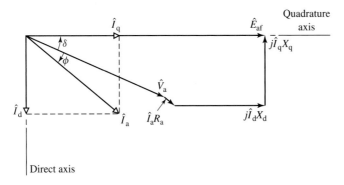

Figure 5.23 Phasor diagram for a synchronous generator showing the relationship between the voltages and the currents.

magnetizing reactance X_φ of cylindrical-rotor theory. The direct- and quadrature-axis synchronous reactances are then given by

$$X_d = X_{al} + X_{\varphi d} \tag{5.55}$$

$$X_q = X_{al} + X_{\varphi q} \tag{5.56}$$

where X_{al} is the armature leakage reactance, assumed to be the same for direct- and quadrature-axis currents. Compare Eqs. 5.55 and 5.56 with Eq. 5.25 for the nonsalient-pole case. As shown in the generator phasor diagram of Fig. 5.23, the generated voltage $\hat{E}_{af}$ equals the phasor sum of the terminal voltage $\hat{V}_a$ plus the armature-resistance drop $\hat{I}_a R_a$ and the component synchronous-reactance drops $j\hat{I}_d X_d + j\hat{I}_q X_q$.

As we have discussed, the quadrature-axis synchronous reactance X_q is less than of the direct axis X_d because of the greater reluctance of the air gap in the quadrature axis. Typically, X_q is between $0.6X_d$ and $0.7X_d$. Note that a small salient-pole effect is also present in turbo-alternators, even though they are cylindrical-rotor machines, because of the effect of the rotor slots on the quadrature-axis reluctance.

Just as for the synchronous reactance X_s of a cylindrical-rotor machine, these reactances are not constant with flux density but rather saturate as the machine flux density increases. It is common to find both unsaturated and saturated values specified for each of these parameters.[4] The saturated values apply to typical machine operating conditions where the terminal voltage is near its rated value. For our purposes in this chapter and elsewhere in this book, we will not focus attention on this issue and, unless specifically stated, the reader may assume that the values of X_d and X_q given are the saturated values.

In using the phasor diagram of Fig. 5.23, the armature current must be resolved into its direct- and quadrature-axis components. This resolution assumes that the phase angle $\phi + \delta$ of the armature current with respect to the generated voltage is known.

---

[4] See, for example, IEEE Std. 115-1995, "IEEE Guide: Test Procedures for Synchronous Machines," Institute of Electrical and Electronic Engineers, Inc., 345 East 47th Street, New York, New York, 10017.

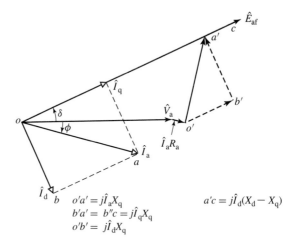

$o'a' = j\hat{I}_a X_q$

$b'a' = b''c = j\hat{I}_q X_q$

$o'b' = j\hat{I}_d X_q$

$a'c = j\hat{I}_d(X_d - X_q)$

Figure 5.24 Relationships between component voltages in a phasor diagram.

Often, however, the power-factor angle ϕ at the machine terminals is explicitly known, rather than the angle $\phi + \delta$. It then becomes necessary to locate the quadrature axis and to compute the power angle δ. This can be done with the aid of the construction of Fig. 5.24.

The phasor diagram of Fig. 5.23 is repeated by the solid-line phasors in Fig. 5.24. Study of this phasor diagram shows that the dashed phasor $o'a'$, perpendicular to $\hat{I}_a$, equals $j\hat{I}_a X_q$. This result follows geometrically from the fact that triangles $o'a'b'$ and oab are similar because their corresponding sides are perpendicular. Thus

$$\frac{o'a'}{oa} = \frac{b'a'}{ba} \tag{5.57}$$

or

$$o'a' = \left(\frac{b'a'}{ba}\right) oa = \frac{|\hat{I}_q| X_q}{|\hat{I}_q|}|\hat{I}_a| = X_q|\hat{I}_a| \tag{5.58}$$

Thus, line $o'a'$ is the phasor $jX_q\hat{I}_a$ and the phasor sum $\hat{V}_a + \hat{I}_a R_a + j\hat{I}_a X_q$ then locates the angular position of the generated voltage $\hat{E}_{af}$ (which in turn lies along the quadrature axis) and therefore the direct and quadrature axes. Physically this must be so, because all the field excitation in a normal machine is in the direct axis. Once the quadrature axis (and hence δ) is known, $\hat{E}_{af}$ can be found as shown in Fig. 5.23

$$\hat{E}_{af} = \hat{V}_a + R_a\hat{I}_a + jX_d\hat{I}_d + jX_q\hat{I}_q \tag{5.59}$$

One use of these relations in determining the excitation requirements for specified operating conditions at the terminals of a salient-pole machine is illustrated in Example 5.9.

EXAMPLE 5.9

The reactances X_d and X_q of a salient-pole synchronous generator are 1.00 and 0.60 per unit, respectively. The armature resistance may be considered to be negligible. Compute the generated voltage when the generator delivers its rated kVA at 0.80 lagging power factor and rated terminal voltage.

■ **Solution**

First, the phase of $\hat{E}_{af}$ must be found so that $\hat{I}_a$ can be resolved into its direct- and quadrature-axis components. The phasor diagram is shown in Fig. 5.25. As is commonly done for such problems, the terminal voltage $\hat{V}_a$ will be used as the reference phasor, i.e., $\hat{V}_a = V_a e^{j0.0°} = V_a$. In per unit

$$\hat{I}_a = I_a e^{j\phi} = 0.80 - j0.60 = 1.0 \, e^{-j36.9°}$$

The quadrature axis is located by the phasor

$$\hat{E}' = \hat{V}_a + jX_q\hat{I}_a = 1.0 + j0.60(1.0 \, e^{-j36.9°}) = 1.44 \, e^{j19.4°}$$

Thus, $\delta = 19.4°$, and the phase angle between $\hat{E}_{af}$ and $\hat{I}_a$ is $\delta - \phi = 19.4° - (-36.9°) = 56.3°$. Note, that although it would appear from Fig. 5.25 that the appropriate angle is $\delta + \phi$, this is not correct because the angle ϕ as drawn in Fig. 5.25 is a negative angle. In general, the desired angle is equal to the difference between the power angle and the phase angle of the terminal current.

The armature current can now be resolved into its direct- and quadrature-axis components. Their magnitudes are

$$I_d = |\hat{I}_a| \sin(\delta - \phi) = 1.00 \sin(56.3°) = 0.832$$

and

$$I_q = |\hat{I}_a| \cos(\delta - \phi) = 1.00 \cos(56.3°) = 0.555$$

As phasors,

$$\hat{I}_d = 0.832 \, e^{j(-90° + 19.4°)} = 0.832 \, e^{j70.6°}$$

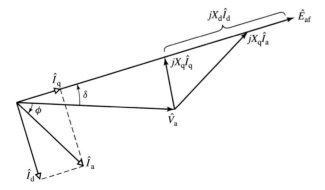

Figure 5.25 Generator phasor diagram for Example 5.9.

and

$$\hat{I}_q = 0.555\, e^{j19.4°}$$

We can now find E_{af} from Eq. 5.59

$$\hat{E}_{af} = \hat{V}_a + jX_d\hat{I}_d + jX_qj\hat{I}_q$$

$$= 1.0 + j1.0(0.832\, e^{j70.6°}) + j0.6(0.555\, e^{j19.4°})$$

$$= 1.77\, e^{j19.4°}$$

and we see that $E_{af} = 1.77$ per unit. Note that, as expected, $\angle \hat{E}_{af} = 19.4° = \delta$, thus confirming that $\hat{E}_{af}$ lies along the quadrature axis.

Practice Problem 5.8

Find the generated voltage for the generator of Example 5.9 if it is loaded to (a) 0.73 per-unit kVA, unity power factor at a terminal voltage of 0.98 per unit and (b) 0.99 per-unit kVA, 0.94 leading power factor and rated terminal voltage.

Solution

   a. $\hat{E}_{af} = 1.20\, e^{j24.5°}$ per unit

   b. $\hat{E}_{af} = 1.08\, e^{j35.0°}$ per unit

EXAMPLE 5.10

In the simplified theory of Section 5.2, the synchronous machine is assumed to be representable by a single reactance, the synchronous reactance X_s of Eq. 5.25. The question naturally arises: How serious an approximation is involved if a salient-pole machine is treated in this simple fashion? Suppose that a salient-pole machine were treated by cylindrical-rotor theory as if it had a single synchronous reactance equal to its direct-axis value X_d? To investigate this question, we will repeat Example 5.9 under this assumption.

■ Solution

In this case, under the assumption that

$$X_q = X_d = X_s = 1.0 \text{ per unit}$$

the generated voltage can be found simply as

$$\hat{E}_{af} = V_a + jX_s\hat{I}_a$$

$$= 1.0 + j1.0(1.0\, e^{-j36.9°}) = 1.79\, e^{j26.6°} \text{ per unit}$$

Comparing this result with that of Example 5.9 (in which we found that $E_{af} = 1.77\, e^{j19.4°}$), we see that the magnitude of the predicted generated voltage is relatively close to the correct value. As a result, we see that the calculation of the field current required for this operating condition will be relatively accurate under the simplifying assumption that the effects of saliency can be neglected.

However, the calculation of the power angle δ (19.4° versus a value of 26.6° if saliency is neglected) shows a considerably larger error. In general, such errors in the calculation of generator steady-state power angles may be of significance when studying the transient behavior of a system including a number of synchronous machines. Thus, although saliency can perhaps be ignored when doing "back-of-the-envelope" system calculations, it is rarely ignored in large-scale, computer-based system studies.

5.7 POWER-ANGLE CHARACTERISTICS OF SALIENT-POLE MACHINES

For the purposes of this discussion, it is sufficient to limit our discussion to the simple system shown in the schematic diagram of Fig. 5.26a, consisting of a salient-pole synchronous machine SM connected to an infinite bus of voltage $\hat{V}_{EQ}$ through a series impedance of reactance X_{EQ}. Resistance will be neglected because it is usually small. Consider that the synchronous machine is acting as a generator. The phasor diagram is shown by the solid-line phasors in Fig. 5.26b. The dashed phasors show the external reactance drop resolved into components due to $\hat{I}_d$ and $\hat{I}_q$. The effect of the external impedance is merely to add its reactance to the reactances of the machine; the total values of the reactance between the excitation voltage $\hat{E}_{af}$ and the bus voltage $\hat{V}_{EQ}$ is therefore

$$X_{dT} = X_d + X_{EQ} \tag{5.60}$$

$$X_{qT} = X_q + X_{EQ} \tag{5.61}$$

If the bus voltage $\hat{V}_{EQ}$ is resolved into components its direct-axis component $V_d = V_{EQ} \sin \delta$ and quadrature-axis component $V_q = V_{EQ} \cos \delta$ in phase with $\hat{I}_d$ and $\hat{I}_q$, respectively, the power P delivered to the bus per phase (or in per unit) is

$$P = I_d V_d + I_q V_q = I_d V_{EQ} \sin \delta + I_q V_{EQ} \cos \delta \tag{5.62}$$

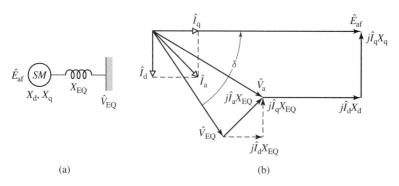

(a)   (b)

Figure 5.26 Salient-pole synchronous machine and series impedance: (a) single-line diagram and (b) phasor diagram.

Also, from Fig. 5.26b,

$$I_d = \frac{E_{af} - V_{EQ} \cos \delta}{X_{dT}} \tag{5.63}$$

and

$$I_q = \frac{V_{EQ} \sin \delta}{X_{qT}} \tag{5.64}$$

Substitution of Eqs. 5.63 and 5.64 in Eq. 5.62 gives

$$P = \frac{E_{af} V_{EQ}}{X_{dT}} \sin \delta + \frac{V_{EQ}^2 (X_{dT} - X_{qT})}{2 X_{dT} X_{qT}} \sin 2\delta \tag{5.65}$$

Equation 5.65 is directly analogous to Eq. 5.47 which applies to the case of a nonsalient machine. It gives the power per phase when E_{af} and V_{EQ} are expressed as line-neutral voltages and the reactances are in Ω/phase, in which case the result must be multiplied by three to get three-phase power. Alternatively, expressing E_{af} and V_{EQ} as line-to-line voltages will result in three-phase power directly. Similarly, Eq. 5.65 can be applied directly if the various quantites are expressed in per unit.

The general form of this power-angle characteristic is shown in Fig. 5.27. The first term is the same as the expression obtained for a cylindrical-rotor machine (Eq. 5.47). The second term includes the effect of salient poles. It represents the fact that the air-gap flux wave creates torque, tending to align the field poles in the position of minimum

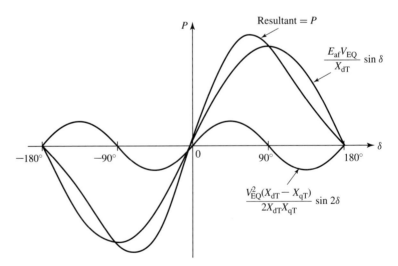

Figure 5.27   Power-angle characteristic of a salient-pole synchronous machine showing the fundamental component due to field excitation and the second-harmonic component due to reluctance torque.

reluctance. This term is the power corresponding to the *reluctance torque* and is of the same general nature as the reluctance torque discussed in Section 3.5. Note that the reluctance torque is independent of field excitation. Also note that, if $X_{dT} = X_{qT}$ as in a uniform-air-gap machine, there is no preferential direction of magnetization, the reluctance torque is zero and Eq. 5.65 reduces to the power-angle equation for a cylindrical-rotor machine (Eq. 5.47).

Notice that the characteristic for negative values of δ is the same except for a reversal in the sign of P. That is, the generator and motor regions are alike if the effects of resistance are negligible. For generator action $\hat{E}_{af}$ leads $\hat{V}_{EQ}$; for motor action $\hat{E}_{af}$ lags $\hat{V}_{EQ}$. Steady-state operation is stable over the range where the slope of the power-angle characteristic is positive. Because of the reluctance torque, a salient-pole machine is "stiffer" than one with a cylindrical rotor; i.e., for equal voltages and equal values of X_{dT}, a salient-pole machine develops a given torque at a smaller value of δ, and the maximum torque which can be developed is somewhat greater.

EXAMPLE 5.11

The 2000-hp, 2300-V synchronous motor of Example 5.7 is assumed to have a synchronous reactance $X_s = 1.95\ \Omega$/phase. In actual fact, it is a salient-pole machine with reactances $X_d = 1.95\ \Omega$/phase and $X_q = 1.40\ \Omega$/phase. Neglecting all losses, compute the maximum mechanical power in kilowatts which this motor can deliver if it is supplied with electric power from an infinite bus (Fig. 5.28a) at rated voltage and frequency and if its field excitation is held constant at that value which would result in unity-power-factor operation at rated load. The shaft load is assumed to be increased gradually so that transient swings are negligible and the steady-state power limit applies. Also, compute the value of δ corresponding to this maximum power operation.

■ **Solution**
The first step is to compute the synchronous motor excitation at rated voltage, full load, and unity power factor. As in Example 5.7, the full-load terminal voltage and current are 1330 V

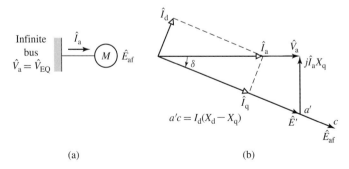

(a)   (b)

Figure 5.28 (a) Single-line diagram and (b) phasor diagram for motor of Example 5.11.

line-to-neutral and 374 A/phase, respectively. The phasor diagram for the specified full-load conditions is shown in Fig. 5.28b. The only essential difference between this phasor diagram and the generator phasor diagram of Fig. 5.25 is that $\hat{I}_a$ in Fig. 5.28 represents motor input current; i.e., we have switched to the motor reference direction for $\hat{I}_a$. Thus, switching the sign of the current to account for the choice of the motor reference direction and neglecting the effects of armature resistance, Eq. 5.59 becomes

$$\hat{E}_{af} = \hat{V}_a - j\hat{I}_d X_d - j\hat{I}_q X_q$$

As in Fig. 5.28b, the quadrature axis can now be located by the phasor

$$\hat{E}' = \hat{V}_a - j\hat{I}_a X_q = 1330 - j374(1.40) = 1429\,e^{-21.5°}$$

That is, $\delta = -21.5°$, with $\hat{E}_{af}$ lagging $\hat{V}_a$. The magnitude of $\hat{I}_d$ is

$$I_d = I_a \sin|\delta| = 374 \sin(21.5°) = 137\,\text{A}$$

With reference to the phasor element labeled $a'c$ in Fig. 5.28b, the magnitude of $\hat{E}_{af}$ can be found by adding the length $a'c = I_d(X_d - X_q)$ numerically to the magnitude of $\hat{E}'$; thus

$$E_{af} = E' + I_d(X_d - X_q) = 1429 + 137(0.55) = 1504\,\text{V line-to-neutral}$$

(Alternatively, E_{af} could have been determined as $\hat{E}_{af} = \hat{V}_a - j\hat{I}_d X_d - j\hat{I}_q X_q$.)

From Eq. 5.65 the power-angle characteristic for this motor is

$$P = \frac{E_{af}V_{EQ}}{X_d}\sin|\delta| + V_{EQ}^2\,\frac{X_d - X_q}{2X_d X_q}\sin(2|\delta|)$$

$$= 1030\sin|\delta| + 178\sin(2|\delta|)\quad\text{kW/phase}$$

Note that we have used $|\delta|$ in this equation. That is because Eq. 5.65 as written applies to a generator and calculates the electrical power output from the generator. For our motor, δ is negative and direct use of Eq. 5.65 will give a value of power $P < 0$ which is of course correct for motor operation. Since we know that this is a motor and that we are calculating electric power into the motor terminals, we ignore the sign issue here entirely and calculate the motor power directly as a positive number.

The maximum motor input power occurs when $dP/d\delta = 0$

$$\frac{dP}{d\delta} = 1030\cos\delta + 356\cos2\delta$$

Setting this equal to zero and using the trigonometric identity

$$\cos2\alpha = 2\cos^2\alpha - 1$$

permit us to solve for the angle δ at which the maximum power occurs:

$$\delta = 73.2°$$

Therefore the maximum power is

$$P_{max} = 1080\,\text{kW/phase} = 3240\,\text{kW, three-phase}$$

We can compare this value with $P_{max} = 3090$ kW found in part (a) of Example 5.7, where the effects of salient poles were neglected. We see that the error caused by neglecting saliency is slightly less than five percent in this case.

Practice Problem 5.9

A 325-MVA, 26-kV, 60-Hz, three-phase, salient-pole synchronous generator is observed to operating at a power output of 250-MW and a lagging power factor of 0.89 at a terminal voltage of 26 kV. The generator synchronous reactances are $X_d = 1.95$ and $X_q = 1.18$, both in per unit. The generator achieves rated-open-circuit voltage at a field current AFNL = 342 A.

Calculate (a) the angle δ between the generator terminal voltage and the generated voltage, (b) the magnitude of the generated voltage E_{af} in per unit, and (c) the required field current in amperes.

Solution

  a. $31.8°$
  b. $E_{af} = 2.29$ per unit
  c. $I_f = 783$ A

The effect, as seen in Example 5.11, of salient poles on the maximum power capability of a synchronous machine increases as the excitation voltage is decreased, as can be seen from Eq. 5.65. Under typical operating conditions, the effect of salient poles usually amounts to a few percent at most. Only at small excitations does the reluctance torque become important. Thus, except at small excitations or when very accurate results are required, a salient-pole machine usually can be adequately treated by simple cylindrical-rotor theory.

5.8 PERMANENT-MAGNET AC MOTORS

Permanent-magnet ac motors are polyphase synchronous motors with permanent-magnet rotors. Thus they are similar to the synchronous machines discussed up to this point in this chapter with the exception that the field windings are replaced by permanent magnets.

Figure 5.29 is a schematic diagram of a three-phase permanent-magnet ac machine. Comparison of this figure with Fig. 5.2 emphasizes the similarities between the permanent-magnet ac machine and the conventional synchronous machine. In fact, the permanent-magnet ac machine can be readily analyzed with the techniques of this chapter simply by assuming that the machine is excited by a field current of constant value, making sure to calculate the various machine inductances based on the effective permeability of the permanent-magnet rotor.

Figure 5.30 shows a cutaway view of a typical permanent-magnet ac motor. This figure also shows a speed and position sensor mounted on the rotor shaft. This sensor is used for control of the motor, as is discussed in Section 11.2. A number

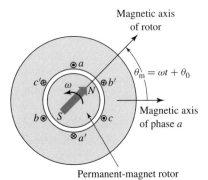

Figure 5.29 Schematic diagram of a three-phase permanent-magnet ac machine. The arrow indicates the direction of rotor magnetization.

Figure 5.30 Cutaway view of a permanent-magnet ac motor. Also shown is the shaft speed and position sensor used to control the motor. (*EG&G Torque Systems.*)

of techniques may be used for shaft-position sensing, including Hall-effect devices, light-emitting diodes and phototransistors in combination with a pulsed wheel, and inductance pickups.

Permanent-magnet ac motors are typically operated from variable-frequency motor drives. Under conditions of constant-frequency, sinusiodal polyphase excitation, a permanent-magnet ac motor behaves similarly to a conventional ac synchronous machine with constant field excitation.

An alternate viewpoint of a permanent-magnet ac motor is that it is a form of permanent-magnet stepping motor with a nonsalient stator (see Section 8.5). Under this viewpoint, the only difference between the two is that there will be little, if any, saliency (cogging) torque in the permanent-magnet ac motor. In the simplest operation, the phases can be simply excited with stepped waveforms so as to cause the rotor to step sequentially from one equilibrium position to the next. Alternatively, using rotor-position feedback from a shaft-position sensor, the motor phase windings can be continuously excited in such a fashion as to control the torque and speed of the motor. As with the stepping motor, the frequency of the excitation determines the motor speed, and the angular position between the rotor magnetic axis and a given phase and the level of excitation in that phase determines the torque which will be produced.

Permanent-magnet ac motors are frequently referred to as *brushless motors* or *brushless dc motors*. This terminology comes about both because of the similarity, when combined with a variable-frequency, variable-voltage drive system, of their speed-torque characteristics to those of dc motors and because of the fact that one can view these motors as inside-out dc motors, with their field winding on the rotor and with their armature electronically commutated by the shaft-position sensor and by switches connected to the armature windings.

5.9 SUMMARY

Under steady-state operating conditions, the physical picture of the operation of a polyphase synchronous machine is simply seen in terms of the interaction of two magnetic fields as discussed in Section 4.7.2. Polyphase currents on the stator produce a rotating magnetic flux wave while dc currents on the rotor produce a flux wave which is stationary with respect to the rotor. Constant torque is produced only when the rotor rotates in synchronism with the stator flux wave. Under these conditions, there is a constant angular displacement between the rotor and stator flux waves and the result is a torque which is proportional to the sine of the displacement angle.

We have seen that a simple set of tests can be used to determine the significant parameters of a synchronous machine including the synchronous reactance X_s or X_d. Two such tests are an open-circuit test, in which the machine terminal voltage is measured as a function of field current, and a short-circuit test, in which the armature is short-circuited and the short-circuit armature current is measured as a function of field current. These test methods are a variation of a testing technique applicable not only to synchronous machines but also to any electrical system whose behavior can be approximated by a linear equivalent circuit to which Thevenin's theorem applies. From a Thevenin-theorem viewpoint, an open-circuit test gives the internal voltage, and a short-circuit test gives information regarding the internal impedance. From the more specific viewpoint of electromechanical machinery, an open-circuit test gives information regarding excitation requirements, core losses, and (for rotating machines) friction and windage losses; a short-circuit test gives information regarding the magnetic reactions of the load current, leakage impedances, and losses associated with the load current such as I^2R and stray losses. The only real complication arises from the effects of magnetic nonlinearity, effects which can be taken into account approximately by considering the machine to be equivalent to an unsaturated one whose magnetization curve is the straight line Op of Fig. 5.9 and whose synchronous reactance is empirically adjusted for saturation as in Eq. 5.29.

In many cases, synchronous machines are operated in conjunction with an external system which can be represented as a constant-frequency, constant-voltage source known as an *infinite bus*. Under these conditions, the synchronous speed is determined by the frequency of the infinite bus, and the machine output power is proportional to the product of the bus voltage, the machine internal voltage (which is, in turn, proportional to the field excitation), and the sine of the phase angle between them (the power angle), and it is inversely proportional to the net reactance between them.

While the real power at the machine terminals is determined by the shaft power input to the machine (if it is acting as a generator) or the shaft load (if it is a motor), varying the field excitation varies the reactive power. For low values of field current, the machine will absorb reactive power from the system and the power angle will be large. Increasing the field current will reduce the reactive power absorbed by the machine as well as the power angle. At some value of field current, the machine power factor will be unity and any further increase in field current will cause the machine to supply reactive power to the system.

Once brought up to synchronous speed, synchronous motors can be operated quite efficiently when connected to a constant-frequency source. However, as we have seen, a synchronous motor develops torque only at synchronous speed and hence has no starting torque. To make a synchronous motor self-starting, a squirrel-cage winding, called an *amortisseur* or *damper winding,* can be inserted in the rotor pole faces, as shown in Fig. 5.31. The rotor then comes up almost to synchronous speed by induction-motor action with the field winding unexcited. If the load and inertia are not too great, the motor will pull into synchronism when the field winding is energized from a dc source.

Alternatively, as we will see in Chapter 11, synchronous motors can be operated from polyphase variable-frequency drive systems. In this case they can be easily started and operated quite flexibly. Small permanent-magnet synchronous machines operated under such conditions are frequently referred to as *brushless motors* or *brushless-dc motors,* both because of the similarity of their speed-torque characteristics to those of dc motors and because of the fact that one can view these motors as inside-out dc motors, with the commutation of the stator windings produced electronically by the drive electronics.

Figure 5.31 Rotor of a six-pole 1200 r/min synchronous motor showing field coils, pole-face damper winding, and construction. (*General Electric Company.*)

5.10 PROBLEMS

5.1 The full-load torque angle of a synchronous motor at rated voltage and frequency is 35 electrical degrees. Neglect the effects of armature resistance and leakage reactance. If the field current is held constant, how would the full-load torque angle be affected by the following changes in operating condition?

   a. Frequency reduced 10 percent, load torque and applied voltage constant.

   b. Frequency reduced 10 percent, load power and applied voltage constant.

   c. Both frequency and applied voltage reduced 10 percent, load torque constant.

   d. Both frequency and applied voltage reduced 10 percent, load power constant.

5.2 The armature phase windings of a two-phase synchronous machine are displaced by 90 electrical degrees in space.

      a.   What is the mutual inductance between these two windings?

      b.   Repeat the derivation leading to Eq. 5.17 and show that the synchronous inductance is simply equal to the armature phase inductance; that is, $L_s = L_{aa0} + L_{al}$, where L_{aa0} is the component of the armature phase inductance due to space-fundamental air-gap flux and L_{al} is the armature leakage inductance.

5.3 Design calculations show the following parameters for a three-phase, cylindrical-rotor synchronous generator:

$$\text{Phase-}a \text{ self-inductance } L_{aa} = 4.83 \text{ mH}$$

$$\text{Armature leakage inductance } L_{al} = 0.33 \text{ mH}$$

Calculate the phase-phase mutual inductance and the machine synchronous inductance.

5.4 The open-circuit terminal voltage of a three-phase, 60-Hz synchronous generator is found to be 15.4 kV rms line-to-line when the field current is 420 A.

      a.   Calculate the stator-to-rotor mutual inductance L_{af}.

      b.   Calculate the open-circuit terminal voltage if the field current is held constant while the generator speed is reduced so that the frequency of the generated voltage is 50 Hz.

5.5 A 460-V, 50-kW, 60-Hz, three-phase synchronous motor has a synchronous reactance of $X_s = 4.15 \ \Omega$ and an armature-to-field mutual inductance, $L_{af} = 83$ mH. The motor is operating at rated terminal voltage and an input power of 40 kW. Calculate the magnitude and phase angle of the line-to-neutral generated voltage $\hat{E}_{af}$ and the field current I_f if the motor is operating at (a) 0.85 power factor lagging, (b) unity power factor, and (c) 0.85 power factor leading.

5.6 The motor of Problem 5.5 is supplied from a 460-V, three-phase source through a feeder whose impedance is $Z_f = 0.084 + j0.82 \ \Omega$. Assuming the system (as measured at the source) to be operating at an input power of 40 kW, calculate the magnitude and phase angle of the line-to-neutral generated voltage $\hat{E}_{af}$ and the field current I_f for power factors of (a) 0.85 lagging, (b) unity, and (c) 0.85 leading.

 5.7 A 50-Hz, two-pole, 750 kVA, 2300 V, three-phase synchronous machine has a synchronous reactance of 7.75 Ω and achieves rated open-circuit terminal voltage at a field current of 120 A.

      a.   Calculate the armature-to-field mutual inductance.

      b.   The machine is to be operated as a motor supplying a 600 kW load at its rated terminal voltage. Calculate the internal voltage E_{af} and the corresponding field current if the motor is operating at unity power factor.

      c.   For a constant load power of 600 kW, write a MATLAB script to plot the terminal current as a function of field current. For your plot, let the field current vary between a minimum value corresponding to a machine loading of 750 kVA at leading power factor and a maximum value corresponding to a machine loading of 750 kVA at lagging power factor. What value of field current produces the minimum terminal current? Why?

5.8 The manufacturer's data sheet for a 26-kV, 750-MVA, 60-Hz, three-phase synchronous generator indicates that it has a synchronous reactance $X_s = 2.04$ and a leakage reactance $X_{al} = 0.18$, both in per unit on the generator base. Calculate (a) the synchronous inductance in mH, (b) the armature leakage inductance in mH, and (c) the armature phase inductance L_{aa} in mH and per unit.

5.9 The following readings are taken from the results of an open- and a short-circuit test on an 800-MVA, three-phase, Y-connected, 26-kV, two-pole, 60-Hz turbine generator driven at synchronous speed:

Field current, A	1540	2960
Armature current, short-circuit test, kA	9.26	17.8
Line voltage, open-circuit characteristic, kV	26.0	(31.8)
Line voltage, air-gap line, kV	29.6	(56.9)

The number in parentheses are extrapolations based upon the measured data. Find (a) the short-circuit ratio, (b) the unsaturated value of the synchronous reactance in ohms per phase and per unit, and (c) the saturated synchronous reactance in per unit and in ohms per phase.

5.10 The following readings are taken from the results of an open- and a short-circuit test on a 5000-kW, 4160-V, three-phase, four-pole, 1800-rpm synchronous motor driven at rated speed:

Field current, A	169	192
Armature current, short-circuit test, A	694	790
Line voltage, open-circuit characteristic, V	3920	4160
Line voltage, air-gap line, V	4640	5270

The armature resistance is 11 mΩ/phase. The armature leakage reactance is estimated to be 0.12 per unit on the motor rating as base. Find (a) the short-circuit ratio, (b) the unsaturated value of the synchronous reactance in ohms per phase and per unit, and (c) the saturated synchronous reactance in per unit and in ohms per phase.

5.11 Write a MATLAB script which automates the calculations of Problems 5.9 and 5.10. The following minimum set of data is required:

■ AFNL: The field current required to achieve rated open-circuit terminal voltage.

■ The corresponding terminal voltage on the air gap line.

■ AFSC: The field current required to achieve rated short-circuit current on the short-circuit characteristic.

Your script should calculate (a) the short-circuit ratio, (b) the unsaturated value of the synchronous reactance in ohms per phase and per unit, and (c) the saturated synchronous reactance in per unit and in ohms per phase.

5.12 Consider the motor of Problem 5.10.

    a. Compute the field current required when the motor is operating at rated voltage, 4200 kW input power at 0.87 power factor leading. Account for saturation under load by the method described in the paragraph relating to Eq. 5.29.

    b. In addition to the data given in Problem 5.10, additional points on the open-circuit characteristic are given below:

Field current, A	200	250	300	350
Line voltage, V	4250	4580	4820	5000

    If the circuit breaker supplying the motor of part (a) is tripped, leaving the motor suddenly open-circuited, estimate the value of the motor terminal voltage following the trip (before the motor begins to slow down and before any protection circuitry reduces the field current).

5.13 Using MATLAB, plot the field current required to achieve unity-power-factor operation for the motor of Problem 5.10 as the motor load varies from zero to full load. Assume the motor to be operating at rated terminal voltage.

5.14 Loss data for the motor of Problem 5.10 are as follows:

    Open-circuit core loss at 4160 V = 37 kW
    Friction and windage loss = 46 kW
    Field-winding resistance at 75°C = 0.279 Ω

Compute the output power and efficiency when the motor is operating at rated input power, unity power factor, and rated voltage. Assume the field-winding to be operating at a temperature of 125°C.

5.15 The following data are obtained from tests on a 145-MVA, 13.8-kV, three-phase, 60-Hz, 72-pole hydroelectric generator.

Open-circuit characteristic:

I_f, A	100	200	300	400	500	600	700	775	800
Voltage, kV	2.27	4.44	6.68	8.67	10.4	11.9	13.4	14.3	14.5

Short-circuit test:

    $I_f = 710$ A, $I_a = 6070$ A

    a. Draw (or plot using MATLAB) the open-circuit saturation curve, the air-gap line, and the short-circuit characteristic.

    b. Find AFNL and AFSC. (Note that if you use MATLAB for part (a), you can use the MATLAB function 'polyfit' to fit a second-order polynomial to the open-circuit saturation curve. You can then use this fit to find AFNL.)

c. Find (*i*) the short-circuit ratio, (*ii*) the unsaturated value of the synchronous reactance in ohms per phase and per unit and (*iii*) the saturated synchronous reactance in per unit and in ohms per phase.

5.16 What is the maximum per-unit reactive power that can be supplied by a synchronous machine operating at its rated terminal voltage whose synchronous reactance is 1.6 per unit and whose maximum field current is limited to 2.4 times that required to achieve rated terminal voltage under open-circuit conditions?

5.17 A 25-MVA, 11.5 kV synchronous machine is operating as a synchronous condenser, as discussed in Appendix D (section D.4.1). The generator short-circuit ratio is 1.68 and the field current at rated voltage, no load is 420 A. Assume the generator to be connected directly to an 11.5 kV source.

a. What is the saturated synchronous reactance of the generator in per unit and in ohms per phase?

The generator field current is adjusted to 150 A.

b. Draw a phasor diagram, indicating the terminal voltage, internal voltage, and armature current.

c. Calculate the armature current magnitude (per unit and amperes) and its relative phase angle with respect to the terminal voltage.

d. Under these conditions, does the synchronous condenser appear inductive or capacitive to the 11.5 kV system?

e. Repeat parts (b) through (d) for a field current of 700 A.

5.18 The synchronous condenser of Problem 5.17 is connected to a 11.5 kV system through a feeder whose series reactance is 0.12 per unit on the machine base. Using MATLAB, plot the voltage (kV) at the synchronous-condenser terminals as the synchronous-condenser field current is varied between 150 A and 700 A.

5.19 A synchronous machine with a synchronous reactance of 1.28 per unit is operating as a generator at a real power loading of 0.6 per unit connected to a system with a series reactance of 0.07 per unit. An increase in its field current is observed to cause a decrease in armature current.

a. Before the increase, was the generator supplying or absorbing reactive power from the power system?

b. As a result of this increase in excitation, did the generator terminal voltage increase or decrease?

c. Repeat parts (a) and (b) if the synchronous machine is operating as a motor.

5.20 Superconducting synchronous machines are designed with superconducting fields windings which can support large current densities and create large magnetic flux densities. Since typical operating magnetic flux densities exceed the saturation flux densities of iron, these machines are typically designed without iron in the magnetic circuit; as a result, these machines exhibit no saturation effects and have low synchronous reactances.

Consider a two-pole, 60-Hz, 13.8-kV, 10-MVA superconducting generator which achieves rated open-circuit armature voltage at a field current of 842 A. It achieves rated armature current into a three-phase terminal short circuit for a field current of 226 A.

a. Calculate the per-unit synchronous reactance.

Consider the situation in which this generator is connected to a 13.8 kV distribution feeder of negligible impedance and operating at an output power of 8.75 MW at 0.9 pf lagging. Calculate:

b. the field current in amperes, the reactive-power output in MVA, and the rotor angle for this operating condition.

c. the resultant rotor angle and reactive-power output in MVA if the field current is reduced to 842 A while the shaft-power supplied by the prime mover to the generator remains constant.

5.21 For a synchronous machine with constant synchronous reactance X_s operating at a constant terminal voltage V_t and a constant excitation voltage E_{af}, show that the locus of the tip of the armature-current phasor is a circle. On a phasor diagram with terminal voltage shown as the reference phasor, indicate the position of the center of this circle and its radius. Express the coordinates of the center and the radius of the circle in terms of V_t, E_{af} and X_s.

5.22 A four-pole, 60-Hz, 24-kV, 650-MVA synchronous generator with a synchronous reactance of 1.82 per unit is operating on a power system which can be represented by a 24-kV infinite bus in series with a reactive impedance of $j0.21\ \Omega$. The generator is equipped with a voltage regulator that adjusts the field excitation such that the generator terminal voltage remains at 24 kV independent of the generator loading.

a. The generator output power is adjusted to 375 MW.

   (i) Draw a phasor diagram for this operating condition.

   (ii) Find the magnitude (in kA) and phase angle (with respect to the generator terminal voltage) of the terminal current.

   (iii) Determine the generator terminal power factor.

   (iv) Find the magnitude (in per unit and kV) of the generator excitation voltage E_{af}.

b. Repeat part (a) if the generator output power is increased to 600 MW.

5.23 The generator of Problem 5.22 achieves rated open-circuit armature voltage at a field current of 850 A. It is operating on the system of Problem 5.22 with its voltage regulator set to maintain the terminal voltage at 0.99 per unit (23.8 kV).

a. Use MATLAB to plot the generator field current (in A) as a function of load (in MW) as the load on the generator output power is varied from zero to full load.

b. Plot the corresponding reactive output power in MVAR as a function of output load.

c. Repeat the plots of parts (a) and (b) if the voltage regulator is set to regulate the terminal voltage to 1.01 per unit (24.2 kV).

5.24 The 145 MW hydroelectric generator of Problem 5.15 is operating on a 13.8-kV power system. Under normal operating procedures, the generator is operated under automatic voltage regulation set to maintain its terminal voltage at 13.8 kV. In this problem you will investigate the possible consequences should the operator forget to switch over to the automatic voltage regulator and instead leave the field excitation constant at AFNL, the value corresponding to rated open-circuit voltage. For the purposes of this problem, neglect the effects of saliency and assume that the generator can be represented by the saturated synchronous reactance found in Problem 5.15.

a. If the power system is represented simply by a 13.8 kV infinite (ignoring the effects of any equivalent impedance), can the generator be loaded to full load? If so, what is the power angle δ corresponding to full load? If not, what is the maximum load that can be achieved?

b. Repeat part (a) with the power system now represented by a 13.8 kV infinite bus in series with a reactive impedance of $j0.14$ Ω.

5.25 Repeat Example 5.9 assuming the generator is operating at one-half of its rated kVA at a lagging power factor of 0.8 and rated terminal voltage.

5.26 Repeat Problem 5.24 assuming that the saturated direct-axis synchronous inductance X_d is equal to that found in Problem 5.15 and that the saturated quadrature-axis synchronous reactance X_q is equal to 75 percent of this value. Compare your answers to those found in Problem 5.24.

5.27 Write a MATLAB script to plot a set of per-unit power-angle curves for a salient-pole synchronous generator connected to an infinite bus ($V_{bus} = 1.0$ per unit). The generator reactances are $X_d = 1.27$ per unit and $X_q = 0.95$ per unit. Assuming $E_{af} = 1.0$ per unit, plot the following curves:

a. Generator connected directly to the infinite bus.

b. Generator connected to the infinite bus through a reactance $X_{bus} = 0.1$ per unit.

c. Generator connected directly to the infinite bus. Neglect saliency effects, setting $X_q = X_d$.

d. Generator connected to the infinite bus through a reactance $X_{bus} = 0.1$ per unit. Neglect saliency effects, setting $X_q = X_d$.

5.28 Draw the steady-state, direct- and quadrature-axis phasor diagram for a salient-pole synchronous motor with reactances X_d and X_q and armature resistance R_a. From this phasor diagram, show that the torque angle δ between the generated voltage $\hat{E}_{af}$ (which lies along the quadrature axis) and the terminal voltage $\hat{V}_t$ is given by

$$\tan \delta = \frac{I_a X_q \cos \phi + I_a R_a \sin \phi}{V_t + I_a X_q \sin \phi - I_a R_a \cos \phi}$$

Here ϕ is the phase angle of the armature current $\hat{I}_a$ and V_t, considered to be negative when $\hat{I}_a$ lags $\hat{V}_t$.

5.29 Repeat Problem 5.28 for synchronous generator operation, in which case the equation for δ becomes

$$\tan \delta = \frac{I_a X_q \cos \phi + I_a R_a \sin \phi}{V_t - I_a X_q \sin \phi + I_a R_a \cos \phi}$$

5.30 What maximum percentage of its rated output power will a salient-pole motor deliver without loss of synchronism when operating at its rated terminal voltage with zero field excitation ($E_{af} = 0$) if $X_d = 0.90$ per unit and $X_q = 0.65$ per unit? Compute the per-unit armature current and reactive power for this operating condition.

5.31 If the synchronous motor of Problem 5.30 is now operated as a synchronous generator connected to an infinite bus of rated voltage, find the minimum per-unit field excitation (where 1.0 per unit is the field current required to achieve rated open-circuit voltage) for which the generator will remain synchronized at (a) half load and (b) full load.

5.32 A salient-pole synchronous generator with saturated synchronous reactances $X_d = 1.57$ per unit and $X_q = 1.34$ per unit is connected to an infinite bus of rated voltage through an external impedance $X_{bus} = 0.11$ per unit. The generator is supplying its rated MVA at 0.95 power factor lagging, as measured at the generator terminals.

   a. Draw a phasor diagram, indicating the infinite-bus voltage, the armature current, the generator terminal voltage, the excitation voltage, and the rotor angle (measured with respect to the infinite bus).

   b. Calculate the per-unit terminal and excitation voltages, and the rotor angle in degrees.

5.33 A salient-pole synchronous generator with saturated synchronous reactances $X_d = 0.78$ per unit and $X_q = 0.63$ per unit is connected to a rated-voltage infinite bus through an external impedance $X_{bus} = 0.09$ per unit.

   a. Assuming the generator to be supplying only reactive power

      (i) Find minimum and maximum per-unit field excitations (where 1.0 per unit is the field current required to achieve rated open-circuit voltage) such that the generator does not exceed its rated terminal current.

      (ii) Using MATLAB, plot the armature current as a function of field excitation as the field excitation is varied between the limits determined in part (i).

   b. Now assuming the generator to be supplying 0.25 per unit rated real power, on the same axes add a plot of the per-unit armature current as a function of field excitation as the field current is varied in the range for which the per-unit armature current is less than 1.0 per unit.

   c. Repeat part (b) for generator output powers of 0.50 and 0.75 per unit. The final result will be a plot of V-curves for this generator in this configuration.

5.34 A two-phase permanent-magnet ac motor has a rated speed of 3000 r/min and a six-pole rotor. Calculate the frequency (in Hz) of the armature voltage required to operate at this speed.

5.35 A 5-kW, three-phase, permanent-magnet synchronous generator produces an open-circuit voltage of 208 V line-to-line, 60-Hz, when driven at a speed of 1800 r/min. When operating at rated speed and supplying a resistive load, its terminal voltage is observed to be 192 V line-to-line for a power output of 4.5 kW.

   a. Calculate the generator phase current under this operating condition.

   b. Assuming the generator armature resistance to be negligible, calculate the generator 60-Hz synchronous reactance.

   c. Calculate the generator terminal voltage which will result if the motor generator load is increased to 5 kW (again purely resistive) while the speed is maintained at 1800 r/min.

5.36 Small single-phase permanent-magnet ac generators are frequently used to generate the power for lights on bicycles. For this application, these generators are typically designed with a significant amount of leakage inductance in their armature winding. A simple model for these generators is an ac voltage source $e_a(t) = \omega K_a \cos \omega t$ in series with the armature leakage inductance L_a and the armature resistance R_a. Here ω is the electrical frequency of the generated voltage which is determined by the speed of the generator as it rubs against the bicycle wheel.

   Assuming that the generator is running a light bulb which can be modeled as a resistance R_b, write an expression for the minimum frequency ω_{min} which must be achieved in order to insure that the light operates at constant brightness, independent of the speed of the bicycle.

6

CHAPTER

Polyphase Induction Machines

T he objective of this chapter is to study the behavior of polyphase induction machines. Our analysis will begin with the development of single-phase equivalent circuits, the general form of which is suggested by the similarity of an induction machine to a transformer. These equivalent circuits can be used to study the electromechanical characteristics of an induction machine as well as the loading presented by the machine on its supply source, whether it is a fixed-frequency source such as a power system or a variable-frequency, variable-voltage motor drive.

6.1 INTRODUCTION TO POLYPHASE INDUCTION MACHINES

As indicated in Section 4.2.1, an *induction motor* is one in which alternating current is supplied to the stator directly and to the rotor by induction or transformer action from the stator. As in the synchronous machine, the stator winding is of the type discussed in Section 4.5. When excited from a balanced polyphase source, it will produce a magnetic field in the air gap rotating at synchronous speed as determined by the number of stator poles and the applied stator frequency f_e (Eq. 4.41).

The rotor of a polyphase induction machine may be one of two types. A *wound rotor* is built with a polyphase winding similar to, and wound with the same number of poles as, the stator. The terminals of the rotor winding are connected to insulated slip rings mounted on the shaft. Carbon brushes bearing on these rings make the rotor terminals available external to the motor, as shown in the cutaway view in Fig. 6.1. Wound-rotor induction machines are relatively uncommon, being found only in a limited number of specialized applications.

On the other hand, the polyphase induction motor shown in cutaway in Fig. 6.2 has a *squirrel-cage rotor* with a winding consisting of conducting bars embedded in slots in the rotor iron and short-circuited at each end by conducting end rings. The

Figure 6.1 Cutaway view of a three-phase induction motor with a wound rotor and slip rings connected to the three-phase rotor winding. (*General Electric Company.*)

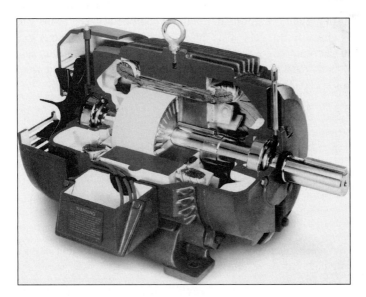

Figure 6.2 Cutaway view of a three-phase squirrel-cage motor. The rotor cutaway shows the squirrel-cage laminations. (*Rockwell Automation/Reliance Electric.*)

 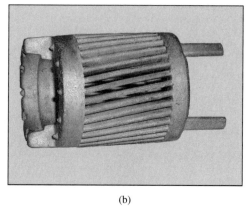

(a) (b)

Figure 6.3 (a) The rotor of a small squirrel-cage motor. (b) The squirrel-cage structure after the rotor laminations have been chemically etched away. (*Athens Products.*)

extreme simplicity and ruggedness of the squirrel-cage construction are outstanding advantages of this type of induction motor and make it by far the most commonly used type of motor in sizes ranging from fractional horsepower on up. Figure 6.3a shows the rotor of a small squirrel-cage motor while Fig. 6.3b shows the squirrel cage itself after the rotor laminations have been chemically etched away.

Let us assume that the rotor is turning at the steady speed of n r/min in the same direction as the rotating stator field. Let the synchronous speed of the stator field be n_s r/min as given by Eq. 4.41. This difference between synchronous speed and the rotor speed is commonly referred to as the *slip* of the rotor; in this case the rotor slip is $n_s - n$, as measured in r/min. Slip is more usually expressed as a fraction of synchronous speed. The *fractional slip s* is

$$s = \frac{n_s - n}{n_s} \tag{6.1}$$

The slip is often expressed in percent, simply equal to 100 percent times the fractional slip of Eq. 6.1.

The rotor speed in r/min can be expressed in terms of the slip and the synchronous speed as

$$n = (1 - s)n_s \tag{6.2}$$

Similarly, the mechanical angular velocity ω_m can be expressed in terms of the synchronous angular velocity ω_s and the slip as

$$\omega_m = (1 - s)\omega_s \tag{6.3}$$

The relative motion of the stator flux and the rotor conductors induces voltages of frequency f_r

$$f_r = sf_e \tag{6.4}$$

called the *slip frequency,* in the rotor. Thus, the electrical behavior of an induction machine is similar to that of a transformer but with the additional feature of frequency transformation produced by the relative motion of the stator and rotor windings. In fact, a wound-rotor induction machine can be used as a frequency changer.

The rotor terminals of an induction motor are short circuited; by construction in the case of a squirrel-cage motor and externally in the case of a wound-rotor motor. The rotating air-gap flux induces slip-frequency voltages in the rotor windings. The rotor currents are then determined by the magnitudes of the induced voltages and the rotor impedance at slip frequency. At starting, the rotor is stationary ($n = 0$), the slip is unity ($s = 1$), and the rotor frequency equals the stator frequency f_e. The field produced by the rotor currents therefore revolves at the same speed as the stator field, and a starting torque results, tending to turn the rotor in the direction of rotation of the stator-inducing field. If this torque is sufficient to overcome the opposition to rotation created by the shaft load, the motor will come up to its operating speed. The operating speed can never equal the synchronous speed however, since the rotor conductors would then be stationary with respect to the stator field; no current would be induced in them, and hence no torque would be produced.

With the rotor revolving in the same direction of rotation as the stator field, the frequency of the rotor currents is $s f_e$ and they will produce a rotating flux wave which will rotate at $s n_s$ r/min *with respect to the rotor* in the forward direction. But superimposed on this rotation is the mechanical rotation of the rotor at n r/min. Thus, with respect to the stator, the speed of the flux wave produced by the rotor currents is the sum of these two speeds and equals

$$sn_s + n = sn_s + n_s(1 - s) = n_s \qquad (6.5)$$

From Eq. 6.5 we see that the rotor currents produce an air-gap flux wave which rotates at synchronous speed and hence in synchronism with that produced by the stator currents. Because the stator and rotor fields each rotate synchronously, they are stationary with respect to each other and produce a steady torque, thus maintaining rotation of the rotor. Such torque, which exists for any mechanical rotor speed n other than synchronous speed, is called an *asynchronous torque.*

Figure 6.4 shows a typical polyphase squirrel-cage induction motor torque-speed curve. The factors influencing the shape of this curve can be appreciated in terms of the torque equation, Eq. 4.81. Note that the resultant air-gap flux Φ_{sr} in this equation is approximately constant when the stator-applied voltage and frequency are constant. Also, recall that the rotor mmf F_r is proportional to the rotor current I_r. Equation 4.81 can then be expressed in the form

$$T = -K I_r \sin \delta_r \qquad (6.6)$$

where K is a constant and δ_r is the angle by which the rotor mmf wave leads the resultant air-gap mmf wave.

The rotor current is equal to the negative of the voltage induced by the air-gap flux divided by the rotor impedance, both at slip frequency. The minus sign is required because the induced rotor current is in the direction to demagnetize the air-gap flux, whereas the rotor current is defined in Chapter 4 as being in the direction to magnetize

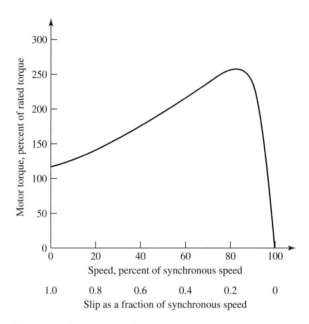

Figure 6.4 Typical induction-motor torque-speed curve for constant-voltage, constant-frequency operation.

the air gap. Under normal running conditions the slip is small: 2 to 10 percent at full load in most squirrel-cage motors. The rotor frequency ($f_r = sf_e$) therefore is very low (of the order of 1 to 6 Hz in 60-Hz motors). In this range the rotor impedance is largely resistive and hence independent of slip. The rotor-induced voltage, on the other hand, is proportional to slip and leads the resultant air-gap flux by 90°. Thus the rotor current is very nearly proportional to the slip, and proportional to and 180° out of phase with the rotor voltage. As a result, the rotor-mmf wave lags the resultant air-gap flux by approximately 90 electrical degrees, and therefore $\sin \delta_r \approx -1$.

Approximate proportionality of torque with slip is therefore to be expected in the range where the slip is small. As slip increases, the rotor impedance increases because of the increasing contribution of the rotor leakage inductance. Thus the rotor current is less than proportional to slip. Also the rotor current lags farther behind the induced voltage, and the magnitude of $\sin \delta_r$ decreases.

The result is that the torque increases with increasing slip up to a maximum value and then decreases, as shown in Fig. 6.4. The maximum torque, or *breakdown torque,* which is typically a factor of two larger than the rated motor torque, limits the short-time overload capability of the motor.

We will see that the slip at which the peak torque occurs is proportional to the rotor resistance. For squirrel-cage motors this peak-torque slip is relatively small, much as is shown in Fig. 6.4. Thus, the squirrel-cage motor is substantially a constant-speed motor having a few percent drop in speed from no load to full load. In the case of a wound-rotor motor, the rotor resistance can be increased by inserting external resistance, hence increasing the slip at peak-torque, and thus decreasing the motor speed for a specified value of torque. Since wound-rotor induction machines are larger, more expensive and require significantly more maintenance than squirrel-cage machines, this method of speed control is rarely used, and induction machines driven from constant-frequency sources tend to be limited to essentially constant-speed applications. As we will see in Chapter 11, use of solid-state, variable-voltage, variable-frequency drive systems makes it possible to readily control the speed of squirrel-cage induction machines and, as a result, they are now widely used in a wide-range of variable-speed applications.

6.2 CURRENTS AND FLUXES IN POLYPHASE INDUCTION MACHINES

For a coil-wound rotor, the flux-mmf situation can be seen with the aid of Fig. 6.5. This sketch shows the development of a simple two-pole, three-phase rotor winding in a two-pole field. It therefore conforms with the restriction that a wound rotor must have the same number of poles as the stator (although the number of phases need not be the same). The rotor flux-density wave is moving to the right at angular velocity ω_s and at slip angular velocity $s\omega_s$ with respect to the rotor winding, which in turn is rotating to the right at angular velocity $(1 - s)\omega_s$. It is shown in Fig. 6.5 in the position of maximum instantaneous voltage in phase a.

If the rotor leakage reactance, equal to $s\omega_s$ times the rotor leakage inductance, is very small compared with the rotor resistance (which is typically the case at the small slips corresponding to normal operation), the phase-a current will also be a maximum. As shown in Section 4.5, the rotor-mmf wave will then be centered on phase a; it is so shown in Fig. 6.5a. The displacement angle, or torque angle, δ_r under these conditions is at its optimum value of $-90°$.

If the rotor leakage reactance is appreciable however, the phase-a current lags the induced voltage by the power-factor angle ϕ_2 of the rotor leakage impedance. The phase-a current will not be at maximum until a correspondingly later time. The rotor-mmf wave will then not be centered on phase a until the flux wave has traveled ϕ_2 degrees farther down the gap, as shown in Fig. 6.5b. The angle δ_r is now $-(90° + \phi_2)$. In general, therefore, the torque angle of an induction motor is

$$\delta_r = -(90° + \phi_2) \tag{6.7}$$

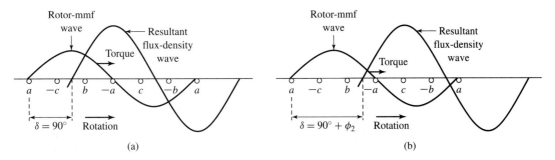

Figure 6.5 Developed rotor winding of an induction motor with its flux-density and mmf waves in their relative positions for (a) zero and (b) nonzero rotor leakage reactance.

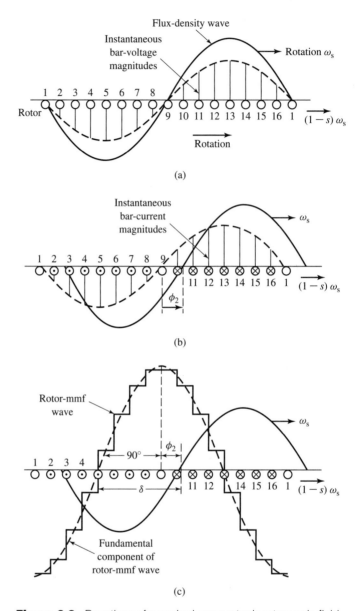

Figure 6.6 Reactions of a squirrel-cage rotor in a two-pole field.

It departs from the optimum value of $-90°$ by the power-factor angle of the rotor leakage impedance at slip frequency. The electromagnetic rotor torque is directed toward the right in Fig. 6.5, or in the direction of the rotating flux wave.

The comparable picture for a squirrel-cage rotor is given in Fig. 6.6. A 16-bar rotor placed in a two-pole field is shown in developed form. To simplify the drawing, only a relatively small number of rotor bars has been chosen and the number is an integral multiple of the number of poles, a choice normally avoided in order to prevent harmful harmonic effects. In Fig. 6.6a the sinusoidal flux-density wave induces a voltage in each bar which has an instantaneous value indicated by the solid vertical lines.

At a somewhat later instant of time, the bar currents assume the instantaneous values indicated by the solid vertical lines in Fig. 6.6b, the time lag corresponding to the rotor power-factor angle ϕ_2. In this time interval, the flux-density wave has traveled in its direction of rotation with respect to the rotor through a space angle ϕ_2 and is then in the position shown in Fig. 6.6b. The corresponding rotor-mmf wave is shown by the step wave of Fig. 6.6c. The fundamental component is shown by the dashed sinusoid and the flux-density wave by the solid sinusoid. Study of these figures confirms the general principle that the number of rotor poles in a squirrel-cage rotor is determined by the inducing flux wave.

6.3 INDUCTION-MOTOR EQUIVALENT CIRCUIT

The foregoing considerations of flux and mmf waves can readily be translated to a steady-state equivalent circuit for a polyphase induction machine. In this derivation, only machines with symmetric polyphase windings excited by balanced polyphase voltages are considered. As in many other discussions of polyphase devices, it is helpful to think of three-phase machines as being Y-connected, so that currents are always line values and voltages always line-to-neutral values. In this case, we can derive the equivalent circuit for one phase, with the understanding that the voltages and currents in the remaining phases can be found simply by an appropriate phase shift of those of the phase under study ($\pm 120°$ in the case of a three-phase machine).

First, consider conditions in the stator. The synchronously-rotating air-gap flux wave generates balanced polyphase counter emfs in the phases of the stator. The stator terminal voltage differs from the counter emf by the voltage drop in the stator leakage impedance $Z_1 = R_1 + jX_1$. Thus

$$\hat{V}_1 = \hat{E}_2 + \hat{I}_1(R_1 + jX_1) \tag{6.8}$$

where

$\hat{V}_1 =$ Stator line-to-neutral terminal voltage

$\hat{E}_2 =$ Counter emf (line-to-neutral) generated by the resultant air-gap flux

$\hat{I}_1 =$ Stator current

$R_1 =$ Stator effective resistance

$X_1 =$ Stator leakage reactance

The polarity of the voltages and currents are shown in the equivalent circuit of Fig. 6.7.

The resultant air-gap flux is created by the combined mmf's of the stator and rotor currents. Just as in the case of a transformer, the stator current can be resolved into two components: a load component and an exciting (magnetizing) component. The load component $\hat{I}_2$ produces an mmf that corresponds to the mmf of the rotor current. The exciting component $\hat{I}_\varphi$ is the additional stator current required to create the resultant

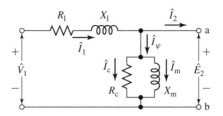

Figure 6.7 Stator equivalent circuit for a polyphase induction motor.

air-gap flux and is a function of the emf $\hat{E}_2$. The exciting current can be resolved into a core-loss component $\hat{I}_c$ in phase with $\hat{E}_2$ and a magnetizing component $\hat{I}_m$ lagging $\hat{E}_2$ by 90°. In the equivalent circuit, the exciting current can be accounted for by means of a shunt branch, formed by a *core-loss resistance* R_c and a *magnetizing reactance* X_m in parallel, connected across $\hat{E}_2$, as in Fig. 6.7. Both R_c and X_m are usually determined at rated stator frequency and for a value of E_2 close to the expected operating value; they are then assumed to remain constant for the small departures of E_2 associated with normal operation of the motor.

The equivalent circuit representing stator phenomena is exactly like that used to represent the primary of a transformer. To complete our model, the effects of the rotor must be incorporated. From the point of view of the stator equivalent circuit of Fig. 6.7, the rotor can be represented by an equivalent impedance Z_2

$$Z_2 = \frac{\hat{E}_2}{\hat{I}_2} \tag{6.9}$$

corresponding to the leakage impedance of an equivalent stationary secondary. To complete the equivalent circuit, we must determine Z_2 by representing the stator and rotor voltages and currents in terms of rotor quantities as referred to the stator.

As we saw in Section 2.3, from the point of view of the primary, the secondary winding of a transformer can be replaced by an equivalent secondary winding having the same number of turns as the primary winding. In a transformer where the turns ratio and the secondary parameters are known, this can be done by referring the secondary impedance to the primary by multiplying it by the square of the primary-to-secondary turns ratio. The resultant equivalent circuit is perfectly general from the point of view of primary quantities.

Similarly, in the case of a polyphase induction motor, if the rotor were to be replaced by an equivalent rotor with a polyphase winding with the same number of phases and turns as the stator but producing the same mmf and air gap flux as the actual rotor, the performance as seen from the stator terminals would be unchanged. This concept, which we will adopt here, is especially useful in modeling squirrel-cage rotors for which the identity of the rotor "phase windings" is in no way obvious.

The rotor of an induction machine is short-circuited, and hence the impedance seen by induced voltage is simply the rotor short-circuit impedance. Consequently the relation between the slip-frequency leakage impedance Z_{2s} of the equivalent rotor and the slip-frequency leakage impedance Z_{rotor} of the actual rotor must be

$$Z_{2s} = \frac{\hat{E}_{2s}}{\hat{I}_{2s}} = N_{\text{eff}}^2 \left(\frac{\hat{E}_{\text{rotor}}}{\hat{I}_{\text{rotor}}} \right) = N_{\text{eff}}^2 Z_{\text{rotor}} \qquad (6.10)$$

where N_{eff} is the effective turns ratio between the stator winding and that of the actual rotor winding. Here the subscript 2s refers to quantities associated with the referred rotor. Thus $\hat{E}_{2s}$ is the voltage induced in the equivalent rotor by the resultant air-gap flux, and $\hat{I}_{2s}$ is the corresponding induced current.

When one is concerned with the actual rotor currents and voltages, the turns ratio N_{eff} must be known in order to convert back from equivalent-rotor quantities to those of the actual rotor. However, for the purposes of studying induction-motor performance as seen from the stator terminals, there is no need for this conversion and a representation in terms of equivalent-rotor quantities is fully adequate. Thus an equivalent circuit based upon equivalent-rotor quantities can be used to represent both coil-wound and squirrel-cage rotors.

Having taken care of the effects of the stator-to-rotor turns ratio, we next must take into account the relative motion between the stator and the rotor with the objective of replacing the actual rotor and its slip-frequency voltages and currents with an equivalent stationary rotor with stator-frequency voltages and currents. Consider first the slip-frequency leakage impedance of the referred rotor.

$$Z_{2s} = \frac{\hat{E}_{2s}}{\hat{I}_{2s}} = R_2 + jsX_2 \qquad (6.11)$$

where

$R_2 = $ Referred rotor resistance

$sX_2 = $ Referred rotor leakage reactance at slip frequency

Note that here X_2 has been defined as the referred rotor leakage reactance at stator frequency f_e. Since the actual rotor frequency $f_r = sf_e$, it has been converted to the slip-frequency reactance simply by multiplying by the slip s. The slip-frequency equivalent circuit of one phase of the referred rotor is shown in Fig. 6.8. This is the equivalent circuit of the rotor as seen in the slip-frequency rotor reference frame.

We next observe that the resultant air-gap mmf wave is produced by the combined effects of the stator current $\hat{I}_1$ and the equivalent load current $\hat{I}_2$. Similarly, it can be

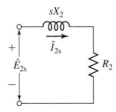

Figure 6.8 Rotor equivalent circuit for a polyphase induction motor at slip frequency.

expressed in terms of the stator current and the equivalent rotor current $\hat{I}_{2s}$. These two currents are equal in magnitude since $\hat{I}_{2s}$ is defined as the current in an equivalent rotor with the same number of turns per phase as the stator. Because the resultant air-gap mmf wave is determined by the phasor sum of the stator current and the rotor current of either the actual or equivalent rotor, $\hat{I}_2$ and $\hat{I}_{2s}$ must also be equal in phase (at their respective electrical frequencies) and hence we can write

$$\hat{I}_{2s} = \hat{I}_2 \tag{6.12}$$

Finally, consider that the resultant flux wave induces both the slip-frequency emf induced in the referred rotor $\hat{E}_{2s}$ and the stator counter emf $\hat{E}_2$. If it were not for the effect of speed, these voltages would be equal in magnitude since the referred rotor winding has the same number of turns per phase as the stator winding. However, because the relative speed of the flux wave with respect to the rotor is s times its speed with respect to the stator, the relation between these emfs is

$$E_{2s} = s E_2 \tag{6.13}$$

We can furthermore argue that since the phase angle between each of these voltages and the resultant flux wave is $90°$, then these two voltages must also be equal in a phasor sense at their respective electrical frequencies. Hence

$$\hat{E}_{2s} = s \hat{E}_2 \tag{6.14}$$

Division of Eq. 6.14 by Eq. 6.12 and use of Eq. 6.11 then gives

$$\frac{\hat{E}_{2s}}{\hat{I}_{2s}} = \frac{s \hat{E}_2}{\hat{I}_2} = Z_{2s} = R_2 + j s X_2 \tag{6.15}$$

Division by the slip s then gives

$$Z_2 = \frac{\hat{E}_2}{\hat{I}_2} = \frac{R_2}{s} + j X_2 \tag{6.16}$$

We have achieved our objective. Z_2 is the impedance of the equivalent stationary rotor which appears across the load terminals of the stator equivalent circuit of Fig. 6.7. The final result is shown in the single-phase equivalent circuit of Fig. 6.9. The combined effect of shaft load and rotor resistance appears as a reflected resistance R_2/s, a function of slip and therefore of the mechanical load. The current in the reflected

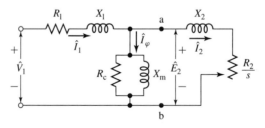

Figure 6.9   Single-phase equivalent circuit for a polyphase induction motor.

rotor impedance equals the load component $\hat{I}_2$ of stator current; the voltage across this impedance equals the stator voltage $\hat{E}_2$. Note that when rotor currents and voltages are reflected into the stator, their frequency is also changed to stator frequency. All rotor electrical phenomena, when viewed from the stator, become stator-frequency phenomena, because the stator winding simply sees mmf and flux waves traveling at synchronous speed.

6.4 ANALYSIS OF THE EQUIVALENT CIRCUIT

The single-phase equivalent circuit of Fig. 6.9 can be used to determine a wide variety of steady-state performance characteristics of polyphase induction machines. These include variations of current, speed, and losses as the load-torque requirements change, as well as the starting torque, and the maximum torque.

The equivalent circuit shows that the total power P_{gap} transferred across the air gap from the stator is

$$P_{gap} = n_{ph} \, I_2^2 \left(\frac{R_2}{s} \right) \tag{6.17}$$

where n_{ph} is the number of stator phases.

The total rotor $I^2 R$ loss, P_{rotor}, can be calculated from the $I^2 R$ loss in the equivalent rotor as

$$P_{rotor} = n_{ph} \, I_{2s}^2 \, R_2 \tag{6.18}$$

Since $I_{2s} = I_2$, we can write Eq. 6.18 as

$$P_{rotor} = n_{ph} \, I_2^2 \, R_2 \tag{6.19}$$

The electromagnetic power P_{mech} developed by the motor can now be determined by subtracting the rotor power dissipation of Eq. 6.19 from the air-gap power of Eq. 6.17.

$$P_{mech} = P_{gap} - P_{rotor} = n_{ph} \, I_2^2 \left(\frac{R_2}{s} \right) - n_{ph} \, I_2^2 \, R_2 \tag{6.20}$$

or equivalently

$$P_{mech} = n_{ph} \, I_2^2 \, R_2 \left(\frac{1 - s}{s} \right) \tag{6.21}$$

Comparing Eq. 6.17 with Eq. 6.21 gives

$$P_{mech} = (1 - s) P_{gap} \tag{6.22}$$

and

$$P_{rotor} = s \, P_{gap} \tag{6.23}$$

We see then that, of the total power delivered across the air gap to the rotor, the fraction $1 - s$ is converted to mechanical power and the fraction s is dissipated as $I^2 R$ loss in the rotor conductors. From this it is evident that an induction motor operating at high slip is an inefficient device. When power aspects are to be emphasized, the

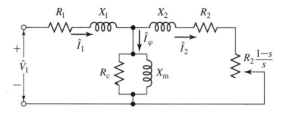

Figure 6.10 Alternative form of equivalent circuit.

equivalent circuit can be redrawn in the manner of Fig. 6.10. The electromechanical power per stator phase is equal to the power delivered to the resistance $R_2(1-s)/s$.

EXAMPLE 6.1

A three-phase, two-pole, 60-Hz induction motor is observed to be operating at a speed of 3502 r/min with an input power of 15.7 kW and a terminal current of 22.6 A. The stator-winding resistance is 0.20 Ω/phase. Calculate the I^2R power dissipated in rotor.

■ Solution

The power dissipated in the stator winding is given by

$$P_{\text{stator}} = 3I_1^2 R_1 = 3(22.6)^2 0.2 = 306 \text{ W}$$

Hence the air-gap power is

$$P_{\text{gap}} = P_{\text{input}} - P_{\text{stator}} = 15.7 - 0.3 = 15.4 \text{ kW}$$

The synchronous speed of this machine can be found from Eq. 4.41

$$n_s = \left(\frac{120}{\text{poles}}\right) f_e = \left(\frac{120}{2}\right) 60 = 3600 \text{ r/min}$$

and hence from Eq. 6.1, the slip is $s = (3600 - 3502)/3600 = 0.0272$. Thus, from Eq. 6.23,

$$P_{\text{rotor}} = s P_{\text{gap}} = 0.0272 \times 15.4 \text{ kW} = 419 \text{ W}$$

Practice Problem 6.1

Calculate the rotor power dissipation for a three-phase, 460-V, 60-Hz, four-pole motor with an armature resistance of 0.056 Ω operating at a speed of 1738 r/min and with an input power of 47.4 kW and a terminal current of 76.2 A.

Solution

1.6 kW

The electromechanical T_{mech} corresponding to the power P_{mech} can be obtained by recalling that mechanical power equals torque times angular velocity. Thus,

$$P_{\text{mech}} = \omega_m T_{\text{mech}} = (1-s)\omega_s T_{\text{mech}} \qquad (6.24)$$

For P_{mech} in watts and ω_s in rad/sec, T_{mech} will be in newton-meters.

Use of Eqs. 6.21 and 6.22 leads to

$$T_{mech} = \frac{P_{mech}}{\omega_m} = \frac{P_{gap}}{\omega_s} = \frac{n_{ph} I_2^2 (R_2/s)}{\omega_s} \qquad (6.25)$$

with the synchronous mechanical angular velocity ω_s being given by

$$\omega_s = \frac{4\pi f_e}{poles} = \left(\frac{2}{poles}\right) \omega_e \qquad (6.26)$$

The mechanical torque T_{mech} and power P_{mech} are not the output values available at the shaft because friction, windage, and stray-load losses remain to be accounted for. It is obviously correct to subtract friction, windage, and other rotational losses from T_{mech} or P_{mech} and it is generally assumed that stray load effects can be subtracted in the same manner. The remainder is available as output power from the shaft for useful work. Thus

$$P_{shaft} = P_{mech} - P_{rot} \qquad (6.27)$$

and

$$T_{shaft} = \frac{P_{shaft}}{\omega_m} = T_{mech} - T_{rot} \qquad (6.28)$$

where P_{rot} and T_{rot} are the power and torque associated with the friction, windage, and remaining rotational losses.

Analysis of the transformer equivalent circuit is often simplified by either neglecting the magnetizing branch entirely or adopting the approximation of moving it out directly to the primary terminals. Such approximations are not used in the case of induction machines under normal running conditions because the presence of the air gap results in a relatively lower magnetizing impedance and correspondingly a relatively higher exciting current—30 to 50 percent of full-load current—and because the leakage reactances are also higher. Some simplification of the induction-machine equivalent circuit results if the core-loss resistance R_c is omitted and the associated core-loss effect is deducted from T_{mech} or P_{mech} at the same time that rotational losses and stray load effects are subtracted. The equivalent circuit then becomes that of Fig. 6.11a or b, and the error introduced is often relatively insignificant. Such

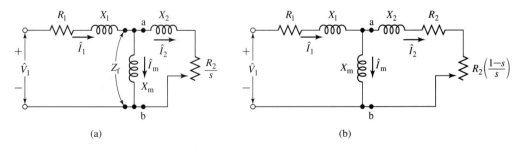

(a)   (b)

Figure 6.11 Equivalent circuits with the core-loss resistance R_c neglected corresponding to (a) Fig. 6.9 and (b) Fig. 6.10.

a procedure also has an advantage during motor testing, for then the no-load core loss need not be separated from friction and windage. These last circuits are used in subsequent discussions.

EXAMPLE 6.2

A three-phase Y-connected 220-V (line-to-line) 7.5-kW 60-Hz six-pole induction motor has the following parameter values in Ω/phase referred to the stator:

$$R_1 = 0.294 \quad R_2 = 0.144$$

$$X_1 = 0.503 \quad X_2 = 0.209 \quad X_m = 13.25$$

The total friction, windage, and core losses may be assumed to be constant at 403 W, independent of load.

For a slip of 2 percent, compute the speed, output torque and power, stator current, power factor, and efficiency when the motor is operated at rated voltage and frequency.

■ **Solution**

Let the impedance Z_f (Fig. 6.11a) represent the per phase impedance presented to the stator by the magnetizing reactance and the rotor. Thus, from Fig. 6.11a

$$Z_f = R_f + jX_f = \left(\frac{R_2}{s} + jX_2 \right) \text{ in parallel with } jX_m$$

Substitution of numerical values gives, for $s = 0.02$,

$$R_f + jX_f = 5.41 + j3.11 \ \Omega$$

The stator input impedance can now be calculated as

$$Z_{in} = R_1 + jX_1 + Z_f = 5.70 + j3.61 = 6.75 \angle 32.3° \ \Omega$$

The line-to-neutral terminal voltage is equal to

$$V_1 = \frac{220}{\sqrt{3}} = 127 \text{ V}$$

and hence the stator current can be calculated as

$$\hat{I}_1 = \frac{V_1}{Z_{in}} = \frac{127}{6.75 \angle 32.3°} = 18.8 \angle{-32.3°} \text{ A}$$

The stator current is thus 18.8 A and the power factor is equal to $\cos(-32.3°) = 0.845$ lagging.

The synchronous speed can be found from Eq. 4.41

$$n_s = \left(\frac{120}{\text{poles}} \right) f_e = \left(\frac{120}{6} \right) 60 = 1200 \text{ r/min}$$

or from Eq. 6.26

$$\omega_s = \frac{4\pi f_e}{\text{poles}} = 125.7 \text{ rad/sec}$$

The rotor speed is

$$n = (1 - s)n_s = (0.98)1200 = 1176 \text{ r/min}$$

or

$$\omega_m = (1 - s)\omega_s = (0.98)125.7 = 123.2 \text{ rad/sec}$$

From Eq. 6.17,

$$P_{gap} = n_{ph} I_2^2 \left(\frac{R_2}{s} \right)$$

Note however that because the only resistance included in Z_f is R_2/s, the power dissipated in Z_f is equal to the power dissipated in R_2/s and hence we can write

$$P_{gap} = n_{ph} I_1^2 R_f = 3(18.8)^2(5.41) = 5740 \text{ W}$$

We can now calculate P_{mech} from Eq. 6.21 and the shaft output power from Eq. 6.27. Thus

$$P_{shaft} = P_{mech} - P_{rot} = (1 - s)P_{gap} - P_{rot}$$

$$= (0.98)5740 - 403 = 5220 \text{ W}$$

and the shaft output torque can be found from Eq. 6.28 as

$$T_{shaft} = \frac{P_{shaft}}{\omega_m} = \frac{5220}{123.2} = 42.4 \text{ N} \cdot \text{m}$$

The efficiency is calculated as the ratio of shaft output power to stator input power. The input power is given by

$$P_{in} = n_{ph}\text{Re}[\hat{V}_1 \hat{I}_1^*] = 3\text{Re}[127(18.8 \angle 32.3°)]$$

$$= 3 \times 127 \times 18.8 \cos(32.2°) = 6060 \text{ W}$$

Thus the efficiency η is equal to

$$\eta = \frac{P_{shaft}}{P_{in}} = \frac{5220}{6060} = 0.861 = 86.1\%$$

The complete performance characteristics of the motor can be determined by repeating these calculations for other assumed values of slip.

Practice Problem 6.2

Find the speed, output power, and efficiency for the motor of Example 6.2 operating at rated voltage and frequency for a slip of 1.5 percent.

Solution

$$\text{Speed} = 1182 \text{ r/min}$$

$$P_{shaft} = 3932 \text{ W}$$

$$\text{Efficiency} = 85.3\%$$

6.5   TORQUE AND POWER BY USE OF THEVENIN'S THEOREM

When torque and power relations are to be emphasized, considerable simplification results from application of Thevenin's network theorem to the induction-motor equivalent circuit. In its general form, Thevenin's theorem permits the replacement of any network of linear circuit elements and complex voltage sources, such as viewed from two terminals a and b (Fig. 6.12a), by a single complex voltage source $\hat{V}_{eq}$ in series with a single impedance Z_{eq} (Fig. 6.12b). The Thevenin-equivalent voltage $\hat{V}_{eq}$ is that appearing across terminals a and b of the original network when these terminals are open-circuited; the Thevenin-equivalent impedance Z_{eq} is that viewed from the same terminals when all voltage sources within the network are set equal to zero. For application to the induction-motor equivalent circuit, points a and b are taken as those so designated in Fig. 6.11a and b. The equivalent circuit then assumes the forms given in Fig. 6.13 where Thevenin's theorem has been used to transform the network to the left of points a and b into an equivalent voltage source $\hat{V}_{1,eq}$ in series with an equivalent impedance $Z_{1,eq} = R_{1,eq} + jX_{1,eq}$.

According to Thevenin's theorem, the equivalent source voltage $\hat{V}_{1,eq}$ is the voltage that would appear across terminals a and b of Fig. 6.11 with the rotor circuits

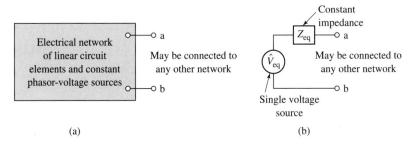

(a)   (b)

Figure 6.12   (a) General linear network and (b) its equivalent at terminals ab by Thevenin's theorem.

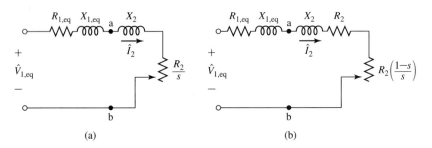

(a)   (b)

Figure 6.13   Induction-motor equivalent circuits simplified by Thevenin's theorem.

removed. The result is a simple voltage divider and thus

$$\hat{V}_{1,\text{eq}} = \hat{V}_1 \left(\frac{jX_\text{m}}{R_1 + j(X_1 + X_\text{m})} \right) \tag{6.29}$$

For most induction motors, negligible error results from neglecting the stator resistance in Eq. 6.29. The Thevenin-equivalent stator impedance $Z_{1,\text{eq}}$ is the impedance between terminals a and b of Fig. 6.11 viewed toward the source with the source voltage set equal to zero (or equivalently replaced by a short circuit) and therefore is

$$Z_{1,\text{eq}} = R_{1,\text{eq}} + jX_{1,\text{eq}} = (R_1 + jX_1) \text{ in parallel with } jX_\text{m} \tag{6.30}$$

or

$$Z_{1,\text{eq}} = \frac{jX_\text{m}(R_1 + jX_1)}{R_1 + j(X_1 + X_\text{m})} \tag{6.31}$$

Note that the core-loss resistance R_c has been neglected in the derivation of Eqs. 6.29 through 6.31. Although this is a very commonly used approximation, its effect can be readily incorporated in the derivations presented here by replacing the magnetizing reactance jX_m by the magnetizing impedance Z_m, equal to the parallel combination of the core-loss resistance R_c and the magnetizing reactance jX_m.

From the Thevenin-equivalent circuit (Fig. 6.13)

$$\hat{I}_2 = \frac{\hat{V}_{1,\text{eq}}}{Z_{1,\text{eq}} + jX_2 + R_2/s} \tag{6.32}$$

and thus from the torque expression (Eq. 6.25)

$$T_\text{mech} = \frac{1}{\omega_\text{s}} \left[\frac{n_\text{ph}V_{1,\text{eq}}^2(R_2/s)}{(R_{1,\text{eq}} + (R_2/s))^2 + (X_{1,\text{eq}} + X_2)^2} \right] \tag{6.33}$$

where ω_s is the synchronous mechanical angular velocity as given by Eq. 6.26. The general shape of the torque-speed or torque-slip curve with the motor connected to a constant-voltage, constant-frequency source is shown in Figs. 6.14 and 6.15.

In normal motor operation, the rotor revolves in the direction of rotation of the magnetic field produced by the stator currents, the speed is between zero and synchronous speed, and the corresponding slip is between 1.0 and 0 (labeled "Motor region" in Fig. 6.14). Motor starting conditions are those of $s = 1.0$.

To obtain operation in the region of s greater than 1 (corresponding to a negative motor speed), the motor must be driven backward, against the direction of rotation of its magnetic field, by a source of mechanical power capable of counteracting the electromechanical torque T_mech. The chief practical usefulness of this region is in bringing motors to a quick stop by a method called *plugging*. By interchanging two stator leads in a three-phase motor, the phase sequence, and hence the direction of rotation of the magnetic field, is reversed suddenly and what was a small slip before the phase reversal becomes a slip close to 2.0 following the reversal; the motor comes to a stop under the influence of torque T_mech and is disconnected from the line before

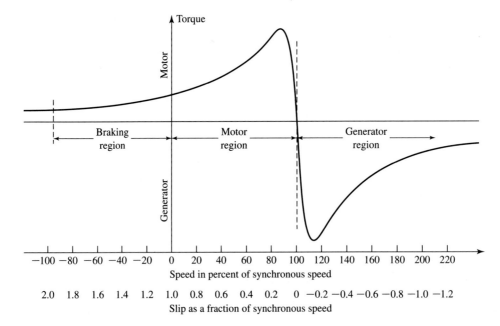

Figure 6.14 Induction-machine torque-slip curve showing braking, motor, and generator regions.

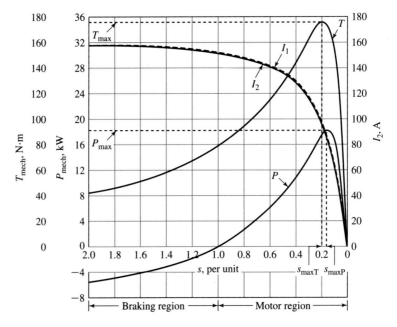

Figure 6.15 Computed torque, power, and current curves for the 7.5-kW motor in Examples 6.2 and 6.3.

it can start in the other direction. Accordingly, the region from $s = 1.0$ to $s = 2.0$ is labeled "Braking region" in Fig. 6.14.

The induction machine will operate as a generator if its stator terminals are connected to a polyphase voltage source and its rotor is driven above synchronous speed (resulting in a negative slip) by a prime mover, as shown in Fig. 6.14. The source fixes the synchronous speed and supplies the reactive power input required to excite the air-gap magnetic field. One such application is that of an induction generator connected to a power system and driven by a wind turbine.

An expression for the *maximum electromechanical torque,* or *breakdown torque,* T_{max}, indicated in Fig. 6.15, can be obtained readily from circuit considerations. As can be seen from Eq. 6.25, the electromechanical torque is a maximum when the power delivered to R_2/s in Fig. 6.13a is a maximum. It can be shown that this power will be greatest when the impedance of R_2/s equals the magnitude of the impedance $R_{1,eq} + j(X_{1,eq} + X_2)$ between it and the constant equivalent voltage $\hat{V}_{1,eq}$. Thus, maximum electromechanical torque will occur at a value of slip (s_{maxT}) for which

$$\frac{R_2}{s_{maxT}} = \sqrt{R_{1,eq}^2 + (X_{1,eq} + X_2)^2} \tag{6.34}$$

The slip s_{maxT} at maximum torque is therefore

$$s_{maxT} = \frac{R_2}{\sqrt{R_{1,eq}^2 + (X_{1,eq} + X_2)^2}} \tag{6.35}$$

and the corresponding torque is, from Eq. 6.33,

$$T_{max} = \frac{1}{\omega_s} \left[\frac{0.5 n_{ph} V_{1,eq}^2}{R_{1,eq} + \sqrt{R_{1,eq}^2 + (X_{1,eq} + X_2)^2}} \right] \tag{6.36}$$

where ω_s is the synchronous mechanical angular velocity as given by Eq. 6.26.

EXAMPLE 6.3

For the motor of Example 6.2, determine (*a*) the load component I_2 of the stator current, the electromechanical torque T_{mech}, and the electromechanical power P_{mech} for a slip $s = 0.03$; (*b*) the maximum electromechanical torque and the corresponding speed; and (*c*) the electromechanical starting torque T_{start} and the corresponding stator load current $I_{2,start}$.

■ **Solution**

First reduce the circuit to its Thevenin-equivalent form. From Eq. 6.29, $V_{1,eq} = 122.3$ V and from Eq. 6.31, $R_{1,eq} + jX_{1,eq} = 0.273 + j0.490 \ \Omega$.

a. At $s = 0.03$, $R_2/s = 4.80$. Then, from Fig. 6.13a,

$$I_2 = \frac{V_{1,eq}}{\sqrt{(R_{1,eq} + R_2/s)^2 + (X_{1,eq} + X_2)^2}} = \frac{122.3}{\sqrt{(5.07)^2 + (0.699)^2}} = 23.9 \text{ A}$$

From Eq. 6.25

$$T_{mech} = \frac{n_{ph} I_2^2 (R_2/s)}{\omega_s} = \frac{3 \times 23.9^2 \times 4.80}{125.7} = 65.4 \text{ N} \cdot \text{m}$$

and from Eq. 6.21

$$P_{mech} = n_{ph} I_2^2 (R_2/s)(1 - s) = 3 \times 23.9^2 \times 4.80 \times 0.97 = 7980 \text{ W}$$

The curves of Fig. 6.15 were computed by repeating these calculations for a number of assumed values of s.

b. At the maximum-torque point, from Eq. 6.35,

$$s_{maxT} = \frac{R_2}{\sqrt{R_{1,eq}^2 + (X_{1,eq} + X_2)^2}}$$

$$= \frac{0.144}{\sqrt{0.273^2 + 0.699^2}} = 0.192$$

and thus the speed at T_{max} is equal to $(1 - s_{maxT})n_s = (1 - 0.192) \times 1200 = 970 \text{ r/min}$
From Eq. 6.36

$$T_{max} = \frac{1}{\omega_s} \left[\frac{0.5 n_{ph} V_{1,eq}^2}{R_{1,eq} + \sqrt{R_{1,eq}^2 + (X_{1,eq} + X_2)^2}} \right]$$

$$= \frac{1}{125.7} \left[\frac{0.5 \times 3 \times 122.3^2}{0.273 + \sqrt{0.273^2 + 0.699^2}} \right] = 175 \text{ N} \cdot \text{m}$$

c. At starting, $s = 1$. Therefore

$$I_{2,start} = \frac{V_{1,eq}}{\sqrt{(R_{1,eq} + R_2)^2 + (X_{1,eq} + X_2)^2}}$$

$$= \frac{122.3}{\sqrt{0.417^2 + 0.699^2}} = 150 \text{ A}$$

From Eq. 6.25

$$T_{start} = \frac{n_{ph} I_2^2 R_2}{\omega_s} = \frac{3 \times 150^2 \times 0.144}{125.7} = 77.3 \text{ N} \cdot \text{m}$$

Practice Problem 6.3

The rotor of the induction motor of Example 6.3 is replaced by a rotor with twice the rotor resistance but which is otherwise identical to the original rotor. Repeat the calculations of Example 6.2.

Solution

a. $I_2 = 12.4 \text{ A}$, $T_{mech} = 35.0 \text{ N} \cdot \text{m}$, $P_{mech} = 4270 \text{ W}$
b. $T_{max} = 175 \text{ N} \cdot \text{m}$ at speed $= 740 \text{ r/min}$
c. At starting, $T_{start} = 128 \text{ N} \cdot \text{m}$, $I_{2,start} = 136 \text{ A}$

For the induction motor of Example 6.3, find (*a*) the rotor resistance required to produce peak electromechanical torque at zero speed (i.e., $s_{maxT} = 1.0$) and (*b*) the corresponding torque T_{max}.

Solution

    a. $R_2 = 0.751\ \Omega$

    b. $T_{max} = 175\ \text{N} \cdot \text{m}$

Under the conditions of constant-frequency operation, a typical conventional induction motor with a squirrel-cage rotor is substantially a constant-speed motor having about 10 percent or less drop in speed from no load to full load. In the case of a wound-rotor induction motor, speed variation can be obtained by inserting external resistance in the rotor circuit; the influence of increased rotor resistance on the torque-speed characteristic is shown by the dashed curves in Fig. 6.16. For such a motor, significant speed variations can be achieved as the rotor resistance is varied. Similarly, the zero-speed torque variations seen in Fig. 6.16 illustrate how the starting torque of a wound-rotor induction motor can be varied by varying the rotor resistance.

Notice from Eqs. 6.35 and 6.36 that the slip at maximum torque is directly proportional to rotor resistance R_2 but the value of the maximum torque is independent of R_2. When R_2 is increased by inserting external resistance in the rotor of a wound-rotor motor, the maximum electromechanical torque is unaffected but the speed at which it occurs can be directly controlled. This result can also be seen by observing

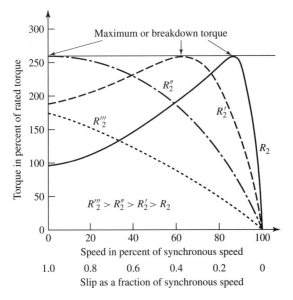

Figure 6.16 Induction-motor torque-slip curves showing effect of changing rotor-circuit resistance.

that the electromechanical torque expression of Eq. 6.33 is a function of the ratio R_2/s. Thus, the torque is unchanged as long as the ratio R_2/s remains constant.

EXAMPLE 6.4

A three-phase, 230-V, 60-Hz, 12-kW, four-pole wound-rotor induction motor has the following parameters expressed in Ω/phase.

$$R_1 = 0.095 \quad X_1 = 0.680 \quad X_2 = 0.672 \quad X_{\mathrm{m}} = 18.7$$

Using MATLAB,[†] plot the electromechanical mechanical torque T_{mech} as a function of rotor speed in r/min for rotor resistances of $R_2 = 0.1, 0.2, 0.5, 1.0$ and $1.5\ \Omega$.

■ **Solution**

The desired plot is shown in Fig. 6.17.

Here is the MATLAB script:

```
clc
clear

%Here are the motor parameters

V1 = 230/sqrt(3);
nph = 3;
poles = 4;
fe = 60;

R1 = 0.095;
X1 = 0.680;
X2 = 0.672;
Xm = 18.7;

%Calculate the synchronous speed

omegas = 4*pi*fe/poles;
ns = 120*fe/poles;

%Calculate stator Thevenin equivalent

Z1eq = j*Xm*(R1+j*X1)/(R1 + j*(X1+Xm));
R1eq = real(Z1eq);
X1eq = imag(Z1eq);

V1eq = abs(V1*j*Xm/(R1 + j*(X1+Xm)));

%Here is the loop over rotor resistance

for m = 1:5

  if m == 1
    R2 = 0.1;
  elseif m==2
    R2 = 0.2;
```

[†] MATLAB is a registered trademark of The MathWorks, Inc.

```
  elseif m==3
    R2 = 0.5;
  elseif m==4
    R2 = 1.0;
  else
    R2 = 1.5;
  end
%Here is the loop over slip
 for n = 1:200
  s(n) = n/200;    %slip
  rpm(n) = ns*(1-s(n));    %rpm
  I2 = abs(V1eq/(Z1eq + j*X2 + R2/s(n)));    %I2
  Tmech(n) = nph*I2^2*R2/(s(n)*omegas); %Electromechanical torque
 end   %End of slip loop
%Now plot
  plot(rpm,Tmech)
  if m ==1
hold
  end
end   %End of resistance loop
hold
xlabel('rpm')
ylabel('Tmech')
```

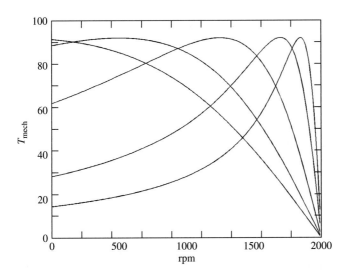

Figure 6.17 Electromechanical torque vs. speed for the wound-rotor induction motor of Example 6.4 for various values of the rotor resistance R_2.

Calculate the electromechanical mechanical torque for the motor of Example 6.4 assuming a rotor resistance of 0.3 Ω and a motor speed of 1719 r/min.

Solution

$36.8 \text{ N} \cdot \text{m}$

In applying the induction-motor equivalent circuit, the idealizations on which it is based should be kept in mind. This is particularly necessary when investigations are carried out over a wide speed range, such as is the case in investigations of motor starting. Saturation under the heavy inrush currents associated with starting conditions has a significant effect on the motor reactances. Moreover, the rotor currents are at slip frequency, which varies from stator frequency at zero speed to a low value at full-load speed. The current distribution in the rotor bars of squirrel-cage motors may vary significantly with frequency, giving rise to significant variations in rotor resistance. In fact, as discussed in Sections 6.7.2 and 6.7.3, motor designers can tailor the shape of the rotor bars in squirrel-cage motors to obtain various speed-torque characteristics. Errors from these causes can be kept to a minimum by using equivalent-circuit parameters corresponding as closely as possible to those of the proposed operating conditions.[1]

6.6   PARAMETER DETERMINATION FROM NO-LOAD AND BLOCKED-ROTOR TESTS

The equivalent-circuit parameters needed for computing the performance of a polyphase induction motor under load can be obtained from the results of a no-load test, a blocked-rotor test, and measurements of the dc resistances of the stator windings. Stray-load losses, which must be taken into account when accurate values of efficiency are to be calculated, can also be measured by tests which do not require loading the motor. The stray-load-loss tests are not described here, however.[2]

6.6.1   No-Load Test

Like the open-circuit test on a transformer, the *no-load test* on an induction motor gives information with respect to exciting current and no-load losses. This test is ordinarily performed at rated frequency and with balanced polyphase voltages applied to the stator terminals. Readings are taken at rated voltage, after the motor has been running long enough for the bearings to be properly lubricated. We will assume that the no-load

---

[1] See, for instance, R. F. Horrell and W. E. Wood, "A Method of Determining Induction Motor Speed–Torque-Current Curves from Reduced Voltage Tests," *Trans. AIEE,* 73(3):670–674 (1954).

[2] For information concerning test methods, see IEEE Std. 112-1996, "Test Procedures for Polyphase Induction Motors and Generators," Institute of Electrical and Electronics Engineers, Inc., 345 East 47th Street, New York, New York, 10017.

test is made with the motor operating at its rated electrical frequency f_r and that the following measurements are available from the no-load test:

$V_{1,nl}$ = The line-to-neutral voltage [V]

$I_{1,nl}$ = The line current [V]

P_{nl} = The total polyphase electrical input power [W]

In polyphase machines it is most common to measure line-to-line voltage, and thus the phase-to-neutral voltage must be then calculated (dividing by $\sqrt{3}$ in the case of a three-phase machine).

At no load, the rotor current is only the very small value needed to produce sufficient torque to overcome the friction and windage losses associated with rotation. The no-load rotor I^2R loss is, therefore, negligibly small. Unlike the continuous magnetic core in a transformer, the magnetizing path in an induction motor includes an air gap which significantly increases the required exciting current. Thus, in contrast to the case of a transformer, whose no-load primary I^2R loss is negligible, the no-load stator I^2R loss of an induction motor may be appreciable because of this larger exciting current.

Neglecting rotor I^2R losses, the *rotational loss* P_{rot} for normal running conditions can be found by subtracting the stator I^2R losses from the no-load input power

$$P_{rot} = P_{nl} - n_{ph}I_{1,nl}^2R_1 \tag{6.37}$$

The total rotational loss at rated voltage and frequency under load usually is considered to be constant and equal to its no-load value. Note that the stator resistance R_1 varies with stator-winding temperature. Hence, when applying Eq. 6.37, care should be taken to use the value corresponding to the temperature of the no-load test.

Note that the derivations presented here ignore the core-loss and the associated core-loss resistance and assign all the no-load losses to friction and windage. Various tests can be performed to separate the friction and windage losses from the core losses. For example, if the motor is not energized, an external drive motor can be used to drive the rotor to the no-load speed and the rotational loss will be equal to the required drive-motor output power.

Alternatively, if the motor is operated at no load and rated speed and if it is then suddenly disconnected from the supply, the decay in motor speed will be determined by the the rotational loss as

$$J\frac{d\omega_m}{dt} = -T_{rot} = -\frac{P_{rot}}{\omega_m} \tag{6.38}$$

Hence, if the rotor inertia J is known, the rotational loss at any speed ω_m can be determined from the resultant speed decay as

$$P_{rot}(\omega_m) = -\omega_m J\frac{d\omega_m}{dt} \tag{6.39}$$

Thus, the rotational losses at rated speed can be determined by evaluating Eq. 6.39 as the motor is first shut off when it is operating at rated speed.

If the no-load rotational losses are determined in this fashion, the core loss can be determined as

$$P_{core} = P_{nl} - P_{rot} - n_{ph} I_{1,nl}^2 R_1 \tag{6.40}$$

Here P_{core} is the total *no-load core loss* corresponding to the voltage of the no-load test (typically rated voltage).

Under no-load conditions, the stator current is relatively low and, to a first approximation, one can neglect the corresponding voltage drop across the stator resistance and leakage reactance. Under this approximation, the voltage across the core-loss resistance will be equal to the no-load line-to-neutral voltage and the core-loss resistance can be determined as

$$R_c = \frac{n_{ph} V_{1,nl}^2}{P_{core}} \tag{6.41}$$

Provided that the machine is operated close to rated speed and rated voltage, the refinement associated with separating out the core loss and specifically incorporating it in the form of a core-loss resistance in the equivalent circuit will not make a significant difference in the results of an analysis. Hence, it is common to ignore the core-loss resistance and to simply include the core losses with the rotational losses. For the purposes of analytical simplicity, this approach will be followed in the remainder of the text. However, if necessary, the reader should find it relatively straight forward to modify the remaining derivations to appropriately include the core-loss resistance.

Because the slip at no load, s_{nl}, is very small, the reflected rotor resistance R_2/s_{nl} is very large. The parallel combination of rotor and magnetizing branches then becomes jX_m shunted by the rotor leakage reactance X_2 in series with a very high resistance, and the reactance of this parallel combination therefore very nearly equals X_m. Consequently the apparent reactance X_{nl} measured at the stator terminals at no load very nearly equals $X_1 + X_m$, which is the self-reactance X_{11} of the stator; i.e.,

$$X_{nl} = X_{11} = X_1 + X_m \tag{6.42}$$

The self-reactance of the stator can therefore be determined from the no-load measurements. The reactive power at no load Q_{nl} can be determined as

$$Q_{nl} = \sqrt{S_{nl}^2 - P_{nl}^2} \tag{6.43}$$

where

$$S_{nl} = n_{ph} V_{1,nl} I_{1,nl} \tag{6.44}$$

is the total apparent power input at no load.

The no-load reactance X_{nl} can then be calculated from Q_{nl} and $I_{1,nl}$ as

$$X_{nl} = \frac{Q_{nl}}{n_{ph} I_{1,nl}^2} \tag{6.45}$$

Usually the no-load power factor is small (i.e., $Q_{nl} \gg P_{nl}$) so that the no-load reactance very nearly equals the no-load impedance.

$$X_{nl} \approx \frac{V_{1,nl}}{I_{1,nl}} \qquad (6.46)$$

6.6.2 Blocked-Rotor Test

Like the short-circuit test on a transformer, the *blocked-rotor test* on an induction motor gives information with respect to the leakage impedances. The rotor is blocked so that it cannot rotate (hence the slip is equal to unity), and balanced polyphase voltages are applied to the stator terminals. We will assume that the following measurements are available from the blocked-rotor test:

$V_{1,bl}$ = The line-to-neutral voltage [V]

$I_{1,bl}$ = The line current [V]

P_{bl} = The total polyphase electrical input power [W]

f_{bl} = The frequency of the blocked-rotor test [Hz]

In some cases, the blocked-rotor torque also is measured.

The equivalent circuit for blocked-rotor conditions is identical to that of a short-circuited transformer. An induction motor is more complicated than a transformer, however, because its leakage impedance may be affected by magnetic saturation of the leakage-flux paths and by rotor frequency. The blocked-rotor impedance may also be affected by rotor position, although this effect generally is small with squirrel-cage rotors.

The guiding principle is that the blocked-rotor test should be performed under conditions for which the current and rotor frequency are approximately the same as those in the machine at the operating condition for which the performance is later to be calculated. For example, if one is interested in the characteristics at slips near unity, as in starting, the blocked-rotor test should be taken at normal frequency and with currents near the values encountered in starting. If, however, one is interested in normal running characteristics, the blocked-rotor test should be taken at a reduced voltage which results in approximately rated current; the frequency also should be reduced, since the values of rotor effective resistance and leakage inductance at the low rotor frequencies corresponding to small slips may differ appreciably from their values at normal frequency, particularly with double-cage or deep-bar rotors, as discussed in Section 6.7.2.

IEEE Standard 112 suggests a blocked-rotor test frequency of 25 percent of rated frequency. The total leakage reactance at normal frequency can be obtained from this test value by considering the reactance to be proportional to frequency. The effects of frequency often are negligible for normal motors of less than 25-hp rating, and the blocked impedance can then be measured directly at normal frequency. The importance of maintaining test currents near their rated value stems from the fact that these leakage reactances are significantly affected by saturation.

Based upon blocked-rotor measurements, the blocked-rotor reactance can be found from the blocked-rotor reactive power

$$Q_{bl} = \sqrt{S_{bl}^2 - P_{bl}^2} \qquad (6.47)$$

where

$$S_{bl} = n_{ph} V_{1,bl} I_{1,bl} \tag{6.48}$$

is the total blocked-rotor apparent power. The blocked-rotor reactance, corrected to rated frequency, can then be calculated as

$$X_{bl} = \left(\frac{f_r}{f_{bl}}\right)\left(\frac{Q_{bl}}{n_{ph} I_{1,bl}^2}\right) \tag{6.49}$$

The blocked-rotor resistance can be calculated from the blocked-rotor input power as

$$R_{bl} = \frac{P_{bl}}{n_{ph} I_{1,bl}^2} \tag{6.50}$$

Once these parameters have been determined, the equivalent circuit parameters can be determined. Under blocked-rotor conditions, an expression for the stator input impedance can be obtained from examination of Fig. 6.11a (with $s = 1$) as

$$Z_{bl} = R_1 + jX_1 + (R_2 + jX_2) \text{ in parallel with } jX_m$$

$$= R_1 + R_2\left(\frac{X_m^2}{R_2^2 + (X_m + X_2)^2}\right)$$

$$+ j\left(X_1 + \frac{X_m(R_2^2 + X_2(X_m + X_2))}{R_2^2 + (X_m + X_2)^2}\right) \tag{6.51}$$

Here we have assumed that the reactances are at their rated-frequency values. Making appropriate approximations (e.g., assuming $R_2 \ll X_m$), Eq. 6.51 can be reduced to

$$Z_{bl} = R_1 + R_2\left(\frac{X_m}{X_2 + X_m}\right)^2 + j\left(X_1 + X_2\left(\frac{X_m}{X_2 + X_m}\right)\right) \tag{6.52}$$

Thus the apparent resistance under blocked-rotor conditions is given by

$$R_{bl} = R_1 + R_2\left(\frac{X_m}{X_2 + X_m}\right)^2 \tag{6.53}$$

and the apparent rated-frequency blocked-rotor reactance by

$$X_{bl} = X_1 + X_2\left(\frac{X_m}{X_2 + X_m}\right) \tag{6.54}$$

From Eqs. 6.54 and 6.53, the rotor leakage reactance X_2 and resistance R_2 can be found as

$$X_2 = (X_{bl} - X_1)\left(\frac{X_m}{X_m + X_1 - X_{bl}}\right) \tag{6.55}$$

and

$$R_2 = (R_{bl} - R_1)\left(\frac{X_2 + X_m}{X_m}\right)^2 \tag{6.56}$$

Table 6.1 Empirical distribution of leakage reactances in induction motors.

Motor class	Description	Fraction of $X_1 + X_2$ X_1	X_2
A	Normal starting torque, normal starting current	0.5	0.5
B	Normal starting torque, low starting current	0.4	0.6
C	High starting torque, low starting current	0.3	0.7
D	High starting torque, high slip	0.5	0.5
Wound rotor	Performance varies with rotor resistance	0.5	0.5

Source: IEEE Standard 112.

In order to achieve maximum accuracy as with the no-load test, if possible the value of the stator resistance R_1 used in Eq. 6.56 should be corrected to the value corresponding to the temperature of the blocked-rotor test.

Substituting for X_m from Eq. 6.42 into Eq. 6.55 gives

$$X_2 = (X_{bl} - X_1) \left(\frac{X_{nl} - X_1}{X_{nl} - X_{bl}} \right) \tag{6.57}$$

Equation 6.57 expresses the rotor leakage reactance X_2 in terms of the measured quantities X_{nl} and X_{bl} and the unknown stator leakage reactance X_1. It is not possible to make an additional measurement from which X_1 and X_2 can be determined uniquely. Fortunately, the performance of the motor is affected relatively little by the way in which the total leakage reactance is distributed between the stator and rotor. IEEE Standard 112 recommends the empirical distribution shown in Table 6.1. If the motor class is unknown, it is common to assume that X_1 and X_2 are equal.

Once the fractional relationship between X_1 and X_2 has been determined, it can be substituted into Eq. 6.57 and X_2 (and hence X_1) can be found in terms of X_{nl} and X_{bl} by solving the resultant quadratic equation.

The magnetizing reactance X_m can then be determined from Eq. 6.42.

$$X_m = X_{nl} - X_1 \tag{6.58}$$

Finally, using the known stator resistance and the values of X_m and X_2 which are now known, the rotor resistance R_2 can now be determined from Eq. 6.56.

EXAMPLE 6.5

The following test data apply to a 7.5-hp, three-phase, 220-V, 19-A, 60-Hz, four-pole induction motor with a double-squirrel-cage rotor of design class C (high-starting-torque, low-starting-current type):

Test 1: No-load test at 60 Hz

$$\text{Applied voltage } V = 219 \text{ V line-to-line}$$

$$\text{Average phase current } I_{1,nl} = 5.70 \text{ A}$$

$$\text{Power } P_{nl} = 380 \text{ W}$$

Test 2: Blocked-rotor test at 15 Hz

$$\text{Applied voltage } V = 26.5 \text{ V line-to-line}$$

$$\text{Average phase current } I_{1,\text{bl}} = 18.57 \text{ A}$$

$$\text{Power } P_{\text{bl}} = 675 \text{ W}$$

Test 3: Average dc resistance per stator phase (measured immediately after test 2)

$$R_1 = 0.262 \ \Omega$$

Test 4: Blocked-rotor test at 60 Hz

$$\text{Applied voltage } V = 212 \text{ V line-to-line}$$

$$\text{Average phase current } I_{1,\text{bl}} = 83.3 \text{ A}$$

$$\text{Power } P_{\text{bl}} = 20.1 \text{ kW}$$

$$\text{Measured starting torque } T_{\text{start}} = 74.2 \text{ N} \cdot \text{m}$$

a. Compute the no-load rotational loss and the equivalent-circuit parameters applying to the normal running conditions. Assume the same temperature as in test 3. Neglect any effects of core loss, assuming that core loss can be lumped in with the rotational losses.
b. Compute the electromechanical starting torque from the input measurements of test 4. Assume the same temperature as in test 3.

■ **Solution**

a. From Eq. 6.37, the rotational losses can be calculated as

$$P_{\text{rot}} = P_{\text{nl}} - n_{\text{ph}} I_{1,\text{nl}}^2 R_1 = 380 - 3 \times 5.70^2 \times 0.262 = 354 \text{ W}$$

The line-to-neutral no-load voltage is equal to $V_{1,\text{nl}} = 219/\sqrt{3} = 126.4$ V and thus, from Eqs. 6.43 and 6.44,

$$Q_{\text{nl}} = \sqrt{(n_{\text{ph}} V_{1,\text{nl}} I_{1,\text{nl}})^2 - P_{\text{nl}}^2} = \sqrt{(3 \times 126.4 \times 5.7)^2 - 380^2} = 2128 \text{ W}$$

and thus from Eq. 6.45

$$X_{\text{nl}} = \frac{Q_{\text{nl}}}{n_{\text{ph}} I_{1,\text{nl}}^2} = \frac{2128}{3 \times 5.7^2} = 21.8 \ \Omega$$

We can assume that the blocked-rotor test at a reduced frequency of 15 Hz and rated current reproduces approximately normal running conditions in the rotor. Thus, from test 2 and Eqs. 6.47 and 6.48 with $V_{1,\text{bl}} = 26.5/\sqrt{3} = 15.3$ V

$$Q_{\text{bl}} = \sqrt{(n_{\text{ph}} V_{1,\text{bl}} I_{1,\text{bl}})^2 - P_{\text{bl}}^2} = \sqrt{(3 \times 15.3 \times 18.57)^2 - 675^2} = 520 \text{ VA}$$

and thus from Eq. 6.49

$$X_{\text{bl}} = \left(\frac{f_{\text{r}}}{f_{\text{bl}}}\right) \left(\frac{Q_{\text{bl}}}{n_{\text{ph}} I_{1,\text{bl}}^2}\right) = \left(\frac{60}{15}\right) \left(\frac{520}{3 \times 18.57^2}\right) = 2.01 \ \Omega$$

Since we are told that this is a Class C motor, we can refer to Table 6.1 and assume that $X_1 = 0.3(X_1 + X_2)$ or $X_1 = kX_2$, where $k = 0.429$. Substituting into Eq. 6.57 results in a quadratic in X_2

$$k^2 X_2^2 + (X_{bl}(1-k) - X_{nl}(1+k))X_2 + X_{nl}X_{bl} = 0$$

or

$$(0.429)^2 X_2 + (2.01(1-0.429) - 22.0(1+0.429))X_2 + 22.0(2.01)$$

$$= 0.184 X_2^2 - 30.29 X_2 + 44.22 = 0$$

Solving gives two roots: 1.48 and 163.1. Clearly, X_2 must be less than X_{nl} and hence it is easy to identify the proper solution as

$$X_2 = 1.48 \ \Omega$$

and thus

$$X_1 = 0.633 \ \Omega$$

From Eq. 6.58,

$$X_m = X_{nl} - X_1 = 21.2 \ \Omega$$

R_{bl} can be found from Eq. 6.50 as

$$R_{bl} = \frac{P_{bl}}{n_{ph} I_{1,bl}^2} = \frac{675}{3 \times 18.57^2} = 0.652 \ \Omega$$

and thus from Eq. 6.56

$$R_2 = (R_{bl} - R_1)\left(\frac{X_2 + X_m}{X_m}\right)^2$$

$$= (0.652 - 0.262)\left(\frac{22.68}{21.2}\right)^2 = 0.447 \ \Omega$$

The parameters of the equivalent circuit for small values of slip have now been calculated.

b. Although we could calculate the electromechanical starting torque from the equivalent-circuit parameters derived in part (a), we recognize that this is a double-squirrel-cage motor and hence these parameters (most specifically the rotor parameters) will differ significantly under starting conditions from their low-slip values calculated in part (a). Hence, we will calculate the electromechanical starting torque from the rated-frequency, blocked-rotor test measurements of test 4.

From the power input and stator $I^2 R$ losses, the air-gap power P_{gap} is

$$P_{gap} = P_{bl} - n_{ph} I_{1,bl}^2 R_1 = 20,100 - 3 \times 83.3^2 \times 0.262 = 14,650 \ \text{W}$$

Since this is a four-pole machine, the synchronous speed can be found from Eq. 6.26 as $\omega_s = 188.5$ rad/sec. Thus, from Eq. 6.25 with $s = 1$

$$T_{start} = \frac{P_{gap}}{\omega_s} = \frac{14,650}{188.5} = 77.7 \ \text{N} \cdot \text{m}$$

The test value, $T_{\text{start}} = 74.2 \, \text{N} \cdot \text{m}$ is a few percent less than the calculated value because the calculations do not account for the power absorbed in the stator core loss or in stray-load losses.

Repeat the equivalent-circuit parameter calculations of Example 6.5 under the assumption that the rotor and stator leakage reactances are equal (i.e., that $X_1 = X_2$).

Solution

$$R_1 = 0.262 \, \Omega \quad R_2 = 0.430 \, \Omega$$

$$X_1 = 1.03 \, \Omega \quad X_m = 20.8 \, \Omega \quad X_2 = 1.03 \, \Omega$$

Calculation of the blocked-rotor reactance can be simplified if one assumes that $X_m \gg X_2$. Under this assumption, Eq. 6.54 reduces to

$$X_{\text{bl}} = X_1 + X_2 \tag{6.59}$$

X_1 and X_2 can then be found from Eq. 6.59 and an estimation of the fractional relationship between X_1 and X_2 (such as from Table 6.1).

Note that one might be tempted to approximate Eq. 6.56, the expression for R_2, in the same fashion. However, because the ratio $(X_2 + X_m)/X_m$ is squared, the approximation tends to result in unacceptably large errors and cannot be justified.

EXAMPLE 6.6

(*a*) Determine the parameters of the motor of Example 6.5 solving for the leakage reactances using Eq. 6.59. (*b*) Assuming the motor to be operating from a 220-V, 60-Hz source at a speed of 1746 r/min, use MATLAB to calculate the output power for the two sets of parameters.

■ Solution

a. As found in Example 6.5,

$$X_{\text{nl}} = 21.8 \, \Omega \quad X_{\text{bl}} = 2.01 \, \Omega$$

$$R_1 = 0.262 \, \Omega \quad R_{\text{bl}} = 0.652 \, \Omega$$

Thus, from Eq. 6.42,

$$X_1 + X_m = X_{\text{nl}} = 21.8 \, \Omega$$

and from Eq. 6.59

$$X_1 + X_2 = X_{\text{bl}} = 2.01 \, \Omega$$

From Table 6.1, $X_1 = 0.3(X_1 + X_2) = 0.603 \, \Omega$ and thus $X_2 = 1.41 \, \Omega$ and $X_m = 21.2 \, \Omega$.

Finally, from Eq. 6.56,

$$R_2 = (R_{bl} - R_1) \left(\frac{X_2 + X_m}{X_m} \right)^2 = 0.444 \, \Omega$$

Comparison with Example 6.5 shows the following

Parameter	Example 6.5	Example 6.6
R_1	0.262 Ω	0.262 Ω
R_2	0.447 Ω	0.444 Ω
X_1	0.633 Ω	0.603 Ω
X_2	1.47 Ω	1.41 Ω
X_m	21.2 Ω	21.2 Ω

b. For the parameters of Example 6.6, $P_{shaft} = 2467$ [W] while for the parameters of part (a) of this example, $P_{shaft} = 2497$ [W]. Thus the approximation associated with Eq. 6.59 results in an error on the order of 1 percent from using the more exact expression of Eq. 6.54. This is a typical result and hence this approximation appears to be justifiable in most cases.

Here is the MATLAB script:

```
clc
clear

% Here are the two sets of parameters
% Set 1 corresponds to the exact solution
% Set 2 corresponds to the approximate solution

R1(1) = 0.262;          R1(2) = 0.262;
R2(1) = 0.447;          R2(2) = 0.444;
X1(1) = 0.633;          X1(2) = 0.603;
X2(1) = 1.47;           X2(2) = 1.41;
Xm(1) = 21.2;           Xm(2) = 21.2;

nph = 3;
poles = 4;
Prot = 354;

%Here is the operating condition

V1 = 220/sqrt(3);
fe = 60;
rpm = 1746;

%Calculate the synchronous speed

ns = 120*fe/poles;
omegas = 4*pi*fe/poles;

slip = (ns-rpm)/ns;
omegam = omegas*(1-slip);
```

```
%Calculate stator Thevenin equivalent

%Loop over the two motors
for m = 1:2

Zgap = j*Xm(m)*(j*X2(m)+R2(m)/slip)/(R2(m)/slip+j*(Xm(m)+X2(m)));
Zin = R1(m) + j*X1(m) + Zgap;
I1 = V1/Zin;
I2 = I1*(j*Xm(m))/(R2(m)/slip+j*(Xm(m)+X2(m)));
Tmech = nph*abs(I2)^2*R2(m)/(slip*omegas); %Electromechanical torque
Pmech = omegam*Tmech;   %Electromechanical power
Pshaft = Pmech - Prot;

if (m == 1)
fprintf('\nExact solution:')
else
fprintf('\nApproximate solution:')
end

fprintf('\n   Pmech = %.1f [W], Pshaft = %.1f [W]',Pmech,Pshaft)
fprintf('\n   I1 = %.1f [A]\n',abs(I1));

end % end of "for m = 1:2" loop
```

6.7 EFFECTS OF ROTOR RESISTANCE; WOUND AND DOUBLE-SQUIRREL-CAGE ROTORS

A basic limitation of induction motors with constant rotor resistance is that the rotor design has to be a compromise. High efficiency under normal running conditions requires a low rotor resistance; but a low rotor resistance results in a low starting torque and high starting current at a low starting power factor.

6.7.1 Wound-Rotor Motors

The use of a *wound rotor* is one effective way of avoiding the need for compromise. The terminals of the rotor winding are connected to slip rings in contact with brushes. For starting, resistors may be connected in series with the rotor windings, the result being increased starting torque and reduced starting current at an improved power factor.

The general nature of the effects on the torque-speed characteristics caused by varying rotor resistance is shown in Fig. 6.16. By use of the appropriate value of rotor resistance, the maximum torque can be made to occur at standstill if high starting torque is needed. As the rotor speeds up, the external resistances can be decreased, making maximum torque available throughout the accelerating range. Since most of the rotor I^2R loss is dissipated in the external resistors, the rotor temperature rise during starting is lower than it would be if the resistance were incorporated in the rotor winding. For normal running, the rotor winding can be short-circuited directly at the brushes. The rotor winding is designed to have low resistance so that running efficiency is high and full-load slip is low. Besides their use when starting requirements are severe, wound-rotor induction motors can be used for adjustable-speed drives. Their chief disadvantage is greater cost and complexity than squirrel-cage motors.

The principal effects of varying rotor resistance on the starting and running characteristics of induction motors can be shown quantitatively by the following example.

EXAMPLE 6.7

A three-phase, 460-V, 60-Hz, four-pole, 500-hp wound-rotor induction motor, with its slip rings short-circuited, has the following properties:

Full-load slip = 1.5 percent
Rotor $I^2 R$ at full-load torque = 5.69 kW
Slip at maximum torque = 6 percent
Rotor current at maximum torque = $2.82 I_{2,fl}$, where $I_{2,fl}$ is the full-load rotor current
Torque at 20 percent slip = $1.20 T_{fl}$, where T_{fl} is the full-load torque
Rotor current at 20 percent slip = $3.95 I_{2,fl}$

If the rotor-circuit resistance is increased to $5 R_{rotor}$ by connecting noninductive resistances in series with each rotor slip ring, determine (a) the slip at which the motor will develop the same full-load torque, (b) total rotor-circuit $I^2 R$ loss at full-load torque, (c) horsepower output at full-load torque, (d) slip at maximum torque, (e) rotor current at maximum torque, (f) starting torque, and (g) rotor current at starting. Express the torques and rotor currents in per unit based on the full-load torque values.

■ Solution

The solution involves recognition of the fact that the effects of changes in the rotor resistance are seen from the stator in terms of changes in the referred resistance R_2/s. Examination of the equivalent circuit shows that, for specified applied voltage and frequency, everything concerning the stator performance is fixed by the value of R_2/s, the other impedance elements being constant. For example, if R_2 is doubled and s is simultaneously doubled, there will be no indication from the stator that anything has changed. The stator current and power factor, the power delivered to the air gap, and the torque will be unchanged as long as the ratio R_2/s remains constant.

Added physical significance can be given to the argument by examining the effects of simultaneously doubling R_2 and s from the viewpoint of the rotor. An observer on the rotor would see the resultant air-gap flux wave traveling past at twice the original slip speed, generating twice the original rotor voltage at twice the original slip frequency. The rotor reactance therefore is doubled, and since the original premise is that the rotor resistance also is doubled, the rotor impedance is doubled while the rotor power factor is unchanged. Since rotor voltage and impedance are both doubled, the effective value of the rotor current remains the same; only its frequency is changed. The air gap still has the same synchronously rotating flux and mmf waves with the same torque angle. An observer on the rotor would then agree with a counterpart on the stator that the torque is unchanged.

An observer on the rotor, however, would be aware of two changes not apparent in the stator: (1) the rotor $I^2 R$ loss will doubled, and (2) the rotor is turning more slowly and therefore developing less mechanical power with the same torque. In other words, more of the power absorbed from the stator goes into $I^2 R$ heat in the rotor, and less is available for mechanical power.

The preceding thought processes can be readily applied to the solution of this example.

a. If the rotor resistance is increased five times, the slip must increase five times for the same value of R_2/s and therefore for the same torque. But the original slip at full load is 0.015. The new slip at full-load torque therefore is $5(0.015) = 0.075$.

b. The effective value of the rotor current is the same as its full-load value before addition of the series resistance, and therefore the rotor R_2/s loss is five times the full-load value of 5.69 kW, or

$$\text{Rotor } I^2 R = 5 \times 5.69 = 28.4 \text{ kW}$$

c. The increased slip has caused the per-unit speed at full-load torque to drop from $1 - s = 0.985$ down to $1 - s = 0.925$. Since the ratio R_2/s is unchanged, the torque is the same and hence the power output has dropped proportionally, or

$$P_{\text{mech}} = \frac{0.925}{0.985}(500) = 470 \text{ hp}$$

Because the air-gap power is unchanged, the decrease in electromechanical mechanical shaft power must be accompanied by a corresponding increase in rotor $I^2 R$ loss.

d. If rotor resistance is increased five times, the slip at maximum torque simply increases five times. But the original slip at maximum torque is 0.060. The new slip at maximum torque with the added rotor resistance therefore is

$$s_{\text{maxT}} = 5(0.060) = 0.30$$

e. The effective value of the rotor current at maximum torque is independent of rotor resistance; only its frequency is changed when rotor resistance is varied. Therefore,

$$I_{2,\text{maxT}} = 2.82 I_{2,\text{fl}}$$

f. With the rotor resistance increased five times, the starting torque will be the same as the original running torque at a slip of 0.20 and therefore equals the running torque without the series resistors, namely,

$$T_{\text{start}} = 1.20 T_{\text{fl}}$$

g. The rotor current at starting with the added rotor resistances will be the same as the rotor current when running at a slip of 0.20 with the slip rings short-circuited, namely,

$$I_{2,\text{start}} = 3.95 I_{2,\text{fl}}$$

Practice Problem 6.7

Consider the motor of Example 6.7. An external resistor is added to the rotor circuits such that the full-load torque is developed at a speed of 1719 r/min. Calculate (*a*) the added resistance in terms of the inherent rotor resistance R_{rotor}, (*b*) the rotor power dissipation at full load, and (*c*) the corresponding electromechanical power.

Solution

    a. Added resistance $= 2R_{rotor}$

    b. Rotor $I^2R = 17.1$ kW

    c. $P_{mech} = 485$ hp

6.7.2 Deep-Bar and Double-Squirrel-Cage Rotors

An ingenious and simple way of obtaining a rotor resistance which will automatically vary with speed makes use of the fact that at standstill the rotor frequency equals the stator frequency; as the motor accelerates, the rotor frequency decreases to a very low value, perhaps 2 or 3 Hz at full load in a 60-Hz motor. With suitable shapes and arrangements for rotor bars, squirrel-cage rotors can be designed so that their effective resistance at 60 Hz is several times their resistance at 2 or 3 Hz. The various schemes all make use of the inductive effect of the slot-leakage flux on the current distribution in the rotor bars. This phenomenon is similar to the skin and proximity effect in any system of conductors carrying alternating current.

    Consider first a squirrel-cage rotor having deep, narrow bars like that shown in cross section in Fig. 6.18. The general character of the slot-leakage field produced by the current in the bar within this slot is shown in the figure. If the rotor iron had infinite permeability, all the leakage-flux lines would close in paths below the slot, as shown. Now imagine the bar to consist of an infinite number of layers of differential depth; one at the bottom and one at the top are indicated crosshatched in Fig. 6.18. The leakage inductance of the bottom layer is greater than that of the top layer because the bottom layer is linked by more leakage flux. Because all the layers are electrically in parallel, under ac conditions, the current in the low-reactance upper layers will be greater than that in the high-reactance lower layers. As a result, the current will be forced toward the top of the slot, and the phase of current in the upper layers will lead that of the current in the lower ones.

    This nonuniform current distribution results in an increase in the effective bar resistance and a smaller decrease in the effective leakage inductance of the bar. Since the distortion in current distribution depends on an inductive effect, the effective resistance is a function of the frequency. It is also a function of the depth of the bar and of the permeability and resistivity of the bar material. Figure 6.19 shows a

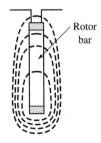

Figure 6.18 Deep rotor bar and slot-leakage flux.

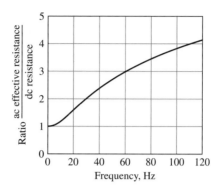

Figure 6.19 Skin effect in a copper rotor bar 2.5 cm deep.

curve of the ratio of effective ac resistance to dc resistance as a function of frequency computed for a copper bar 2.5 cm deep. A squirrel-cage rotor with deep bars can be readily designed to have an effective resistance at stator frequency (corresponding to rotor standstill conditions) several times greater than its dc resistance. As the motor accelerates, the rotor frequency decreases and therefore the effective rotor resistance decreases, approaching its dc value at small slips.

An alternative way of attaining similar results is the double-cage arrangement shown in Fig. 6.20. In this case, the squirrel-cage winding consists of two layers of bars short-circuited by end rings. The upper bars are of smaller cross-sectional area than the lower bars and consequently have higher resistance. The general nature of the slot-leakage field is shown in Fig. 6.20, from which it can be seen that the inductance of the lower bars is greater than that of the upper ones because of the flux crossing the slot between the two layers. The difference in inductance can be made quite large by properly proportioning the constriction in the slot between the two bars. At standstill, when rotor frequency equals stator frequency, there is relatively little current in the lower bars because of their high reactance; the effective resistance of the rotor at standstill is then approximately equal to that of the high-resistance upper layer. At the low rotor frequencies corresponding to small slips, however, reactance effects become negligible, and the rotor resistance then approaches that of the two layers in parallel.

Note that since the effective resistance and leakage inductance of double-cage and deep-bar rotors vary with frequency, the parameters R_2 and X_2, representing the

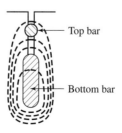

Figure 6.20 Double-squirrel-cage rotor bars and slot-leakage flux.

referred effects of rotor resistance and leakage inductance as viewed from the stator, vary with rotor speed and are not constant. Strictly speaking, a more complicated form of equivalent circuit, with multiple parallel branches, is required in order to represent these cases.

Under steady-state conditions, the simple equivalent circuit derived in Section 6.3 can still be used to represent induction machines in these cases. However R_2 and X_2 must be varied with slip. All the basic relations still apply to the motor if the values of R_2 and X_2 are properly adjusted with changes in slip. For example, in computing the starting performance, R_2 and X_2 should be taken as their effective values at stator frequency, while in computing the running performance at small slips, R_2 should be taken as its effective value at a low frequency, and X_2 should be taken as the stator-frequency value of the reactance corresponding to a low-frequency effective value of the rotor leakage inductance. Over the normal running range of slips, the rotor resistance and leakage inductance usually can be considered constant at substantially their dc values.

6.7.3 Motor-Application Considerations

By use of double-cage and deep-bar rotors, squirrel-cage motors can be designed to have the good starting characteristics resulting from high rotor resistance and, at the same time, the good running characteristics resulting from low rotor resistance. The design is necessarily somewhat of a compromise, however, and such motors lack the flexibility of a wound-rotor machine with external rotor resistance. As a result, wound-rotor motors were commonly preferred when starting requirements were severe. However, as discussed in Section 11.3, when combined with power-electronics, squirrel-cage motors can achieve all the flexibility of wound-rotor motors, and hence wound-rotor motors are becoming increasingly less common even in these cases.

To meet the usual needs of industry, integral-horsepower, three-phase, squirrel-cage motors are available from manufacturers' stock in a range of standard ratings up to 200 hp at various standard frequencies, voltages, and speeds. (Larger motors are generally regarded as special-purpose rather than general-purpose motors.) Several standard designs are available to meet various starting and running requirements. Representative torque-speed characteristics of the four most common designs are shown in Fig. 6.21. These curves are fairly typical of 1800 r/min (synchronous-speed) motors in ratings from 7.5 to 200 hp although it should be understood that individual motors may differ appreciably from these average curves.

Briefly, the characteristic features of these designs are as follows.

Design Class A: Normal Starting Torque, Normal Starting Current, Low Slip
This design usually has a low-resistance, single-cage rotor. It emphasizes good running performance at the expense of starting. The full-load slip is low and the full-load efficiency is high. The maximum torque usually is well over 200 percent of full-load torque and occurs at a small slip (less than 20 percent). The starting torque at full voltage varies from about 200 percent of full-load torque in small motors to about 100 percent in large motors. The high starting current (500 to 800 percent of full-load current when started at rated voltage) is the principal disadvantage of this design.

In sizes below about 7.5 hp these starting currents usually are within the limits on inrush current which the distribution system supplying the motor can withstand, and

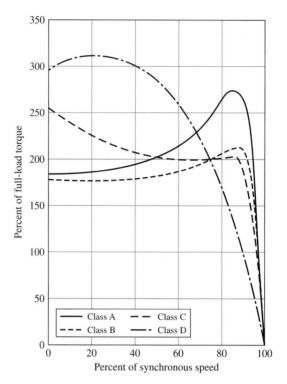

Figure 6.21 Typical torque-speed curves for 1800-r/min general-purpose induction motors.

across-the-line starting at full voltage then can be used. Otherwise, reduced-voltage starting must be used. Reduced-voltage starting results in a decrease in starting torque because the starting torque is proportional to the square of the voltage applied to the motor terminals. The reduced voltage for starting is usually obtained from an autotransformer, called a *starting compensator*, which may be manually operated or automatically operated by relays which cause full voltage to be applied after the motor is up to speed. A circuit diagram of one type of compensator is shown in Fig. 6.22. If a smoother start is necessary, series resistance or reactance in the stator may be used.

The class A motor is the basic standard design in sizes below about 7.5 and above about 200 hp. It is also used in intermediate ratings where design considerations may make it difficult to meet the starting-current limitations of the class-B design. Its field of application is about the same as that of the class-B design described next.

Design Class B: Normal Starting Torque, Low Starting Current, Low Slip   This design has approximately the same starting torque as the class-A design with but 75 percent of the starting current. Full-voltage starting, therefore, may be used with larger sizes than with class A. The starting current is reduced by designing for relatively high leakage reactance, and the starting torque is maintained by use of a double-cage or deep-bar rotor. The full-load slip and efficiency are good, about the same as for the class A design. However, the use of high reactance slightly decreases

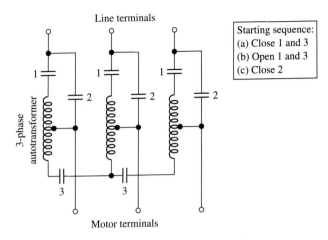

Figure 6.22 Connections of a one-step starting autotransformer.

the power factor and decidedly lowers the maximum torque (usually only slightly over 200 percent of full-load torque being obtainable).

This design is the most common in the 7.5 to 200-hp range of sizes. It is used for substantially constant-speed drives where starting-torque requirements are not severe, such as in driving fans, blowers, pumps, and machine tools.

Design Class C: High Starting Torque, Low Starting Current   This design uses a double-cage rotor with higher rotor resistance than the class-B design. The result is higher starting torque with low starting current but somewhat lower running efficiency and higher slip than the class-A and class-B designs. Typical applications are in driving compressors and conveyers.

Design Class D: High Starting Torque, High Slip   This design usually has a single-cage, high-resistance rotor (frequently brass bars). It produces very high starting torque at low starting current, high maximum torque at 50 to 100 percent slip, but runs at a high slip at full load (7 to 11 percent) and consequently has low running efficiency. Its principal uses are for driving intermittent loads involving high accelerating duty and for driving high-impact loads such as punch presses and shears. When driving high-impact loads, the motor is generally aided by a flywheel which helps supply the impact and reduces the pulsations in power drawn from the supply system. A motor whose speed falls appreciably with an increase in torque is required so that the flywheel can slow down and deliver some of its kinetic energy to the impact.

6.8 SUMMARY

In a polyphase induction motor, slip-frequency currents are induced in the rotor windings as the rotor slips past the synchronously-rotating stator flux wave. These rotor currents, in turn, produce a flux wave which rotates in synchronism with the stator flux wave; torque is produced by the interaction of these two flux waves. For increased load on the motor, the rotor speed decreases, resulting in larger slip, increased induced rotor currents, and greater torque.

Examination of the flux-mmf interactions in a polyphase induction motor shows that, electrically, the machine is a form of transformer. The synchronously-rotating air-gap flux wave in the induction machine is the counterpart of the mutual core flux in the transformer. The rotating field induces emf's of stator frequency in the stator windings and of slip frequency in the rotor windings (for all rotor speeds other than synchronous speed). Thus, the induction machine transforms voltages and at the same time changes frequency. When viewed from the stator, all rotor electrical and magnetic phenomena are transformed to stator frequency. The rotor mmf reacts on the stator windings in the same manner as the mmf of the secondary current in a transformer reacts on the primary. Pursuit of this line of reasoning leads to a single-phase equivalent circuit for polyphase induction machines which closely resemble that of a transformer.

For applications requiring a substantially constant speed without excessively severe starting conditions, the squirrel-cage motor usually is unrivaled because of its ruggedness, simplicity, and relatively low cost. Its only disadvantage is its relatively low power factor (about 0.85 to 0.90 at full load for four-pole, 60-Hz motors and considerably lower at light loads and for lower-speed motors). The low power factor is a consequence of the fact that all the excitation must be supplied by lagging reactive power taken from the ac source.

One of the salient facts affecting induction-motor applications is that the slip at which maximum torque occurs can be controlled by varying the rotor resistance. A high rotor resistance gives optimum starting conditions but poor running performance. A low rotor resistance, however, may result in unsatisfactory starting conditions. However, the design of a squirrel-cage motor is, therefore, quite likely to be a compromise.

Marked improvement in the starting performance with relatively little sacrifice in running performance can be built into a squirrel-cage motor by using a deep-bar or double-cage rotor whose effective resistance increases with slip. A wound-rotor motor can be used for very severe starting conditions or when speed control by rotor resistance is required. Variable-frequency solid-state motor drives lend considerable flexibility to the application of induction motors in variable-speed applications. These issues are discussed in Chapter 11.

6.9 PROBLEMS

6.1 The nameplate on a 460-V, 50-hp, 60-Hz, four-pole induction motor indicates that its speed at rated load is 1755 r/min. Assume the motor to be operating at rated load.

    a. What is the slip of the rotor?

    b. What is the frequency of the rotor currents?

    c. What is the angular velocity of the stator-produced air-gap flux wave with respect to the stator? With respect to the rotor?

d. What is the angular velocity of the rotor-produced air-gap flux wave with respect to the stator? With respect to the rotor?

6.2 Stray leakage fields will induce rotor-frequency voltages in a pickup coil mounted along the shaft of an induction motor. Measurement of the frequency of these induced voltages can be used to determine the rotor speed.

a. What is the rotor speed in r/min of a 50-Hz, six-pole induction motor if the frequency of the induced voltage is 0.89 Hz?

b. Calculate the frequency of the induced voltage corresponding to a four-pole, 60-Hz induction motor operating at a speed of 1740 r/min. What is the corresponding slip?

6.3 A three-phase induction motor runs at almost 1198 r/min at no load and 1112 r/min at full load when supplied from a 60-Hz, three-phase source.

a. How many poles does this motor have?

b. What is the slip in percent at full load?

c. What is the corresponding frequency of the rotor currents?

d. What is the corresponding speed of the rotor field with respect to the rotor? With respect to the stator?

6.4 Linear induction motors have been proposed for a variety of applications including high-speed ground transportation. A linear motor based on the induction-motor principle consists of a car riding on a track. The track is a developed squirrel-cage winding, and the car, which is 4.5 m long and 1.25 m wide, has a developed three-phase, 12-pole-pair armature winding. Power at 75 Hz is fed to the car from arms extending through slots to rails below ground level.

a. What is the synchronous speed in km/hr?

b. Will the car reach this speed? Explain your answer.

c. What is the slip if the car is traveling 95 km/hr? What is the frequency of the track currents under this condition?

d. If the control system controls the magnitude and frequency of the car currents to maintain constant slip, what is the frequency of the armature-winding currents when the car is traveling 75 km/hr? What is the frequency of the track currents under this condition?

6.5 A three-phase, variable-speed induction motor is operated from a variable-frequency, variable-voltage source which is controlled to maintain constant peak air-gap flux density as the frequency of the applied voltage is varied. The motor is to be operated at constant slip frequency while the motor speed is varied between one half rated speed and rated speed.

a. Describe the variation of magnitude and frequency of the applied voltage with speed.

b. Describe how the magnitude and frequency of the rotor currents will vary as the motor speed is varied.

c. How will the motor torque vary with speed?

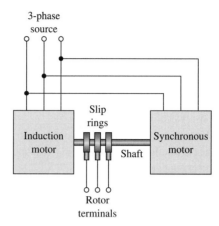

Figure 6.23 Interconnected induction and synchronous machines (Problems 6.7 and 6.8).

6.6 Describe the effect on the torque-speed characteristic of an induction motor produced by (*a*) halving the applied voltage and (*b*) halving both the applied voltage and the frequency. Sketch the resultant torque-speed curves relative to that of rated-voltage and rated-frequency. Neglect the effects of stator resistance and leakage reactance.

6.7 Figure 6.23 shows a system consisting of a three-phase wound-rotor induction machine whose shaft is rigidly coupled to the shaft of a three-phase synchronous motor. The terminals of the three-phase rotor winding of the induction machine are brought out to slip rings as shown. With the system supplied from a three-phase, 60-Hz source, the induction machine is driven by the synchronous motor at the proper speed and in the proper direction of rotation so that three-phase, 120-Hz voltages appear at the slip rings. The induction motor has four-pole stator winding.

  a. How many poles are on the rotor winding of the induction motor?
  b. If the stator field in the induction machine rotates in a clockwise direction, what is the rotation direction of its rotor?
  c. What is the rotor speed in r/min?
  d. How many poles are there on the synchronous motor?
  e. It is proposed that this system can produce dc voltage by reversing two of the phase leads to the induction motor stator. Is this proposal valid?

6.8 A system such at that shown in Fig. 6.23 is used to convert balanced 50-Hz voltages to other frequencies. The synchronous motor has four poles and drives the interconnected shaft in the clockwise direction. The induction machine has six poles and its stator windings are connected to the source in such a fashion as to produce a counterclockwise rotating field (in the direction opposite to the rotation of the synchronous motor). The machine has a wound rotor whose terminals are brought out through slip rings.

a. At what speed does the motor run?

b. What is the frequency of the voltages produced at the slip rings of the induction motor?

c. What will be the frequency of the voltages produced at the slip rings of the induction motor if two leads of the induction-motor stator are interchanged, reversing the direction of rotation of the resultant rotating field?

6.9 A three-phase, eight-pole, 60-Hz, 4160-V, 1000-kW squirrel-cage induction motor has the following equivalent-circuit parameters in ohms per phase Y referred to the stator:

$$R_1 = 0.220 \quad R_2 = 0.207 \quad X_1 = 1.95 \quad X_2 = 2.42 \quad X_m = 45.7$$

Determine the changes in these constants which will result from the following proposed design modifications. Consider each modification separately.

a. Replace the stator winding with an otherwise identical winding with a wire size whose cross-sectional area is increased by 4 percent.

b. Decrease the inner diameter of the stator laminations such that the air gap is decreased by 15 percent.

c. Replace the aluminum rotor bars (conductivity 3.5×10^7 mhos/m) with copper bars (conductivity 5.8×10^7 mhos/m).

d. Reconnect the stator winding, originally connected in Y for 4160-V operation, in Δ for 2.4 kV operation.

6.10 A three-phase, Y-connected, 460-V (line-line), 25-kW, 60-Hz, four-pole induction motor has the following equivalent-circuit parameters in ohms per phase referred to the stator:

$$R_1 = 0.103 \quad R_2 = 0.225 \quad X_1 = 1.10 \quad X_2 = 1.13 \quad X_m = 59.4$$

The total friction and windage losses may be assumed constant at 265 W, and the core loss may be assumed to be equal to 220 W. With the motor connected directly to a 460-V source, compute the speed, output shaft torque and power, input power and power factor and efficiency for slips of 1, 2 and 3 percent. You may choose either to represent the core loss by a resistance connected directly across the motor terminals or by resistance R_c connected in parallel with the magnetizing reactance X_m.

6.11 Consider the induction motor of Problem 6.10.

a. Find the motor speed in r/min corresponding to the rated shaft output power of 25 kW. (Hint: This can be easily done by writing a MATLAB script which searches over the motor slip.)

b. Similarly, find the speed in r/min at which the motor will operate with no external shaft load (assuming the motor load at that speed to consist only of the friction and windage losses).

c. Write a MATLAB script to plot motor efficiency versus output power as the motor output power varies from zero to full load.

d.   Make a second plot of motor efficiency versus output power as the motor output power varies from roughly 5 kW to full load.

6.12 Write a MATLAB script to analyze the performance of a three-phase induction motor operating at its rated frequency and voltage. The inputs should be the rated motor voltage, power and frequency, the number of poles, the equivalent-circuit parameters, and the rotational loss. Given a specific speed, the program should calculate the motor output power, the input power and power factor and the motor efficiency. Exercise your program on a 500-kW, 4160 V, three-phase, 60-Hz, four-pole induction motor operating at 1725 r/min whose rated speed rotational loss is 3.5 kW and whose equivalent-circuit parameters are:

$$R_1 = 0.521 \quad R_2 = 1.32 \quad X_1 = 4.98 \quad X_2 = 5.32 \quad X_m = 136$$

6.13 A 15-kW, 230-V, three-phase, Y-connected, 60-Hz, four-pole squirrel-cage induction motor develops full-load internal torque at a slip of 3.5 percent when operated at rated voltage and frequency. For the purposes of this problem, rotational and core losses can be neglected. The following motor parameters, in ohms per phase, have been obtained:

$$R_1 = 0.21 \quad X_1 = X_2 = 0.26 \quad X_m = 10.1$$

Determine the maximum internal torque at rated voltage and frequency, the slip at maximum torque, and the internal starting torque at rated voltage and frequency.

6.14 The induction motor of Problem 6.13 is supplied from a 230-V source through a feeder of impedance $Z_f = 0.05 + j0.14$ ohms. Find the motor slip and terminal voltage when it is supplying rated load.

6.15 A three-phase induction motor, operating at rated voltage and frequency, has a starting torque of 135 percent and a maximum torque of 220 percent, both with respect to its rated-load torque. Neglecting the effects of stator resistance and rotational losses and assuming constant rotor resistance, determine:

a.   the slip at maximum torque.

b.   the slip at rated load.

c.   the rotor current at starting (as a percentage of rotor current at rated load).

6.16 When operated at rated voltage and frequency, a three-phase squirrel-cage induction motor (of the design classification known as a high-slip motor) delivers full load at a slip of 8.7 percent and develops a maximum torque of 230 percent of full load at a slip of 55 percent. Neglect core and rotational losses and assume that the rotor resistance and inductance remain constant, independent of slip. Determine the torque at starting, with rated voltage and frequency, in per unit based upon its full-load value.

6.17 A 500-kW, 2400-V, four-pole, 60-Hz induction machine has the following equivalent-circuit parameters in ohms per phase Y referred to the stator:

$$R_1 = 0.122 \quad R_2 = 0.317 \quad X_1 = 1.364 \quad X_2 = 1.32 \quad X_m = 45.8$$

It achieves rated shaft output at a slip of 3.35 percent with an efficiency of 94.0 percent. The machine is to be used as a generator, driven by a wind turbine. It will be connected to a distribution system which can be represented by a 2400-V infinite bus.

a. From the given data calculate the total rotational and core losses at rated load.

b. With the wind turbine driving the induction machine at a slip of -3.2 percent, calculate (*i*) the electric power output in kW, (*ii*) the efficiency (electric power output per shaft input power) in percent and (*iii*) the power factor measured at the machine terminals.

c. The actual distribution system to which the generator is connected has an effective impedance of $0.18 + j0.41$ Ω/phase. For a slip of -3.2 percent, calculate the electric power as measured (*i*) at the infinite bus and (*ii*) at the machine terminals.

6.18 Write a MATLAB script to plot the efficiency as a function of electric power output for the induction generator of Problem 6.17 as the slip varies from -0.5 to -3.2 percent. Assume the generator to be operating into the system with the feeder impedance of part (c) of Problem 6.17.

6.19 For a 25-kW, 230-V, three-phase, 60-Hz squirrel-cage motor operating at rated voltage and frequency, the rotor $I^2 R$ loss at maximum torque is 9.0 times that at full-load torque, and the slip at full-load torque is 0.023. Stator resistance and rotational losses may be neglected and the rotor resistance and inductance assumed to be constant. Expressing torque in per unit of the full-load torque, find

a. the slip at maximum torque.

b. the maximum torque.

c. the starting torque.

6.20 A squirrel-cage induction motor runs at a full-load slip of 3.7 percent. The rotor current at starting is 6.0 times the rotor current at full load. The rotor resistance and inductance is independent of rotor frequency and rotational losses, stray-load losses and stator resistance may be neglected. Expressing torque in per unit of the full-load torque, compute

a. the starting torque.

b. the maximum torque and the slip at which the maximum torque occurs.

6.21 A Δ-connected, 25-kW, 230-V, three-phase, six-pole, 50-Hz squirrel-cage induction motor has the following equivalent-circuit parameters in ohms per phase Y:

$$R_1 = 0.045 \quad R_2 = 0.054 \quad X_1 = 0.29 \quad X_2 = 0.28 \quad X_m = 9.6$$

a. Calculate the starting current and torque for this motor connected directly to a 230-V source.

b. To limit the starting current, it is proposed to connect the stator winding in Y for starting and then to switch to the Δ connection for normal operation. (*i*) What are the equivalent-circuit parameters in ohms per phase for the Y connection? (*ii*) With the motor Y-connected and running directly off of a 230-V source, calculate the starting current and torque.

6.22 The following data apply to a 125-kW, 2300-V, three-phase, four pole, 60-Hz squirrel-cage induction motor:

$$\text{Stator-resistance between phase terminals} = 2.23 \ \Omega$$

No-load test at rated frequency and voltage:

$$\text{Line current} = 7.7 \ \text{A} \quad \text{Three-phase power} = 2870 \ \text{W}$$

Blocked-rotor test at 15 Hz:

$$\text{Line voltage} = 268 \ \text{V} \quad \text{Line current} = 50.3 \ \text{A}$$

$$\text{Three-phase power} = 18.2 \ \text{kW}$$

a. Calculate the rotational losses.
b. Calculate the equivalent-circuit parameters in ohms. Assume that $X_1 = X_2$.
c. Compute the stator current, input power and power factor, output power and efficiency when this motor is operating at rated voltage and frequency at a slip of 2.95 percent.

6.23 Two 50-kW, 440-V, three-phase, six-pole, 60-Hz squirrel-cage induction motors have identical stators. The dc resistance measured between any pair of stator terminals is 0.204 Ω. Blocked-rotor tests at 60-Hz produce the following results:

Motor	Volts (line-to-line)	Amperes	Three-phase power, kW
1	74.7	72.9	4.40
2	99.4	72.9	11.6

Determine the ratio of the internal starting torque developed by motor 2 to that of motor 1 (*a*) for the same current and (*b*) for the same voltage. Make reasonable assumptions.

6.24 Write a MATLAB script to calculate the parameters of a three-phase induction motor from open-circuit and blocked-rotor tests.

Input:

Rated frequency
Open-circuit test: Voltage, current and power
Blocked-rotor test: Frequency, voltage, current and power
Stator-resistance measured phase to phase
Assumed ratio X_1/X_2

Output:

> Rotational loss
> Equivalent circuit parameters R_1, R_2, X_1, X_2 and X_m

Exercise your program on a 2300-V, three-phase, 50-Hz, 250-kW induction motor whose test results are:

$$\text{Stator-resistance between phase terminals} = 0.636 \ \Omega$$

No-load test at rated frequency and voltage:

$$\text{Line current} = 20.2 \text{ A} \quad \text{Three-phase power} = 3.51 \text{ kW}$$

Blocked-rotor test at 12.5 Hz:

$$\text{Line voltage} = 142 \text{ V} \quad \text{Line current} = 62.8 \text{ A}$$

$$\text{Three-phase power} = 6.55 \text{ kW}$$

You may assume that $X_1 = 0.4(X_1 + X_2)$.

6.25 A 230-V, three-phase, six-pole, 60-Hz squirrel-cage induction motor develops a maximum internal torque of 288 percent at a slip of 15 percent when operated at rated voltage and frequency. If the effect of stator resistance is neglected, determine the maximum internal torque that this motor would develop if it were operated at 190 V and 50 Hz. Under these conditions, at what speed would the maximum torque be developed?

6.26 A 75-kW, 50-Hz, four-pole, 460-V three-phase, wound-rotor induction motor develops full-load torque at 1438 r/min with the rotor short-circuited. An external non-inductive resistance of 1.1 Ω is placed in series with each phase of the rotor, and the motor is observed to develop its rated torque at a speed of 1405 r/min. Calculate the rotor resistance per phase of the motor itself.

6.27 A 75-kW, 460-V, three-phase, four-pole, 60-Hz, wound-rotor induction motor develops a maximum internal torque of 225 percent at a slip of 16 percent when operated at rated voltage and frequency with its rotor short-circuited directly at the slip rings. Stator resistance and rotational losses may be neglected, and the rotor resistance and inductance may be assumed to be constant, independent of rotor frequency. Determine

a. the slip at full load in percent.

b. the rotor I^2R loss at full load in watts.

c. the starting torque at rated voltage and frequency in per unit and in N · m.

   If the rotor resistance is doubled (by inserting external series resistance at the slip rings) and the motor load is adjusted for such that the line current is equal to the value corresponding to rated load with no external resistance, determine

d. the corresponding slip in percent and

e. the torque in N · m.

6.28 Neglecting any effects of rotational and core losses, use MATLAB to plot the internal torque versus speed curve for the induction motor of Problem 6.10 for rated-voltage, rated-frequency operation. On the same plot, plot curves of internal torque versus speed for this motor assuming the rotor resistance increases by a factor of 2, 5 and 10.

6.29 A 100-kW, three-phase, 60-Hz, 460-V, six-pole wound-rotor induction motor develops its rated full-load output at a speed of 1158 r/min when operated at rated voltage and frequency with its slip rings short-circuited. The maximum torque it can develop at rated voltage and frequency is 310 percent of full-load torque. The resistance of the rotor winding is 0.17 Ω/phase Y. Neglect any effects of rotational and stray-load loss and stator resistance.

   a. Compute the rotor I^2R loss at full load.

   b. Compute the speed at maximum torque in r/min.

   c. How much resistance must be inserted in series with the rotor windings to produce maximum starting torque?

      With the rotor windings short-circuited, the motor is now run from a 50-Hz supply with the applied voltage adjusted so that the air-gap flux wave is essentially equal to that at rated 60-Hz operation.

   d. Compute the 50-Hz applied voltage.

   e. Compute the speed at which the motor will develop a torque equal to its rated 60-Hz value with its slip-rings shorted.

6.30 A 460-V, three-phase, six-pole, 60-Hz, 150-kW, wound-rotor induction motor develops an internal torque of 190 percent with a line current of 200 percent (torque and current expressed as a percentage of their full-load values) at a slip of 5.6 percent when running at rated voltage and frequency with its rotor terminals short-circuited. The rotor resistance is measured to be 90 mΩ between each slip ring and may be assumed to remain constant. A balanced set of Y-connected resistors is to be connected to the slip rings in order to limit the rated-voltage starting current to 200 percent of its rated value. What resistance must be chosen for each leg of the Y connection? What will be the starting torque under these conditions?

6.31 The resistance measured between each pair of slip rings of a three-phase, 60-Hz, 250-kW, 16-pole, wound-rotor induction motor is 49 mΩ. With the slip rings short-circuited, the full-load slip is 0.041. For the purposes of this problem, it may be assumed that the slip-torque curve is a straight line from no load to full load. The motor drives a fan which requires 250 kW at the full-load speed of the motor. Assuming the torque to drive the fan varies as the square of the fan speed, what resistance should be connected in series with the rotor resistance to reduce the fan speed to 400 r/min?

CHAPTER 7

DC Machines

Dc machines are characterized by their versatility. By means of various combinations of shunt-, series-, and separately-excited field windings they can be designed to display a wide variety of volt-ampere or speed-torque characteristics for both dynamic and steady-state operation. Because of the ease with which they can be controlled, systems of dc machines have been frequently used in applications requiring a wide range of motor speeds or precise control of motor output. In recent years, solid-state ac drive system technology has developed sufficiently that these systems are replacing dc machines in applications previously associated almost exclusively with dc machines. However, the versatility of dc machines in combination with the relative simplicity of their drive systems will insure their continued use in a wide variety of applications.

7.1 INTRODUCTION

The essential features of a dc machine are shown schematically in Fig. 7.1. The stator has salient poles and is excited by one or more field coils. The air-gap flux distribution created by the field windings is symmetric about the center line of the field poles. This axis is called the *field axis* or *direct axis*.

As discussed in Section 4.6.2, the ac voltage generated in each rotating armature coil is converted to dc in the external armature terminals by means of a rotating commutator and stationary brushes to which the armature leads are connected. The commutator-brush combination forms a mechanical rectifier, resulting in a dc armature voltage as well as an armature-mmf wave which is fixed in space. Commutator action is discussed in detail in Section 7.2.

The brushes are located so that commutation occurs when the coil sides are in the neutral zone, midway between the field poles. The axis of the armature-mmf wave then is 90 electrical degrees from the axis of the field poles, i.e., in the *quadrature axis*. In the schematic representation of Fig. 7.1a, the brushes are shown in the quadrature axis

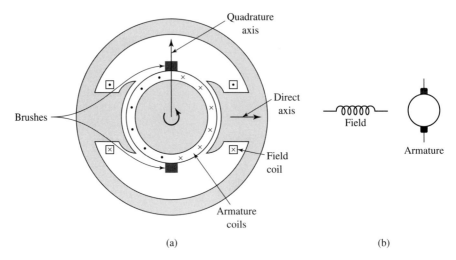

Figure 7.1   Schematic representations of a dc machine.

because this is the position of the coils to which they are connected. The armature-mmf wave then is along the brush axis, as shown. (The geometric position of the brushes in an actual machine is approximately 90 electrical degrees from their position in the schematic diagram because of the shape of the end connections to the commutator. For example, see Fig. 7.7.) For simplicity, the circuit representation usually will be drawn as in Fig. 7.1b.

Although the magnetic torque and the speed voltage appearing at the brushes are somewhat dependent on the spatial waveform of the flux distribution, for convenience we continue to assume a sinusoidal flux-density wave in the air gap as was done in Chapter 4. The torque can then be found from the magnetic field viewpoint of Section 4.7.2.

The electromagnetic torque T_{mech} can be expressed in terms of the interaction of the direct-axis air-gap flux per pole Φ_d and the space-fundamental component F_{a1} of the armature-mmf wave, in a form similar to Eq. 4.81. With the brushes in the quadrature axis, the angle between these fields is 90 electrical degrees, and its sine equals unity. Substitution in Eq. 4.81 then gives

$$T_{mech} = \frac{\pi}{2} \left(\frac{\text{poles}}{2} \right)^2 \Phi_d F_{a1} \tag{7.1}$$

in which the minus sign has been dropped because the positive direction of the torque can be determined from physical reasoning. The peak value of the sawtooth armature-mmf wave is given by Eq. 4.9, and its space fundamental F_{a1} is $8/\pi^2$ times its peak. Substitution in Eq. 7.1 then gives

$$T_{mech} = \left(\frac{\text{poles } C_a}{2\pi m} \right) \Phi_d i_a = K_a \Phi_d i_a \tag{7.2}$$

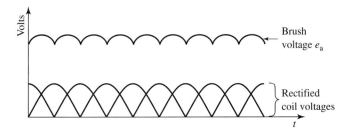

Figure 7.2 Rectified coil voltages and resultant voltage between brushes in a dc machine.

where

i_a = current in external armature circuit

C_a = total number of conductors in armature winding

m = number of parallel paths through winding

and

$$K_a = \frac{\text{poles } C_a}{2\pi m} \tag{7.3}$$

is a constant determined by the design of the winding.

The rectified voltage generated in the armature has already been found in Section 4.6.2 for an elementary single-coil armature, and its waveform is shown in Fig. 4.33. The effect of distributing the winding in several slots is shown in Fig. 7.2, in which each of the rectified sine waves is the voltage generated in one of the coils, with commutation taking place at the moment when the coil sides are in the neutral zone.

The generated voltage as observed from the brushes is the sum of the rectified voltages of all the coils in series between brushes and is shown by the rippling line labeled e_a in Fig. 7.2. With a dozen or so commutator segments per pole, the ripple becomes very small and the average generated voltage observed from the brushes equals the sum of the average values of the rectified coil voltages. From Eq. 4.53 the rectified voltage e_a between brushes, known also as the *speed voltage,* is

$$e_a = \left(\frac{\text{poles } C_a}{2\pi m}\right)\Phi_d\omega_m = K_a\Phi_d\omega_m \tag{7.4}$$

where K_a is the winding constant defined in Eq. 7.3. The rectified voltage of a distributed winding has the same average value as that of a concentrated coil. The difference is that the ripple is greatly reduced.

From Eqs. 7.2 and 7.4, with all variables expressed in SI units,

$$e_a i_a = T_{mech}\omega_m \tag{7.5}$$

Noting that the product of torque and mechanical speed is the mechanical power, this equation simply says that the instantaneous electric power associated with the speed voltage equals the instantaneous mechanical power associated with the magnetic

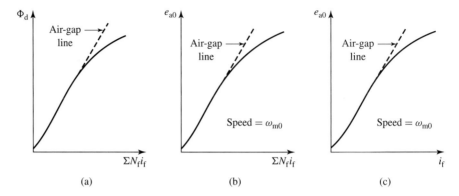

Figure 7.3 Typical form of magnetization curves of a dc machine.

torque, the direction of power flow being determined by whether the machine is acting as a motor or generator.

The direct-axis air-gap flux is produced by the combined mmf $\sum N_f i_f$ of the field windings; the flux-mmf characteristic is referred to as the *magnetization curve* for the machine. The form of a typical magnetization curve is shown in Fig. 7.3a, in which it is assumed that the armature mmf has no effect on the direct-axis flux because the axis of the armature-mmf wave is along the quadrature axis and hence perpendicular to the field axis. It will be necessary to reexamine this assumption later in this chapter, where the effects of saturation are investigated more thoroughly. Note that the magnetization curve of Fig. 7.3a does not pass through the origin. This behaviour will occur in cases where the field structure exhibits *residual magnetism,* i.e., where the magnetic material of the field does not fully demagnetize when the net field mmf is reduced to zero.

Because the armature emf is proportional to flux times speed, it is usually more convenient to express the magnetization curve in terms of the armature emf e_{a0} at a constant speed ω_{m0} as shown in Fig. 7.3b. The voltage e_a for a given flux at any other speed ω_m is proportional to the speed; i.e., from Eq. 7.4

$$\frac{e_a}{\omega_m} = K_a \Phi_d = \frac{e_{a0}}{\omega_{m0}} \tag{7.6}$$

Thus

$$e_a = \left(\frac{\omega_m}{\omega_{m0}}\right) e_{a0} \tag{7.7}$$

or, in terms of rotational speed in r/min

$$e_a = \left(\frac{n}{n_0}\right) e_{a0} \tag{7.8}$$

where n_0 is the rotational speed corresponding to the armature emf of e_{a0}.

Figure 7.3c shows the magnetization curve with only one field winding excited, in this case with the armature voltage plotted against the field current instead of the field ampere-turns. This curve can easily be obtained by test methods; since the field current can be measured directly, no knowledge of any design details is required.

Over a fairly wide range of excitation the reluctance of the electrical steel in the machine is negligible compared with that of the air gap. In this region the flux is linearly proportional to the total mmf of the field windings, the constant of proportionality being the *direct-axis permeance $\mathcal{P}_d$*; thus

$$\Phi_d = \mathcal{P}_d \sum N_f i_f \qquad (7.9)$$

The dashed straight line through the origin coinciding with the straight portion of the magnetization curves in Fig. 7.3 is called the *air-gap line*. This nomenclature refers to the fact that this linear magnetizing characteristic would be found if the reluctance of the magnetic material portion of the flux path remained negligible compared to that of the air gap, independent of the degree of magnetic saturation of the motor steel.

The outstanding advantages of dc machines arise from the wide variety of operating characteristics which can be obtained by selection of the method of excitation of the field windings. Various connection diagrams are shown in Fig. 7.4. The method of excitation profoundly influences both the steady-state characteristics and the dynamic behavior of the machine in control systems.

Consider first dc generators. The connection diagram of a *separately-excited generator* is given in Fig. 7.4a. The required field current is a very small fraction of the rated armature current; on the order of 1 to 3 percent in the average generator. A small amount of power in the field circuit may control a relatively large amount of power in the armature circuit; i.e., the generator is a power amplifier. Separately-excited generators are often used in feedback control systems when control of the armature voltage over a wide range is required.

The field windings of *self-excited generators* may be supplied in three different ways. The field may be connected in series with the armature (Fig. 7.4b), resulting in

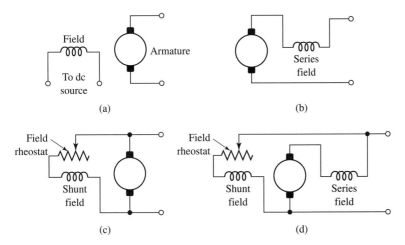

Figure 7.4 Field-circuit connections of dc machines: (a) separate excitation, (b) series, (c) shunt, (d) compound.

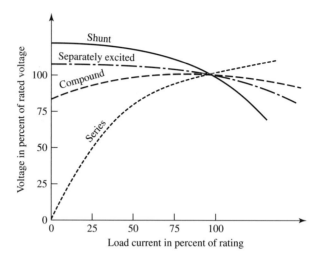

Figure 7.5 Volt-ampere characteristics of dc generators.

a *series generator*. The field may be connected in shunt with the armature (Fig. 7.4c), resulting in a *shunt generator,* or the field may be in two sections (Fig. 7.4d), one of which is connected in series and the other in shunt with the armature, resulting in a *compound generator*. With *self-excited generators,* residual magnetism must be present in the machine iron to get the self-excitation process started. The effects of residual magnetism can be clearly seen in Fig. 7.3, where the flux and voltage are seen to have nonzero values when the field current is zero.

Typical steady-state volt-ampere characteristics of dc generators are shown in Fig. 7.5, constant-speed operation being assumed. The relation between the steady-state generated emf E_a and the armature terminal voltage V_a is

$$V_a = E_a - I_a R_a \qquad (7.10)$$

where I_a is the armature current output and R_a is the armature circuit resistance. In a generator, E_a is larger than V_a, and the electromagnetic torque T_{mech} is a countertorque opposing rotation.

The terminal voltage of a separately-excited generator decreases slightly with an increase in the load current, principally because of the voltage drop in the armature resistance. The field current of a series generator is the same as the load current, so that the air-gap flux and hence the voltage vary widely with load. As a consequence, series generators are not often used. The voltage of shunt generators drops off somewhat with load, but not in a manner that is objectionable for many purposes. Compound generators are normally connected so that the mmf of the series winding aids that of the shunt winding. The advantage is that through the action of the series winding the flux per pole can increase with load, resulting in a voltage output which is nearly constant or which even rises somewhat as load increases. The shunt winding usually contains many turns of relatively small wire. The series winding, wound on the outside, consists of a few turns of comparatively heavy conductor because it must carry the full armature current of the machine. The voltage of both shunt and compound

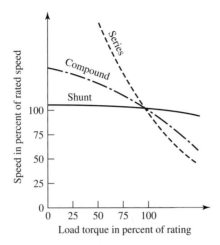

Figure 7.6 Speed-torque characteristics of dc motors.

generators can be controlled over reasonable limits by means of rheostats in the shunt field.

Any of the methods of excitation used for generators can also be used for motors. Typical steady-state dc-motor speed-torque characteristics are shown in Fig. 7.6, in which it is assumed that the motor terminals are supplied from a constant-voltage source. In a motor the relation between the emf E_a generated in the armature and the armature terminal voltage V_a is

$$V_a = E_a + I_a R_a \qquad (7.11)$$

or

$$I_a = \frac{V_a - E_a}{R_a} \qquad (7.12)$$

where I_a is now the armature-current input to the machine. The generated emf E_a is now smaller than the terminal voltage V_a, the armature current is in the opposite direction to that in a generator, and the electromagnetic torque is in the direction to sustain rotation of the armature.

In *shunt-* and *separately-excited motors,* the field flux is nearly constant. Consequently, increased torque must be accompanied by a very nearly proportional increase in armature current and hence by a small decrease in counter emf E_a to allow this increased current through the small armature resistance. Since counter emf is determined by flux and speed (Eq. 7.4), the speed must drop slightly. Like the squirrel-cage induction motor, the shunt motor is substantially a constant-speed motor having about 6 percent drop in speed from no load to full load. A typical speed-torque characteristic is shown by the solid curve in Fig. 7.6. Starting torque and maximum torque are limited by the armature current that can be successfully commutated.

An outstanding advantage of the shunt motor is ease of speed control. With a rheostat in the shunt-field circuit, the field current and flux per pole can be varied at will, and variation of flux causes the inverse variation of speed to maintain counter emf approximately equal to the impressed terminal voltage. A maximum speed range of about 4 or 6 to 1 can be obtained by this method, the limitation again being commutating conditions. By variation of the impressed armature voltage, very wide speed ranges can be obtained.

In the *series motor*, increase in load is accompanied by increases in the armature current and mmf and the stator field flux (provided the iron is not completely saturated). Because flux increases with load, speed must drop in order to maintain the balance between impressed voltage and counter emf; moreover, the increase in armature current caused by increased torque is smaller than in the shunt motor because of the increased flux. The series motor is therefore a varying-speed motor with a markedly drooping speed-torque characteristic of the type shown in Fig. 7.6. For applications requiring heavy torque overloads, this characteristic is particularly advantageous because the corresponding power overloads are held to more reasonable values by the associated speed drops. Very favorable starting characteristics also result from the increase in flux with increased armature current.

In the *compound motor*, the series field may be connected either *cumulatively*, so that its mmf adds to that of the shunt field, or *differentially*, so that it opposes. The differential connection is rarely used. As shown by the broken-dash curve in Fig. 7.6, a cumulatively-compounded motor has speed-load characteristics intermediate between those of a shunt and a series motor, with the drop of speed with load depending on the relative number of ampere-turns in the shunt and series fields. It does not have the disadvantage of very high light-load speed associated with a series motor, but it retains to a considerable degree the advantages of series excitation.

The application advantages of dc machines lie in the variety of performance characteristics offered by the possibilities of shunt, series, and compound excitation. Some of these characteristics have been touched upon briefly in this section. Still greater possibilities exist if additional sets of brushes are added so that other voltages can be obtained from the commutator. Thus the versatility of dc-machine systems and their adaptability to control, both manual and automatic, are their outstanding features.

7.2 COMMUTATOR ACTION

The dc machine differs in several respects from the ideal model of Section 4.2.2. Although the basic concepts of Section 4.2.2 are still valid, a reexamination of the assumptions and a modification of the model are desirable. The crux of the matter is the effect of the commutator shown in Figs. 4.2 and 4.16.

Figure 7.7 shows diagrammatically the armature winding of Figs. 4.22 and 4.23a with the addition of the commutator, brushes, and connections of the coils to the commutator segments. The commutator is represented by the ring of segments in the center of the figure. The segments are insulated from each other and from the shaft. Two stationary brushes are shown by the black rectangles inside the commutator. Actually the brushes usually contact the outer surface, as shown in Fig. 4.16. The coil sides in the slots are shown in cross section by the small circles with dots and crosses in them, indicating currents toward and away from the reader, respectively, as in Fig. 4.22. The connections of the coils to the commutator segments are shown by

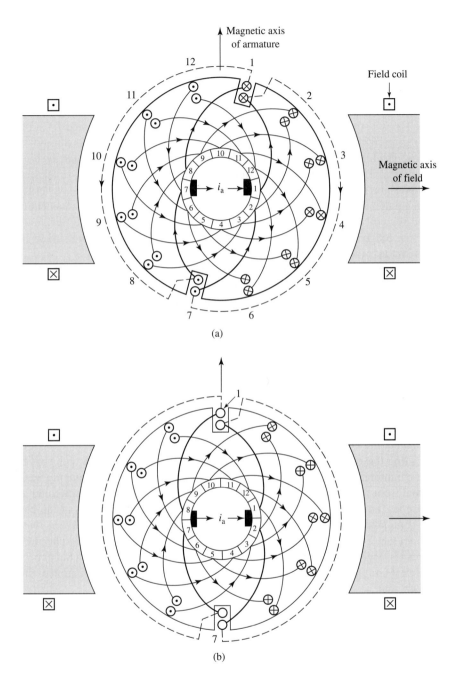

Figure 7.7 Dc machine armature winding with commutator and brushes.
(a), (b) Current directions for two positions of the armature.

the circular arcs. The end connections at the back of the armature are shown dashed for the two coils in slots 1 and 7, and the connections of these coils to adjacent commutator segments are shown by the heavy arcs. All coils are identical. The back end connections of the other coils have been omitted to avoid complicating the figure, but they can easily be traced by remembering that each coil has one side in the top of a slot and the other side in the bottom of the diametrically-opposite slot.

In Fig. 7.7a the brushes are in contact with commutator segments 1 and 7. Current entering the right-hand brush divides equally between two parallel paths through the winding. The first path leads to the inner coil side in slot 1 and finally ends at the brush on segment 7. The second path leads to the outer coil side in slot 6 and also finally ends at the brush on segment 7. The current directions in Fig. 7.7a can readily be verified by tracing these two paths. They are the same as in Fig. 4.22. The effect is identical to that of a coil wrapped around the armature with its magnetic axis vertical, and a clockwise magnetic torque is exerted on the armature, tending to align its magnetic field with that of the field winding.

Now suppose the machine is acting as a generator driven in the counterclockwise direction by an applied mechanical torque. Figure 7.7b shows the situation after the armature has rotated through the angle subtended by half a commutator segment. The right-hand brush is now in contact with both segments 1 and 2, and the left-hand brush is in contact with both segments 7 and 8. The coils in slots 1 and 7 are now short-circuited by the brushes. The currents in the other coils are shown by the dots and crosses, and they produce a magnetic field whose axis again is vertical.

After further rotation, the brushes will be in contact with segments 2 and 8, and slots 1 and 7 will have rotated into the positions which were previously occupied by slots 12 and 6 in Fig. 7.7a. The current directions will be similar to those of Fig. 7.7a except that the currents in the coils in slots 1 and 7 will have reversed. The magnetic axis of the armature is still vertical.

During the time when the brushes are simultaneously in contact with two adjacent commutator segments, the coils connected to these segments are temporarily removed from the main circuit comprising the armature winding, short-circuited by the brushes, and the currents in them are reversed. Ideally, the current in the coils being commutated should reverse linearly with time, a condition referred to as *linear commutation*. Serious departure from linear commutation will result in sparking at the brushes. Means for obtaining sparkless commutation are discussed in Section 7.9. With linear commutation the waveform of the current in any coil as a function of time is trapezoidal, as shown in Fig. 7.8.

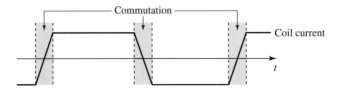

Figure 7.8 Waveform of current in an armature coil with linear commutation.

The winding of Fig. 7.7 is simpler than that used in most dc machines. Ordinarily more slots and commutator segments would be used, and except in small machines, more than two poles are common. Nevertheless, the simple winding of Fig. 7.7 includes the essential features of more complicated windings.

7.3 EFFECT OF ARMATURE MMF

Armature mmf has definite effects on both the space distribution of the air-gap flux and the magnitude of the net flux per pole. The effect on flux distribution is important because the limits of successful commutation are directly influenced; the effect on flux magnitude is important because both the generated voltage and the torque per unit of armature current are influenced thereby. These effects and the problems arising from them are described in this section.

It was shown in Section 4.3.2 and Fig. 4.23 that the armature-mmf wave can be closely approximated by a sawtooth, corresponding to the wave produced by a finely-distributed armature winding or current sheet. For a machine with brushes in the neutral position, the idealized mmf wave is again shown by the dashed sawtooth in Fig. 7.9, in which a positive mmf ordinate denotes flux lines leaving the armature surface. Current directions in all windings other than the main field are indicated by black and cross-hatched bands. Because of the salient-pole field structure found in almost all dc machines, the associated space distribution of flux will not be triangular. The distribution of air-gap flux density with only the armature excited is given by the solid curve of Fig. 7.9. As can readily be seen, it is appreciably decreased by the long air path in the interpolar space.

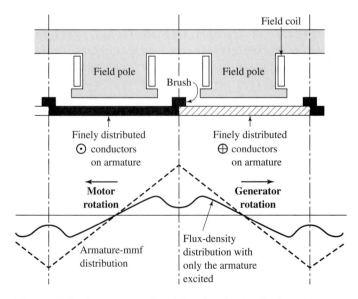

Figure 7.9 Armature-mmf and flux-density distribution with brushes on neutral and only the armature excited.

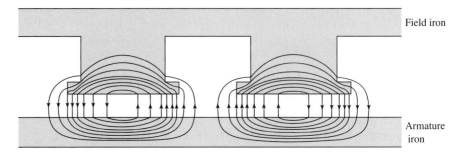

Figure 7.10 Flux with only the armature excited and brushes on neutral.

The axis of the armature mmf is fixed at 90 electrical degrees from the main-field axis by the brush position. The corresponding flux follows the paths shown in Fig. 7.10. The effect of the armature mmf is seen to be that of creating flux crossing the pole faces; thus its path in the pole shoes crosses the path of the main-field flux. For this reason, armature reaction of this type is called *cross-magnetizing armature reaction*. It evidently causes a decrease in the resultant air-gap flux density under one half of the pole and an increase under the other half.

When the armature and field windings are both excited, the resultant air-gap flux-density distribution is of the form given by the solid curve of Fig. 7.11. Superimposed on this figure are the flux distributions with only the armature excited (long-dash curve) and only the field excited (short-dash curve). The effect of cross-magnetizing armature reaction in decreasing the flux under one pole tip and increasing it under the other can be seen by comparing the solid and short-dash curves. In general, the solid curve is not the algebraic sum of the two dashed curves because of the nonlinearity of the iron magnetic circuit. Because of saturation of the iron, the flux density is decreased by a greater amount under one pole tip than it is increased under the other. Accordingly, the resultant flux per pole is lower than would be produced by the field winding alone, a consequence known as the *demagnetizing effect of cross-magnetizing armature reaction*. Since it is caused by saturation, its magnitude is a nonlinear function of both the field current and the armature current. For normal machine operation at the flux densities used commercially, the effect is usually significant, especially at heavy loads, and must often be taken into account in analyses of performance.

The distortion of the flux distribution caused by cross-magnetizing armature reaction may have a detrimental influence on the commutation of the armature current, especially if the distortion becomes excessive. In fact, this distortion is usually an important factor limiting the short-time overload capability of a dc machine. Tendency toward distortion of the flux distribution is most pronounced in a machine, such as a shunt motor, where the field excitation remains substantially constant while the armature mmf may reach very significant proportions at heavy loads. The tendency is least pronounced in a series-excited machine, such as the series motor, for both the field and armature mmf increase with load.

The effect of cross-magnetizing armature reaction can be limited in the design and construction of the machine. The mmf of the main field should exert predominating

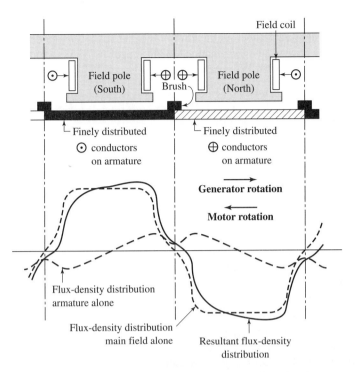

Figure 7.11 Armature, main-field, and resultant flux-density distributions with brushes on neutral.

control on the air-gap flux, so that the condition of weak field mmf and strong armature mmf should be avoided. The reluctance of the cross-flux path (essentially the armature teeth, pole shoes, and the air gap, especially at the pole tips) can be increased by increasing the degree of saturation in the teeth and pole faces, by avoiding too small an air gap, and by using a chamfered or eccentric pole face, which increases the air gap at the pole tips. These expedients affect the path of the main flux as well, but the influence on the cross flux is much greater. The best, but also the most expensive, curative measure is to compensate the armature mmf by means of a winding embedded in the pole faces, a measure discussed in Section 7.9.

If the brushes are not in the neutral position, the axis of the armature mmf wave is not 90° from the main-field axis. The armature mmf then produces not only cross magnetization but also a direct-axis demagnetizing or magnetizing effect, depending on the direction of brush shift. Shifting of the brushes from the neutral position is usually inadvertent due to incorrect positioning of the brushes or a poor brush fit. Before the invention of interpoles, however, shifting the brushes was a common method of securing satisfactory commutation, the direction of the shift being such that demagnetizing action was produced. It can be shown that brush shift in the direction of rotation in a generator or against rotation in a motor produces a direct-axis demagnetizing mmf which may result in unstable operation of a motor or excessive drop in voltage of a generator. Incorrectly placed brushes can be detected by a load test. If the brushes are on neutral, the terminal voltage of a generator or the speed of a motor should be the same for identical conditions of field excitation and armature current when the direction of rotation is reversed.

7.4 ANALYTICAL FUNDAMENTALS: ELECTRIC-CIRCUIT ASPECTS

From Eqs. 7.1 and 7.4, the electromagnetic torque and generated voltage of a dc machine are, respectively,

$$T_{\text{mech}} = K_a \Phi_d I_a \tag{7.13}$$

and

$$E_a = K_a \Phi_d \omega_m \tag{7.14}$$

where

$$K_a = \frac{\text{poles } C_a}{2\pi m} \tag{7.15}$$

Here the capital-letter symbols E_a for generated voltage and I_a for armature current are used to emphasize that we are primarily concerned with steady-state considerations in this chapter. The remaining symbols are as defined in Section 7.1. Equations 7.13 through 7.15 are basic equations for analysis of the machine. The quantity $E_a I_a$ is frequently referred to as the *electromagnetic power;* from Eqs. 7.13 and 7.14 it is related to electromagnetic torque by

$$T_{\text{mech}} = \frac{E_a I_a}{\omega_m} = K_a \Phi_d I_a \tag{7.16}$$

The electromagnetic power differs from the mechanical power at the machine shaft by the rotational losses and differs from the electric power at the machine terminals by the shunt-field and armature $I^2 R$ losses. Once the electromagnetic power $E_a I_a$ has been determined, numerical addition of the rotational losses for generators and subtraction for motors yields the mechanical power at the shaft.

The interrelations between voltage and current are immediately evident from the connection diagram of Fig. 7.12. Thus,

$$V_a = E_a \pm I_a R_a \tag{7.17}$$

$$V_t = E_a \pm I_a (R_a + R_s) \tag{7.18}$$

and

$$I_L = I_a \pm I_f \tag{7.19}$$

where the plus sign is used for a motor and the minus sign for a generator and R_a and R_s are the resistances of the armature and series field, respectively. Here, the voltage V_a

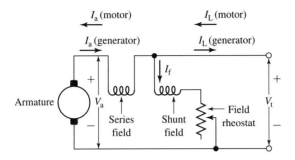

Figure 7.12 Motor or generator connection diagram with current directions.

refers to the terminal voltage of the armature winding and V_t refers to the terminal voltage of the dc machine, including the voltage drop across the series-connected field winding; they are equal if there is no series field winding.

Some of the terms in Eqs. 7.17 to 7.19 are omitted when the machine connections are simpler than those shown in Fig. 7.12. The resistance R_a is to be interpreted as that of the armature plus brushes unless specifically stated otherwise. Sometimes R_a is taken as the resistance of the armature winding alone and the brush-contact voltage drop is accounted for separately, usually assumed to be two volts.

EXAMPLE 7.1

A 25-kW 125-V separately-excited dc machine is operated at a constant speed of 3000 r/min with a constant field current such that the open-circuit armature voltage is 125 V. The armature resistance is 0.02 Ω.

Compute the armature current, terminal power, and electromagnetic power and torque when the terminal voltage is (*a*) 128 V and (*b*) 124 V.

■ **Solution**

a. From Eq. 7.17, with $V_t = 128$ V and $E_a = 125$ V, the armature current is

$$I_a = \frac{V_t - E_a}{R_a} = \frac{128 - 125}{0.02} = 150 \text{ A}$$

in the motor direction, and the power input at the motor terminal is

$$V_t I_a = 128 \times 150 = 19.20 \text{ kW}$$

The electromagnetic power is given by

$$E_a I_a = 125 \times 150 = 18.75 \text{ kW}$$

In this case, the dc machine is operating as a motor and the electromagnetic power is hence smaller than the motor input power by the power dissipated in the armature resistance.

Finally, the electromagnetic torque is given by Eq. 7.16:

$$T_{mech} = \frac{E_a I_a}{\omega_m} = \frac{18.75 \times 10^3}{100\pi} = 59.7 \, \text{N} \cdot \text{m}$$

b. In this case, E_a is larger than V_t and hence armature current will flow out of the machine, and thus the machine is operating as a generator. Hence

$$I_a = \frac{E_a - V_t}{R_a} = \frac{125 - 124}{0.02} = 50 \, \text{A}$$

and the terminal power is

$$V_t I_a = 124 \times 50 = 6.20 \, \text{kW}$$

The electromagnetic power is

$$E_a I_a = 125 \times 50 = 6.25 \, \text{kW}$$

and the electromagnetic torque is

$$T_{mech} = \frac{6.25 \times 10^3}{100\pi} = 19.9 \, \text{N} \cdot \text{m}$$

Practice Problem 7.1

The speed of the separately-excited dc machine of Example 7.1 is observed to be 2950 r/min with the field current at the same value as in Example 7.1. For a terminal voltage of 125 V, calculate the terminal current and power and the electromagnetic power for the machine. Is it acting as a motor or a generator?

Solution

Terminal current: $I_a = 104 \, \text{A}$

Terminal power: $V_t I_a = 13.0 \, \text{kW}$

Electromechanical power: $E_a I_a = 12.8 \, \text{kW}$

The machine is acting as a motor.

EXAMPLE 7.2

Consider again the separately-excited dc machine of Example 7.1 with the field-current maintained constant at the value that would produce a terminal voltage of 125 V at a speed of 3000 r/min. The machine is observed to be operating as a motor with a terminal voltage of 123 V and with a terminal power of 21.9 kW. Calculate the speed of the motor.

■ Solution

The terminal current can be found from the terminal voltage and power as

$$I_a = \frac{\text{Input power}}{V_t} = \frac{21.9 \times 10^3}{123} = 178 \, \text{A}$$

Thus the generated voltage is

$$E_a = V_t - I_a R_a = 119.4 \text{ V}$$

From Eq. 7.8, the rotational speed can be found as

$$n = n_0 \left(\frac{E_a}{E_{a0}} \right) = 3000 \left(\frac{119.4}{125} \right) = 2866 \text{ r/min}$$

Repeat Example 7.2 if the machine is observed to be operating as a generator with a terminal voltage of 124 V and a terminal power of 24 kW.

Solution

3069 r/min

For compound machines, another variation may occur. Figure 7.12 shows a *long-shunt connection* in that the shunt field is connected directly across the line terminals with the series field between it and the armature. An alternative possibility is the *short-shunt connection,* illustrated in Fig. 7.13, with the shunt field directly across the armature and the series field between it and the line terminals. The series-field current is then I_L instead of I_a, and the voltage equations are modified accordingly. There is so little practical difference between these two connections that the distinction can usually be ignored: unless otherwise stated, compound machines will be treated as though they were long-shunt connected.

Although the difference between terminal voltage V_t and armature generated voltage E_a is comparatively small for normal operation, it has a definite bearing on performance characteristics. This voltage difference divided by the armature resistance determines the value of armature current I_a and hence the strength of the armature flux. Complete determination of machine behavior requires a similar investigation of factors influencing the direct-axis flux or, more particularly, the net flux per pole Φ_d.

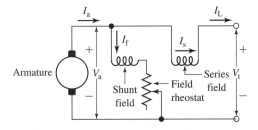

Figure 7.13 Short-shunt compound-generator connections.

7.5 ANALYTICAL FUNDAMENTALS: MAGNETIC-CIRCUIT ASPECTS

The net flux per pole is that resulting from the combined mmf's of the field and armature windings. Although in a idealized, shunt- or separately-excited dc machine the armature mmf produces magnetic flux only along the quadrature axis, in a practical device the armature current produces flux along the direct axis, either directly as produced, for example, by a series field winding or indirectly through saturation effects as discussed in Section 7.3. The interdependence of the generated armature voltage E_a and magnetic circuit conditions in the machine is accordingly a function of the sum of all the mmf's on the polar- or direct-axis flux path. First we consider the mmf intentionally placed on the stator main poles to create the working flux, i.e., the *main-field mmf,* and then we include armature-reaction effects.

7.5.1 Armature Reaction Neglected

With no load on the machine or with armature-reaction effects ignored, the resultant mmf is the algebraic sum of the mmf's acting on the main or direct axis. For the usual compound generator or motor having N_f shunt-field turns per pole and N_s series-field turns per pole,

$$\text{Main-field mmf} = N_f I_f + N_s I_s \tag{7.20}$$

Note that the mmf of the series field can either add to or subtract from that of the shunt field; the sign convention of Eq. 7.20 is such that the mmf's add. For example, in the long-shunt connection of Fig. 7.12, this would correspond to the cumulative series-field connection in which $I_s = I_a$. If the connection of this series field-winding were to be reversed such that $I_s = -I_a$, forming a differential series-field connection, then the mmf of the series field would subtract from that of the shunt field.

Additional terms will appear in Eq. 7.20 when there are additional field windings on the main poles and when, unlike the compensating windings of Section 7.9, they are wound concentric with the normal field windings to permit specialized control. When either the series or the shunt field is absent, the corresponding term in Eq. 7.20 naturally is omitted.

Equation 7.20 thus sums up in ampere-turns per pole the gross mmf of the main-field windings acting on the main magnetic circuit. The magnetization curve for a dc machine is generally given in terms of current in only the principal field winding, which is almost invariably the shunt-field winding when one is present. The mmf units of such a magnetization curve and of Eq. 7.20 can be made the same by one of two rather obvious steps. The field current on the magnetization curve can be multiplied by the turns per pole in that winding, giving a curve in terms of ampere-turns per pole; or both sides of Eq. 7.20 can be divided by N_f, converting the units to the equivalent current in the N_f coil alone which produces the same mmf. Thus

$$\text{Gross mmf} = I_f + \left(\frac{N_s}{N_f}\right) I_s \quad \text{equivalent shunt-field amperes} \tag{7.21}$$

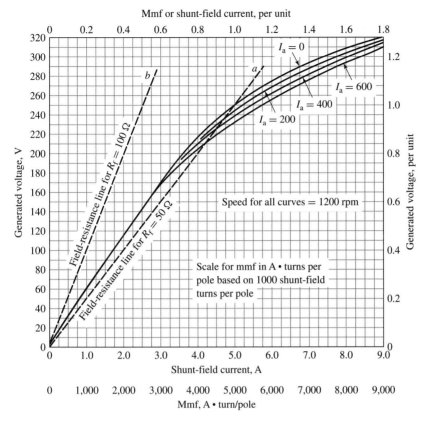

Figure 7.14 Magnetization curves for a 250-V 1200-r/min dc machine. Also shown are field-resistance lines for the discussion of self-excitation in Section 7.6.1.

This latter procedure is often the more convenient and the one more commonly adopted. As discussed in conjunction with Eq. 7.20, the connection of the series field winding will determine whether or not the series-field mmf adds to or subtracts from that of the main field winding.

An example of a *no-load magnetization characteristic* is given by the curve for $I_a = 0$ in Fig. 7.14, with values representative of those for a 100-kW, 250-V, 1200-r/min generator. Note that the mmf scale is given in both shunt-field current and ampere-turns per pole, the latter being derived from the former on the basis of a 1000 turns-per-pole shunt field. The characteristic can also be presented in normalized, or per-unit, form, as shown by the upper mmf and right-hand voltage scales. On these scales, 1.0 per-unit field current or mmf is that required to produce rated voltage at rated speed when the machine is unloaded; similarly, 1.0 per-unit voltage equals rated voltage.

Use of the magnetization curve with generated voltage, rather than flux, plotted on the vertical axis may be somewhat complicated by the fact that the speed of a dc machine need not remain constant and that speed enters into the relation between flux and generated voltage. Hence, generated voltage ordinates correspond to a unique machine speed. The generated voltage E_a at any speed ω_m is given by Eqs. 7.7 and 7.8, repeated here in terms of the steady-state values of generated voltage.

$$E_a = \left(\frac{\omega_m}{\omega_{m0}}\right) E_{a0} \tag{7.22}$$

or, in terms of rotational speed in r/min,

$$E_a = \left(\frac{n}{n_0}\right) E_{a0} \tag{7.23}$$

In these equations, ω_{m0} and n_0 are the magnetizing-curve speed in rad/sec and r/min respectively and E_{a0} is the corresponding generated voltage.

EXAMPLE 7.3

A 100-kW, 250-V, 400-A, long-shunt compound generator has an armature resistance (including brushes) of 0.025 Ω, a series-field resistance of 0.005 Ω, and the magnetization curve of Fig. 7.14. There are 1000 shunt-field turns per pole and three series-field turns per pole. The series field is connected in such a fashion that positive armature current produces direct-axis mmf which adds to that of the shunt field.

Compute the terminal voltage at rated terminal current when the shunt-field current is 4.7 A and the speed is 1150 r/min. Neglect the effects of armature reaction.

■ Solution

As is shown in Fig. 7.12, for a long-shunt connection the armature and series field-currents are equal. Thus

$$I_s = I_a = I_L + I_f = 400 + 4.7 = 405 \text{ A}$$

From Eq. 7.21 the main-field gross mmf is

$$\text{Gross mmf} = I_f + \left(\frac{N_s}{N_f}\right) I_s$$

$$= 4.7 + \left(\frac{3}{1000}\right) 405 = 5.9 \quad \text{equivalent shunt-field amperes}$$

By examining the $I_a = 0$ curve of Fig. 7.14 at this equivalent shunt-field current, one reads a generated voltage of 274 V. Accordingly, the actual emf at a speed of 1150 r/min can be found from Eq. 7.23

$$E_a = \left(\frac{n}{n_0}\right) E_{a0} = \left(\frac{1150}{1200}\right) 274 = 263 \text{ V}$$

Then

$$V_t = E_a - I_a(R_a + R_s) = 263 - 405(0.025 + 0.005) = 251 \text{ V}$$

Repeat Example 7.3 for a terminal current of 375 A and a speed of 1190 r/min.

Solution

257 V

7.5.2 Effects of Armature Reaction Included

As described in Section 7.3, current in the armature winding gives rise to a demagnetizing effect caused by a cross-magnetizing armature reaction. Analytical inclusion of this effect is not straightforward because of the nonlinearities involved. One common approach is to base analyses on the measured performance of the machine in question or for one of similar design and frame size. Data are taken with both the field and armature excited, and the tests are conducted so that the effects on generated emf of varying both the main-field excitation and the armature mmf can be noted.

One form of summarizing and correlating the results is illustrated in Fig. 7.14. Curves are plotted not only for the no-load characteristic ($I_a = 0$) but also for a family of values of I_a. In the analysis of machine performance, the inclusion of armature reaction then becomes simply a matter of using the magnetization curve corresponding to the armature current involved. Note that the ordinates of all these curves give values of armature-generated voltage E_a, not terminal voltage under load. Note also that all the curves tend to merge with the air-gap line as the saturation of the iron decreases.

The load-saturation curves are displaced to the right of the no-load curve by an amount which is a function of I_a. The effect of armature reaction then is approximately the same as a demagnetizing mmf F_{ar} acting on the main-field axis. This additional term can then be included in Eq. 7.20, with the result that the net direct-axis mmf can be assumed to be

$$\text{Net mmf} = \text{gross mmf} - F_{ar} = N_f I_f + N_s I_s - AR \qquad (7.24)$$

The no-load magnetization curve can then be used as the relation between generated emf and net excitation under load with the armature reaction accounted for as a demagnetizing mmf. Over the normal operating range (about 240 to about 300 V for the machine of Fig. 7.14), the demagnetizing effect of armature reaction may be assumed to be approximately proportional to the armature current.

The reader should be aware that the amount of armature reaction present in Fig. 7.14 is chosen so that some of its disadvantageous effects will appear in a pronounced form in subsequent numerical examples and problems illustrating generator and motor performance features. It is definitely more than one would expect to find in a normal, well-designed machine operating at normal currents.

EXAMPLE 7.4

Consider again the long-shunt compound dc generator of Example 7.3. As in Example 7.3, compute the terminal voltage at rated terminal current when the shunt-field current is 4.7 A and the speed is 1150 r/min. In this case however, include the effects of armature reaction.

■ **Solution**

As calculated in Example 7.3, $I_s = I_a = 400$ A and the gross mmf is equal to 5.9 equivalent shunt-field amperes. From the curve labeled $I_a = 400$ in Fig. 7.14 (based upon a rated terminal current of 400 A), the corresponding generated emf is found to be 261 V (as compared to 274 V with armature reaction neglected). Thus from Eq. 7.23, the actual generated voltage at a speed of 1150 r/min is equal to

$$E_a = \left(\frac{n}{n_0}\right) E_{a0} = \left(\frac{1150}{1200}\right) 261 = 250 \text{ V}$$

Then

$$V_t = E_a - I_a(R_a + R_s) = 250 - 405(0.025 + 0.005) = 238 \text{ V}$$

EXAMPLE 7.5

To counter the effects of armature reaction, a fourth turn is added to the series field winding of the dc generator of Examples 7.3 and 7.4, increasing its resistance to 0.007 Ω. Repeat the terminal-voltage calculation of Example 7.4.

■ **Solution**

As in Examples 7.3 and 7.4, $I_s = I_a = 405$ A. The main-field mmf can then be calculated as

$$\text{Gross mmf} = I_f + \left(\frac{N_s}{N_f}\right) I_s = 4.7 + \left(\frac{4}{1000}\right) 405$$
$$= 6.3 \quad \text{equivalent shunt-field amperes}$$

From the $I_a = 400$ curve of Fig. 7.14 with an equivalent shunt-field current of 6.3 A, one reads a generated voltage 269 V which corresponds to an emf at 1150 r/min of

$$E_a = \left(\frac{1150}{1200}\right) 269 = 258 \text{ V}$$

The terminal voltage can now be calculated as

$$V_t = E_a - I_a(R_a + R_s) = 258 - 405(0.025 + 0.007) = 245 \text{ V}$$

Practice Problem 7.4

Repeat Example 7.5 assuming that a fifth turn is added to the series field winding, bringing its total resistance to 0.009 Ω.

Solution

250 V

7.6 ANALYSIS OF STEADY-STATE PERFORMANCE

Although exactly the same principles apply to the analysis of a dc machine acting as a generator as to one acting as a motor, the general nature of the problems ordinarily encountered is somewhat different for the two methods of operation. For a generator, the speed is usually fixed by the prime mover, and problems often encountered are to determine the terminal voltage corresponding to a specified load and excitation or to find the excitation required for a specified load and terminal voltage. For a motor, however, problems frequently encountered are to determine the speed corresponding to a specific load and excitation or to find the excitation required for specified load and speed conditions; terminal voltage is often fixed at the value of the available source. The routine techniques of applying the common basic principles therefore differ to the extent that the problems differ.

7.6.1 Generator Analysis

Since the main-field current is independent of the generator voltage, separately-excited generators are the simplest to analyze. For a given load, the equivalent main-field excitation is given by Eq. 7.21 and the associated armature-generated voltage E_a is determined by the appropriate magnetization curve. This voltage, together with Eq. 7.17 or 7.18, fixes the terminal voltage.

Shunt-excited generators will be found to self-excite under properly chosen operating conditions. Under these conditions, the generated voltage will build up spontaneously (typically initiated by the presence of a small amount of residual magnetism in the field structure) to a value ultimately limited by magnetic saturation. In self-excited generators, the shunt-field excitation depends on the terminal voltage and the series-field excitation depends on the armature current. Dependence of shunt-field current on terminal voltage can be incorporated graphically in an analysis by drawing the *field-resistance line,* the line 0a in Fig. 7.14, on the magnetization curve. The field-resistance line 0a is simply a graphical representation of Ohm's law applied to the shunt field. It is the locus of the terminal voltage versus shunt-field-current operating point. Thus, the line 0a is drawn for $R_f = 50\ \Omega$ and hence passes through the origin and the point (1.0 A, 50 V).

The tendency of a shunt-connected generator to self-excite can be seen by examining the buildup of voltage for an unloaded shunt generator. When the field circuit is closed, the small voltage from residual magnetism (the 6-V intercept of the magnetization curve, Fig. 7.14) causes a small field current. If the flux produced by the resulting ampere-turns adds to the residual flux, progressively greater voltages and field currents are obtained. If the field ampere-turns opposes the residual magnetism, the shunt-field terminals must be reversed to obtain buildup.

This process can be seen with the aid of Fig. 7.15. In Fig. 7.15, the generated voltage e_a is shown in series with the armature inductance L_a and resistance R_a. The shunt-field winding, shown connected across the armature terminals, is represented by its inductance L_f and resistance R_f. Recognizing that since there is no load current

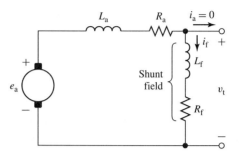

Figure 7.15 Equivalent circuit for analysis of voltage buildup in a self-excited dc generator.

on the generator ($i_L = 0$), $i_a = i_f$ and thus the differential equation describing the buildup of the field current i_f is

$$(L_a + L_f)\frac{di_f}{dt} = e_a - (R_a + R_f)i_f \tag{7.25}$$

From this equation it is clear that as long as the net voltage across the winding inductances $e_a - i_f(R_a + R_f)$ is positive, the field current and the corresponding generated voltage will increase. Buildup continues until the volt-ampere relations represented by the magnetization curve and the field-resistance line are simultaneously satisfied, which occurs at their intersection $e_a = (R_a + R_f) i_f$; in this case at $e_a = 250$ V for the line $0a$ in Fig. 7.14. From Eq. 7.25, it is clear that the field resistance line should also include the armature resistance. However, this resistance is in general much less than the field and is typically neglected.

Notice that if the field resistance is too high, as shown by line $0b$ for $R_f = 100 \, \Omega$ in Fig. 7.14, the intersection is at very low voltage and buildup is not obtained. Notice also that if the field-resistance line is essentially tangent to the lower part of the magnetization curve, corresponding to a field resistance of 57 Ω in Fig. 7.14, the intersection may be anywhere from about 60 to 170 V, resulting in very unstable conditions. The corresponding resistance is the *critical field resistance,* above which buildup will not be obtained. The same buildup process and the same conclusions apply to compound generators; in a long-shunt compound generator, the series-field mmf created by the shunt-field current is entirely negligible.

For a shunt generator, the magnetization curve for the appropriate value of I_a is the locus of E_a versus I_f. The field-resistance line is the locus V_t versus I_f. Under steady-state operating conditions, at any value of I_f, the vertical distance between the line and the curve must be the $I_a R_a$ drop at the load corresponding to that condition. Determination of the terminal voltage for a specified armature current is then simply a matter of finding where the line and curve are separated vertically by the proper amount; the ordinate of the field-resistance line at that field current is then the terminal voltage. For a compound generator, however, the series-field mmf causes corresponding points on the line and curve to be displaced horizontally as well as vertically. The horizontal displacement equals the series-field mmf measured in equivalent shunt-field amperes, and the vertical displacement is still the $I_a R_a$ drop.

Great precision is evidently not obtained from the foregoing computational process. The uncertainties caused by magnetic hysteresis in dc machines make high precision unattainable in any event. In general, the magnetization curve on which the machine operates on any given occasion may range from the rising to the falling part of the rather fat hysteresis loop for the magnetic circuit of the machine, depending essentially on the magnetic history of the iron. The curve used for analysis is usually the mean magnetization curve, and thus the results obtained are substantially correct on the average. Significant departures from the average may be encountered in the performance of any dc machine at a particular time, however.

EXAMPLE 7.6

A 100-kW, 250-V, 400-A, 1200-r/min dc shunt generator has the magnetization curves (including armature-reaction effects) of Fig. 7.14. The armature-circuit resistance, including brushes, is 0.025 Ω. The generator is driven at a constant speed of 1200 r/min, and the excitation is adjusted (by varying the shunt-field rheostat) to give rated voltage at no load.

(a) Determine the terminal voltage at an armature current of 400 A. (b) A series field of four turns per pole having a resistance of 0.005 Ω is to be added. There are 1000 turns per pole in the shunt field. The generator is to be *flat-compounded* so that the full-load voltage is 250 V when the shunt-field rheostat is adjusted to give a no-load voltage of 250 V. Show how a resistance across the series field (referred to as a *series-field diverter*) can be adjusted to produce the desired performance.

■ **Solution**

a. The 50 Ω field-resistance line 0a (Fig. 7.14) passes through the 250-V, 5.0-A point of the no-load magnetization curve. At $I_a = 400$ A

$$I_a R_a = 400 \times 0.025 = 10 \text{ V}$$

Thus the operating point under this condition corresponds to a condition for which the terminal voltage V_t (and hence the shunt-field voltage) is 10 V less than the generated voltage E_a.

A vertical distance of 10 V exists between the magnetization curve for $I_a = 400$ A and the field-resistance line at a field current of 4.1 A, corresponding to $V_t = 205$ V. The associated line current is

$$I_L = I_a - I_f = 400 - 4 = 396 \text{ A}$$

Note that a vertical distance of 10 V also exists at a field current of 1.2 A, corresponding to $V_t = 60$ V. The voltage-load curve is accordingly double-valued in this region. It can be shown that this operating point is unstable and that the point for which $V_t = 205$ V is the normal operating point.

b. For the no-load voltage to be 250 V, the shunt-field resistance must be 50 Ω and the field-resistance line is 0a (Fig. 7.14). At full load, $I_f = 5.0$ A because $V_t = 250$ V. Then

$$I_a = 400 + 5.0 = 405 \text{ A}$$

and

$$E_a = V_t + I_a(R_a + R_p) = 250 + 405(0.025 + R_p)$$

where R_p is the parallel combination of the series-field resistance $R_s = 0.005\ \Omega$ and the diverter resistance R_d

$$R_p = \frac{R_s R_d}{(R_s + R_d)}$$

The series field and the diverter resistor are in parallel, and thus the shunt-field current can be calculated as

$$I_s = 405\left(\frac{R_d}{R_s + R_d}\right) = 405\left(\frac{R_p}{R_s}\right)$$

and the equivalent shunt-field amperes can be calculated from Eq. 7.21 as

$$I_{net} = I_f + \frac{4}{1000}I_s = 5.0 + \frac{4}{1000}I_s$$

$$= 5.0 + 1.62\left(\frac{R_p}{R_s}\right)$$

This equation can be solved for R_p which can be, in turn, substituted (along with $R_s = 0.005\ \Omega$) in the equation for E_a to yield

$$E_a = 253.9 + 1.25I_{net}$$

This can be plotted on Fig. 7.14 (E_a on the vertical axis and I_{net} on the horizontal axis). Its intersection with the magnetization characteristic for $I_a = 400$ A (strictly speaking, of course, a curve for $I_a = 405$ A should be used, but such a small distinction is obviously meaningless here) gives $I_{net} = 6.0$ A.
     Thus

$$R_p = \frac{R_s(I_{net} - 5.0)}{1.62} = 0.0031\ \Omega$$

and

$$R_d = 0.0082\ \Omega$$

Practice Problem 7.5

Repeat part (b) of Example 7.6, calculating the diverter resistance which would give a full-load voltage of 240-V if the excitation is adjusted for a no-load voltage of 250-V.

Solution

$$R_d = 1.9\ m\Omega$$

7.6.2 Motor Analysis

The terminal voltage of a motor is usually held substantially constant or controlled to a specific value. Hence, motor analysis most nearly resembles that for separately-excited generators, although speed is now an important variable and often the one whose value is to be found. Analytical essentials include Eqs. 7.17 and 7.18 relating terminal voltage and generated voltage (counter emf); Eq. 7.21 for the main-field excitation; the magnetization curve for the appropriate armature current as the graphical relation between counter emf and excitation; Eq. 7.13 showing the dependence of electromagnetic torque on flux and armature current; and Eq. 7.14 relating counter emf to flux and speed. The last two relations are particularly significant in motor analysis. The former is pertinent because the interdependence of torque and the stator and rotor field strengths must often be examined. The latter is the usual medium for determining motor speed from other specified operating conditions.

Motor speed corresponding to a given armature current I_a can be found by first computing the actual generated voltage E_a from Eq. 7.17 or 7.18. Next the main-field excitation can be obtained from Eq. 7.21. Since the magnetization curve will be plotted for a constant speed ω_{m0}, which in general will be different from the actual motor speed ω_m, the generated voltage read from the magnetization curve at the foregoing main-field excitation will correspond to the correct flux conditions but to speed ω_{m0}. Substitution in Eq. 7.22 then yields the actual motor speed.

Note that knowledge of the armature current is postulated at the start of this process. When, as is frequently the case, the speed at a stated shaft power or torque output is to be found, an iterative procedure based on assumed values of I_a usually forms the basis for finding the solution.

EXAMPLE 7.7

A 100-hp, 250-V dc shunt motor has the magnetization curves (including armature-reaction effects) of Fig. 7.14. The armature circuit resistance, including brushes, is 0.025 Ω. No-load rotational losses are 2000 W and the stray-load losses equal 1.0% of the output. The field rheostat is adjusted for a no-load speed of 1100 r/min.

a. As an example of computing points on the speed-load characteristic, determine the speed in r/min and output in horsepower (1 hp = 746 W) corresponding to an armature current of 400 A.

b. Because the speed-load characteristic observed to in part (a) is considered undesirable, a *stabilizing winding* consisting of 1-1/2 cumulative series turns per pole is to be added. The resistance of this winding is assumed negligible. There are 1000 turns per pole in the shunt field. Compute the speed corresponding to an armature current of 400 A.

■ Solution

a. At no load, $E_a = 250$ V. The corresponding point on the 1200-r/min no-load saturation curve is

$$E_{a0} = 250 \left(\frac{1200}{1100} \right) = 273 \text{ V}$$

for which $I_f = 5.90$ A. The field current remains constant at this value.

At $I_a = 400$ A, the actual counter emf is

$$E_a = 250 - 400 \times 0.025 = 240 \text{ V}$$

From Fig. 7.14 with $I_a = 400$ and $I_f = 5.90$, the value of E_a would be 261 V if the speed were 1200 r/min. The actual speed is then found from Eq. 7.23

$$n = 1200 \left(\frac{240}{261} \right) = 1100 \text{ r/min}$$

The electromagnetic power is

$$E_a I_a = 240 \times 400 = 96 \text{ kW}$$

Deduction of the rotational losses leaves 94 kW. With stray load losses accounted for, the power output P_0 is given by

$$94 \text{ kW} - 0.01 P_0 = P_0$$

or

$$P_0 = 93.1 \text{ kW} = 124.8 \text{ hp}$$

Note that the speed at this load is the same as at no load, indicating that armature-reaction effects have caused an essentially flat speed-load curve.

b. With $I_f = 5.90$ A and $I_s = I_a = 400$ A, the main-field mmf in equivalent shunt-field amperes is

$$5.90 + \left(\frac{1.5}{1000} \right) 400 = 6.50 \text{ A}$$

From Fig. 7.14 the corresponding value of E_a at 1200 r/min would be 271 V. Accordingly, the speed is now

$$n = 1200 \left(\frac{240}{271} \right) = 1063 \text{ r/min}$$

The power output is the same as in part (a). The speed-load curve is now drooping, due to the effect of the stabilizing winding.

Practice Problem 7.6

Repeat Example 7.7 for an armature current of $I_a = 200$ A.

Solution

a. Speed $= 1097$ r/min and $P_0 = 46.5$ kW $= 62.4$ hp
b. Speed $= 1085$ r/min

7.7 PERMANENT-MAGNET DC MACHINES

Permanent-magnet dc machines are widely found in a wide variety of low-power applications. The field winding is replaced by a permanent magnet, resulting in simpler construction. Permanent magnets offer a number of useful benefits in these applications. Chief among these is that they do not require external excitation and its associated power dissipation to create magnetic fields in the machine. The space required for the permanent magnets may be less than that required for the field winding, and thus permanent-magnet machines may be smaller, and in some cases cheaper, than their externally-excited counterparts.

Alternatively, permanent-magnet dc machines are subject to limitations imposed by the permanent magnets themselves. These include the risk of demagnetization due to excessive currents in the motor windings or due to overheating of the magnet. In addition, permanent magnets are somewhat limited in the magnitude of air-gap flux density that they can produce. However, with the development of new magnetic materials such as samarium-cobalt and neodymium-iron-boron (Section 1.6), these characteristics are becoming less and less restrictive for permanent-magnet machine design.

Figure 7.16 shows a disassembled view of a small permanent-magnet dc motor. Notice that the rotor of this motor consists of a conventional dc armature with commutator segments and brushes. There is also a small permanent magnet on one end which constitutes the field of an ac tachometer which can be used in applications where precise speed control is required.

Unlike the salient-pole field structure characteristic of a dc machine with external field excitation (see Fig. 7.23), permanent-magnet motors such as that of Fig. 7.16 typically have a smooth stator structure consisting of a cylindrical shell (or fraction thereof) of uniform thickness permanent-magnet material magnetized in the radial direction. Such a structure is illustrated in Fig. 7.17, where the arrows indicate the

Figure 7.16 Disassembled permanent-magnet dc motor. A permanent-magnet ac tachometer is also included in the same housing for speed control. (*Buehler Products Inc.*)

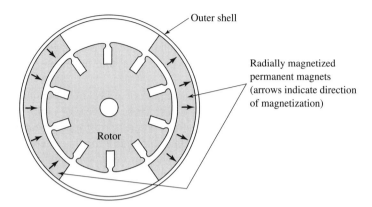

Figure 7.17 Cross section of a typical permanent-magnet motor. Arrows indicate the direction of magnetization in the permanent magnets.

direction of magnetization. The rotor of Fig. 7.17 has winding slots and has a commutator and brushes, as in all dc machines. Notice also that the outer shell in these motors serves a dual purpose: it is made up of a magnetic material and thus serves as a return path for magnetic flux as well as a support for the magnets.

EXAMPLE 7.8

Figure 7.18a defines the dimensions of a permanent-magnet dc motor similar to that of Fig. 7.17. Assume the following values:

> Rotor radius $R_r = 1.2$ cm
> Gap length $t_g = 0.05$ cm
> Magnet thickness $t_m = 0.35$ cm

Also assume that both the rotor and outer shell are made of infinitely permeable magnetic material ($\mu \to \infty$) and that the magnet is neodymium-iron-boron (see Fig. 1.19).

Ignoring the effects of rotor slots, estimate the magnetic flux density B in the air gap of this motor.

■ **Solution**

Because the rotor and outer shell are assumed to be made of material with infinite magnetic permeability, the motor can be represented by a magnetic equivalent circuit consisting of an air gap of length $2t_g$ in series with a section of neodymium-iron-boron of length $2t_m$ (see Fig. 7.18b). Note that this equivalent circuit is approximate because the cross-sectional area of the flux path in the motor increases with increasing radius, whereas it is assumed to be constant in the equivalent circuit.

The solution can be written down by direct analogy with Example 1.9. Replacing the air-gap length g with $2t_g$ and the magnet length l_m with $2t_m$, the equation for the load line can be written as

$$B_m = -\mu_0 \left(\frac{t_m}{t_g}\right) H_m = -7\mu_0 H_m$$

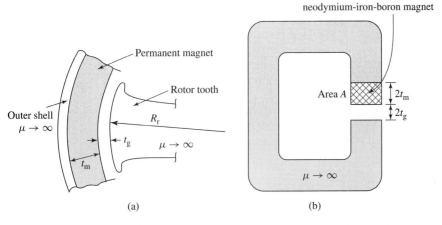

Figure 7.18 (a) Dimension definitions for the motor of Fig. 7.17, (b) approximate magnetic equivalent circuit.

This relationship can be plotted on Fig. 1.19 to find the operating point from its intersection with the dc magnetization curve for neodymium-iron-boron. Alternatively, recognizing that, in SI units, the dc magnetization curve for neodymium-iron-boron is a straight line of the form

$$B_m = 1.06\mu_0 H_m + 1.25$$

we find that

$$B_m = B_g = 1.09 \text{ T}$$

Practice Problem 7.7

Estimate the magnetic flux density in the motor of Example 7.8 if the rotor radius is increased to $R_r = 1.3$ cm and the magnetic thickness is decreased to $t_m = 0.25$ cm.

Solution

$$B_m = B_g = 1.03 \text{ T}$$

Figure 7.19 shows an exploded view of an alternate form of permanent-magnet dc motor. In this motor, the armature windings are made into the form of a thin disk (with no iron in the armature). As in any dc motor, brushes are used to commutate the armature current, contacting the commutator portion of the armature which is at its inner radius. Currents in the disk armature flow radially, and the disk is placed between two sets of permanent magnets which create axial flux through the armature winding. The combination of axial magnetic flux and radial currents produces a torque which results in rotation, as in any dc motor. This motor configuration can be shown to produce large acceleration (due to low rotor inertia), no cogging torque (due to the fact that the rotor is nonmagnetic), and long brush life and high-speed capability (due

Figure 7.19 Exploded view of a disk armature permanent-magnet servomotor. Magnets are Alnico. (*PMI Motion Technologies.*)

to the fact that the armature inductance is low and thus there will be little arcing at the commutator segments).

The principal difference between permanent-magnet dc machines and those discussed previously in this chapter is that they have a fixed source of field-winding flux which is supplied by a permanent magnet. As a result, the equivalent circuit for a permanent-magnet dc motor is identical to that of the externally-excited dc motor except that there are no field-winding connections. Figure 7.20 shows the equivalent circuit for a permanent-magnet dc motor.

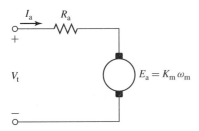

Figure 7.20 Equivalent circuit of a permanent-magnet dc motor.

From Eq. 7.14, the speed-voltage term for a dc motor can be written in the form $E_a = K_a \Phi_d \omega_m$ where Φ_d is the net flux along the field-winding axis and K_a is a geometric constant. In a permanent-magnet dc machine, Φ_d is constant and thus Eq. 7.14 can be reduced to

$$E_a = K_m \omega_m \qquad (7.26)$$

where

$$K_m = K_a \Phi_d \qquad (7.27)$$

is known as the *torque constant* of the motor and is a function of motor geometry and magnet properties.

Finally the torque of the machine can be easily found from Eq. 7.16 as

$$T_{mech} = \frac{E_a I_a}{\omega_m} = K_m I_a \qquad (7.28)$$

In other words, the torque of a permanent magnet motor is given by the product of the torque constant and the armature current.

EXAMPLE 7.9

A permanent-magnet dc motor is known to have an armature resistance of $1.03\ \Omega$. When operated at no load from a dc source of 50 V, it is observed to operate at a speed of 2100 r/min and to draw a current of 1.25 A. Find (a) the torque constant K_m, (b) the no-load rotational losses of the motor and (c) the power output of the motor when it is operating at 1700 r/min from a 48-V source.

■ **Solution**

a. From the equivalent circuit of Fig. 7.20, the generated voltage E_a can be found as

$$E_a = V_t - I_a R_a$$

$$= 50 - 1.25 \times 1.03 = 48.7\ \text{V}$$

At a speed of 2100 r/min,

$$\omega_m = \left(\frac{2100\ \text{r}}{\text{min}}\right) \times \left(\frac{2\pi\ \text{rad}}{\text{r}}\right) \times \left(\frac{1\ \text{min}}{60\ \text{s}}\right)$$

$$= 220\ \text{rad/sec}$$

Therefore, from Eq. 7.26,

$$K_m = \frac{E_a}{\omega_m} = \frac{48.7}{220} = 0.22 \ \text{V/(rad/sec)}$$

b. At no load, all the power supplied to the generated voltage E_a is used to supply rotational losses. Therefore

$$\text{Rotational losses} = E_a I_a = 48.7 \times 1.25 = 61 \ \text{W}$$

c. At 1700 r/min,

$$\omega_m = 1700 \left(\frac{2\pi}{60} \right) = 178 \ \text{rad/sec}$$

and

$$E_a = K_m \omega_m = 0.22 \times 178 = 39.2 \ \text{V}$$

The input current can now be found as

$$I_a = \frac{V_t - E_a}{R_a} = \frac{48 - 39.2}{1.03} = 8.54 \ \text{A}$$

The electromagnetic power can be calculated as

$$P_{\text{mech}} = E_a I_a = 39.2 \times 8.54 = 335 \ \text{W}$$

Assuming the rotational losses to be constant at their no-load value (certainly an approximation), the output shaft power can be calculated:

$$P_{\text{shaft}} = P_{\text{mech}} - \text{rotational losses} = 274 \ \text{W}$$

Practice Problem 7.8

The armature resistance of a small dc motor is measured to be 178 mΩ. With an applied voltage of 9 V, the motor is observed to operate at a no-load speed of 14,600 r/min while drawing a current of 437 mA. Calculate (*a*) the rotational loss and (*b*) the motor torque constant K_m.

Solution

a. Rotational loss = 3.90 W
b. $K_m = 5.84 \times 10^{-3} \ \text{V/(rad/sec)}$

7.8 COMMUTATION AND INTERPOLES

One of the most important limiting factors on the satisfactory operation of a dc machine is the ability to transfer the necessary armature current through the brush contact at the commutator without sparking and without excessive local losses and heating of the brushes and commutator. Sparking causes destructive blackening, pitting, and wear of both the commutator and the brushes, conditions which rapidly become worse and burn away the copper and carbon. Sparking may be caused by faulty mechanical conditions, such as chattering of the brushes or a rough, unevenly worn commutator, or, as in any switching problem, by electrical conditions. The latter conditions are seriously influenced by the armature mmf and the resultant flux wave.

As indicated in Section 7.2, a coil undergoing commutation is in transition between two groups of armature coils: at the end of the commutation period, the coil current must be equal but opposite to that at the beginning. Figure 7.7b shows the armature in an intermediate position during which the coils in slots 1 and 7 are being commutated. The commutated coils are short-circuited by the brushes. During this period the brushes must continue to conduct the armature current I_a from the armature winding to the external circuit. The short-circuited coil constitutes an inductive circuit with time-varying resistances at the brush contact, with rotational voltages induced in the coil, and with both conductive and inductive coupling to the rest of the armature winding.

The attainment of good commutation is more an empirical art than a quantitative science. The principal obstacle to quantitative analysis lies in the electrical behavior of the carbon-copper (brush-commutator) contact film. Its resistance is nonlinear and is a function of current density, current direction, temperature, brush material, moisture, and atmospheric pressure. Its behavior in some respects is like that of an ionized gas or plasma. The most significant fact is that an unduly high current density in a portion of the brush surface (and hence an unduly high energy density in that part of the contact film) results in sparking and a breakdown of the film at that point. The boundary film also plays an important part in the mechanical behavior of the rubbing surfaces. At high altitudes, definite steps must be taken to preserve it, or extremely-rapid brush wear takes place.

The empirical basis of securing sparkless commutation, then, is to avoid excessive current densities at any point in the copper-carbon contact. This basis, combined with the principle of utilizing all material to the fullest extent, indicates that optimum conditions are obtained when the current density is uniform over the brush surface during the entire commutation period. A linear change of current with time in the commutated coil, corresponding to linear commutation as shown in Fig. 7.8, brings about this condition and is accordingly the optimum.

The principal factors tending to produce linear commutation are changes in brush-contact resistance resulting from the linear decrease in area at the trailing brush edge and linear increase in area at the leading edge. Several electrical factors mitigate against linearity. Resistance in the commutated coil is one example. Usually, however, the voltage drop at the brush contacts is sufficiently large (of the order of 1.0 V) in comparison with the resistance drop in a single armature coil to permit the latter to be ignored. Coil inductance is a much more serious factor. Both the voltage of self-induction in the commutated coil and the voltage of mutual-induction from other coils (particularly those in the same slot) undergoing commutation at the same time oppose changes in current in the commutated coil. The sum of these two voltages is often referred to as the *reactance voltage*. Its result is that current values in the short-circuited coil lag in time the values dictated by linear commutation. This condition is known as *undercommutation* or *delayed commutation*.

Armature inductance thus tends to produce high losses and sparking at the trailing brush tip. For best commutation, inductance must be held to a minimum by using the fewest possible number of turns per armature coil and by using a multipolar design with a short armature. The effect of a given reactance voltage in delaying commutation is minimized when the resistive brush-contact voltage drop is significant compared with it. This fact is one of the main reasons for the use of carbon brushes with their appreciable contact drop. When good commutation is secured by virtue of resistance drops, the process is referred to as *resistance commutation*. It is typically used as the exclusive means only in fractional-horsepower machines.

Another important factor in the commutation process is the rotational voltage induced in the short-circuited coil. Depending on its sign, this voltage may hinder or aid commutation. In Fig. 7.11, for example, cross-magnetizing armature reaction creates a definite flux in the interpolar region. The direction of the corresponding rotational voltage in the commutated coil is the same as the current under the immediately preceding pole face. This voltage then encourages the continuance of current in the old direction and, like the resistance voltage, opposes its reversal. To aid commutation, the rotational voltage must oppose the reactance voltage. The general principle is to produce in the coil undergoing commutation a rotational voltage which approximately compensates for the reactance voltage, a principle called *voltage commutation*. Voltage commutation is used in almost all modern integral-horsepower commutating machines. The appropriate flux density is introduced in the commutating zone by means of small, narrow poles located between the main poles. These auxiliary poles are called *interpoles* or *commutating poles*.

The general appearance of interpoles and an approximate map of the flux produced when they alone are excited are shown in Fig. 7.21. The interpoles are the smaller poles between the larger main poles in the dc-motor section shown in Fig. 7.23. The polarity of a commutating pole must be that of the main pole just ahead of it, i.e., in the direction of rotation for a generator, and just behind it for a motor. The interpole mmf must be sufficient to neutralize the cross-magnetizing armature mmf in the interpolar region and enough more to furnish the flux density required for the rotational voltage in the short-circuited armature coil to cancel the reactance voltage. Since both the armature mmf and the reactance voltage are proportional to the armature current, the commutating winding must be connected in series with the armature. To preserve the desired linearity, the commutating pole should operate at a relatively low flux level. By the use of commutating fields sparkless commutation can be obtained over a wide range in large dc machines. In accordance with the performance standards

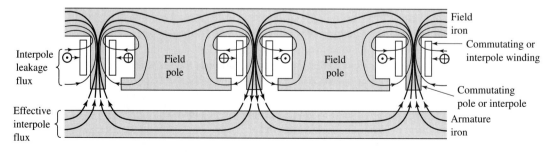

Figure 7.21   Interpoles and their associated component flux.

of NEMA,[1] general-purpose dc machines must be capable of carrying for one minute, with successful commutation, loads of 150 percent of the current corresponding to their continuous rating when operating with a field current equal to their rated-load excitation.

7.9 COMPENSATING WINDINGS

For machines subjected to heavy overloads, rapidly changing loads, or operation with a weak main field, there is the possibility of trouble other than simply sparking at the brushes. At the instant when an armature coil is located at the peak of a badly distorted flux wave, the coil voltage may be high enough to break down the air between the adjacent segments to which the coil is connected and result in flashover, or arcing, between segments. The breakdown voltage here is not high, because the air near the commutator is in a condition favorable to breakdown, due to the presence of the plasma carrying the armature current between the brushes and the commutator. The maximum allowable voltage between segments is of the order of 30 to 40 V, a fact which limits the average voltage between segments to lower values and thus determines the minimum number of segments which can be used in a proposed design. Under transient conditions, high voltages between segments may result from the induced voltages associated with growth and decay of armature flux. Inspection of Fig. 7.10, for instance, may enable one to visualize very appreciable voltages of this nature being induced in a coil under the pole centers by the growth or decay of the armature flux shown in the sketch. Consideration of the sign of this induced voltage will show that it adds to the normal rotational emf when load is dropped from a generator or added to a motor. Flashing between segments may quickly spread around the entire commutator and, in addition to its possibly destructive effects on the commutator, constitutes a direct short circuit on the line. Even with interpoles present, therefore, armature reaction under the poles definitely limits the conditions under which a machine can operate.

These limitations can be considerably extended by compensating or neutralizing the armature mmf under the pole faces. Such compensation can be achieved by means of a *compensating* or *pole-face winding* (Fig. 7.22) embedded in slots in the pole face and having a polarity opposite to that of the adjoining armature winding. The physical appearance of such a winding can be seen in the stator section of Fig. 7.23. Since the axis of the compensating winding is the same as that of the armature, it will almost completely neutralize the armature reaction of the armature conductors under the pole faces when it is given the proper number of turns. It must be connected in series with the armature in order to carry a proportional current. The net effect of the main field, armature, commutating winding, and compensating winding on the air-gap flux is that, except for the commutation zone, the resultant flux-density distribution is substantially the same as that produced by the main field alone (Fig. 7.11). Furthermore, the addition

---

[1] NEMA Standards Publication No. MG1-1998, Motors and Generators, Sections 23 and 24, National Electrical Manufactures Association, 300 North 17th Street, Suite 1847, Rosslyn, Virginia, 22209.

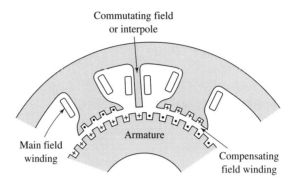

Figure 7.22 Section of dc machine showing compensating winding.

of a compensating winding improves the speed of response of the machine because it reduces the armature-circuit time constant.

The main disadvantage of pole-face windings is their expense. They are used in machines designed for heavy overloads or rapidly-changing loads (steel-mill motors are a good example of machines subjected to severe duty cycles) or in motors intended to operate over wide speed ranges by shunt-field control. By way of a schematic summary, Fig. 7.24 shows the circuit diagram of a compound machine with a compensating winding. The relative position of the coils in this diagram indicates that the commutating and compensating fields act along the armature axis, and the shunt and

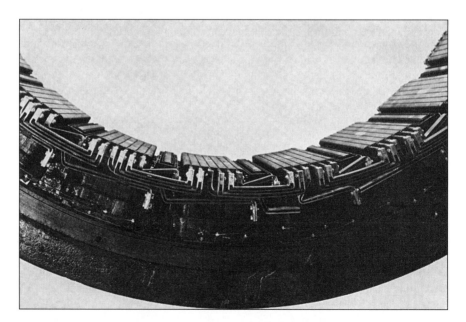

Figure 7.23 Section of a dc motor stator showing shunt and series coils, interpoles and pole-face compensating winding. (*Westinghouse Electric Company.*)

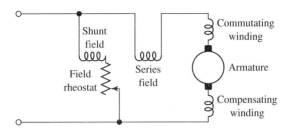

Figure 7.24 Schematic connection diagram of a dc machine.

series fields act along the axis of the main poles. Rather complete control of air-gap flux around the entire armature periphery is thus achieved.

7.10 SERIES UNIVERSAL MOTORS

Figure 7.25 shows a dc machine connection with a series-connected field winding. For this connection the direct-axis flux Φ_d is proportional to the armature current. Hence from Eq. 7.14, the generated voltage E_a is proportional to the product of the armature current and the motor speed, and from Eq. 7.16 we see that the torque will be proportional to the square of the armature current.

The dashed line in Fig. 7.26 shows a typical speed-torque characteristic for such a series-connected motor under dc operating conditions. Note that because the torque is proportional to the square of the armature current, the torque depends only upon the magnitude of the armature voltage and not its polarity; reversing the polarity of the applied voltage will not change the magnitude or direction of the applied torque.

If the rotor and stator structures of a series connected motor are properly laminated to reduce ac eddy-current losses, the resultant motor is referred to as a *series universal motor*. The series universal motor has the convenient ability to run on either alternating or direct current and with similar characteristics. Such a single-phase series motor therefore is commonly called a *universal motor*. The torque angle is fixed by the brush position and is normally at its optimum value of 90°. If alternating current is supplied to a series universal motor, the torque will always be in the same direction, although it will pulsate in magnitude at twice line frequency. Average torque will be produced, and the performance of the motor will be generally similar to that with direct current.

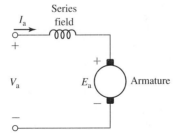

Figure 7.25 Series-connected universal machine.

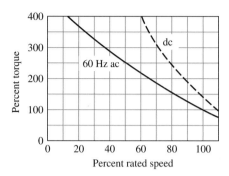

Figure 7.26 Typical torque-speed characteristics of a series universal motor.

Small universal motors are used where light weight is important, as in vacuum cleaners, kitchen appliances, and portable tools, and usually operate at high speeds (1500 to 15,000 r/min). Typical characteristics are shown in Fig. 7.26. The ac and dc characteristics differ somewhat for two reasons: (1) With alternating current, reactance-voltage drops in the field and armature absorb part of the applied voltage; therefore for a specified current and torque the rotational counter emf generated in the armature is less than with direct current, and the speed tends to be lower. (2) With alternating current, the magnetic circuit may be appreciably saturated at the peaks of the current wave. Thus the rms value of the flux may be appreciably less with alternating current than with the same rms value of direct current. The torque therefore tends to be less with alternating than with direct current. The universal motor provides the highest horsepower per dollar in the fractional-horsepower range, at the expense of noise, relatively short life, and high speed.

To obtain control of the speed and torque of a series universal motor, the applied ac voltage may be varied by the use of a Triac, such as is discussed in Chapter 10. The firing angle of the Triac can be manually adjusted, as in a trigger-controlled electric drill, or it can be controlled by a speed control circuit, as in some portable tools and appliances. The combination of a series motor and a solid-state device provides an economical, controllable motor package.

7.11   SUMMARY

This chapter has discussed the significant operating characteristics of dc machines. In general, the outstanding advantage of dc machines lies in their flexibility and versatility. Before the widespread availability of ac motor drives, dc machines were essentially the only choice available for many applications which required a high degree of control. Their principal disadvantages stem from the complexity associated with the armature winding and the commutator/brush system. Not only does this additional complexity increase the cost over competing ac machines, it also increases the need for maintenance and reduces the potential reliability of these machines. Yet the advantages of dc motors remain, and they continue to retain a strong competitive position in both large sizes for industrial applications and in smaller sizes for a wide variety of applications.

Dc generators are a simple solution to the problem of converting mechanical energy to electric energy in dc form, although ac generators feeding rectifier systems are certainly an option which must be considered. Among dc generators themselves, separately-excited and cumulatively-compounded, self-excited machines are the most common. Separately-excited generators have the advantage of permitting a wide range of output voltages, whereas self-excited machines may produce unstable voltages at lower output voltages where the field-resistance line becomes essentially tangent to the magnetization curve. Cumulatively-compounded generators may produce a substantially flat voltage characteristic or one which rises with load, whereas shunt- or separately-excited generators may produce a drooping voltage characteristic unless external regulating means (such as a series field winding) are added.

Among dc motors, the outstanding characteristics of each type are as follows. The series motor operates with a decidedly drooping speed as load is added, the no-load speed usually being prohibitively high; the torque is proportional to almost the square of the current at low flux levels and to some power between 1 and 2 as saturation increases. The shunt motor at constant field current operates at a slightly drooping but almost constant speed as load is added, the torque being almost proportional to armature current; equally important, however, is the fact that its speed can be controlled over wide ranges by shunt-field control, armature-voltage control, or a combination of both. Depending on the relative strengths of the shunt and series field, the cumulatively-compounded motor is intermediate between the other two and may be given essentially the advantages of one or the other.

In a wide variety of low-power applications in systems which are run from a dc source (automotive applications, portable electronics, etc.), dc machines are the most cost-effective choice. These dc machines are constructed in a wide-range of configurations, and many of them are based upon permanent-magnet excitation. In spite of the wide variety of dc machines which can be found in these various applications, their performance can readily be determined using the models and techniques presented in this chapter.

7.12 PROBLEMS

7.1 Consider a separately-excited dc motor. Describe the speed variation of the motor operating unloaded under the following conditions:

   a. The armature terminal voltage is varied while the field current is held constant.

   b. The field current is varied while the armature terminal voltage is held constant.

   c. The field winding is connected in shunt directly to the armature terminals, and the armature terminal voltage is then varied.

7.2 A dc shunt motor operating at an armature terminal voltage of 125 V is observed to be operating at a speed of 1180 r/min. When the motor is operated unloaded at the same armature terminal voltage but with an additional resistance of 5 Ω in series with the shunt field, the motor speed is observed to be 1250 r/min.

a. Calculate the resistance of the series field.

b. Calculate the motor speed which will result if the series resistance is increased from 5 Ω to 15 Ω.

7.3 For each of the following changes in operating condition for a dc shunt motor, describe how the armature current and speed will vary:

a. Halving the armature terminal voltage while the field flux and load torque remain constant.

b. Halving the armature terminal voltage while the field current and load power remain constant.

c. Doubling the field flux while the armature terminal voltage and load torque remain constant.

d. Halving both the field flux and armature terminal voltage while the load power remains constant.

e. Halving the armature terminal voltage while the field flux remains constant and the load torque varies as the square of the speed.

Only brief quantitative statements describing the general nature of the effect are required, for example, "speed approximately doubled."

7.4 The constant-speed magnetization curve for a 25-kW, 250-V dc machine at a speed of 1200 r/min is shown in Fig. 7.27. This machine is separately excited and has an armature resistance of 0.14 Ω. This machine is to be operated as a dc generator while driven from a synchronous motor at constant speed.

a. What is the rated armature current of this machine?

b. With the generator speed held at 1200 r/min and if the armature current is limited to its rated value, calculate the maximum power output of the

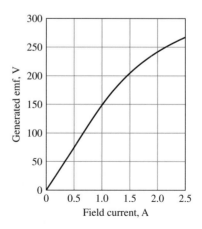

Figure 7.27 1200 r/min magnetization curve for the dc generator of Problem 7.4.

generator and the corresponding armature voltage for constant field currents of (*i*) 1.0 A, (*ii*) 2.0 A and (*iii*) 2.5 A.

c.   Repeat part (b) if the speed of the synchronous generator is reduced to 900 r/min.

7.5 The dc generator of Problem 7.4 is to be operated at a constant speed of 1200 r/min into a load resistance of 2.5 Ω.

a.   Using the "spline()" function of MATLAB and the points of the magnetization curve of Fig. 7.27 at 0, 0.5, 1.0, 1.5, 2.0, and 2.5 A, create a MATLAB plot of the magnetization curve of Fig. 7.27.

b.   Using the "spline()" function as in part (a), use MATLAB to plot (*i*) the terminal voltage and (*ii*) the power delivered to the load as the generator field current is varied from 0 to 2.5 A.

7.6 The dc machine of Problem 7.4 is to be operated as a motor supplied by a constant armature terminal voltage of 250 V. If saturation effects are ignored, the magnetization curve of Fig. 7.27 becomes a straight line with a constant slope of 150 volts per ampere of field current. For the purposes of this problem, you may assume that saturation effects can be neglected.

a.   Assuming that the field current is held constant at 1.67 A, plot the motor speed as a function of motor shaft power as the shaft power varies from 0 to 25 kW.

b.   Now assuming that the field current can be adjusted in order to maintain the motor speed constant at 1200 r/min, plot the required field current as a function of motor shaft power as the shaft power varies from 0 to 25 kW.

7.7 Repeat Problem 7.6 including the saturation effects represented by the saturation curve of Fig. 7.27. For part (a), set the field current equal to the value required to produce an open-circuit armature terminal voltage of 250 V at 1200 r/min. (Hint: This problem is most easily solved using MATLAB and its "spline()" function as in Problem 7.5.)

7.8 A 15-kW, 250-V, 1150 r/min shunt generator is driven by a prime mover whose speed is 1195 r/min when the generator delivers no load. The speed falls to 1140 r/min when the generator delivers 15 kW and may be assumed to decrease in proportion to the generator output. The generator is to be changed into a short-shunt compound generator by equipping it with a series field winding which will cause its voltage to rise from 230 V at no load to 250 V for a load of 61.5 A. It is estimated that the series field winding will have a resistance of 0.065 Ω. The armature resistance (including brushes) is 0.175 Ω. The shunt field winding has 500 turns per pole.

     To determine the necessary series-field turns, the machine is run as a separately-excited generator and the following load data are obtained:

    Armature terminal voltage = 254 V

    Armature current = 62.7 A

Field current $= 1.95$ A

Speed $= 1140$ r/min

The magnetization curve at 1195 r/min is as follows:

E_a, V	230	240	250	260	270	280
I_f, A	1.05	1.13	1.25	1.44	1.65	1.91

Determine

a. the armature reaction in equivalent demagnetizing ampere-turns per pole for $I_a = 62.7$ A and

b. the necessary number of series-field turns per pole.

(Hint: This problem can be solved either graphically or by use of the MATLAB "spline()" function to represent the magnetization curve.)

7.9 When operated from a 230-V dc supply, a dc series motor operates at 975 r/min with a line current of 90 A. Its armature-circuit resistance is 0.11 Ω and its series-field resistance is 0.08 Ω. Due to saturation effects, the flux produced by an armature current of 30 A is 48 percent of that at an armature current of 90 A. Find the motor speed when the armature voltage is 230 V and the armature current is 30 A.

7.10 Consider the long-shunt, 250-V, 100-kW dc machine of Example 7.3. Assuming the machine is operated as a motor at a constant supply voltage of 250-V with a constant shunt-field current of 5.0 A, use MATLAB to plot the motor speed as a function of load. Use the MATLAB "spline()" function to represent the magnetization curve of Fig. 7.14. Neglect armature-reaction effects. Include two plots, one for the case where the series-field ampere-turns add to those of the shunt field and the second for the case where the series-field ampere-turns oppose those of the shunt field.

7.11 A 250-V dc shunt-wound motor is used as an adjustable-speed drive over the range from 0 to 2000 r/min. Speeds from 0 to 1200 r/min are obtained by adjusting the armature terminal voltage from 0 to 250 V with the field current kept constant. Speeds from 1200 r/min to 2000 r/min are obtained by decreasing the field current with the armature terminal voltage remaining at 250 V. Over the entire speed range, the torque required by the load remains constant.

a. Sketch the general form of the curve of armature current versus speed over the entire range. Ignore machine losses and armature-reaction effects.

b. Suppose that, instead of operating with constant torque, the load torque at any given speed is adjusted to maintain the armature current at its rated value. Sketch the general form of the allowable torque as a function of speed assuming the motor is controlled as described above.

7.12 Two adjustable-speed dc shunt motors have maximum speeds of 1800 r/min and minimum speeds of 500 r/min. Speed adjustment is obtained by field-rheostat control. Motor A drives a load requiring constant power over the speed range; motor B drives one requiring constant torque. All losses and armature reaction may be neglected.

a. If the power outputs of the two motors are equal at 1800 r/min and the armature currents are each 125 A, what will the armature currents be at 500 r/min?

b. If the power outputs of the two motors are equal at 500 r/min and the armature currents are each 125 A, what will the armature current be at 1800 r/min?

c. Answer parts (a) and (b) with speed adjustment by armature-voltage control with conditions otherwise the same.

7.13 Consider a dc shunt motor connected to a constant-voltage source and driving a load requiring constant electromagnetic torque. Show that if $E_a > 0.5V_t$ (the normal situation), increasing the resultant air-gap flux decreases the speed, whereas if $E_a < 0.5V_t$ (as might be brought about by inserting a relatively high resistance in series with the armature), increasing the resultant air-gap flux increases the speed.

7.14 A separately-excited dc motor is mechanically coupled to a three-phase, four-pole, 30-kVA, 460-V, cylindrical-pole synchronous generator. The dc motor is connected to a constant 230-V dc supply, and the ac generator is connected to a 460-V, fixed-voltage, fixed-frequency, three-phase supply. The synchronous reactance of the synchronous generator is 5.13 Ω/phase. The armature resistance of the dc motor is 30 mA. The four-pole dc machine is rated 30 kW at 230 V. All unspecified losses are to be neglected.

a. If the two machines act as a motor-generator set receiving power from the dc source and delivering power to the ac supply, what is the excitation voltage of the ac machine in volts per phase (line-to-neutral) when it delivers 30 kW at unity power factor? What is the internal voltage of the dc motor?

b. Leaving the field current of the ac machine at the value corresponding to the condition of part (a), what adjustment can be made to reduce the power transfer between the two machines to zero? Under this condition of zero power transfer, what is the armature current of the dc machine? What is the armature current of the ac machine?

c. Leaving the field current of the ac machine as in parts (a) and (b), what adjustment can be made to cause the transfer of 30 kW from the ac source to the dc source? Under these conditions what are the armature current and internal voltage of the dc machine? What will be the magnitude and phase of the current of the ac machine?

7.15 A 150-kW, 600-V, 600 r/min dc series-wound railway motor has a combined field and armature resistance (including brushes) of 0.125 Ω. The full-load current at rated voltage and speed is 250 A. The magnetization curve at 400 r/min is as follows:

Generated emf, V	375	400	425	450	475
Series-field current, A	227	260	301	350	402

Determine the internal starting torque when the starting current is limited to 460 A. Assume the armature reaction to be equivalent to a demagnetizing mmf which varies as the square of the current. (Hint: This problem can be solved either graphically or by use of the MATLAB "spline()" function to represent the magnetization curve.)

7.16 A 25-kW, 230-V shunt motor has an armature resistance of 0.11 Ω and a field resistance of 117 Ω. At no load and rated voltage, the speed is 2150 r/min and the armature current is 6.35 A. At full load and rated voltage, the armature current is 115 A and, because of armature reaction, the flux is 6 percent less than its no-load value. What is the full-load speed?

7.17 A 91-cm axial-flow fan is to deliver air at 16.1 m³/sec against a static pressure of 120 Pa when rotating at a speed of 1165 r/min. The fan has the following speed-load characteristic

Speed, r/min	700	800	900	1000	1100	1200
Power, kW	3.6	4.9	6.5	8.4	10.8	13.9

It is proposed to drive the fan by a 12.5 kW, 230-V, 46.9-A, four-pole dc shunt motor. The motor has an armature winding with two parallel paths and $C_a = 666$ active conductors. The armature-circuit resistance is 0.215 Ω. The armature flux per pole is $\Phi_d = 10^{-2}$ Wb and armature reaction effects can be neglected. No-load rotational losses (to be considered constant) are estimated to be 750 W. Determine the shaft power output and the operating speed of the motor when it is connected to the fan load and operated from a 230-V source. (Hint: This problem can be easily solved using MATLAB with the fan characteristic represented by the MATLAB "spline()" function.)

7.18 A shunt motor operating from a 230-V line draws a full-load armature current of 46.5 A and runs at a speed of 1300 r/min at both no load and full load. The following data is available on this motor:

Armature-circuit resistance (including brushes) = 0.17 Ω

Shunt-field turns per pole = 1500 turns

The magnetization curve taken with the machine operating as a motor at no load and 1300 r/min is

E_a, V	180	200	220	240	250
I_f, A	0.98	1.15	1.46	1.93	2.27

a. Determine the shunt-field current of this motor at no load and 1300 r/min when connected to a 230-V line. Assume negligible armature-circuit resistance and armature reaction at no load.

b. Determine the effective armature reaction at full load in ampere-turns per pole.

c. How many series-field turns should be added to make this machine into a long-shunt cumulatively compounded motor whose speed will be 1210 r/min when the armature current is 46.5 A and the applied voltage is 230 V? Assume that the series field has a resistance of 0.038 Ω.

d. If a series-field winding having 20 turns per pole and a resistance of 0.038 Ω is installed, determine the speed when the armature current is 46.5 A and the applied voltage is 230 V.

(Hint: This problem can be solved either graphically or by use of the MATLAB "spline()" function to represent the magnetization curve.)

7.19 A 7.5-kW, 230-V shunt motor has 2000 shunt-field turns per pole, an armature resistance (including brushes) of 0.21 Ω, and a commutating-field resistance of 0.035 Ω. The shunt-field resistance (exclusive of rheostat) is 310 Ω. When the motor is operated at no load with rated terminal voltage and varying shunt-field resistance, the following data are obtained:

Speed, r/min	1110	1130	1160	1200	1240
I_f, A	0.672	0.634	0.598	0.554	0.522

The no-load armature current is negligible. When the motor is operating at full load and rated terminal voltage with a field current of 0.554 A, the armature current is 35.2 A and the speed is 1185 r/min.

a. Calculate the full-load armature reaction in equivalent demagnetizating ampere-turns per pole.

b. Calculate the full-load electromagnetic torque at this operating condition.

c. What starting torque will the motor produce with maximum field current if the starting armature current is limited to 65 A? Assume that the armature reaction under these conditions is equal to 160 ampere-turns per pole.

d. Design a series field winding to give a speed of 1050 r/min when the motor is loaded to an armature current of 35.2 A and when the shunt field current is adjusted to give a no-load speed of 1200 r/min. Assume the series field will have a resistance of 0.05 Ω.

(Hint: This problem can be solved either graphically or by use of the MATLAB "spline()" function to represent the magnetization curve.)

7.20 When operated at rated voltage, a 230-V shunt motor runs at 1750 r/min at full load and at no load. The full-load armature current is 70.8 A. The shunt field winding has 2000 turns per pole. The resistance of the armature circuit (including brushes and interpoles) is 0.15 Ω. The magnetization curve at 1750 r/min is

E_a, V	200	210	220	230	240	250
I_f, A	0.40	0.44	0.49	0.55	0.61	0.71

a. Compute the demagnetizing effect of the armature reaction at full load.

b. A long-shunt cumulative series field winding having four turns per pole and a resistance of 0.038 Ω is added to the machine. Compute the speed at full-load current and rated voltage. The shunt field current will remain equal to that of part (a).

c. With the series-field winding of part (b) installed, compute the internal starting torque in N · m if the starting armature current is limited to 125 A. Assume that the corresponding demagnetizating effect of armature reaction is 230 ampere-turns per pole.

(Hint: This problem can be solved either graphically or by use of the MATLAB "spline()" function to represent the magnetization curve.)

7.21 A 230-V dc shunt motor has an armature-circuit resistance of 0.23 Ω. When operating from a 230-V supply and driving a constant-torque load, the motor is observed to be drawing an armature current of 60 A. An external resistance of 1.0 Ω is now inserted in series with the armature while the shunt field current is unchanged. Neglecting the effects of rotational losses and armature reaction, calculate

a. the resultant armature current and

b. the fractional speed change of the motor.

7.22 A common industrial application of dc series motors is in crane and hoist drives. This problem relates to the computation of selected motor performance characterstics for such a drive. The specific motor concerned is a series-wound, 230-V, totally-enclosed motor having a 1/2-hour crane rating of 100 kW with a 75°C temperature rise. The performance characteristics

Table 7.1 Motor characteristics for Problem 7.22.

Line current A	Shaft torque N · m	Speed r/min
100	217	940
200	570	630
300	1030	530
400	1480	475
500	1980	438
600	2470	407
700	3000	385
800	3430	370

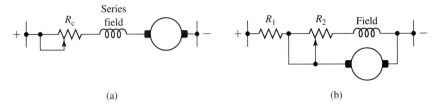

Figure 7.28 Series crane motor (Problem 7.22): (a) hoisting connection and (b) lowering connection.

of the motor alone at 230 V as found in the manufacturer's catalog are listed in Table 7.1. The resistance of the armature (including brushes) plus commutating winding is 0.065 Ω and that of the series field winding is 0.027 Ω. Armature reaction effects can be ignored.

The motor is to be connected as in Fig. 7.28a for hoisting and Fig. 7.28b for lowering. The former connection consists simply of series-resistance control. The latter connection provides dynamic breaking with the field reconnected in shunt with the addition of an adjustable series resistance.

You will use MATLAB to plot some sample speed-torque curves (speed as a function of torque) to determine the suitability of the motor and control for the specified application. Plot all of the curves on a single set of axes covering roughly the torque-magnitude range found in Table 7.1. Provide for both positive and negative values of speed, corresponding respectively to hoisting and lowering, as well as for both positive and negative values of torque, corresponding respectively to torque in the direction of raising the load and torque in the direction of lowering the load.

a. For the hoisting connection, plot speed-torque curves for the control resistor R_c set at 0, 0.3 and 0.6 Ω. If any of these curves extend into the fourth quadrant within the range of torques covered, interpret physically what operation in that regime means.

b. For the lowering connection, plot a speed-torque curve for $R_1 = 0.3\ \Omega$ and $R_2 = 0.3\ \Omega$. The most important portion of this curve is in the fourth quadrant, but if it extends into the third quadrant, that region should also be plotted and interpreted physically.

c. In part (b), what is the lowering speed corresponding to a torque of 1500 N · m? (Hint: This can be found easily using the MATLAB "spline()" function.)

7.23 A 25-kW, 230-V shunt motor has an armature resistance of 0.064 Ω and a field-circuit resistance of 95 Ω. The motor delivers rated output power at rated voltage when its armature current is 122 A. When the motor is operating at rated voltage, the speed is observed to be 1150 r/min when the machine is loaded such that the armature current is 69.5 A.

   a.   Calculate the rated-load speed of this motor.

   In order to protect both the motor and the dc supply under starting conditions, an external resistance will be connected in series with the armature winding (with the field winding remaining directly across the 230-V supply). The resistance will then be automatically adjusted in steps so that the armature current does not exceed 200 percent of rated current. The step size will be determined such that, until all the external resistance is switched out, the armature current will not be permitted to drop below rated value. In other words, the machine is to start with 200 percent of rated armature current and as soon as the current falls to rated value, sufficient series resistance is to be cut out to restore the current to 200 percent. This process will be repeated until all of the series resistance has been eliminated.

   b.   Find the maximum value of the series resistance.

   c.   How much resistance should be cut out at each step in the starting operation and at what speed should each step change occur?

7.24 The manufacturer's data sheet for a permanent-magnet dc motor indicates that it has a torque constant $K_m = 0.21$ V/(rad/sec) and an armature resistance of 1.9 Ω. For a constant applied armature voltage of 85 V dc, calculate

   a.   the no-load speed of the motor in r/min and

   b.   its stall (zero-speed) current and torque (in N · m).

   c.   Plot the motor torque as a function of speed.

7.25 Measurements on a small permanent-magnet dc motor indicate that it has an armature resistance of 4.6 Ω. With an applied armature voltage of 5 V, the motor is observed to achieve a no-load speed of 11,210 r/min while drawing an armature current of 12.5 mA.

   a.   Calculate the motor torque constant K_m in V/(rad/sec).

   b.   Calculate the no-load rotational losses in mW.

   Assume the motor to be operating from an applied armature voltage of 5 V.

   c.   Find the stall current and torque of the motor.

   d.   At what speeds will the motor achieve an output power of 1 W? Estimate the motor efficiency under these operating conditions. Assume that the rotational loss varies as the cube of the speed.

7.26 Write a MATLAB script to calculate the parameters of a dc motor. The inputs will be the armature resistance and the no-load armature voltage, speed, and armature current. The output should be the no-load rotational loss and the torque constant K_m.

7.27 The dc motor of Problem 7.25 will be used to drive a load which requires a power of 0.75 W at a speed of 8750 r/min. Calculate the armature voltage which must be applied to achieve this operating condition.

CHAPTER 8

Variable-Reluctance Machines and Stepping Motors

V ariable-reluctance machines[1] (often abbreviated as *VRMs*) are perhaps the simplest of electrical machines. They consist of a stator with excitation windings and a magnetic rotor with saliency. Rotor conductors are not required because torque is produced by the tendency of the rotor to align with the stator-produced flux wave in such a fashion as to maximize the stator flux linkages that result from a given applied stator current. Torque production in these machines can be evaluated by using the techniques of Chapter 3 and the fact that the stator winding inductances are functions of the angular position of the rotor.

Although the concept of the VRM has been around for a long time, only in the past few decades have these machines begun to see widespread use in engineering applications. This is due in large part to the fact that although they are simple in construction, they are somewhat complicated to control. For example, the position of the rotor must be known in order to properly energize the phase windings to produce torque. It is the widespread availability and low cost of micro and power electronics that has made the VRM competitive with other motor technologies in a wide range of applications.

By sequentially exciting the phases of a VRM, the rotor will rotate in a step-wise fashion, rotating through a specific angle per step. *Stepper motors* are designed to take advantage of this characteristic. Such motors often combine the use of a variable-reluctance geometry with permanent magnets to produce increased torque and precision position accuracy.

---

[1] Variable-reluctance machines are often referred to as *switched-reluctance machines* (*SRMs*) to indicate the combination of a VRM and the switching inverter required to drive it. This term is popular in the technical literature.

8.1 BASICS OF VRM ANALYSIS

Common variable-reluctance machines can be categorized into two types: singly-salient and doubly-salient. In both cases, their most noticeable features are that there are no windings or permanent magnets on their rotors and that their only source of excitation consists of stator windings. This can be a significant feature because it means that all the resistive winding losses in the VRM occur on the stator. Because the stator can typically be cooled much more effectively and easily than the rotor, the result is often a smaller motor for a given rating and frame size.

As is discussed in Chapter 3, to produce torque, VRMs must be designed such that the stator-winding inductances vary with the position of the rotor. Figure 8.1a shows a cross-sectional view of a *singly-salient VRM,* which can be seen to consist of a nonsalient stator and a two-pole salient rotor, both constructed of high-permeability magnetic material. In the figure, a two-phase stator winding is shown although any number of phases are possible.

Figure 8.2a shows the form of the variation of the stator inductances as a function of rotor angle θ_m for a singly-salient VRM of the form of Fig. 8.1a. Notice that the inductance of each stator phase winding varies with rotor position such that the inductance is maximum when the rotor axis is aligned with the magnetic axis of that phase and minimum when the two axes are perpendicular. The figure also shows that the mutual inductance between the phase windings is zero when the rotor is aligned with the magnetic axis of either phase but otherwise varies periodically with rotor position.

Figure 8.1b shows the cross-sectional view of a two-phase *doubly-salient VRM* in which both the rotor and stator have salient poles. In this machine, the stator has four poles, each with a winding. However, the windings on opposite poles are of the same phase; they may be connected either in series or in parallel. Thus this machine is quite similar to that of Fig. 8.1a in that there is a two-phase stator winding and a two-pole salient rotor. Similarly, the phase inductance of this configuration varies from a maximum value when the rotor axis is aligned with the axis of that phase to a minimum when they are perpendicular.

Unlike the singly-salient machine of Fig. 8.1a, under the assumption of negligible iron reluctance the mutual inductances between the phases of the doubly-salient VRM of Fig. 8.1b will be zero, with the exception of a small, essentially-constant component associated with leakage flux. In addition, the saliency of the stator enhances the difference between the maximum and minimum inductances, which in turn enhances the torque-producing characteristics of the doubly-salient machine. Figure 8.2b shows the form of the variation of the phase inductances for the doubly-salient VRM of Fig. 8.1b.

The relationship between flux linkage and current for the singly-salient VRM is of the form

$$\begin{bmatrix} \lambda_1 \\ \lambda_2 \end{bmatrix} = \begin{bmatrix} L_{11}(\theta_m) & L_{12}(\theta_m) \\ L_{12}(\theta_m) & L_{22}(\theta_m) \end{bmatrix} \begin{bmatrix} i_1 \\ i_2 \end{bmatrix} \tag{8.1}$$

Here $L_{11}(\theta_m)$ and $L_{22}(\theta_m)$ are the self-inductances of phases 1 and 2, respectively, and $L_{12}(\theta_m)$ is the mutual inductances. Note that, by symmetry

$$L_{22}(\theta_m) = L_{11}(\theta_m - 90°) \tag{8.2}$$

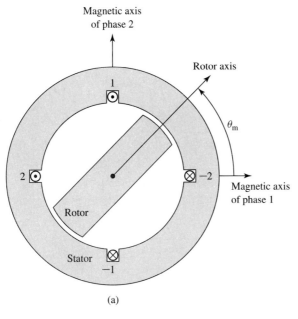

(a)

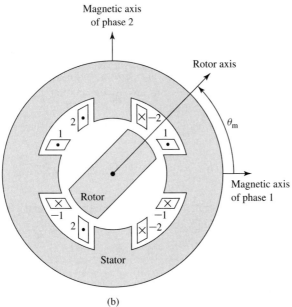

(b)

Figure 8.1 Basic two-phase VRMs: (a) singly-salient
and (b) doubly-salient.

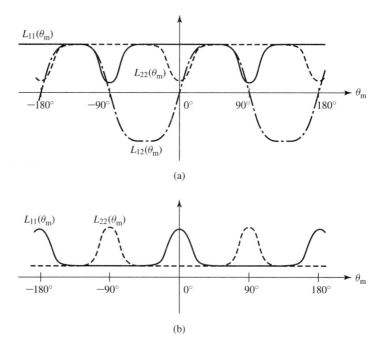

(a)

(b)

Figure 8.2 Plots of inductance versus θ_m for (a) the singly-salient VRM of Fig. 8.1a and (b) the doubly-salient VRM of Fig. 8.1b.

Note also that all of these inductances are periodic with a period of 180° because rotation of the rotor through 180° from any given angular position results in no change in the magnetic circuit of the machine.

From Eq. 3.68 the electromagnetic torque of this system can be determined from the coenergy as

$$T_{mech} = \frac{\partial W'_{fld}(i_1, i_2, \theta_m)}{\partial \theta_m} \qquad (8.3)$$

where the partial derivative is taken while holding currents i_1 and i_2 constant. Here, the coenergy can be found from Eq. 3.70,

$$W'_{fld} = \frac{1}{2}L_{11}(\theta_m)i_1^2 + L_{12}(\theta_m)i_1 i_2 + \frac{1}{2}L_{22}(\theta_m)i_2^2 \qquad (8.4)$$

Thus, combining Eqs. 8.3 and 8.4 gives the torque as

$$T_{mech} = \frac{1}{2}i_1^2\frac{dL_{11}(\theta_m)}{d\theta_m} + i_1 i_2 \frac{dL_{12}(\theta_m)}{d\theta_m} + \frac{1}{2}i_2^2\frac{dL_{22}(\theta_m)}{d\theta_m} \qquad (8.5)$$

For the double-salient VRM of Fig. 8.1b, the mutual-inductance term $dL_{12}(\theta_m)/d\theta_m$ is zero and the torque expression of Eq. 8.5 simplifies to

$$T_{mech} = \frac{1}{2}i_1^2\frac{dL_{11}(\theta_m)}{d\theta_m} + \frac{1}{2}i_2^2\frac{dL_{22}(\theta_m)}{d\theta_m} \qquad (8.6)$$

Substitution of Eq. 8.2 then gives

$$T_{\text{mech}} = \frac{1}{2}i_1^2 \frac{dL_{11}(\theta_{\text{m}})}{d\theta_{\text{m}}} + \frac{1}{2}i_2^2 \frac{dL_{11}(\theta_{\text{m}} - 90°)}{d\theta_{\text{m}}} \tag{8.7}$$

Equations 8.6 and 8.7 illustrate an important characteristic of VRMs in which mutual-inductance effects are negligible. In such machines the torque expression consists of a sum of terms, each of which is proportional to the square of an individual phase current. As a result, the torque depends only on the magnitude of the phase currents and not on their polarity. Thus the electronics which supply the phase currents to these machines can be unidirectional; i.e., bidirectional currents are not required.

Since the phase currents are typically switched on and off by solid-state switches such as transistors or thyristors and since each switch need only handle currents in a single direction, this means that the motor drive requires only half the number of switches (as well as half the corresponding control electronics) that would be required in a corresponding bidirectional drive. The result is a drive system which is less complex and may be less expensive. Typical VRM motor drives are discussed in Section 11.4.

The assumption of negligible mutual inductance is valid for the doubly-salient VRM of Fig. 8.1b both due to symmetry of the machine geometry and due to the assumption of negligible iron reluctance. In practice, even in situations where symmetry might suggest that the mutual inductances are zero or can be ignored because they are independent of rotor position (e.g., the phases are coupled through leakage fluxes), significant nonlinear and mutual-inductance effects can arise due to saturation of the machine iron. In such cases, although the techniques of Chapter 3, and indeed torque expressions of the form of Eq. 8.3, remain valid, analytical expressions are often difficult to obtain (see Section 8.4).

At the design and analysis stage, the winding flux-current relationships and the motor torque can be determined by using numerical-analysis packages which can account for the nonlinearity of the machine magnetic material. Once a machine has been constructed, measurements can be made, both to validate the various assumptions and approximations which were made as well as to obtain an accurate measure of actual machine performance.

From this point on, we shall use the symbol p_s to indicate the number of stator poles and p_r to indicate the number of rotor poles, and the corresponding machine is called a p_s/p_r machine. Example 8.1 examines a 4/2 VRM.

EXAMPLE 8.1

A 4/2 VRM is shown in Fig. 8.3. Its dimensions are

$$R = 3.8 \text{ cm} \quad \alpha = \beta = 60° = \pi/3 \text{ rad}$$

$$g = 2.54 \times 10^{-2} \text{ cm} \quad D = 13.0 \text{ cm}$$

and the poles of each phase winding are connected in series such that there are a total of $N = 100$ turns (50 turns per pole) in each phase winding. Assume the rotor and stator to be of infinite magnetic permeability.

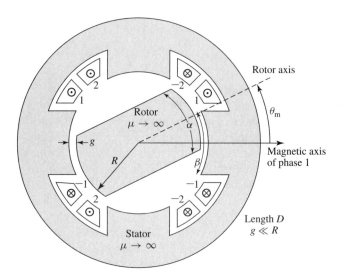

Figure 8.3   4/2 VRM for Example 8.1.

a. Neglecting leakage and fringing fluxes, plot the phase-1 inductance $L(\theta_m)$ as a function of θ_m.
b. Plot the torque, assuming (*i*) $i_1 = I_1$ and $i_2 = 0$ and (*ii*) $i_1 = 0$ and $i_2 = I_2$.
c. Calculate the net torque (in N · m) acting on the rotor when both windings are excited such that $i_1 = i_2 = 5$ A and at angles (*i*) $\theta_m = 0°$, (*ii*) $\theta_m = 45°$, (*iii*) $\theta_m = 75°$.

■ Solution

a. Using the magnetic circuit techniques of Chapter 1, we see that the maximum inductance L_{max} for phase 1 occurs when the rotor axis is aligned with the phase-1 magnetic axis. From Eq. 1.31, we see that L_{max} is equal to

$$L_{max} = \frac{N^2 \mu_0 \alpha R D}{2g}$$

where $\alpha R D$ is the cross-sectional area of the air gap and $2g$ is the total gap length in the magnetic circuit. For the values given,

$$
\begin{aligned}
L_{max} &= \frac{N^2 \mu_0 \alpha R D}{2g} \\
&= \frac{(100)^2 (4\pi \times 10^{-7})(\pi/3)(3.8 \times 10^{-2})(0.13)}{2 \times (2.54 \times 10^{-4})} \\
&= 0.128 \text{ H}
\end{aligned}
$$

Neglecting fringing, the inductance $L(\theta_m)$ will vary linearly with the air-gap cross-sectional area as shown in Fig. 8.4a. Note that this idealization predicts that the inductance is zero when there is no overlap when in fact there will be some small value of inductance, as shown in Fig. 8.2.

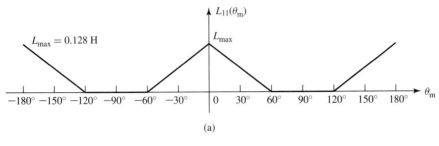

(a)

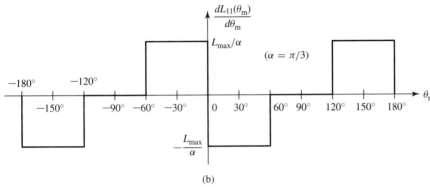

(b)

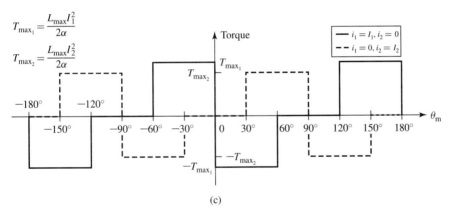

(c)

Figure 8.4 (a) $L_{11}(\theta_m)$ versus θ_m, (b) $dL_{11}(\theta_m)/d\theta_m$ versus θ_m, and (c) torque versus θ_m.

b. From Eq. 8.7, the torque consists of two terms

$$T_{mech} = \frac{1}{2}i_1^2 \frac{dL_{11}(\theta_m)}{d\theta_m} + \frac{1}{2}i_2^2 \frac{dL_{11}(\theta_m - 90°)}{d\theta_m}$$

and $dL_{11}/d\theta_m$ can be seen to be the stepped waveform of Fig. 8.4b whose maximum values are given by $\pm L_{max}/\alpha$ (with α expressed in radians!). Thus the torque is as shown in Fig. 8.4c.

c. The peak torque due to each of the windings is given by

$$T_{max} = \left(\frac{L_{max}}{2\alpha}\right) i^2 = \left(\frac{0.128}{2(\pi/3)}\right) 5^2 = 1.53 \text{ N} \cdot \text{m}$$

(i) From the plot in Fig. 8.4c, at $\theta_m = 0°$, the torque contribution from phase 2 is clearly zero. Although the phase-1 contribution appears to be indeterminate, in an actual machine the torque change from T_{max_1} to $-T_{max_1}$ at $\theta_m = 0°$ would have a finite slope and the torque would be zero at $\theta = 0°$. Thus the net torque from phases 1 and 2 at this position is zero.

Notice that the torque at $\theta_m = 0$ is zero independent of the current levels in phases 1 and 2. This is a problem with the 4/2 configuration of Fig. 8.3 since the rotor can get "stuck" at this position (as well as at $\theta_m = \pm 90°, \pm 180°$), and there is no way that electrical torque can be produced to move it.

(ii) At $\theta_m = 45°$ both phases are providing torque. That of phase 1 is negative while that of phase 2 is positive. Because the phase currents are equal, the torques are thus equal and opposite and the net torque is zero. However, unlike the case of $\theta_m = 0°$, the torque at this point can be made either positive or negative simply by appropriate selection of the phase currents.

(iii) At $\theta_m = 75°$ phase 1 produces no torque while phase 2 produces a positive torque of magnitude T_{max_2}. Thus the net torque at this position is positive and of magnitude 1.53 N · m. Notice that there is no combination of phase currents that will produce a negative torque at this position since the phase-1 torque is always zero while that of phase 2 can be only positive (or zero).

Practice Problem 8.1

Repeat the calculation of Example 8.1, part (c), for the case in which $\alpha = \beta = 70°$.

Solution

(i) $T = 0 \text{ N} \cdot \text{m}$
(ii) $T = 0 \text{ N} \cdot \text{m}$
(iii) $T = 1.59 \text{ N} \cdot \text{m}$

Example 8.1 illustrates a number of important considerations for the design of VRMs. Clearly these machines must be designed to avoid the occurrence of rotor positions for which none of the phases can produce torque. This is of concern in the design of 4/2 machines which will always have such positions if they are constructed with uniform, symmetric air gaps.

It is also clear that to operate VRMs with specified torque characteristics, the phase currents must be applied in a fashion consistent with the rotor position. For example, positive torque production from each phase winding in Example 8.1 can be seen from Fig. 8.4c to occur only for specific values of θ_m. Thus operation of VRMs must include some sort of rotor-position sensing as well as a controller which determines both the sequence and the waveform of the phase currents to achieve the desired operation. This is typically implemented by using electronic switching devices (transistors, thyristors, gate-turn-off devices, etc.) under the supervision of a microprocessor-based controller.

Although a 4/2 VRM such as in Example 8.1 can be made to work, as a practical matter it is not particularly useful because of undesirable characteristics such as its zero-torque positions and the fact that there are angular locations at which it is not possible to achieve a positive torque. For example, because of these limitations, this machine cannot be made to generate a constant torque independent of rotor angle; certainly no combination of phase currents can result in torque at the zero-torque positions or positive torque in the range of angular locations where only negative torque can be produced. As discussed in Section 8.2, these difficulties can be eliminated by 4/2 designs with asymmetric geometries, and so practical 4/2 machines can be constructed.

As has been seen in this section, the analysis of VRMs is conceptually straightforward. In the case of linear machine iron (no magnetic saturation), finding the torque is simply a matter of finding the stator-phase inductances (self and mutual) as a function of rotor position, expressing the coenergy in terms of these inductances, and then calculating the derivative of the coenergy with respect to angular position (holding the phase currents constant when taking the derivative). Similarly, as discussed in Section 3.8, the electric terminal voltage for each of the phases can be found from the sum of the time derivative of the phase flux linkage and the iR drop across the phase resistance.

In the case of nonlinear machine iron (where saturation effects are important) as is discussed in Section 8.4, the coenergy can be found by appropriate integration of the phase flux linkages, and the torque can again be found from the derivative of the coenergy with respect to the angular position of the rotor. In either case, there are no rotor windings and typically no other rotor currents in a well-designed variable-reluctance motor; hence, unlike other ac machine types (synchronous and induction), there are no electrical dynamics associated with the machine rotor. This greatly simplifies their analysis.

Although VRMs are simple in concept and construction, their operation is somewhat complicated and requires sophisticated control and motor-drive electronics to achieve useful operating characteristics. These issues and others are discussed in Sections 8.2 to 8.5.

8.2 PRACTICAL VRM CONFIGURATIONS

Practical VRM drive systems (the motor and its inverter) are designed to meet operating criteria such as

- Low cost.
- Constant torque independent of rotor angular position.
- A desired operating speed range.
- High efficiency.
- A large torque-to-mass ratio.

As in any engineering situation, the final design for a specific application will involve a compromise between the variety of options available to the designer. Because VRMs require some sort of electronics and control to operate, often the designer is concerned with optimizing a characteristic of the complete drive system, and this will impose additional constraints on the motor design.

VRMs can be built in a wide variety of configurations. In Fig. 8.1, two forms of a 4/2 machine are shown: a singly-salient machine in Fig. 8.1a and a doubly-salient machine in Fig. 8.1b. Although both types of design can be made to work, a doubly-salient design is often the superior choice because it can generally produce a larger torque for a given frame size.

This can be seen qualitatively (under the assumption of a high-permeability, nonsaturating magnetic structure) by reference to Eq. 8.7, which shows that the torque is a function of $dL_{11}(\theta_m)/d\theta_m$, the derivative of the phase inductance with respect to angular position of the rotor. Clearly, all else being equal, the machine with the largest derivative will produce the largest torque.

This derivative can be thought of as being determined by the ratio of the maximum to minimum phase inductances L_{max}/L_{min}. In other words, we can write,

$$\frac{dL_{11}(\theta_m)}{d\theta_m} \simeq \frac{L_{max} - L_{min}}{\Delta\theta_m}$$

$$= \frac{L_{max}}{\Delta\theta_m}\left(1 - \frac{L_{min}}{L_{max}}\right) \tag{8.8}$$

where $\Delta\theta_m$ is the angular displacement of the rotor between the positions of maximum and minimum phase inductance. From Eq. 8.8, we see that, for a given L_{max} and $\Delta\theta_m$, the largest value of L_{max}/L_{min} will give the largest torque. Because of its geometry, a doubly-salient structure will typically have a lower minimum inductance and thus a larger value of L_{max}/L_{min}; hence it will produce a larger torque for the same rotor structure.

For this reason doubly-salient machines are the predominant type of VRM, and hence for the remainder of this chapter we consider only doubly-salient VRMs. In general, doubly-salient machines can be constructed with two or more poles on each of the stator and rotor. It should be pointed out that once the basic structure of a VRM is determined, L_{max} is fairly well determined by such quantities as the number of turns, air-gap length, and basic pole dimensions. The challenge to the VRM designer is to achieve a small value of L_{min}. This is a difficult task because L_{min} is dominated by leakage fluxes and other quantities which are difficult to calculate and analyze.

As shown in Example 8.1, the geometry of a symmetric 4/2 VRM with a uniform air gap gives rise to rotor positions for which no torque can be developed for any combination of excitation of the phase windings. These torque zeros can be seen to occur at rotor positions where all the stator phases are simultaneously at a position of either maximum or minimum inductance. Since the torque depends on the derivative of inductance with respect to angular position, this simultaneous alignment of maximum and minimum inductance points necessarily results in zero net torque.

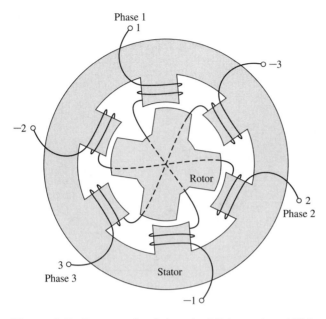

Figure 8.5 Cross-sectional view of a 6/4 three-phase VRM.

Figure 8.5 shows a 6/4 VRM from which we see that a fundamental feature of the 6/4 machine is that no such simultaneous alignment of phase inductances is possible. As a result, this machine does not have any zero-torque positions. This is a significant point because it eliminates the possibility that the rotor might get stuck in one of these positions at standstill, requiring that it be mechanically moved to a new position before it can be started. In addition to the fact that there are not positions of simultaneous alignment for the 6/4 VRM, it can be seen that there also are no rotor positions at which only a torque of a single sign (either positive or negative) can be produced. Hence by proper control of the phase currents, it should be possible to achieve constant-torque, independent of rotor position.

In the case of a symmetric VRM with p_s stator poles and p_r rotor poles, a simple test can be used to determine if zero-torque positions exist. If the ratio p_s/p_r (or alternatively p_r/p_s if p_r is larger than p_s) is an integer, there will be zero-torque positions. For example, for a 6/4 machine the ratio is 1.5, and there will be no zero-torque positions. However, the ratio is 2.0 for a 6/3 machine, and there will be zero-torque positions.

In some instances, design constraints may dictate that a machine with an integral pole ratio is desirable. In these cases, it is possible to eliminate the zero-torque positions by constructing a machine with an asymmetric rotor. For example, the rotor radius can be made to vary with angle as shown in grossly exaggerated fashion in Fig. 8.6a. This design, which also requires that the width of the rotor pole be wider than that of the stator, will not produce zero torque at positions of alignment because $dL(\theta_m)/d\theta_m$ is not zero at these points, as can be seen with reference to Fig. 8.6b.

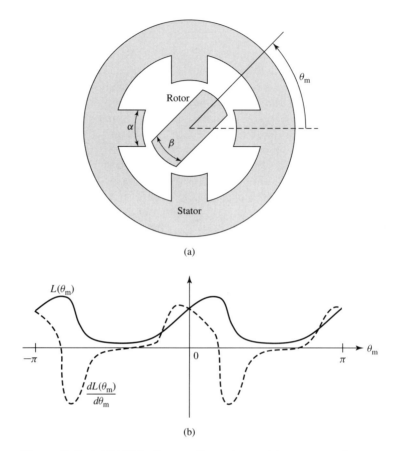

Figure 8.6   A 4/2 VRM with nonuniform air gap: (a) exaggerated
schematic view and (b) plots of $L(\theta_m)$ and $dL(\theta_m)/d\theta_m$ versus θ_m.

An alternative procedure for constructing an integral-pole-ratio VRM without
zero-torque positions is to construct a stack of two or more VRMs in series, aligned
such that each of the VRMs is displaced in angle from the others and with all rotors
sharing a common shaft. In this fashion, the zero-torque positions of the individual
machines will not align, and thus the complete machine will not have any torque zeros.
For example, a series stack of two two-phase, 4/2 VRMs such as that of Example 8.1
(Fig. 8.3) with a 45° angular displacement between the individual VRMs will result
in a four-phase VRM without zero-torque positions.

Generally VRMs are wound with a single coil on each pole. Although it is possible
to control each of these windings separately as individual phases, it is common practice
to combine them into groups of poles which are excited simultaneously. For example,
the 4/2 VRM of Fig. 8.3 is shown connected as a two-phase machine. As shown in
Fig. 8.5, a 6/4 VRM is commonly connected as a three-phase machine with opposite
poles connected to the same phase and in such a fashion that the windings drive flux
in the same direction through the rotor.

In some cases, VRMs are wound with a parallel set of windings on each phase. This configuration, known as a *bifilar winding,* in some cases can result in a simple inverter configuration and thus a simple, inexpensive motor drive. The use of a bifilar winding in VRM drives is discussed in Section 11.4.

In general, when a given phase is excited, the torque is such that the rotor is pulled to the nearest position of maximum flux linkage. As excitation is removed from that phase and the next phase is excited, the rotor "follows" as it is then pulled to a new maximum flux-linkage position. Thus, the rotor speed is determined by the frequency of the phase currents. However, unlike the case of a synchronous machine, the relationship between the rotor speed and the frequency and sequence of the phase-winding excitation can be quite complex, depending on the number of rotor poles and the number of stator poles and phases. This is illustrated in Example 8.2.

Consider a four-phase, 8/6 VRM. If the stator phases are excited sequentially, with a total time of T_0 sec required to excite the four phases (i.e., each phase is excited for a time of $T_0/4$ sec), find the angular velocity of the stator flux wave and the corresponding angular velocity of the rotor. Neglect any system dynamics and assume that the rotor will instantaneously track the stator excitation.

■ Solution

Figure 8.7 shows in schematic form an 8/6 VRM. The details of the pole shapes are not of importance for this example and thus the rotor and stator poles are shown simply as arrows indicating their locations. The figure shows the rotor aligned with the stator phase-1 poles. This position corresponds to that which would occur if there were no load on the rotor and the stator phase-1 windings were excited, since it corresponds to a position of maximum phase-1 flux linkage.

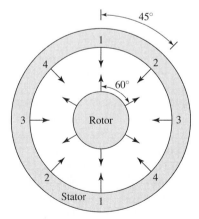

Figure 8.7 Schematic view of a four-phase 8/6 VRM. Pole locations are indicated by arrows.

Consider next that the excitation on phase 1 is removed and phase 2 is excited. At this point, the stator flux wave has rotated 45° in the clockwise direction. Similarly, as the excitation on phase 2 is removed and phase 3 is excited, the stator flux wave will move an additional 45° clockwise. Thus the angular velocity ω_s of the stator flux wave can be calculated quite simply as $\pi/4$ rad (45°) divided by $T_0/4$ sec, or $\omega_s = \pi/T_0$ rad/sec.

Note, however, that this is not the angular velocity of the rotor itself. As the phase-1 excitation is removed and phase 2 is excited, the rotor will move in such a fashion as to maximize the phase-2 flux linkages. In this case, Fig. 8.7 shows that the rotor will move 15° counterclockwise since the nearest rotor poles to phase 2 are actually 15° ahead of the phase-2 poles. Thus the angular velocity of the rotor can be calculated as $-\pi/12$ rad (15°, with the minus sign indicating counterclockwise rotation) divided by $T_0/4$ sec, or $\omega_m = -\pi/(3T_0)$ rad/sec.

In this case, the rotor travels at one-third the angular velocity of the stator excitation and in the opposite direction!

Practice Problem 8.2

Repeat the calculation of Example 8.2 for the case of a four-phase, 8/10 VRM.

Solution

$$\omega_m = \pi/(5T_0) \text{ rad/sec}$$

Example 8.2 illustrates the complex relationship that can exist between the excitation frequency of a VRM and the "synchronous" rotor frequency. This relationship is directly analogous to that between two mechanical gears for which the choice of different gear shapes and configurations gives rise to a wide range of speed ratios. It is difficult to derive a single rule which will describe this relationship for the immense variety of VRM configurations which can be envisioned. It is, however, a fairly simple matter to follow a procedure similar to that shown in Example 8.2 to investigate any particular configuration of interest.

Further variations on VRM configurations are possible if the main stator and rotor poles are subdivided by the addition of individual teeth (which can be thought of as a set of small poles excited simultaneously by a single winding). The basic concept is illustrated in Fig. 8.8, which shows a schematic view of three poles of a three-phase VRM with a total of six main stator poles. Such a machine, with the stator and rotor poles subdivided into teeth, is known as a *castleated* VRM, the name resulting from the fact that the stator teeth appear much like the towers of a medieval castle.

In Fig. 8.8 each stator pole has been divided into four subpoles by the addition of four teeth of width $6\frac{3}{7}^\circ$ (indicated by the angle β in the figure), with a slot of the same width between each tooth. The same tooth/slot spacing is chosen for the rotor, resulting in a total of 28 teeth on the rotor. Notice that this number of rotor teeth and the corresponding value of β were chosen so that when the rotor teeth are aligned with those of the phase-1 stator pole, they are not aligned with those of phases 2 and 3. In this fashion, successive excitation of the stator phases will result in a rotation of the rotor.

Castleation further complicates the relationship between the rotor speed and the frequency and sequence of the stator-winding excitation. For example, from Fig. 8.8

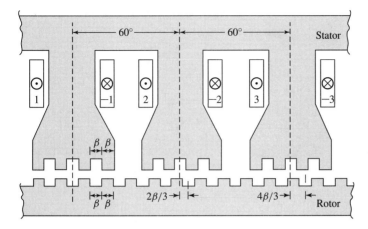

Figure 8.8 Schematic view of a three-phase castleated VRM with six stator poles and four teeth per pole and 28 rotor poles.

it can be seen that for this configuration, when the excitation of phase 1 is removed and phase 2 is excited (corresponding to a rotation of the stator flux wave by 60° in the clockwise direction), the rotor will rotate by an angle of $(2\beta/3) = 4\frac{2}{7}^\circ$ in the counterclockwise direction.

From the preceding analysis, we see that the technique of castleation can be used to create VRMs capable of operating at low speeds (and hence producing high torque for a given stator power input) and with very precise rotor position accuracy. For example, the machine of Fig. 8.8 can be rotated precisely by angular increments of $(2\beta/3)$. The use of more teeth can further increase the position resolution of these machines. Such machines can be found in applications where low speed, high torque, and precise angular resolution are required. This castleated configuration is one example of a class of VRMs commonly referred to as *stepping motors* because of their capability to produce small steps in angular resolution.

8.3 CURRENT WAVEFORMS FOR TORQUE PRODUCTION

As is seen in Section 8.1, the torque produced by a VRM in which saturation and mutual-inductance effects can be neglected is determined by the summation of terms consisting of the derivatives of the phase inductances with respect to the rotor angular position, each multiplied by the square of the corresponding phase current. For example, we see from Eqs. 8.6 and 8.7 that the torque of the two-phase, 4/2 VRM of Fig. 8.1b is given by

$$
\begin{aligned}
T_{\text{mech}} &= \frac{1}{2}i_1^2 \frac{dL_{11}(\theta_m)}{d\theta_m} + \frac{1}{2}i_2^2 \frac{dL_{22}(\theta_m)}{d\theta_m} \\
&= \frac{1}{2}i_1^2 \frac{dL_{11}(\theta_m)}{d\theta_m} + \frac{1}{2}i_2^2 \frac{dL_{11}(\theta_m - 90°)}{d\theta_m}
\end{aligned}
\tag{8.9}
$$

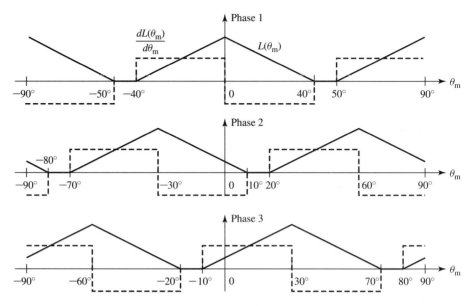

Figure 8.9 Idealized inductance and $dL/d\theta_m$ curves for a three-phase 6/4 VRM with 40° rotor and stator poles.

For each phase of a VRM, the phase inductance is periodic in rotor angular position, and thus the area under the curve of $dL/d\theta_m$ calculated over a complete period of $L(\theta_m)$ is zero, i.e.,

$$\int_0^{2\pi/p_r} \frac{dL(\theta_m)}{d\theta_m} \, d\theta_m = L(2\pi/p_r) - L(0) = 0 \qquad (8.10)$$

where p_r is the number of rotor poles.

The average torque produced by a VRM can be found by integrating the torque equation (Eq. 8.9) over a complete period of rotation. Clearly, if the stator currents are held constant, Eq. 8.10 shows that the average torque will be zero. Thus, to produce a time-averaged torque, the stator currents must vary with rotor position. The desired average output torque for a VRM depends on the nature of the application. For example, motor operation requires a positive time-averaged shaft torque. Similarly, braking or generator action requires negative time-averaged torque.

Positive torque is produced when a phase is excited at angular positions with positive $dL/d\theta_m$ for that phase, and negative torque is produced by excitation at positions at which $dL/d\theta_m$ is negative. Consider a three-phase, 6/4 VRM (similar to that shown in Fig. 8.5) with 40° rotor and stator poles. The inductance versus rotor position for this machine will be similar to the idealized representation shown in Fig. 8.9.

Operation of this machine as a motor requires a net positive torque. Alternatively, it can be operated as a generator under conditions of net negative torque. Noting that

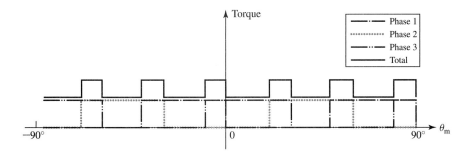

Figure 8.10 Individual phase torques and total torque for the motor of Fig. 8.9. Each phase is excited with a constant current I_0 only at positions where $dL/d\theta_m > 0$.

positive torque is generated when excitation is applied at rotor positions at which $dL/d\theta_m$ is positive, we see that a control system is required that determines rotor position and applies the phase-winding excitations at the appropriate time. It is, in fact, the need for this sort of control that makes VRM drive systems more complex than might perhaps be thought, considering only the simplicity of the VRM itself.

One of the reasons that VRMs have found application in a wide variety of situations is because the widespread availability and low cost of microprocessors and power electronics have brought the cost of the sensing and control required to successfully operate VRM drive systems down to a level where these systems can be competitive with competing technologies. Although the control of VRM drives is more complex than that required for dc, induction, and permanent-magnet ac motor systems, in many applications the overall VRM drive system turns out to be less expensive and more flexible than the competition.

Assuming that the appropriate rotor-position sensor and control system is available, the question still remains as to how to excite the armature phases. From Fig. 8.9, one possible excitation scheme would be to apply a constant current to each phase at those angular positions at which $dL/d\theta_m$ is positive and zero current otherwise.

If this is done, the resultant torque waveform will be that of Fig. 8.10. Note that because the torque waveforms of the individual phases overlap, the resultant torque will not be constant but rather will have a pulsating component on top of its average value. In general, such pulsating torques are to be avoided both because they may produce damaging stresses in the VRM and because they may result in the generation of excessive vibration and noise.

Consideration of Fig. 8.9 shows that there are alternative excitation strategies which can reduce the torque pulsations of Fig. 8.10. Perhaps the simplest strategy is to excite each phase for only 30° of angular position instead of the 40° which resulted in Fig. 8.9. Thus, each phase would simply be turned off as the next phase is turned on, and there would be no torque overlap between phases.

Although this strategy would be an ideal solution to the problem, as a practical matter it is not possible to implement. The problem is that because each phase winding has a self-inductance, it is not possible to instantaneously switch on or off the phase currents. Specifically, for a VRM with independent (uncoupled) phases,[2] the voltage-current relationship of the jth phase is given by

$$v_j = R_j i_j + \frac{d\lambda_j}{dt} \tag{8.11}$$

where

$$\lambda_j = L_{jj}(\theta_m) i_j \tag{8.12}$$

Thus,

$$v_j = R_j i_j + \frac{d}{dt}[L_{jj}(\theta_m) i_j] \tag{8.13}$$

Equation 8.13 can be rewritten as

$$v_j = \left\{ R_j + \frac{d}{dt}[L_{jj}(\theta_m)] \right\} i_j + L_{jj}(\theta_m) \frac{di_j}{dt} \tag{8.14}$$

or

$$v_j = \left[R_j + \frac{dL_{jj}(\theta_m)}{d(\theta_m)} \frac{d\theta_m}{dt} \right] i_j + L_{jj}(\theta_m) \frac{di_j}{dt} \tag{8.15}$$

Although Eqs. 8.13 through 8.15 are mathematically complex and often require numerical solution, they clearly indicate that some time is required to build up currents in the phase windings following application of voltage to that phase. A similar analysis can be done for conditions associated with removal of the phase currents. The delay time associated with current build up can limit the maximum achievable torque while the current decay time can result in negative torque if current is still flowing when $dL(\theta_m)/d\theta_m$ reverses sign. These effects are illustrated in Example 8.3 which also shows that in cases where winding resistance can be neglected, an approximate solution to these equations can be found.

EXAMPLE 8.3

Consider the idealized 4/2 VRM of Example 8.1. Assume that it has a winding resistance of $R = 1.5\ \Omega$/phase and a leakage inductance $L_l = 5$ mH in each phase. For a constant rotor speed of 4000 r/min, calculate (a) the phase-1 current as a function of time during the interval $-60° \le \theta_m \le 0°$, assuming that a constant voltage of $V_0 = 100$ V is applied to phase 1 just as $dL_{11}(\theta_m)/d\theta_m$ becomes positive (i.e., at $\theta_m = -60° = -\pi/3$ rad), and (b) the decay of phase-1 current if a negative voltage of -200 V is applied at $\theta_m = 0°$ and maintained until the current reaches zero. (c) Using MATLAB[†], plot these currents as well as the corresponding torque. Also calculate the integral under the torque-versus-time plot and compare it to the integral under the torque-versus-time curve for the time period during which the torque is positive.

---

[2] The reader is reminded that in some cases the assumption of independent phases is not justified, and then a more complex analysis of the VRM is required (see the discussion following the derivation of Eq. 8.5).

[†] MATLAB is a registered trademark of The MathWorks, Inc.

■ Solution

a. From Eq. 8.15, the differential equation governing the current buildup in phase 1 is given by

$$v_1 = \left[R + \frac{dL_{11}(\theta_m)}{d\theta_m} \frac{d\theta_m}{dt} \right] i_1 + L_{11}(\theta_m) \frac{di_1}{dt}$$

At 4000 r/min,

$$\omega_m = \frac{d\theta_m}{dt} = 4000 \text{ r/min} \times \frac{\pi}{30} \left[\frac{\text{rad/sec}}{\text{r/min}} \right] = \frac{400\pi}{3} \text{ rad/sec}$$

From Fig. 8.4 (for $-60° \le \theta_m \le 0°$)

$$L_{11}(\theta_m) = L_l + \frac{L_{max}}{\pi/3} \left(\theta_m + \frac{\pi}{3} \right)$$

$$= 0.005 + 0.122(\theta_m + \pi/3)$$

Thus

$$\frac{dL_{11}(\theta_m)}{d\theta_m} = 0.122 \text{ H/rad}$$

and

$$\frac{dL_{11}(\theta_m)}{d\theta_m} \frac{d\theta_m}{dt} = 51.1 \ \Omega$$

which is much greater than the resistance $R = 1.5 \ \Omega$

This will enable us to obtain an approximate solution for the current by neglecting the Ri term in Eq. 8.13. We must then solve

$$\frac{d(L_{11}i_1)}{dt} = v_1$$

for which the solution is

$$i_1(t) = \frac{\int_0^t v_1 dt}{L_{11}(t)} = \frac{V_0 t}{L_{11}(t)}$$

Substituting

$$\theta_m = -\frac{\pi}{3} + \omega_m t$$

into the expression for $L_{11}(\theta_m)$ then gives

$$i_1(t) = \frac{100t}{0.005 + 51.1t} \text{ A}$$

which is valid until $\theta_m = 0°$ at $t = 2.5$ msec, at which point $i_1(t) = 1.88$ A.

b. During the period of current decay the solution proceeds as in part (a). From Fig. 8.4, for $0° \le \theta_m \le 60°$, $dL_{11}(\theta_m)/dt = -51.1 \ \Omega$ and the Ri term can again be ignored in Eq. 8.13.

Thus, since the applied voltage is -200 V for this time period ($t \ge 2.5$ msec until $i_1(t) = 0$) in an effort to bring the current rapidly to zero, since the current must be continuous at time $t_0 = 2.5$ msec, and since, from Fig. 8.4 (for $0° \leq \theta_m \leq 60°$)

$$L_{11}(\theta_m) = L_l + \frac{L_{max}}{\pi/3}\left(\frac{\pi}{3} - \theta_m\right)$$

$$= 0.005 + 0.122(\pi/3 - \theta_m)$$

we see that the solution becomes

$$i_1(t) = \frac{L_{11}(t_0)i_1(t_0) + \int_{t_0}^{t} v_1 \, dt}{L_{11}(t)}$$

$$= \frac{0.25 - 200(t - 2.5 \times 10^{-3})}{0.005 + 51.1(5 \times 10^{-3} - t)}$$

From this equation, we see that the current reaches zero at $t = 3.75$ msec.

c. The torque can be found from Eq. 8.9 by setting $i_2 = 0$. Thus

$$T_{mech} = \frac{1}{2}i_1^2\frac{dL_{11}}{d\theta_m}$$

Using MATLAB and the results of parts (a) and (b), the current waveform is plotted in Fig. 8.11a and the torque in Fig. 8.11b. The integral under the torque curve is 3.35×10^{-4} N·m·sec while that under the positive portion of the torque curve corresponding to positive torque is 4.56×10^{-4} N·m·sec. Thus we see that the negative torque produces a 27 percent reduction in average torque from that which would otherwise be available if the current could be reduced instantaneously to zero.

Notice first from the results of part (b) and from Fig. 8.11a that, in spite of applying a negative voltage of twice the magnitude of the voltage used to build up the current, current continues to flow in the winding for 1.25 ms after reversal of the applied voltage. From Fig. 8.11b, we see that the result is a significant period of negative torque production. In practice, this may, for example, dictate a control scheme which reverses the phase current in advance of the time that the sign of $dL(\theta_m)/d\theta_m$ reverses, achieving a larger average torque by trading off some reduction in average positive torque against a larger decrease in average negative torque.

This example also illustrates another important aspect of VRM operation. For a system of resistance of 1.5 Ω and constant inductance, one would expect a steady-state current of $100/1.5 = 66.7$ A. Yet in this system the steady-state current is less than 2 A. The reason for this is evident from Eqs. 8.14 and 8.15 where we see that $dL_{11}(\theta_m)/dt = 51.1$ Ω appears as an apparent resistance in series with the winding resistance which is much larger than the winding resistance itself. The corresponding voltage drop (the speed voltage) is of sufficient magnitude to limit the steady-state current to a value of $100/51.1 = 1.96$ A.

Here is the MATLAB script:

```
clc
clear

% Here are the inductances
Lmax = 0.128;
Lleak = 0.005;

Posintegral = 0;
integral = 0;
```

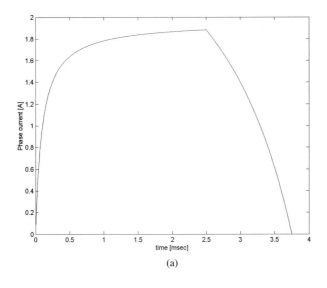

(a)

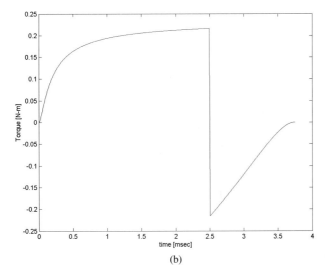

(b)

Figure 8.11 Example 8.3: (a) phase-1 current and (b) torque profile.

```
N = 500;
tmax = 3.75e-3;
deltat = tmax/N;

% Now do the calculations
for n = 1:(N+1)

    t(n) = tmax*(n-1)/N;
    thetam(n) = -(pi/3) + (400*pi/3) * t(n);
```

```
   if (thetam(n) <= 0)
      i(n) = 100*t(n)/(0.005 + 51.1 *t(n));
      dld11dtheta = 0.122;
      Torque(n) = 0.5*i(n)^2*dld11dtheta;
      Posintegral = Posintegral + Torque(n)*deltat;
      integral = Posintegral;
   else
      i(n) = (0.25 - 200*(t(n) - 2.5e-3))/(0.005+51.1*(5e-3 - t(n)));
      dld11dtheta = -0.122;
      Torque(n) = 0.5*i(n)^2*dld11dtheta;
      integral = integral + Torque(n)*deltat;
   end

end

fprintf('\nPositive torque integral = %g [N-m-sec]',Posintegral)
fprintf('\nTorque integral = %g [N-m-sec]\n',integral)

plot(t*1000,i)
xlabel('time [msec]')
ylabel('Phase current [A]')

pause

plot(t*1000,Torque)
xlabel('time [msec]')
ylabel('Torque [N-m]')
```

Practice Problem 8.3

 Reconsider Example 8.3 under the condition that a voltage of -250 V is applied to turn off the phase current. Use MATLAB to calculate the integral under the torque-versus-time plot and compare it to the integral under the torque-versus-time curve for the time period during which the torque is positive.

Solution

The current returns to zero at $t = 3.5$ msec. The integral under the torque curve is 3.67×10^{-4} N $\cdot$ m $\cdot$ s while that under the positive portion of the torque curve corresponding to positive torque remains equal to 4.56×10^{-4} N $\cdot$ m $\cdot$ s. In this case, the negative torque produces a 20 percent reduction in torque from that which would otherwise be available if the current could be reduced instantaneously to zero.

Example 8.3 illustrates important aspects of VRM performance which do not appear in an idealized analysis such as that of Example 8.1 but which play an extremely important role in practical applications. It is clear that it is not possible to readily apply phase currents of arbitrary waveshapes. Winding inductances (and their time derivatives) significantly affect the current waveforms that can be achieved for a given applied voltage.

In general, the problem becomes more severe as the rotor speed is increased. Consideration of Example 8.3 shows, for a given applied voltage, (1) that as the speed is increased, the current will take a larger fraction of the available time during which $dL(\theta_m)/d\theta_m$ is positive to achieve a given level and (2) that the steady-state current which can be achieved is progressively lowered. One common method for maximizing the available torque is to apply the phase voltage somewhat in advance of the time when $dL(\theta_m)/d\theta_m$ begins to increase. This gives the current time to build up to a significant level before torque production begins.

Yet a more significant difficulty (also illustrated in Example 8.3) is that just as the currents require a significant amount of time to increase at the beginning of a turn-on cycle, they also require time to decrease at the end. As a result, if the phase excitation is removed at or near the end of the positive $dL(\theta_m)/d\theta_m$ period, it is highly likely that there will be phase current remaining as $dL(\theta_m)/d\theta_m$ becomes negative, so there will be a period of negative torque production, reducing the effective torque-producing capability of the VRM.

One way to avoid such negative torque production would be to turn off the phase excitation sufficiently early in the cycle that the current will have decayed essentially to zero by the time that $dL(\theta_m)/d\theta_m$ becomes negative. However, there is clearly a point of diminishing returns, because turning off the phase current while $dL(\theta_m)/d\theta_m$ is positive also reduces positive torque production. As a result, it is often necessary to accept a certain amount of negative torque (to get the required positive torque) and to compensate for it by the production of additional positive torque from another phase.

Another possibility is illustrated in Fig. 8.12. Figure 8.12a shows the cross-sectional view of a 4/2 VRM similar to that of Fig. 8.3 with the exception that the rotor pole angle has been increased from 60° to 75°, with the result that the rotor pole overhangs that of the stator by 15°. As can be seen from Fig. 8.12b, this results in a region of constant inductance separating the positive and negative $dL(\theta_m)/d\theta_m$ regions, which in turn provides additional time for the phase current to be turned off before the region of negative torque production is reached.

Although Fig. 8.12 shows an example with 15° of rotor overhang, in any particular design the amount of overhang would be determined as part of the overall design process and would depend on such issues as the amount of time required for the phase current to decay and the operating speed of the VRM. Also included in this design process must be recognition that the use of wider rotor poles will result in a larger value of L_{min}, which itself tends to reduce torque production (see the discussion of Eq. 8.8) and to increase the time for current buildup.

Under conditions of constant-speed operation, it is often desirable to achieve constant torque independent of rotor position. Such operation will minimize pulsating torques which may cause excessive noise and vibration and perhaps ultimately lead to component failure due to material fatigue. This means that as the torque production of one phase begins to decrease, that of another phase must increase to compensate. As can be seen from torque waveforms such as those found in Fig. 8.11, this represents a

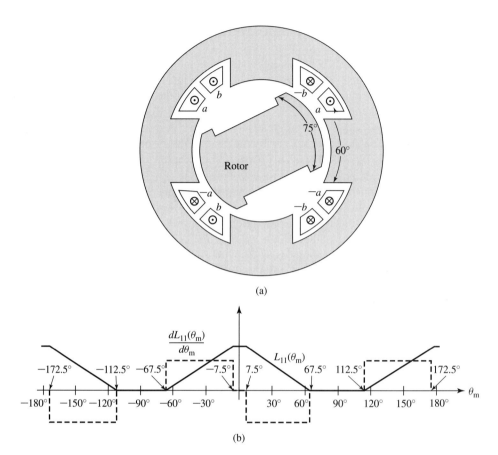

Figure 8.12   A 4/2 VRM with 15° rotor overhang: (a) cross-sectional view and (b) plots of $L_{11}(\theta_m)$ and $dL_{11}(\theta_m)/d\theta_m$ versus θ_m.

complex control problem for the phase excitation, and totally ripple-free torque will be difficult to achieve in many cases.

8.4 NONLINEAR ANALYSIS

Like most electric machines, VRMs employ magnetic materials both to direct and shape the magnetic fields in the machine and to increase the magnetic flux density that can be achieved from a given amplitude of current. To obtain the maximum benefit from the magnetic material, practical VRMs are operated with the magnetic flux density high enough so that the magnetic material is in saturation under normal operating conditions.

As with the synchronous, induction, and dc machines discussed in Chapters 5–7, the actual operating flux density is determined by trading off such quantities as cost, efficiency, and torque-to-mass ratio. However, because the VRM and its drive electronics are quite closely interrelated, VRMs design typically involves additional trade-offs that in turn affect the choice of operating flux density.

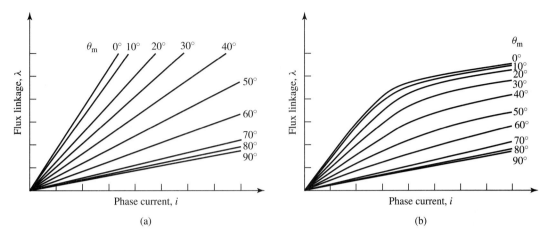

Figure 8.13 Plots of λ versus *i* for a VRM with (a) linear and (b) nonlinear magnetics.

Figure 8.2 shows typical inductance-versus-angle curves for the VRMs of Fig. 8.1. Such curves are characteristic of all VRMs. It must be recognized that the use of the concept of inductance is strictly valid only under the condition that the magnetic circuit in the machine is linear so that the flux density (and hence the winding flux linkage) is proportional to the winding current. This linear analysis is based on the assumption that the magnetic material in the motor has constant magnetic permeability. This assumption was used for all the analyses earlier in this chapter.

An alternate representation of the flux-linkage versus current characteristic of a VRM is shown in Fig. 8.13. This representation consists of a series of plots of the flux linkage versus current at various rotor angles. In this figure, the curves correspond to a machine with a two-pole rotor such as in Fig. 8.1, and hence a plot of curves from $0°$ to $90°$ is sufficient to completely characterize the machine.

Figure 8.13a shows the set of λ-i characteristics which would be measured in a machine with linear magnetics, i.e., constant magnetic permeability and no magnetic saturation. For each rotor angle, the curve is a straight line whose slope corresponds to the inductance $L(\theta_m)$ at that angular position. In fact, a plot of $L(\theta_m)$ versus θ_m such as in Fig. 8.2 is an equivalent representation to that of Fig. 8.13a.

In practice, VRMs do operate with their magnetic material in saturation and their λ-i characteristics take on the form of Fig. 8.13b. Notice that for low current levels, the curves are linear, corresponding to the assumption of linear magnetics of Fig. 8.13a. However, for higher current levels, saturation begins to occur and the curves bend over steeply, with the result that there is significantly less flux linkage for a given current level. Finally, note that saturation effects are maximum at $\theta_m = 0°$ (for which the rotor and stator poles are aligned) and minimal for higher angles as the rotor approaches the nonaligned position.

Saturation has two important, somewhat contradictory effects on VRM performance. On the one hand, saturation limits flux densities for a given current level and thus tends to limit the amount of torque available from the VRM. On the other hand, it can be shown that saturation tends to lower the required inverter volt-ampere rating

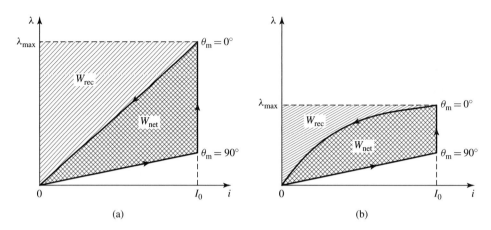

Figure 8.14   (a) Flux-linkage-current trajectory for the (a) linear and (b) nonlinear machines of Fig. 8.13.

for a given VRM output power and thus tends to make the inverter smaller and less costly. A well-designed VRM system will be based on a trade-off between the two effects.[3]

These effects of saturation can be investigated by considering the two machines of Figs. 8.13a and b operating at the same rotational speed and under the same operating condition. For the sake of simplicity, we assume a somewhat idealized condition in which the phase-1 current is instantaneously switched on to a value I_0 at $\theta_m = -90°$ (the unaligned position for phase 1) and is instantaneously switched off at $\theta_m = 0°$ (the aligned position). This operation is similar to that discussed in Example 8.1 in that we will neglect the complicating effects of the current buildup and decay transients which are illustrated in Example 8.3.

Because of rotor symmetry, the flux linkages for negative rotor angles are identical to those for positive angles. Thus, the flux linkage-current trajectories for one current cycle can be determined from Figs. 8.13a and b and are shown for the two machines in Figs. 8.14a and b.

As each trajectory is traversed, the power input to the winding is given by its volt-ampere product

$$p_{in} = iv = i\frac{d\lambda}{dt} \tag{8.16}$$

The net electric energy input to the machine (the energy that is converted to mechanical work) in a cycle can be determined by integrating Eq. 8.16 around the

[3] For a discussion of saturation effects in VRM drive systems, see T. J. E. Miller, "Converter Volt-Ampere Requirements of the Switched Reluctance Motor," *IEEE Trans. Ind. Appl.,* IA-21:1136–1144 (1985).

trajectory

$$\text{Net work} = \int p_{\text{in}} \, dt = \oint i \, d\lambda \tag{8.17}$$

This can be seen graphically as the area enclosed by the trajectory, labeled W_{net} in Figs. 8.14a and b. Note that the saturated machine converts less useful work per cycle than the unsaturated machine. As a result, to get a machine of the same power output, the saturated machine will have to be larger than a corresponding (hypothetical) unsaturated machine. This analysis demonstrates the effects of saturation in lowering torque and power output.

The peak energy input to the winding from the inverter can also be calculated. It is equal to the integral of the input power from the start of the trajectory to the point $(I_0, \lambda_{\text{max}})$:

$$\text{Peak energy} = \int_0^{\lambda_{\text{max}}} i \, d\lambda \tag{8.18}$$

This is the total area under the λ-i curve, shown in Fig. 8.14a and b as the sum of the areas labeled W_{rec} and W_{net}.

Since we have seen that the energy represented by the area W_{net} corresponds to useful output energy, it is clear that the energy represented by the area W_{rec} corresponds to energy input that is required to make the VRM operate (i.e., it goes into creating the magnetic fields in the VRM). This energy produced no useful work; rather it must be recycled back into the inverter at the end of the trajectory.

The inverter volt-ampere rating is determined by the average power per phase processed by the inverter as the motor operates, equal to the peak energy input to the VRM divided by the time T between cycles. Similarly, the average output power per phase of the VRM is given by the net energy input per cycle divided by T. Thus the ratio of the inverter volt-ampere rating to power output is

$$\frac{\text{Inverter volt-ampere rating}}{\text{Net output area}} = \frac{\text{area}(W_{\text{rec}} + W_{\text{net}})}{\text{area}(W_{\text{net}})} \tag{8.19}$$

In general, the inverter volt-ampere rating determines its cost and size. Thus, for a given power output from a VRM, a smaller ratio of inverter volt-ampere rating to output power means that the inverter will be both smaller and cheaper. Comparison of Figs. 8.14a and b shows that this ratio is smaller in the machine which saturates; the effect of saturation is to lower the amount of energy which must be recycled each cycle and hence the volt-ampere rating of the inverter required to supply the VRM.

EXAMPLE 8.4

Consider a symmetrical two-phase 4/2 VRM whose λ-i characteristic can be represented by the following λ-i expression (for phase 1) as a function of θ_{m} over the range $0 \le \theta_{\text{m}} \le 90°$

$$\lambda_1 = \left(0.005 + 0.09 \left(\frac{90° - \theta_{\text{m}}}{90°} \right) \left(\frac{8.0}{8.0 + i_1} \right) \right) i_1$$

Phase 2 of this motor is identical to that of phase 1, and there is no significant mutual inductance between the phases. Assume that the winding resistance is negligible.

a. Using MATLAB, plot a family of λ_1-i_1 curves for this motor as θ_m varies from 0 to 90° in 10° increments and as i_1 is varied from 0 to 30 A.

b. Again using MATLAB, use Eq. 8.19 and Fig. 8.14 to calculate the ratio of the inverter volt-ampere rating to the VRM net power output for the following idealized operating cycle:

  (*i*) The current is instantaneously raised to 25 A when $\theta_m = -90°$.
  (*ii*) The current is then held constant as the rotor rotates to $\theta_m = 0°$.
  (*iii*) At $\theta_m = 0°$, the current is reduced to zero.

c. Assuming the VRM to be operating as a motor using the cycle described in part (b) and rotating at a constant speed of 2500 r/min, calculate the net electromechanical power supplied to the rotor.

■ Solution

a. The λ_1-i_1 curves are shown in Fig. 8.15a.

b. Figure 8.15b shows the areas W_{net} and W_{rec}. Note that, as pointed out in the text, the λ-i curves are symmetrical around $\theta_m = 0°$ and thus the curves for negative values of θ_m are identical to those for the corresponding positive values. The area W_{net} is bounded by the λ_1-i_1 curves corresponding to $\theta_m = 0°$ and $\theta_m = 90°$ and the line $i_1 = 25$ A. The area W_{rec} is bounded by the line $\lambda_1 = \lambda_{max}$ and the λ_1-i_1 curve corresponding to $\theta_m = 0°$, where $\lambda_{max} = \lambda_1(25$ A, $0°)$.

Using MATLAB to integrate the areas, the desired ratio can be calculated from Eq. 8.19 as

$$\frac{\text{Inverter volt-ampere rating}}{\text{Net output power}} = \frac{\text{area}(W_{rec} + W_{net})}{\text{area}(W_{net})} = 1.55$$

c. Energy equal to area(W_{net}) is supplied by each phase to the rotor twice during each revolution of the rotor. If area(W_{net}) is measured in joules, the power in watts supplied per phase is thus equal to

$$P_{phase} = 2\left(\frac{\text{area}(W_{net})}{T}\right) \text{ W}$$

where T is the time for one revolution (in seconds).

From MATLAB, area(W_{net}) = 9.91 joules and for 2500 r/min, $T = 60/2500 = 0.024$ sec,

$$P_{phase} = 2\left(\frac{9.91}{0.024}\right) = 825 \text{ W}$$

and thus

$$P_{mech} = 2P_{phase} = 1650 \text{ W}$$

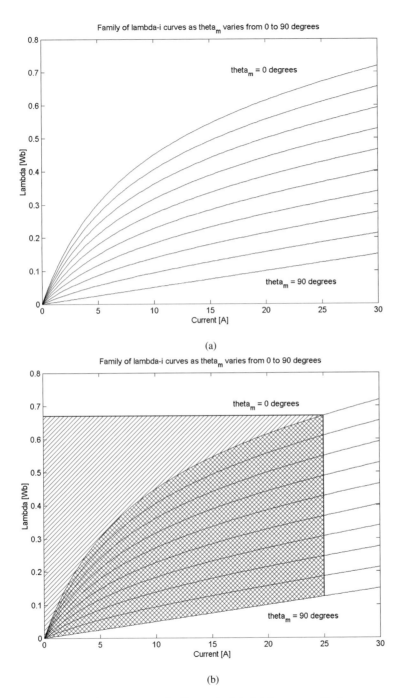

(a)

(b)

Figure 8.15 (a) λ_1-i_1 curves for Example 8.4. (b) Areas used in the calculation of part (b).

Here is the MATLAB script:

```
clc
clear

%(a) First plot the lambda-i characteristics

for m = 1:10
theta(m) = 10*(m-1);
   for n=1:101
i(n) = 30*(n-1)/100;
Lambda(n) = i(n)*(0.005 + 0.09*((90-theta(m))/90)*(8/(i(n)+8)));
   end

   plot(i,Lambda)
   if m==1
hold
   end
end

hold
xlabel('Current [A]')
ylabel('Lambda [Wb]')
title('Family of lambda-i curves as theta_m varies from 0 to 90 degrees')
text(17,.7,'theta_m = 0 degrees')
text(20,.06,'theta_m = 90 degrees')
%(b) Now integrate to find the areas.

%Peak lambda at 0 degrees, 25 Amps
lambdamax = 25*(0.005+0.09*(8/(25+8)));

AreaWnet = 0;
AreaWrec = 0;

% 100 integration step
deli = 25/100;

for n=1:101
  i(n) = 25*(n-1)/100;
  AreaWnet = AreaWnet + deli*i(n)*(0.09)*(8/(i(n)+8));
  AreaWrec = AreaWrec + deli*(lambdamax - i(n)*(0.005+0.09*(8/(i(n)+8))));
end

Ratio = (AreaWrec + AreaWnet)/AreaWnet;

fprintf('\nPart(b) Ratio = %g',Ratio)

%(c) Calculate the power

rpm = 2500;
rps = 2500/60;
T = 1/rps;
Pphase = 2*AreaWnet/T;
Ptot = 2*Pphase;
```

```
fprintf('\n\nPart(c) AreaWnet = %g [Joules]',AreaWnet)
fprintf('\n        Pphase = %g [W] and Ptot = %g [W]\n',Pphase,Ptot)
```

Consider a two-phase VRM which is identical to that of Example 8.4 with the exception of an additional 5 mH of leakage inductance in each phase. (*a*) Calculate the ratio of the inverter volt-ampere rating to the VRM net power output for the following idealized operating cycle:

(*i*) The current is instantaneously raised to 25 A when $\theta_m = -90°$.

(*ii*) The current is then held constant as the rotor rotates to $\theta_m = 10°$.

(*iii*) At $\theta_m = 10°$, the current is reduced to zero.

(*b*) Assuming the VRM to be operating as a motor using the cycle described in part (a) and rotating at a constant speed of 2500 r/min, calculate the net electromechanical power supplied to the rotor.

Solution

a.

$$\frac{\text{Inverter volt-ampere rating}}{\text{Net output power}} = 1.75$$

b. $P_{\text{mech}} = 1467$ W

Saturation effects clearly play a significant role in the performance of most VRMs and must be taken into account. In addition, the idealized operating cycle illustrated in Example 8.4 cannot, of course, be achieved in practice since some rotor motion is likely to take place over the time scale over which current changes occur. As a result, it is often necessary to resort to numerical-analysis packages such as finite-element programs as part of the design process for practical VRM systems. Many of these programs incorporate the ability to model the nonlinear effects of magnetic saturation as well as mechanical (e.g., rotor motion) and electrical (e.g., current buildup) dynamic effects.

As we have seen, the design of a VRM drive system typically requires that a trade-off be made. On the one hand, saturation tends to increase the size of the VRM for a given power output. On the other hand, on comparing two VRM systems with the same power output, the system with the higher level of saturation will typically require an inverter with a lower volt-ampere rating. Thus the ultimate design will be determined by a trade-off between the size, cost, and efficiency of the VRM and of the inverter.

8.5 STEPPING MOTORS

As we have seen, when the phases of a VRM are energized sequentially in an appropriate step-wise fashion, the VRM will rotate a specific angle for each step. Motors designed specifically to take advantage of this characteristic are referred to as *stepping motors* or *stepper motors*. Frequently stepping motors are designed to produce a large number of steps per revolution, for example 50, 100, or 200 steps per revolution (corresponding to a rotation of 7.2°, 3.6° and 1.8° per step).

An important characteristic of the stepping motor is its compatibility with digital-electronic systems. These systems are common in a wide variety of applications and continue to become more powerful and less expensive. For example, the stepping motor is often used in digital control systems where the motor receives open-loop commands in the form of a train of pulses to turn a shaft or move an object a specific distance. Typical applications include paper-feed and print-head-positioning motors in printers and plotters, drive and head-positioning motors in disk drives and CD players, and worktable and tool positioning in numerically controlled machining equipment. In many applications, position information can be obtained simply by keeping count of the pulses sent to the motor, in which case position sensors and feedback control are not required.

The angular resolution of a VRM is determined by the number of rotor and stator teeth and can be greatly enhanced by techniques such as *castleation,* as is discussed in Section 8.2. Stepping motors come in a wide variety of designs and configurations. In addition to variable-reluctance configurations, these include permanent-magnet and hybrid configurations. The use of permanent magnets in combination with a variable-reluctance geometry can significantly enhance the torque and positional accuracy of a stepper motor.

The VRM configurations discussed in Sections 8.1 through 8.3 consist of a single rotor and stator with multiple phases. A stepping motor of this configuration is called a *single-stack, variable-reluctance stepping motor.* An alternate form of variable-reluctance stepping motor is known as a *multistack variable-reluctance stepping motor.* In this configuration, the motor can be considered to be made up of a set of axially displaced, single-phase VRMs mounted on a single shaft.

Figure 8.16 shows a multistack variable-reluctance stepping motor. This type of motor consists of a series of stacks, each axially displaced, of identical geometry and each excited by a single phase winding, as shown in Fig. 8.17. The motor of Fig. 8.16 has three stacks and three phases, although motors with additional phases and stacks are common. For an n_s-stack motor, the rotor or stator (but not both) on each stack is displaced by $1/n_s$ times the pole-pitch angle. In Fig. 8.16, the rotor poles are aligned, but the stators are offset in angular displacement by one-third of the pole pitch. By successively exciting the individual phases, the rotor can be turned in increments of this displacement angle.

A schematic diagram of a two-phase stepping motor with a permanent-magnet, two-pole rotor is shown in Fig. 8.18. Note that this machine is in fact a two-phase synchronous machine, similar for example to the three-phase permanent-magnet ac machine of Fig. 5.29. The distinction between such a stepping motor and a synchronous motor arises not from the construction of the motor but rather from how the motor is operated. The synchronous motor is typically intended to drive a load at a specified speed, and the stepping motor is typically intended to control the position of a load.

Figure 8.16 Cutaway view of a three-phase, three-stack variable-reluctance stepping motor. (*Warner Electric.*)

The rotor of the stepping motor of Fig. 8.18 assumes the angles $\theta_m = 0, 45°$, $90°, \ldots$ as the windings are excited in the sequence:

1. Positive current in phase 1 alone.
2. Equal-magnitude positive currents in phase 1 and phase 2.
3. Positive current in phase 2 alone.
4. Equal-magnitude negative current in phase 1 and positive current in phase 2.
5. Negative current in phase 1 alone.
6. And so on.

Note that if a ferromagnetic rotor were substituted for the permanent-magnet rotor, the rotor would move in a similar fashion.

The stepping motor of Fig. 8.18 can also be used for 90° steps by exciting the coils singly. In the latter case, only a permanent-magnet rotor can be used. This can be seen from the torque-angle curves for the two types of rotors shown in Fig. 8.19. Whereas the permanent-magnet rotor produces peak torque when the excitation is shifted 90°, the ferromagnetic rotor produces zero torque and may move in either direction.

The rotor position in the permanent-magnet stepping motor of Fig. 8.18 is defined by the winding currents with no ambiguity and depends on the direction of the phase

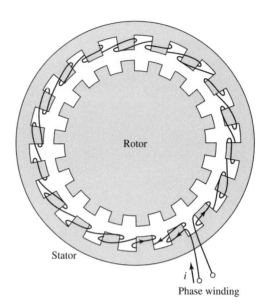

Figure 8.17 Diagram of one stack and phase of a multiphase, multistack variable-reluctance stepping motor, such as that in Fig. 8.16. For an n_s-stack motor, the rotor or stator (but not both) on each stack is displaced by $1/n_s$ times the pole pitch.

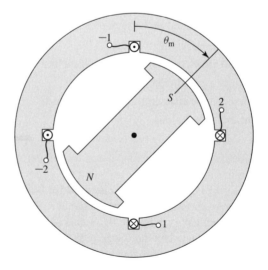

Figure 8.18 Schematic diagram of a two-phase permanent-magnet stepping motor.

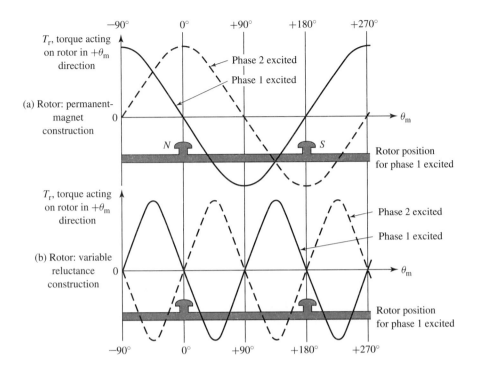

Figure 8.19 Torque-angle curves for the stepping motor of Fig. 8.18: (a) permanent-magnet rotor and (b) variable-reluctance rotor.

currents. Reversing the phase currents will cause the rotor to reverse its orientation. This is in contrast to VRM configuration with a ferromagnetic rotor, in which two rotor positions are equally stable for any particular set of phase currents, and hence the rotor position cannot be determined uniquely. Permanent-magnet stepping motors are also unlike their VRM counterparts in that torque tending to align the rotor with the stator poles will be generated even when there is no excitation applied to the phase windings. Thus the rotor will have preferred unexcited rest positions, a fact which can be used to advantage in some applications.

EXAMPLE 8.5

Using the techniques of Chapter 3 and neglecting saturation effects, the torque of a two-phase, permanent-magnet stepping motor of the form of Fig. 8.18 can be expressed as

$$T_{mech} = T_0 \, (i_1 \cos \theta_m + i_2 \sin \theta_m)$$

where T_0 is a positive constant that depends upon the motor geometry and the properties of the permanent magnet.

Calculate the rest (zero-torque) positions which will result if the motor is driven by a drive such that each phase current can be set equal to three values $-I_0$, 0, and I_0. Using such a drive, what is the motor step size?

■ Solution

In general, the zero-torque positions of the motor can be found by setting the torque expression to zero and solving for the resultant rotor position. Thus setting

$$T_{mech} = T_0 \, (i_1 \sin \theta_m - i_2 \cos \theta_m) = 0$$

gives

$$i_1 \sin \theta_m - i_2 \cos \theta_m = 0$$

or

$$\theta_m = \tan^{-1} \left(\frac{i_2}{i_1} \right)$$

Note that not all of these zero-torque positions correspond to stable equilibrium positions. For example, operation with $i_1 = I_0$ and $i_2 = 0$ gives two zero-torque positions: $\theta_m = 0°$ and $\theta_m = 180°$. Yet only the position $\theta_m = 0°$ is stable. This is directly analogous to the case of a hanging pendulum which sees zero torque both when it is hanging downward ($\theta = 0°$) and when it is sitting inverted ($\theta = 180°$). Yet, it is clear that the slightest perturbation of the position of the inverted pendulum will cause it to rotate downwards and that it will eventually come to rest in the stable hanging position.

Stable rest positions of the rotor are determined by the requirement that a restoring torque is produced as the rotor moves from that position. Thus, a negative torque should result if the rotor moves in the $+\theta_m$ direction, and a positive torque should result for motion in the $-\theta_m$ direction. Mathematically, this can be expressed as an additional constraint on the torque at the rest position

$$\left. \frac{\partial T_{mech}}{\partial \theta_m} \right|_{i_1, i_2} < 0$$

where the partial derivative is evaluated at the zero-torque position and is taken with the phase currents held constant. Thus, in this case, the rest position must satisfy the additional constraint that

$$\left. \frac{\partial T_{mech}}{\partial \theta_m} \right|_{i_1, i_2} = -T_0 \, (i_1 \cos \theta_m + i_2 \sin \theta_m) < 0$$

From this equation, we see for example that with $i_1 = I_0$ and $i_2 = 0$, at $\theta_m = 0°$, $\partial T_{mech}/\partial \theta_m < 0$ and thus $\theta_m = 0°$ is a stable rest position. Similarly, at $\theta_m = 180°$, $\partial T_{mech}/\partial \theta_m > 0$ and thus $\theta_m = 180°$ is not a stable rest position.

Using these relationships, Table 8.1 lists the stable rest positions of the rotor for the various combinations of phase currents.

From this table we see that this drive results in a step size of $45°$.

Table 8.1 Rotor rest positions for Example 8.5.

i_1	i_2	θ_m
0	0	-
0	$-I_0$	270°
0	I_0	90°
$-I_0$	0	180°
$-I_0$	$-I_0$	225°
$-I_0$	I_0	135°
I_0	0	0°
I_0	$-I_0$	315°
I_0	I_0	45°

Practice Problem 8.5

In order to achieve a step size of 22.5°, the motor drive of Example 8.5 is modified so that each phase can be driven by currents of magnitude 0, $\pm k I_0$, and $\pm I_0$. Find the required value of the constant k.

Solution

$$k = \tan^{-1}(22.5°) = 0.4142$$

In Example 8.5 we see that stable equilibrium positions of an unloaded stepping motor satisfy the conditions that there is zero torque, i.e.,

$$T_{mech} = 0 \tag{8.20}$$

and that there is positive restoring torque, i.e.,

$$\left.\frac{\partial T_{mech}}{\partial \theta_m}\right|_{i_1,i_2} < 0 \tag{8.21}$$

In practice, there will of course be a finite load torque tending to perturb the stepping motor from these idealized positions. For open-loop control systems (i.e., control systems in which there is no mechanism for position feedback), a high-degree of position control can be achieved by designing the stepping motor to produce large restoring torque (i.e., a large magnitude of $\partial T_{mech}/\partial \theta_m$). In such a stepping motor, load torques will result in only a small movement of the rotor from the idealized positions which satisfies Eqs. 8.20 and 8.21.

Example 8.5 also shows how carefully controlled combinations of phase currents can enhance the resolution of a stepper motor. This technique, referred to as *microstepping,* can be used to achieve increased step resolution of a wide variety of stepper motors. As the following example shows, microstepping can be used to produce extremely fine position resolution. The increased resolution comes, however, at the expense of an increase in complexity of the stepping-motor drive electronics and control algorithms, which must accurately control the distribution of current to multiple phases simultaneously.

EXAMPLE 8.6

Consider again the two-phase, permanent-magnet stepping motor of Example 8.5. Calculate the rotor position which will result if the phase currents are controlled to be sinusoidal functions of a reference angle θ_{ref} in the form

$$i_1 = I_0 \cos \theta_{ref}$$

$$i_2 = I_0 \sin \theta_{ref}$$

■ **Solution**

Substitution of the current expressions into the torque expression of Example 8.5 gives

$$T_{mech} = T_0 (i_1 \cos \theta_m + i_2 \sin \theta_m) = T_0 I_0 (\cos \theta_{ref} \cos \theta_m + \sin \theta_{ref} \sin \theta_m)$$

Use of the trigonometric identity $\cos(\alpha - \beta) = \cos \alpha \cos \beta + \sin \alpha \sin \beta$ gives

$$T_{mech} = T_0 I_0 \; \cos(\theta_{ref} - \theta_m)$$

From this expression and using the analysis of Example 8.5, we see that the rotor equilibrium position will be equal to the reference angle, i.e., $\theta_m = \theta_{ref}$. In a practical implementation, a digital controller is likely to be used to increment θ_{ref} in finite steps, which will result in finite steps in the position of the stepping-motor.

The *hybrid stepping motor* combines characteristics of the variable-reluctance and permanent-magnet stepping motors. A photo of a hybrid stepping motor is shown in Fig. 8.20, and a schematic view of a hybrid stepping motor is shown in Fig. 8.21. The hybrid-stepping-motor rotor configuration appears much like that of a multistack variable-reluctance stepping motor. In the rotor of Fig. 8.21a, two identical rotor stacks are displaced axially along the rotor and displaced in angle by one-half the rotor pole pitch, while the stator pole structure is continuous along the length of the rotor. Unlike the multistack variable-reluctance stepping motor, in the hybrid stepping motor, the rotor stacks are separated by an axially-directed permanent magnet. As a result, in Fig. 8.21a one end of the rotor can be considered to have a north magnetic pole and

Figure 8.20 Disassembled 1.8°/step hybrid stepping motor. (*Oriental Motor.*)

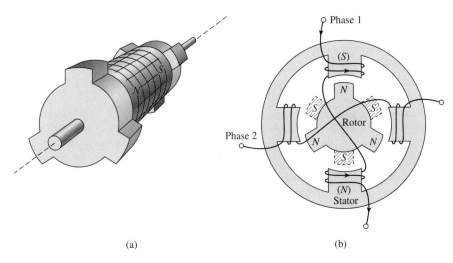

(a)

(b)

Figure 8.21 Schematic view of a hybrid stepping motor. (a) Two-stack rotor showing the axially-directed permanent magnet and the pole pieces displaced by one-half the pole pitch. (b) End view from the rotor north poles and showing the rotor south poles at the far end (shown crosshatched). Phase 1 of the stator is energized to align the rotor as shown.

the other end a south magnetic pole. Figure 8.21b shows a schematic end view of a hybrid stepping motor. The stator has four poles with the phase-1 winding wound on the vertical poles and the phase-2 winding wound on the horizontal poles. The rotor is shown with its north-pole end at the near end of the motor and the south-pole end (shown crosshatched) at the far end.

In Fig. 8.21b, phase 1 is shown excited such that the top stator pole is a south pole while the bottom pole is a north pole. This stator excitation interacts with the permanent-magnet flux of the rotor to align the rotor with a pole on its north-pole end vertically upward and a pole on its south-pole end vertically downward, as shown in the figure. Note that if the stator excitation is removed, there will still be a permanent-magnet torque tending to maintain the rotor in the position shown.

To turn the rotor, excitation is removed from phase 1, and phase 2 is excited. If phase 2 is excited such that the right-hand stator pole is a south pole and the left-hand one is a north pole, the rotor will rotate 30° counterclockwise. Similarly, if the opposite excitation is applied to the phase-2 winding, a 30° rotation in the clockwise direction will occur. Thus, by alternately applying phase-1 and phase-2 excitation of the appropriate polarity, the rotor can be made to rotate in either direction by a specified angular increment.

Practical hybrid stepping motors are generally built with more rotor poles than are indicated in the schematic motor of Fig. 8.21, in order to give much better angular resolution. Correspondingly, the stator poles are often castleated (see Fig. 8.8) to further increase the angular resolution. In addition, they may be built with more than two stacks per rotor.

The hybrid stepping motor design offers advantages over the permanent-magnet design discussed earlier. It can achieve small step sizes easily and with a simple magnet structure while a purely permanent-magnet motor would require a multipole permanent magnet. In comparison with the variable-reluctance stepping motor, the hybrid design may require less excitation to achieve a given torque because some of the excitation is supplied by the permanent magnet. In addition, the hybrid stepping motor will tend to maintain its position when the stator excitation is removed, as does the permanent-magnet design.

The actual choice of a stepping-motor design for a particular application is determined based on the desired operating characteristics, availability, size, and cost. In addition to the three classifications of stepping motors discussed in this chapter, a number of other different and often quite clever designs have been developed. Although these encompass a wide range of configurations and construction techniques, the operating principles remain the same.

Stepping motors may be driven by electronic drive components similar to those discussed in Section 11.4 in the context of VRM drives. Note that the issue of controlling a stepping motor to obtain the desired response under dynamic, transient conditions is quite complex and remains the subject of considerable investigation.[4]

8.6 SUMMARY

Variable-reluctance machines are perhaps the simplest of electrical machines. They consist of a stator with excitation windings and a magnetic rotor with saliency. Torque is produced by the tendency of the salient-pole rotor to align with excited magnetic poles on the stator.

VRMs are synchronous machines in that they produce net torque only when the rotor motion is in some sense synchronous with the applied stator mmf. This synchronous relationship may be complex, with the rotor speed being some specific fraction of the applied electrical frequency as determined not only by the number of stator and rotor poles but also by the number of stator and rotor teeth on these poles. In fact, in some cases, the rotor will be found to rotate in the direction opposite to the rotation direction of the applied stator mmf.

Successful operation of a VRM depends on exciting the stator phase windings in a specific fashion correlated to the instantaneous position of the rotor. Thus, rotor position must be measured, and a controller must be employed to determine the appropriate excitation waveforms and to control the output of the inverter. Typically chopping is required to obtain these waveforms. The net result is that although the VRM is itself a simple device, somewhat complex electronics are typically required to make a complete drive system.

The significance of VRMs in engineering applications stems from their low cost, reliability, and controllability. Because their torque depends only on the square of the applied stator currents and not on their direction, these machines can be operated from

---

[4] For further information on stepping motors, see P. Acarnley, *Stepping Motors: A Guide to Modern Theory and Practice,* 2nd ed., Peter Peregrinus Ltd., London, 1982; Takashi Kenjo, *Stepping Motors and Their Microprocessor Controls,* Clarendon Press, Oxford, 1984; and Benjamin C. Kuo, *Theory and Applications of Step Motors,* West Publishing Co., St. Paul, Minnesota, 1974.

unidirectional drive systems, reducing the cost of the power electronics. However, it is only recently, with the advent of low-cost, flexible power electronic circuitry and microprocessor-based control systems, that VRMs have begun to see widespread application in systems ranging from traction drives to high-torque, precision position control systems for robotics applications.

Practical experience with VRMs has shown that they have the potential for high reliability. This is due in part to the simplicity of their construction and to the fact that there are no windings on their rotors. In addition, VRM drives can be operated successfully (at a somewhat reduced rating) following the failure of one or more phases, either in the machine or in the inverter. VRMs typically have a large number of stator phases (four or more), and significant output can be achieved even if some of these phases are out of service. Because there is no rotor excitation, there will be no voltage generated in a phase winding which fails open-circuited or current generated in a phase winding which fails short-circuited, and thus the machine can continue to be operated without risk of further damage or additional losses and heating.

Because VRMs can be readily manufactured with a large number of rotor and stator teeth (resulting in large inductance changes for small changes in rotor angle), they can be constructed to produce very large torque per unit volume. There is, however, a trade-off between torque and velocity, and such machines will have a low rotational velocity (consistent with the fact that only so much power can be produced by a given machine frame size). On the opposite extreme, the simple configuration of a VRM rotor and the fact that it contains no windings suggest that it is possible to build extremely rugged VRM rotors. These rotors can withstand high speeds, and motors which operate in excess of 200,000 r/min have been built.

Finally, we have seen that saturation plays a large role in VRM performance. As recent advances in power electronic and microelectronic circuitry have brought VRM drive systems into the realm of practicality, so have advances in computer-based analytical techniques for magnetic-field analysis. Use of these techniques now makes it practical to perform optimized designs of VRM drive systems which are competitive with alternative technologies in many applications.

Stepping motors are closely related to VRMs in that excitation of each successive phase of the stator results in a specific angular rotation of the rotor. Stepping motors come in a wide variety of designs and configurations. These include variable-reluctance, permanent-magnet, and hybrid configurations. The rotor position of a variable-reluctance stepper motor is not uniquely determined by the phase currents since the phase inductances are not unique functions of the rotor angle. The addition of a permanent magnet changes this situation and the rotor position of a permanent-magnet stepper motor is a unique function of the phase currents.

Stepping motors are the electromechanical companions to digital electronics. By proper application of phase currents to the stator windings, these motors can be made to rotate in well-defined steps ranging down to a fraction of a degree per pulse. They are thus essential components of digitally controlled electromechanical systems where a high degree of precision is required. They are found in a wide range of applications including numerically controlled machine tools, in printers and plotters, and in disk drives.

8.7 PROBLEMS

8.1 Repeat Example 8.1 for a machine identical to that considered in the example except that the stator pole-face angle is $\beta = 45°$.

8.2 In the paragraph preceeding Eq. 8.1, the text states that "under the assumption of negligible iron reluctance the mutual inductances between the phases of the doubly-salient VRM of Fig. 8.1b will be zero, with the exception of a small, essentially constant component associated with leakage flux." Neglect any leakage flux effects and use magnetic circuit techniques to show that this statement is true.

8.3 Use magnetic-circuit techniques to show that the phase-to-phase mutual inductance in the 6/4 VRM of Fig. 8.5 is zero under the assumption of infinite rotor- and stator-iron permeability. Neglect any contributions of leakage flux.

8.4 A 6/4 VRM of the form of Fig. 8.5 has the following properties:

> Stator pole angle $\beta = 30°$
>
> Rotor pole angle $\alpha = 30°$
>
> Air-gap length $g = 0.35$ mm
>
> Rotor outer radius $R = 5.1$ cm
>
> Active length $D = 7$ cm

This machine is connected as a three-phase motor with opposite poles connected in series to form each phase winding. There are 40 turns per pole (80 turns per phase). The rotor and stator iron can be considered to be of infinite permeability and hence mutual-inductance effects can be neglected.

a. Defining the zero of rotor angle θ_m at the position when the phase-1 inductance is maximum, plot and label the inductance of phase 1 as a function of rotor angle.

b. On the plot of part (a), plot the inductances of phases 2 and 3.

c. Find the phase-1 current I_0 which results in a magnetic flux density of 1.0 T in the air gap under the phase-1 pole face when the rotor is in a position of maximum phase-1 inductance.

d. Assuming that the phase-1 current is held constant at the value found in part (c) and that there is no current in phases 2 and 3, plot the torque as a function of rotor position.

   The motor is to be driven from a three-phase current-source inverter which can be switched on or off to supply either zero current or a constant current of magnitude I_0 in phases 2 and 3; plot the torque as a function of rotor position.

e. Under the idealized assumption that the currents can be instantaneously switched, determine the sequence of phase currents (as a function of rotor position) that will result in constant positive motor torque, independent of rotor position.

f. If the frequency of the stator excitation is such that a time $T_0 = 35$ msec is required to sequence through all three phases under the excitation conditions of part (e), find the rotor angular velocity and its direction of rotation.

8.5 In Section 8.2, when discussing Fig. 8.5, the text states: "In addition to the fact that there are not positions of simultaneous alignment for the 6/4 VRM, it can be seen that there also are no rotor positions at which only a torque of a single sign (either positive or negative) can be produced." Show that this statement is true.

8.6 Consider a three-phase 6/8 VRM. The stator phases are excited sequentially, requiring a total time of 15 msec. Find the angular velocity of the rotor in r/min.

8.7 The phase windings of the castleated machine of Fig. 8.8 are to be excited by turning the phases on and off individually (i.e., only one phase can be on at any given time).

    a. Describe the sequence of phase excitations required to move the rotor to the right (clockwise) by an angle of approximately $21.4°$.

    b. The stator phases are to be excited as a regular sequence of pulses. Calculate the phase order and the time between pulses required to produce a steady-state rotor rotation of 125 r/min in the counterclockwise direction.

8.8 Replace the 28-tooth rotor of Problem 8.7 with a rotor with 26 teeth.

    a. Phase 1 is excited, and the rotor is allowed to come to rest. If the excitation on phase 1 is removed and excitation is applied to phase 2, calculate the resultant direction and magnitude (in degrees) of rotor rotation.

    b. The stator phases are to be excited as a regular sequence of pulses. Calculate the phase order and the time between pulses required to produce a steady-state rotor rotation of 80 r/min in the counterclockwise direction.

8.9 Repeat Example 8.3 for a rotor speed of 4500 r/min.

8.10 Repeat Example 8.3 under the condition that the rotor speed is 4500 r/min and that a negative voltage of -250 V is used to turn off the phase current.

8.11 The three-phase 6/4 VRM of Problem 8.4 has a winding resistance of 0.15 Ω/phase and a leakage inductance of 4.5 mH in each phase. Assume that the rotor is rotating at a constant angular velocity of 1750 r/min.

    a. Plot the phase-1 inductance as a function of the rotor angle θ_m.

    b. A voltage of 75 V is applied to phase 1 as the rotor reaches the position $\theta_m = -30°$ and is maintained constant until $\theta_m = 0°$. Calculate and plot the phase-1 current as a function of time during this period.

    c. When the rotor reaches $\theta = 0°$, the applied voltage is reversed so that a voltage of -75 V is applied to the winding. This voltage is maintained until the winding current reaches zero, at which point the winding is open-circuited. Calculate and plot the current decay during the time until the current decays to zero.

    d. Calculate and plot the torque during the time periods investigated in parts (b) and (c).

8.12 Assume that the VRM of Examples 8.1 and 8.3 is modified by replacing its rotor with a rotor with 75° pole-face angles as shown in Fig. 8.12a. All other dimensions and parameters of the VRM are unchanged.

a. Calculate and plot $L(\theta_m)$ for this machine.

b. Repeat Example 8.3 except that the constant voltage 100 V is first applied at $\theta_m = -67.5°$ when $dL(\theta_m)/d\theta_m$ becomes positive and the constant voltage of -100 V is then applied at $\theta_m = -7.5°$ (i.e., when $dL(\theta_m)/d\theta_m$ becomes zero) and is maintained until the winding current reaches zero.

c. Plot the corresponding torque.

8.13 Repeat Example 8.4 for a symmetrical two-phase 4/2 VRM whose λ-i characteristic can be represented by the following expression (for phase 1) as a function of θ_m over the range $0 \le \theta_m \le 90°$:

$$\lambda_1 = \left(0.01 + 0.15 \left(\frac{90° - \theta_m}{90°} \right) \left(\frac{12.0}{12.0 + i_1} \right)^{1.2} \right) i_1$$

8.14 Consider a two-phase stepper motor with a permanent-magnet rotor such as shown in Fig. 8.18 and whose torque-angle curve is as shown in Fig. 8.19a. This machine is to be excited by a four-bit digital sequence corresponding to the following winding excitation:

bit			bit		
1	2	i_1	3	4	i_2
0	0	0	0	0	0
0	1	$-I_0$	0	1	$-I_0$
1	0	I_0	1	0	I_0
1	1	0	1	1	0

a. Make a table of 4-bit patterns which will produce rotor angular positions of 0, 45°, ..., 315°.

b. By sequencing through the bit pattern found in part (a) the motor can be made to rotate. What time interval (in milliseconds) between bit-pattern changes will result in a rotor speed of 1200 r/min?

8.15 Figure 8.22 shows a two-phase hybrid stepping motor with castleated poles on the stator. The rotor is shown in the position it occupies when current is flowing into the positive lead of phase 1.

a. If phase one is turned off and phase 2 is excited with current flowing into its positive lead, calculate the corresponding angular rotation of the rotor. Is it in the clockwise or counterclockwise direction?

b. Describe an excitation sequence for the phase windings which will result in a steady rotation of the rotor in the clockwise direction.

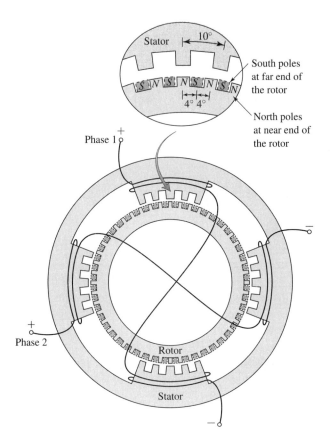

Figure 8.22 Castleated hybrid stepping motor for
Problem 8.15.

   c. Determine the frequency of the phase currents required to achieve a rotor
speed of 8 r/min.

8.16 Consider a multistack, multiphase variable-reluctance stepping motor, such
as that shown schematically in Fig. 8.17, with 14 poles on each of the rotor
and stator stacks and three stacks with one phase winding per stack. The
motor is built such that the stator poles of each stack are aligned.

   a. Calculate the angular displacement between the rotor stacks.

   b. Determine the frequency of phase currents required to achieve a rotor
speed of 900 r/min.

9 CHAPTER

Single- and Two-Phase Motors

This chapter discusses single-phase motors. While focusing on induction motors, synchronous-reluctance, hysteresis, and shaded-pole induction motors are also discussed. Note that another common single-phase motor, the series universal motor, is discussed in Section 7.10. Most induction motors of fractional-kilowatt (fractional horsepower) rating are single-phase motors. In residential and commercial applications, they are found in a wide range of equipment including refrigerators, air conditioners and heat pumps, fans, pumps, washers, and dryers.

In this chapter, we will describe these motors qualitatively in terms of rotating-field theory and will begin with a rigorous analysis of a single-phase motor operating off of a single winding. However, most single-phase induction motors are actually two-phase motors with unsymmetrical windings; the two windings are typically quite different, with different numbers of turns and/or winding distributions. Thus this chapter also discusses two-phase motors and includes a development of a quantitative theory for the analysis of single-phase induction motors when operating off both their main and auxiliary windings.

9.1 SINGLE-PHASE INDUCTION MOTORS: QUALITATIVE EXAMINATION

Structurally, the most common types of single-phase induction motors resemble polyphase squirrel-cage motors except for the arrangement of the stator windings. An induction motor with a squirrel-cage rotor and a single-phase stator winding is represented schematically in Fig. 9.1. Instead of being a concentrated coil, the actual stator winding is distributed in slots to produce an approximately sinusoidal space distribution of mmf. As we saw in Section 4.5.1, a single-phase winding produces equal forward- and backward-rotating mmf waves. By symmetry, it is clear that such a motor inherently will produce no starting torque since at standstill, it will produce equal torque in both directions. However, we will show that if it is started by auxiliary

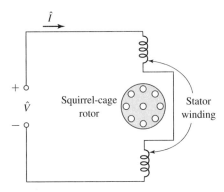

Figure 9.1 Schematic view of a single-phase induction motor.

means, the result will be a net torque in the direction in which it is started, and hence the motor will continue to run.

Before we consider auxiliary starting methods, we will discuss the basic properties of the schematic motor of Fig. 9.1. If the stator current is a cosinusoidal function of time, the resultant air-gap mmf is given by Eq. 4.18

$$\mathcal{F}_{ag1} = F_{max} \cos(\theta_{ae}) \cos \omega_e t \tag{9.1}$$

which, as shown in Section 4.5.1, can be written as the sum of positive- and negative-traveling mmf waves of equal magnitude. The positive-traveling wave is given by

$$\mathcal{F}_{ag1}^{+} = \frac{1}{2} F_{max} \cos(\theta_{ae} - \omega_e t) \tag{9.2}$$

and the negative-traveling wave is given by

$$\mathcal{F}_{ag1}^{-} = \frac{1}{2} F_{max} \cos(\theta_{ae} + \omega_e t) \tag{9.3}$$

Each of these component mmf waves produces induction-motor action, but the corresponding torques are in opposite directions. With the rotor at rest, the forward and backward air-gap flux waves created by the combined mmf's of the stator and rotor currents are equal, the component torques are equal, and no starting torque is produced. If the forward and backward air-gap flux waves were to remain equal when the rotor revolves, each of the component fields would produce a torque-speed characteristic similar to that of a polyphase motor with negligible stator leakage impedance, as illustrated by the dashed curves f and b in Fig. 9.2a. The resultant torque-speed characteristic, which is the algebraic sum of the two component curves, shows that if the motor were started by auxiliary means, it would produce torque in whatever direction it was started.

The assumption that the air-gap flux waves remain equal when the rotor is in motion is a rather drastic simplification of the actual state of affairs. First, the effects of stator leakage impedance are ignored. Second, the effects of induced rotor currents are not properly accounted for. Both these effects will ultimately be included in the

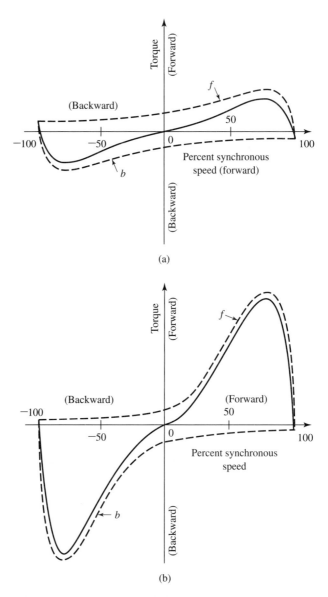

Figure 9.2 Torque-speed characteristic of a single-phase induction motor (a) on the basis of constant forward and backward flux waves, (b) taking into account changes in the flux waves.

detailed quantitative theory of Section 9.3. The following qualitative explanation shows that the performance of a single-phase induction motor is considerably better than would be predicted on the basis of equal forward and backward flux waves.

When the rotor is in motion, the component rotor currents induced by the backward field are greater than at standstill, and their power factor is lower. Their mmf, which opposes that of the stator current, results in a reduction of the backward flux wave. Conversely, the magnetic effect of the component currents induced by the forward field is less than at standstill because the rotor currents are less and their power factor is higher. As speed increases, therefore, the forward flux wave increases while the backward flux wave decreases. The sum of these flux waves must remain roughly constant since it must induce the stator counter emf, which is approximately constant if the stator leakage-impedance voltage drop is small.

Hence, with the rotor in motion, the torque of the forward field is greater and that of the backward field less than in Fig. 9.2a, the true situation being about that shown in Fig. 9.2b. In the normal running region at a few percent slip, the forward field is several times greater than the backward field, and the flux wave does not differ greatly from the constant-amplitude revolving field in the air gap of a balanced polyphase motor. In the normal running region, therefore, the torque-speed characteristic of a single-phase motor is not too greatly inferior to that of a polyphase motor having the same rotor and operating with the same maximum air-gap flux density.

In addition to the torques shown in Fig. 9.2, double-stator-frequency torque pulsations are produced by the interactions of the oppositely rotating flux and mmf waves which rotate past each other at twice synchronous speed. These interactions produce no average torque, but they tend to make the motor noisier than a polyphase motor. Such torque pulsations are unavoidable in a single-phase motor because of the pulsations in instantaneous power input inherent in a single-phase circuit. The effects of the pulsating torque can be minimized by using an elastic mounting for the motor. The torque referred to on the torque-speed curves of a single-phase motor is the time average of the instantaneous torque.

9.2 STARTING AND RUNNING PERFORMANCE OF SINGLE-PHASE INDUCTION AND SYNCHRONOUS MOTORS

Single-phase induction motors are classified in accordance with their starting methods and are usually referred to by names descriptive of these methods. Selection of the appropriate motor is based on the starting- and running-torque requirements of the load, the duty cycle of the load, and the limitations on starting and running current from the supply line for the motor. The cost of single-phase motors increases with their rating and with their performance characteristics such as starting-torque-to-current ratio. Typically, in order to minimize cost, an application engineer will select the motor with the lowest rating and performance that can meet the specifications of the application. Where a large number of motors are to be used for a specific purpose, a special motor may be designed in order to ensure the least cost. In the fractional-kilowatt motor business, small differences in cost are important.

Starting methods and the resulting torque-speed characteristics are considered qualitatively in this section. A quantitative theory for analyzing these motors is developed in Section 9.4.2.

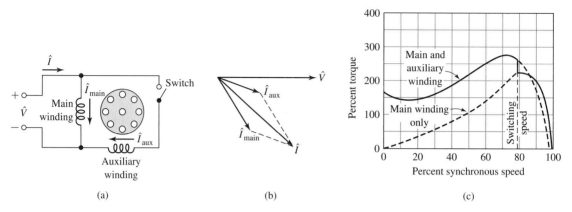

Figure 9.3 Split-phase motor: (a) connections, (b) phasor diagram at starting, and (c) typical torque-speed characteristic.

9.2.1 Split-Phase Motors

Split-phase motors have two stator windings, a *main winding* (also referred to as the *run winding*) which we will refer to with the subscript 'main' and an *auxiliary winding* (also referred to as the *start winding*) which we will refer to with the subscript 'aux'. As in a two-phase motor, the axes of these windings are displaced 90 electrical degrees in space, and they are connected as shown in Fig. 9.3a. The auxiliary winding has a higher resistance-to-reactance ratio than the main winding, with the result that the two currents will be out of phase, as indicated in the phasor diagram of Fig. 9.3b, which is representative of conditions at starting. Since the auxiliary-winding current $\hat{I}_{aux}$ leads the main-winding current $\hat{I}_{main}$, the stator field first reaches a maximum along the axis of the auxiliary winding and then somewhat later in time reaches a maximum along the axis of the main winding.

The winding currents are equivalent to unbalanced two-phase currents, and the motor is equivalent to an unbalanced two-phase motor. The result is a rotating stator field which causes the motor to start. After the motor starts, the auxiliary winding is disconnected, usually by means of a centrifugal switch that operates at about 75 percent of synchronous speed. The simple way to obtain the high resistance-to-reactance ratio for the auxiliary winding is to wind it with smaller wire than the main winding, a permissible procedure because this winding operates only during starting. Its reactance can be reduced somewhat by placing it in the tops of the slots. A typical torque-speed characteristic for such a motor is shown in Fig. 9.3c.

Split-phase motors have moderate starting torque with low starting current. Typical applications include fans, blowers, centrifugal pumps, and office equipment. Typical ratings are 50 to 500 watts; in this range they are the lowest-cost motors available.

9.2.2 Capacitor-Type Motors

Capacitors can be used to improve motor starting performance, running performance, or both, depending on the size and connection of the capacitor. The *capacitor-start*

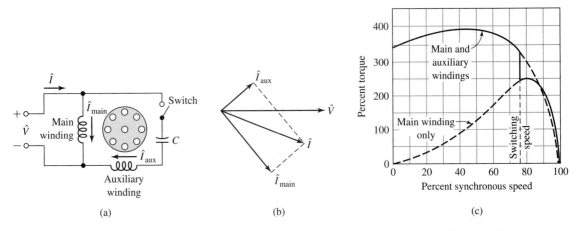

Figure 9.4 Capacitor-start motor: (a) connections, (b) phasor diagram at starting, and (c) typical torque-speed characteristic.

motor is also a split-phase motor, but the time-phase displacement between the two currents is obtained by means of a capacitor in series with the auxiliary winding, as shown in Fig. 9.4a. Again the auxiliary winding is disconnected after the motor has started, and consequently the auxiliary winding and capacitor can be designed at minimum cost for intermittent service.

By using a *starting capacitor* of appropriate value, the auxiliary-winding current $\hat{I}_{aux}$ at standstill can be made to lead the main-winding current $\hat{I}_{main}$ by 90 electrical degrees, as it would in a balanced two-phase motor (see Fig. 9.4b). In practice, the best compromise between starting torque, starting current, and cost typically results with a phase angle somewhat less than 90°. A typical torque-speed characteristic is shown in Fig. 9.4c, high starting torque being an outstanding feature. These motors are used for compressors, pumps, refrigeration and air-conditioning equipment, and other hard-to-start loads. A cutaway view of a capacitor-start motor is shown in Fig. 9.5.

In the *permanent-split-capacitor motor,* the capacitor and auxiliary winding are not cut out after starting; the construction can be simplified by omission of the switch, and the power factor, efficiency, and torque pulsations improved. For example, the capacitor and auxiliary winding could be designed for perfect two-phase operation (i.e., no backwards flux wave) at any one desired load. The losses due to the backward field at this operating point would then be eliminated, with resulting improvement in efficiency. The double-stator-frequency torque pulsations would also be eliminated, with the capacitor serving as an energy storage reservoir for smoothing out the pulsations in power input from the single-phase line, resulting in quieter operation. Starting torque must be sacrificed because the choice of capacitance is necessarily a compromise between the best starting and running values. The resulting torque-speed characteristic and a schematic diagram are given in Fig. 9.6.

If two capacitors are used, one for starting and one for running, theoretically optimum starting and running performance can both be obtained. One way of accomplishing this result is shown in Fig. 9.7a. The small value of capacitance required for

Figure 9.5   Cutaway view of a capacitor-start induction motor. The starting switch is at the right of the rotor. The motor is of drip-proof construction. (*General Electric Company.*)

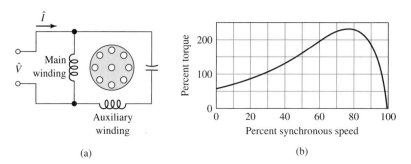

(a)

(b)

Figure 9.6   Permanent-split-capacitor motor and typical torque-speed characteristic.

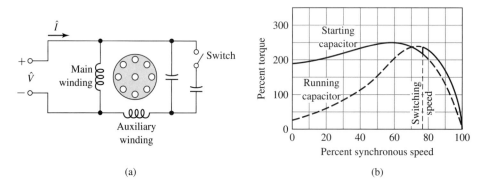

(a)

(b)

Figure 9.7   Capacitor-start, capacitor-run motor and typical torque-speed characteristic.

optimum running conditions is permanently connected in series with the auxiliary winding, and the much larger value required for starting is obtained by a capacitor connected in parallel with the running capacitor via a switch with opens as the motor comes up to speed. Such a motor is known as a *capacitor-start, capacitor-run motor*.

The capacitor for a capacitor-start motor has a typical value of 300 μF for a 500-W motor. Since it must carry current for just the starting time, the capacitor is a special compact ac electrolytic type made for motor-starting duty. The capacitor for the same motor permanently connected has a typical rating of 40 μF, and since it operates continuously, the capacitor is an ac paper, foil, and oil type. The cost of the various motor types is related to performance: the capacitor-start motor has the lowest cost, the permanent-split-capacitor motor next, and the capacitor-start, capacitor-run motor the highest cost.

EXAMPLE 9.1

A 2.5-kW 120-V 60-Hz capacitor-start motor has the following impedances for the main and auxiliary windings (at starting):

$$Z_{main} = 4.5 + j3.7 \ \Omega \quad \text{main winding}$$

$$Z_{aux} = 9.5 + j3.5 \ \Omega \quad \text{auxiliary winding}$$

Find the value of starting capacitance that will place the main and auxiliary winding currents in quadrature at starting.

■ **Solution**

The currents $\hat{I}_{main}$ and $\hat{I}_{aux}$ are shown in Fig. 9.4a and b. The impedance angle of the main winding is

$$\phi_{main} = \tan^{-1}\left(\frac{3.7}{4.5}\right) = 39.6°$$

To produce currents in time quadrature with the main winding, the impedance angle of the auxiliary winding circuit (including the starting capacitor) must be

$$\phi = 39.6° - 90.0° = -50.4°$$

The combined impedance of the auxiliary winding and starting capacitor is equal to

$$Z_{total} = Z_{aux} + jX_c = 9.5 + j(3.5 + X_c) \ \Omega$$

where $X_c = -\frac{1}{\omega C}$ is the reactance of the capacitor and $\omega = 2\pi 60 \approx 377$ rad/sec. Thus

$$\tan^{-1}\left(\frac{3.5 + X_c}{9.5}\right) = -50.4°$$

$$\frac{3.5 + X_c}{9.5} = \tan(-50.4°) = -1.21$$

and hence

$$X_c = -1.21 \times 9.5 - 3.5 = -15.0 \ \Omega$$

The capacitance C is then

$$C = \frac{-1}{\omega X_c} = \frac{-1}{377 \times (-15.0)} = 177 \ \mu\text{F}$$

Consider the motor of Example 9.1. Find the phase angle between the main- and auxiliary-winding currents if the 177-μF capacitor is replaced by a 200-μF capacitor.

Solution

85.2°

9.2.3 Shaded-Pole Induction Motors

As illustrated schematically in Fig. 9.8a, the *shaded-pole induction motor* usually has salient poles with one portion of each pole surrounded by a short-circuited turn of copper called a *shading coil*. Induced currents in the shading coil cause the flux in the shaded portion of the pole to lag the flux in the other portion. The result is similar to a rotating field moving in the direction from the unshaded to the shaded portion of the pole; currents are induced in the squirrel-cage rotor and a low starting torque is produced. A typical torque-speed characteristic is shown in Fig. 9.8b. Their efficiency is low, but shaded-pole motors are the least expensive type of subfractional-kilowatt motor. They are found in ratings up to about 50 watts.

9.2.4 Self-Starting Synchronous-Reluctance Motors

Any one of the induction-motor types described above can be made into a *self-starting synchronous-reluctance motor*. Anything which makes the reluctance of the air gap a function of the angular position of the rotor with respect to the stator coil axis will produce reluctance torque when the rotor is revolving at synchronous speed. For

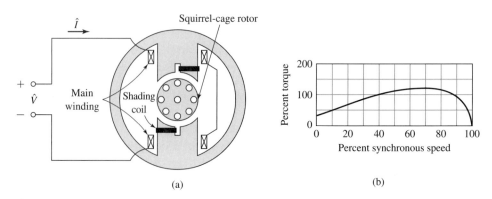

(a)   (b)

Figure 9.8   Shaded-pole induction motor and typical torque-speed characteristic.

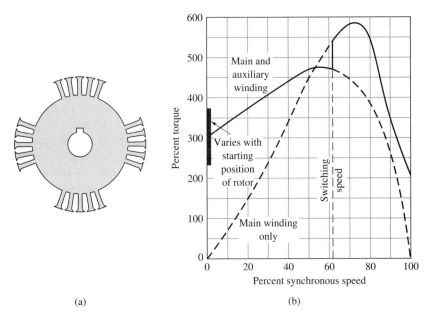

(a)

(b)

Figure 9.9 Rotor punching for four-pole synchronous-reluctance motor and typical torque-speed characteristic.

example, suppose some of the teeth are removed from a squirrel-cage rotor, leaving the bars and end rings intact, as in an ordinary squirrel-cage induction motor. Figure 9.9a shows a lamination for such a rotor designed for use with a four-pole stator. The stator may be polyphase or any one of the single-phase types described above.

The motor will start as an induction motor and at light loads will speed up to a small value of slip. The reluctance torque arises from the tendency of the rotor to try to align itself in the minimum-reluctance position with respect to the synchronously revolving forward air-gap flux wave, in accordance with the principles discussed in Chapter 3. At a small slip, this torque alternates slowly in direction; the rotor is accelerated during a positive half cycle of the torque variation and decelerated during the succeeding negative half cycle. If the moment of inertia of the rotor and its mechanical load are sufficiently small, the rotor will be accelerated from slip speed up to synchronous speed during an accelerating half cycle of the reluctance torque. The rotor will then pull into synchronism and continue to run at synchronous speed. The presence of any backward-revolving stator flux wave will produce torque ripple and additional losses, but synchronous operation will be maintained provided the load torque is not excessive.

A typical torque-speed characteristic for a split-phase-start synchronous-reluctance motor is shown in Fig. 9.9b. Notice the high values of induction-motor torque. The reason for this is that in order to obtain satisfactory synchronous-motor characteristics, it has been found necessary to build synchronous-reluctance motors in frames which would be suitable for induction motors of two or three times their synchronous-motor rating. Also notice that the principal effect of the salient-pole rotor on the induction-motor characteristic is at standstill, where considerable "cogging" is evident; i.e., the torque varies considerably with rotor position.

9.2.5 Hysteresis Motors

The phenomenon of hysteresis can be used to produce mechanical torque. In its simplest form, the rotor of a *hysteresis motor* is a smooth cylinder of magnetically hard steel, without windings or teeth. It is placed inside a slotted stator carrying distributed windings designed to produce as nearly as possible a sinusoidal space distribution of flux, since undulations in the flux wave greatly increase the losses. In single-phase motors, the stator windings usually are of the permanent-split-capacitor type, as in Fig. 9.6. The capacitor is chosen so as to result in approximately balanced two-phase conditions within the motor windings. The stator then produces a primarily space-fundamental air-gap field revolving at synchronous speed.

Instantaneous magnetic conditions in the air gap and rotor are indicated in Fig. 9.10a for a two-pole stator. The axis SS' of the stator-mmf wave revolves at synchronous speed. Because of hysteresis, the magnetization of the rotor lags behind the inducing mmf wave, and therefore the axis RR' of the rotor flux wave lags behind the axis of the stator-mmf wave by the hysteretic lag angle δ (Fig. 9.10a). If the rotor is stationary, starting torque is produced proportional to the product of the fundamental components of the stator mmf and rotor flux and the sine of the torque angle δ. The rotor then accelerates if the torque of the load is less than the developed torque of the motor.

As long as the rotor is turning at less than synchronous speed, each region of the rotor is subjected to a repetitive hysteresis cycle at slip frequency. While the rotor accelerates, the lag angle δ remains constant if the flux is constant, since the angle δ

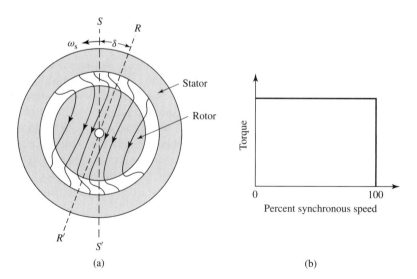

(a)   (b)

Figure 9.10   (a) General nature of the magnetic field in the air gap and rotor of a hysteresis motor; (b) idealized torque-speed characteristic.

depends merely on the hysteresis loop of the rotor material and is independent of the rate at which the loop is traversed. The motor therefore develops constant torque right up to synchronous speed, as shown in the idealized torque-speed characteristic of Fig. 9.10b. This feature is one of the advantages of the hysteresis motor. In contrast with a reluctance motor, which must "snap" its load into synchronism from an induction-motor torque-speed characteristic, a hysteresis motor can synchronize any load which it can accelerate, no matter how great the inertia. After reaching synchronism, the motor continues to run at synchronous speed and adjusts its torque angle so as to develop the torque required by the load.

The hysteresis motor is inherently quiet and produces smooth rotation of its load. Furthermore, the rotor takes on the same number of poles as the stator field. The motor lends itself to multispeed synchronous operation when the stator is wound with several sets of windings and utilizes pole-changing connections. The hysteresis motor can accelerate and synchronize high-inertia loads because its torque is uniform from standstill to synchronous speed.

9.3 REVOLVING-FIELD THEORY OF SINGLE-PHASE INDUCTION MOTORS

As discussed in Section 9.1, the stator-mmf wave of a single-phase induction motor can be shown to be equivalent to two constant-amplitude mmf waves revolving at synchronous speed in opposite directions. Each of these component stator-mmf waves induces its own component rotor currents and produces induction-motor action just as in a balanced polyphase motor. This double-revolving-field concept not only is useful for qualitative visualization but also can be developed into a quantitative theory applicable to a wide variety of induction-motor types. We will not discuss the full quantitative theory here.[1] However, we will consider the simpler, but important case of a single-phase induction motor running on only its main winding.

Consider conditions with the rotor stationary and only the main stator winding excited. The motor then is equivalent to a transformer with its secondary short-circuited. The equivalent circuit is shown in Fig. 9.11a, where $R_{1,\text{main}}$ and $X_{1,\text{main}}$ are, respectively, the resistance and leakage reactance of the main winding, $X_{\text{m,main}}$ is the magnetizing reactance, and $R_{2,\text{main}}$ and $X_{2,\text{main}}$ are the standstill values of the rotor resistance and leakage reactance referred to the main stator winding by use of the appropriate turns ratio. Core loss, which is omitted here, will be accounted for later as if it were a rotational loss. The applied voltage is $\hat{V}$, and the main-winding current is $\hat{I}_{\text{main}}$. The voltage $\hat{E}_{\text{main}}$ is the counter emf generated in the main winding by the stationary pulsating air-gap flux wave produced by the combined action of the stator and rotor currents.

In accordance with the double-revolving-field concept of Section 9.1, the stator mmf can be resolved into half-amplitude forward and backward rotating fields. At

---

[1] For an extensive treatment of single-phase motors, see, for example, C. B. Veinott, *Fractional- and Subfractional-Horsepower Electric Motors,* McGraw-Hill, New York, 1970.

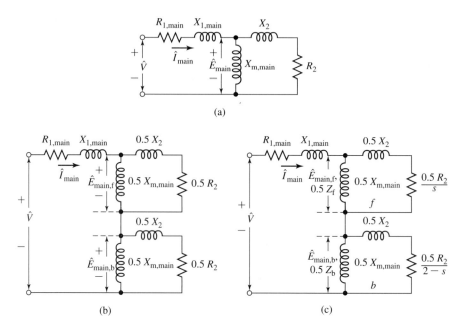

Figure 9.11 Equivalent circuits for a single-phase induction motor: (a) rotor blocked; (b) rotor blocked, showing effects of forward and backward fields; (c) running conditions.

standstill the amplitudes of the forward and backward resultant air-gap flux waves both equal half the amplitude of the pulsating field. In Fig. 9.11b the portion of the equivalent circuit representing the effects of the air-gap flux is split into two equal portions, representing the effects of the forward and backward fields, respectively.

Now consider conditions after the motor has been brought up to speed by some auxiliary means and is running on only its main winding in the direction of the forward field at a per-unit slip s. The rotor currents induced by the forward field are of slip frequency sf_e, where f_e is the stator applied electrical frequency. Just as in any polyphase motor with a symmetric polyphase or squirrel-cage rotor, these rotor currents produce an mmf wave traveling forward at slip speed with respect to the rotor and therefore at synchronous speed with respect to the stator. The resultant of the forward waves of stator and rotor mmf creates a resultant forward wave of air-gap flux, which generates a counter emf $\hat{E}_{\mathrm{main,f}}$ in the main winding of the stator. The reflected effect of the rotor as viewed from the stator is like that in a polyphase motor and can be represented by an impedance $0.5R_{2,\mathrm{main}}/s + j0.5X_{2,\mathrm{main}}$ in parallel with $j0.5X_{\mathrm{m,main}}$ as in the portion of the equivalent circuit of Fig. 9.11c labeled 'f'. The factors of 0.5 come from the resolution of the pulsating stator mmf into forward and backward components.

Now consider conditions with respect to the backward field. The rotor is still turning at a slip s with respect to the forward field, and its per-unit speed n in the direction of the forward field is $n = 1 - s$. The relative speed of the rotor with respect to the backward field is $1 + n$, or its slip with respect to the backward field is $1 + n = 2 - s$. The backward field then induces rotor currents whose frequency is $(2 - s)f_e$. For small slips, these rotor currents are of almost twice stator frequency.

At a small slip, an oscilloscope trace of rotor current will therefore show a high-frequency component from the backward field superposed on a low-frequency component from the forward field. As viewed from the stator, the rotor-mmf wave of the backward-field induced rotor current travels at synchronous speed but in the backward direction. The equivalent-circuit representing these internal reactions from the viewpoint of the stator is like that of a polyphase motor whose slip is $2 - s$ and is shown in the portion of the equivalent circuit (Fig. 9.11c) labeled 'b'. As with the forward field, the factors of 0.5 come from the resolution of the pulsating stator mmf into forward and backward components. The voltage $\hat{E}_{\text{main,b}}$ across the parallel combination representing the backward field is the counter emf generated in the main winding of the stator by the resultant backward field.

By use of the equivalent circuit of Fig. 9.11c, the stator current, power input, and power factor can be computed for any assumed value of slip when the applied voltage and the motor impedances are known. To simplify the notation, let

$$Z_f \equiv R_f + jX_f \equiv \left(\frac{R_{2,\text{main}}}{s} + jX_{2,\text{main}} \right) \text{ in parallel with } jX_{m,\text{main}} \qquad (9.4)$$

and

$$Z_b \equiv R_b + jX_b \equiv \left(\frac{R_{2,\text{main}}}{2 - s} + jX_{2,\text{main}} \right) \text{ in parallel with } jX_{m,\text{main}} \qquad (9.5)$$

The impedances representing the reactions of the forward and backward fields from the viewpoint of the single-phase main stator winding are $0.5Z_f$ and $0.5Z_b$, respectively, in Fig. 9.11c.

Examination of the equivalent circuit (Fig. 9.11c) confirms the conclusion, reached by qualitative reasoning in Section 9.1 (Fig. 9.2b), that the forward air-gap flux wave increases and the backward wave decreases when the rotor is set in motion. When the motor is running at a small slip, the reflected effect of the rotor resistance in the forward field, $0.5R_{2,\text{main}}/s$, is much larger than its standstill value, while the corresponding effect in the backward field, $0.5R_{2,\text{main}}/(2 - s)$, is smaller. The forward-field impedance therefore is larger than its standstill value, while that of the backward field is smaller. The forward-field counter emf $\hat{E}_{\text{main,f}}$ therefore is larger than its standstill value, while the backward-field counter emf $\hat{E}_{\text{main,b}}$ is smaller; i.e., the forward air-gap flux wave increases, while the backward flux wave decreases.

Mechanical power and torque can be computed by application of the torque and power relations developed for polyphase motors in Chapter 6. The torques produced by the forward and backward fields can each be treated in this manner. The interactions of the oppositely rotating flux and mmf waves cause torque pulsations at twice stator frequency but produce no average torque.

As in Eq. 6.25, the electromagnetic torque $T_{\text{main,f}}$ of the forward field in newton-meters equals $1/\omega_s$ times the power $P_{\text{gap,f}}$ in watts delivered by the stator winding to the forward field, where ω_s is the synchronous angular velocity in mechanical radians per second; thus

$$T_{\text{main,f}} = \frac{1}{\omega_s} P_{\text{gap,f}} \tag{9.6}$$

When the magnetizing impedance is treated as purely inductive, $P_{\text{gap,f}}$ is the power absorbed by the impedance $0.5Z_f$; that is,

$$P_{\text{gap,f}} = I^2(0.5R_f) \tag{9.7}$$

where R_f is the resistive component of the forward-field impedance defined in Eq. 9.4. Similarly, the internal torque $T_{\text{main,b}}$ of the backward field is

$$T_{\text{main,b}} = \frac{1}{\omega_s} P_{\text{gap,b}} \tag{9.8}$$

where $P_{\text{gap,b}}$ is the power delivered by the stator winding to the backward field, or

$$P_{\text{gap,b}} = I^2(0.5R_b) \tag{9.9}$$

where R_b is the resistive component of the backward-field impedance Z_b defined in Eq. 9.5.

The torque of the backward field is in the opposite direction to that of the forward field, and therefore the net internal torque T_{mech} is

$$T_{\text{mech}} = T_{\text{main,f}} - T_{\text{main,b}} = \frac{1}{\omega_s}(P_{\text{gap,f}} - P_{\text{gap,b}}) \tag{9.10}$$

Since the rotor currents produced by the two component air-gap fields are of different frequencies, the total rotor I^2R loss is the numerical sum of the losses caused by each field. In general, as shown by comparison of Eqs. 6.17 and 6.19, the rotor I^2R loss caused by a rotating field equals the slip of the field times the power absorbed from the stator. Thus

$$\text{Forward-field rotor } I^2R = s\, P_{\text{gap,f}} \tag{9.11}$$

$$\text{Backward-field rotor } I^2R = (2 - s)\, P_{\text{gap,b}} \tag{9.12}$$

$$\text{Total rotor } I^2R = s\, P_{\text{gap,f}} + (2 - s)\, P_{\text{gap,b}} \tag{9.13}$$

Since power is torque times angular velocity and the angular velocity of the rotor is $(1 - s)\omega_s$, using Eq. 9.10, the internal power P_{mech} converted to mechanical form, in watts, is

$$P_{\text{mech}} = (1 - s)\omega_s T_{\text{mech}} = (1 - s)(P_{\text{gap,f}} - P_{\text{gap,b}}) \tag{9.14}$$

As in the polyphase motor, the internal torque T_{mech} and internal power P_{mech} are not the output values because rotational losses remain to be accounted for. It is obviously correct to subtract friction and windage losses from T_{mech} or P_{mech} and it is usually assumed that core losses can be treated in the same manner. For the small changes in speed encountered in normal operation, the rotational losses are often assumed to be constant.[2]

EXAMPLE 9.2

A $\frac{1}{4}$-hp, 110-V, 60-Hz, four-pole, capacitor-start motor has the following equivalent circuit parameter values (in Ω) and losses:

$$R_{1,main} = 2.02 \qquad X_{1,main} = 2.79 \qquad R_{2,main} = 4.12$$

$$X_{2,main} = 2.12 \qquad X_{m,main} = 66.8$$

$$Core\ loss = 24\ W \qquad Friction\ and\ windage\ loss = 13\ W$$

For a slip of 0.05, determine the stator current, power factor, power output, speed, torque, and efficiency when this motor is running as a single-phase motor at rated voltage and frequency with its starting winding open.

■ Solution

The first step is to determine the values of the forward- and backward-field impedances at the assigned value of slip. The following relations, derived from Eq. 9.4, simplify the computations of the forward-field impedance Z_f:

$$R_f = \left(\frac{X^2_{m,main}}{X_{22}} \right) \frac{1}{sQ_{2,main} + 1/(sQ_{2,main})} \qquad X_f = \frac{X_{2,main}X_{m,main}}{X_{22}} + \frac{R_f}{sQ_{2,main}}$$

where

$$X_{22} = X_{2,main} + X_{m,main} \qquad and \qquad Q_{2,main} = \frac{X_{22}}{R_{2,main}}$$

Substitution of numerical values gives, for $s = 0.05$,

$$Z_f = R_f + jX_f = 31.9 + j40.3\ \Omega$$

Corresponding relations for the backward-field impedance Z_b are obtained by substituting $2 - s$ for s in these equations. When $(2 - s)Q_{2,main}$ is greater than 10, as is usually the case, less than 1 percent error results from use of the following approximate forms:

$$R_b = \frac{R_{2,main}}{2 - s} \left(\frac{X_{m,main}}{X_{22}} \right)^2 \qquad X_b = \frac{X_{2,main}X_{m,main}}{X_{22}} + \frac{R_b}{(2 - s)Q_{2,main}}$$

Substitution of numerical values gives, for $s = 0.05$,

$$Z_b = R_b + jX_b = 1.98 + j2.12\ \Omega$$

---

[2] For a treatment of the experimental determination of motor constants and losses, see Veinott, op. cit., Chapter 18.

Addition of the series elements in the equivalent circuit of Fig. 9.11c gives

$$R_{1,\text{main}} + jX_{1,\text{main}} = 2.02 + j2.79$$

$$0.5(R_f + jX_f) = 15.95 + j20.15$$

$$\underline{0.5(R_b + jX_b) = 0.99 + j1.06}$$

$$\text{Total Input } Z = 18.96 + j24.00 = 30.6 \angle 51.7°$$

$$\text{Stator current } I = \frac{V}{Z} = \frac{110}{30.6} = 3.59 \text{ A}$$

$$\text{Power factor} = \cos(51.7°) = 0.620$$

$$\text{Power input} = P_{\text{in}} = VI \times \text{power factor} = 110 \times 3.59 \times 0.620 = 244 \text{ W}$$

The power absorbed by the forward field (Eq. 9.7) is

$$P_{\text{gap,f}} = I^2(0.5R_f) = 3.59^2 \times 15.95 = 206 \text{ W}$$

The power absorbed by the backward field (Eq. 9.9) is

$$P_{\text{gap,b}} = I^2(0.5R_b) = 3.59^2 \times 0.99 = 12.8 \text{ W}$$

The internal mechanical power (Eq. 9.14) is

$$P_{\text{mech}} = (1-s)(P_{\text{gap,f}} - P_{\text{gap,b}}) = 0.95(206 - 13) = 184 \text{ W}$$

Assuming that the core loss can be combined with the friction and windage loss, the rotational loss becomes $24 + 13 = 37$ W and the shaft output power is the difference. Thus

$$P_{\text{shaft}} = 184 - 37 = 147 \text{ W} = 0.197 \text{ hp}$$

From Eq. 4.40, the synchronous speed in rad/sec is given by

$$\omega_s = \left(\frac{2}{\text{poles}}\right)\omega_e = \left(\frac{2}{4}\right)120\pi = 188.5 \text{ rad/sec}$$

or in terms of r/min from Eq. 4.41

$$n_s = \left(\frac{120}{\text{poles}}\right)f_e = \left(\frac{120}{4}\right)60 = 1800 \text{ r/min}$$

$$\text{Rotor speed} = (1-s)(\text{synchronous speed})$$

$$= 0.95 \times 1800 = 1710 \text{ r/min}$$

and

$$\omega_m = 0.95 \times 188.5 = 179 \text{ rad/sec}$$

The torque can be found from Eq. 9.14.

$$T_{\text{shaft}} = \frac{P_{\text{shaft}}}{\omega_m} = \frac{147}{179} = 0.821 \text{ N} \cdot \text{m}$$

and the efficiency is

$$\eta = \frac{P_{\text{shaft}}}{P_{\text{in}}} = \frac{147}{244} = 0.602 = 60.2\%$$

As a check on the power bookkeeping, compute the losses:

$$I^2 R_{1,\text{main}} = (3.59)^2 (2.02) = 26.0$$

$$\text{Forward-field rotor } I^2 R \text{ (Eq. 9.11)} = 0.05 \times 206 = 10.3$$

$$\text{Backward-field rotor } I^2 R \text{ (Eq. 9.12)} = 1.95 \times 12.8 = 25.0$$

$$\text{Rotational losses} = \underline{37.0}$$

$$98.3 \text{ W}$$

From $P_{\text{in}} - P_{\text{shaft}}$, the total losses $= 97$ W which checks within accuracy of computations.

Assume the motor of Example 9.2 to be operating at a slip of 0.065 and at rated voltage and frequency. Determine (a) the stator current and power factor and (b) the power output.

Solution

    a. 4.0 A, power factor $= 0.70$ lagging
    b. 190 W

Examination of the order of magnitude of the numerical values in Example 9.2 suggests approximations which usually can be made. These approximations pertain particularly to the backward-field impedance. Note that the impedance $0.5(R_b + jX_b)$ is only about 5 percent of the total motor impedance for a slip near full load. Consequently, an approximation as large as 20 percent of this impedance would cause only about 1 percent error in the motor current. Although, strictly speaking, the backward-field impedance is a function of slip, very little error usually results from computing its value at any convenient slip in the normal running region, e.g., 5 percent, and then assuming R_b and X_b to be constants.

Corresponding to a slightly greater approximation, the shunting effect of $jX_{m,\text{main}}$ on the backward-field impedance can often be neglected, whence

$$Z_b \approx \frac{R_{2,\text{main}}}{2 - s} + jX_{2,\text{main}} \tag{9.15}$$

This equation gives values of the backward-field resistance that are a few percent high, as can be seen by comparison with the exact expression given in Example 9.2. Neglecting s in Eq. 9.15 would tend to give values of the backward-field resistance that would be too low, and therefore such an approximation would tend to counteract the error in Eq. 9.15. Consequently, for small slips

$$Z_b \approx \frac{R_{2,\text{main}}}{2} + jX_{2,\text{main}} \tag{9.16}$$

In a polyphase motor (Section 6.5), the maximum internal torque and the slip at which it occurs can easily be expressed in terms of the motor parameters; the maximum internal torque is independent of rotor resistance. No such simple expressions exist for a single-phase motor. The single-phase problem is much more involved because of the presence of the backward field, the effect of which is twofold: (1) it absorbs some of the applied voltage, thus reducing the voltage available for the forward field and decreasing the forward torque developed; and (2) the backward field produces negative torque, reducing the effective developed torque. Both of these effects depend on rotor resistance as well as leakage reactance. Consequently, unlike the polyphase motor, the maximum internal torque of a single-phase motor is influenced by rotor resistance; increasing the rotor resistance decreases the maximum torque and increases the slip at which maximum torque occurs.

Principally because of the effects of the backward field, a single-phase induction motor is somewhat inferior to a polyphase motor using the same rotor and the same stator core. The single-phase motor has a lower maximum torque which occurs at a lower slip. For the same torque, the single-phase motor has a higher slip and greater losses, largely because of the backward-field rotor I^2R loss. The volt-ampere input to the single-phase motor is greater, principally because of the power and reactive volt-amperes consumed by the backward field. The stator I^2R loss also is somewhat higher in the single-phase motor, because one phase, rather than several, must carry all the current. Because of the greater losses, the efficiency is lower, and the temperature rise for the same torque is higher. A larger frame size must be used for a single-phase motor than for a polyphase motor of the same power and speed rating. Because of the larger frame size, the maximum torque can be made comparable with that of a physically smaller but equally rated polyphase motor. In spite of the larger frame size and the necessity for auxiliary starting arrangements, general-purpose single-phase motors in the standard fractional-kilowatt ratings cost approximately the same as correspondingly rated polyphase motors because of the much greater volume of production of the former.

9.4 TWO-PHASE INDUCTION MOTORS

As we have seen, most single-phase induction motors are actually constructed in the form of two-phase motors, with two stator windings in space quadrature. The main and auxiliary windings are typically quite different, with a different number of turns, wire size, and turns distribution. This difference, in combination with the capacitor that is typically used in series with the auxiliary winding, guarantees that the mmfs produced by the two winding currents will be quite unbalanced; at best they may be balanced at one specific operating point. We will thus discuss various analytical techniques for two-phase motors, both to expand our understanding and insight into machine performance and also to develop techniques for the analysis of single- and two-phase motors.

Under balanced operating conditions, a symmetrical two-phase motor can be analyzed using techniques developed in Chapter 6 for three-phase motors, modified only slightly to take into account the fact that there are two phases rather than three. In this section, we will first discuss one technique that can be used to analyze a symmetrical two-phase motor operating under unbalanced operating conditions. We will then formally derive an analytical model for an unsymmetrical two-phase motor that can be applied to the general case single-phase motors operating off both their main and auxiliary windings.

9.4.1 Unbalanced Operation of Symmetrical Two-Phase Machines; The Symmetrical-Component Concept

When operating from the main winding alone, the single-phase motor is the extreme case of a motor operating under unbalanced stator-current conditions. In some cases, unbalanced voltages or currents are produced in the supply network to a motor, e.g., when a line fuse is blown. In other cases, unbalanced voltages are produced by the starting impedances of single-phase motors, as described in Section 9.2. The purpose of this section is to develop the symmetrical-component theory of two-phase induction motors from the double-revolving-field concept and to show how the theory can be applied to a variety of problems involving induction motors having two stator windings in space quadrature.

First consider in review what happens when balanced two-phase voltages are applied to the stator terminals of a two-phase machine having a uniform air gap, a symmetrical polyphase or cage rotor, and two identical stator windings α and β in space quadrature. The stator currents are equal in magnitude and in time quadrature. When the current in winding α is at its instantaneous maximum, the current in winding β is zero and the stator-mmf wave is centered on the axis of winding α. Similarly, the stator-mmf wave is centered on the axis of winding β at the instant when the current in winding β is at its instantaneous maximum. The stator-mmf wave therefore travels 90 electrical degrees in space in a time interval corresponding to a 90° phase change of the applied voltage, with the direction of its travel depending on the phase sequence of the currents. A more complete analysis in the manner of Section 4.5 shows that the traveling wave has constant amplitude and constant angular velocity. This fact is, of course, the basis for the theory of the balanced operation of induction machines.

The behavior of the motor for balanced two-phase applied voltages of either phase sequence can be readily determined. Thus, if the rotor is turning at a slip s in the direction from winding α toward winding β, the terminal impedance per phase is given by the equivalent circuit of Fig. 9.12a when the applied voltage $\hat{V}_\beta$ lags the applied voltage $\hat{V}_\alpha$ by 90°. Throughout the rest of this treatment, this phase sequence is called *positive sequence* and is designated by the subscript 'f' since positive-sequence currents result in a forward field. With the rotor running at the same speed and in the same direction, the terminal impedance per phase is given by the equivalent circuit of Fig. 9.12b when $\hat{V}_\beta$ leads $\hat{V}_\alpha$ by 90°. This phase sequence is called *negative sequence* and is designated by subscript 'b', since negative-sequence currents produce a backward field.

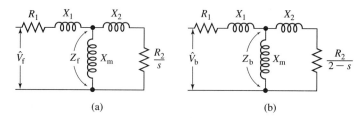

(a)   (b)

Figure 9.12 Single-phase equivalent circuits for a two-phase motor under unbalanced conditions: (a) forward field and (b) backward field.

Suppose now that *two* balanced two-phase voltage sources of *opposite phase sequence* are connected in series and applied simultaneously to the motor, as indicated in Fig. 9.13a, where phasor voltages $\hat{V}_f$ and $j\hat{V}_f$ applied, respectively, to windings α and β form a balanced system of positive sequence, and phasor voltages $\hat{V}_b$ and $-j\hat{V}_b$ form another balanced system but of negative sequence.

The resultant voltage V_α applied to winding α is, as a phasor,

$$\hat{V}_\alpha = \hat{V}_f + \hat{V}_b \tag{9.17}$$

and that applied to winding β is

$$\hat{V}_\beta = j\hat{V}_f - j\hat{V}_b \tag{9.18}$$

Fig. 9.13b shows a generalized phasor diagram in which the forward, or positive-sequence, system is given by the phasors $\hat{V}_f$ and $j\hat{V}_f$ and the backward, or negative-sequence, system is given by the phasors $\hat{V}_b$ and $-j\hat{V}_b$. The resultant voltages, given by the phasors $\hat{V}_\alpha$ and $\hat{V}_\beta$ are not, in general, either equal in magnitude or in time

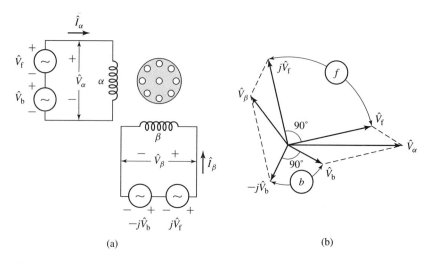

(a)   (b)

Figure 9.13 Synthesis of an unbalanced two-phase system from the sum of two balanced systems of opposite phase sequence.

quadrature. From this discussion we see that an unbalanced two-phase system of applied voltages V_α and V_β can be synthesized by combining two balanced voltage sets of opposite phase sequence.

The symmetrical-component systems are, however, much easier to work with than their unbalanced resultant system. Thus, it is easy to compute the component currents produced by each symmetrical-component system of applied voltages because the induction motor operates as a balanced two-phase motor for each component system. By superposition, the actual current in a winding then is the sum of its components. Thus, if $\hat{I}_f$ and $\hat{I}_b$ are, respectively, the positive- and negative-sequence component phasor currents in winding α, then the corresponding positive- and negative-sequence component phasor currents in winding β are, respectively, $j\hat{I}_f$ and $-j\hat{I}_b$, and the actual winding currents $\hat{I}_\alpha$ and $\hat{I}_\beta$ are

$$\hat{I}_\alpha = \hat{I}_f + \hat{I}_b \tag{9.19}$$

$$\hat{I}_\beta = j\hat{I}_f - j\hat{I}_b \tag{9.20}$$

The inverse operation of finding the symmetrical components of specified voltages or currents must be performed often. Solution of Eqs. 9.17 and 9.18 for the phasor components $\hat{V}_f$ and $\hat{V}_b$ in terms of known phasor voltages $\hat{V}_\alpha$ and $\hat{V}_\beta$ gives

$$\hat{V}_f = \frac{1}{2}(\hat{V}_\alpha - j\hat{V}_\beta) \tag{9.21}$$

$$\hat{V}_b = \frac{1}{2}(\hat{V}_\alpha + j\hat{V}_\beta) \tag{9.22}$$

These operations are illustrated in the phasor diagram of Fig. 9.14. Obviously, similar relations give the phasor symmetrical components $\hat{I}_f$ and $\hat{I}_b$ of the current in winding α in terms of specified phasor currents $\hat{I}_m$ and $\hat{I}_a$ in the two phases; thus

$$\hat{I}_f = \frac{1}{2}(\hat{I}_\alpha - j\hat{I}_\beta) \tag{9.23}$$

$$\hat{I}_b = \frac{1}{2}(\hat{I}_\alpha + j\hat{I}_\beta) \tag{9.24}$$

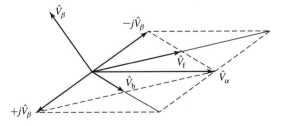

Figure 9.14 Resolution of unbalanced two-phase voltages into symmetrical components.

EXAMPLE 9.3

The equivalent-circuit parameters of a 5-hp 220-V 60-Hz four-pole two-phase squirrel-cage induction motor in ohms per phase are

$$R_1 = 0.534 \qquad X_1 = 2.45 \qquad X_m = 70.1 \qquad R_2 = 0.956 \qquad X_2 = 2.96$$

This motor is operated from an unbalanced two-phase 60-Hz source whose phase voltages are, respectively, 230 and 210 V, the smaller voltage leading the larger by 80°. For a slip of 0.05, find (a) the positive- and negative-sequence components of the applied voltages, (b) the positive- and negative-sequence components of the stator-phase currents, (c) the effective values of the phase currents, and (d) the internal mechanical power.

■ Solution

We will solve this example using MATLAB.[†]

a. Let $\hat{V}_\alpha$ and $\hat{V}_\beta$ denote the voltages applied to the two phases, respectively. Then

$$\hat{V}_\alpha = 230\angle 0° = 230 + j0 \text{ V}$$

$$\hat{V}_\beta = 210\angle 80° = 36.4 + j207 \text{ V}$$

From Eqs. 9.21 and 9.22 the forward and backward components of voltages are, respectively,

$$\hat{V}_f = 218.4 - j18.2 = 219.2\angle -4.8° \text{ V}$$

$$\hat{V}_b = 11.6 + j18.2 = 21.6\angle 57.5° \text{ V}$$

b. Because of the ease with which MATLAB handles complex numbers, there is no need to use approximations such as are derived in Example 9.2. Rather, the forward- and backward-field input impedances of the motor can be calculated from the equivalent circuits of Figs. 9.12a and b. Dividing the forward-field voltage by the forward-field input impedance gives

$$\hat{I}_f = \frac{\hat{V}_f}{R_1 + jX_1 + Z_f} = 9.3 - j6.3 = 11.2\angle -34.2° \text{ A}$$

Similarly, dividing the backward-field voltage by the backward-field input impedance gives

$$\hat{I}_b = \frac{\hat{V}_b}{R_1 + jX_1 + Z_b} = 3.7 - j1.5 = 4.0\angle -21.9° \text{ A}$$

c. The winding currents can be calculated from Eqs. 9.19 and 9.20

$$\hat{I}_\alpha = \hat{I}_f + \hat{I}_b = 13.0 - j7.8 = 15.2\angle -31.0° \text{ A}$$

$$\hat{I}_\beta = j\hat{I}_f - j\hat{I}_b = 4.8 + j5.6 = 7.4\angle 49.1° \text{ A}$$

Note that the winding currents are much more unbalanced than the applied voltages. Even though the motor is not overloaded insofar as shaft load is concerned, the losses are

---

[†] MATLAB is a registered trademark of The MathWorks, Inc.

appreciably increased by the current unbalance, and the stator winding with the greatest current may overheat.

d. The power delivered across the air gap by the forward field is equal to the forward-field equivalent-circuit input power minus the corresponding stator loss

$$P_{gap,f} = 2\left(\text{Re}[\hat{V}_f \hat{I}_f^*] - I_f^2 R_1\right) = 4149 \text{ W}$$

where the factor of 2 accounts for the fact that this is a two-phase motor. Similarly, the power delivered to the backward field is

$$P_{gap,b} = 2\left(\text{Re}[\hat{V}_b \hat{I}_b^*] - I_b^2 R_1\right) = 14.5 \text{ W}$$

Here, the symbol Re[ ] indicates the real part of a complex number, and the superscript * indicates the complex conjugate.

Finally, from Eq. 9.14, the internal mechanical power developed is equal to $(1 - s)$ times the total air-gap power or

$$P_{mech} = (1 - s)(P_{gap,f} - P_{gap,b}) = 3927 \text{ W}$$

If the core losses, friction and windage, and stray load losses are known, the shaft output can be found by subtracting them from the internal power. The friction and windage losses depend solely on the speed and are the same as they would be for balanced operation at the same speed. The core and stray load losses, however, are somewhat greater than they would be for balanced operation with the same positive-sequence voltage and current. The increase is caused principally by the $(2 - s)$-frequency core and stray losses in the rotor caused by the backward field.

Here is the MATLAB script:

```
clc
clear

% Useful constants
f = 60; %60 Hz system
omega = 2*pi*f;
s = 0.05; % slip

% Parameters

R1 = 0.534;
X1 = 2.45;
Xm = 70.1;
R2 = 0.956;
X2 = 2.96;

% Winding voltages

Valpha = 230;
Vbeta = 210 * exp(j*80*pi/180);

%(a) Calculate Vf and Vb from Equations and 9-21 and 9-22

Vf = 0.5*(Valpha - j*Vbeta);
Vb = 0.5*(Valpha + j*Vbeta);
```

```
magVf = abs(Vf);
angleVf = angle(Vf)*180/pi;

magVb = abs(Vb);
angleVb = angle(Vb)*180/pi;

fprintf('\n(a)')
fprintf('\n Vf = %.1f + j %.1f = %.1f at angle %.1f degrees V', ...
 real(Vf),imag(Vf),magVf,angleVf);
fprintf('\n Vb = %.1f + j %.1f = %.1f at angle %.1f degrees V\n', ...
 real(Vb),imag(Vb),magVb,angleVb);

%(b) First calculate the forward-field input impedance of the motor from
% the equivalent circuit of Fig. 9-12(a).

Zforward = R1 + j*X1 + j*Xm*(R2/s+j*X2)/(R2/s+j*(X2+Xm));

%Now calculate the forward-field current.

If = Vf/Zforward;

magIf = abs(If);
angleIf = angle(If)*180/pi;

% Next calculate the backward-field input impedance of the motor from
% Fig. 9-12(b).

Zback = R1 + j*X1 + j*Xm*(R2/(2-s)+j*X2)/(R2/(2-s)+j*(X2+Xm));

%Now calculate the backward-field current.

Ib = Vb/Zback;

magIb = abs(Ib);
angleIb = angle(Ib)*180/pi;

fprintf('\n(b)')
fprintf('\n If = %.1f + j %.1f = %.1f at angle %.1f degrees A', ...
 real(If),imag(If),magIf,angleIf);
fprintf('\n Ib = %.1f + j %.1f = %.1f at angle %.1f degrees A\n', ...
 real(Ib),imag(Ib),magIb,angleIb);

%(c) Calculate the winding currents from Eqs. 9-19 and 9-20

Ialpha = If + Ib;

Ibeta = j*(If - Ib);

magIalpha = abs(Ialpha);
angleIalpha = angle(Ialpha)*180/pi;

magIbeta = abs(Ibeta);
angleIbeta = angle(Ibeta)*180/pi;

fprintf('\n(c)')
fprintf('\n Ialpha = %.1f + j %.1f = %.1f at angle %.1f degrees A', ...
 real(Ialpha),imag(Ialpha),magIalpha,angleIalpha);
fprintf('\n Ibeta = %.1f + j %.1f = %.1f at angle %.1f degrees A\n', ...
 real(Ibeta),imag(Ibeta),magIbeta,angleIbeta);
```

```
%(d) Power delivered to the forward field is equal to the
% forward-field input power less the stator-winding I^2R loss

Pgf = 2*(real(Vf*conj(If)) - R1*magIf^2);

% Power delivered to the backward field is equal to the
% backward-field input power less the stator-winding I^2R loss

Pgb = 2*(real(Vb*conj(Ib)) - R1*magIb^2);

% The electromagnetic power is equal to (1-s) times the
% net air-gap power

Pmech = (1-s)*(Pgf - Pgb);

fprintf('\n(d)')
fprintf('\n Power to forward field = %.1f W',Pgf)
fprintf('\n Power to backward field = %.1f W',Pgb)
fprintf('\n Pmech = %.1f W\n',Pmech)
fprintf('\n')
```

Practice Problem 9.3

For the motor of Example 9.3, use MATLAB to produce a plot of the internal mechanical power as a function of slip as the slip varies from $s = 0.04$ to $s = 0.05$ for the unbalanced voltages assumed in the example. On the same axes (using dashed lines), plot the internal mechanical power for balanced two-phase voltages of 220-V magnitude and 90° phase shift.

Solution

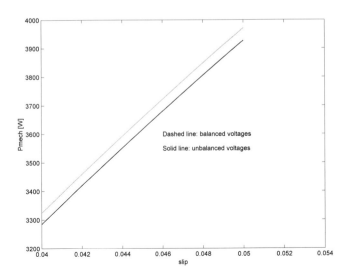

Figure 9.15 MATLAB plot for Practice Problem 9.3.

9.4.2 The General Case: Unsymmetrical Two-Phase Induction Machines

As we have discussed, a single-phase induction motor with a main and auxiliary winding is an example of an unsymmetrical two-phase induction motor. In this section we will develop a model for such a two-phase motor, using notation appropriate to the single-phase motor. We will assume, as is commonly the case, that the windings are in space quadrature but that they are unsymmetrical in that they may have a different number of turns, a different winding distribution, and so on.

Our analytical approach is to represent the rotor by an equivalent two-phase winding as shown in schematic form in Fig. 9.16 and to start with flux-linkage/current relationships for the rotor and stator of the form

$$
\begin{bmatrix} \lambda_{\text{main}} \\ \lambda_{\text{aux}} \\ \lambda_{\text{r1}} \\ \lambda_{\text{r2}} \end{bmatrix} = \begin{bmatrix} L_{\text{main}} & 0 & \mathcal{L}_{\text{main,r1}}(\theta_{\text{me}}) & \mathcal{L}_{\text{main,r2}}(\theta_{\text{me}}) \\ 0 & L_{\text{aux}} & \mathcal{L}_{\text{aux,r1}}(\theta_{\text{me}}) & \mathcal{L}_{\text{aux,r2}}(\theta_{\text{me}}) \\ \mathcal{L}_{\text{main,r1}}(\theta_{\text{me}}) & \mathcal{L}_{\text{aux,r1}}(\theta_{\text{me}}) & L_{\text{r}} & 0 \\ \mathcal{L}_{\text{main,r2}}(\theta_{\text{me}}) & \mathcal{L}_{\text{aux,r2}}(\theta_{\text{me}}) & 0 & L_{\text{r}} \end{bmatrix} \begin{bmatrix} i_{\text{main}} \\ i_{\text{aux}} \\ i_{\text{r1}} \\ i_{\text{r2}} \end{bmatrix}
$$

$$(9.25)$$

where θ_{me} is the rotor angle measured in electrical radians.

$$L_{\text{main}} = \text{Self-inductance of the main winding}$$

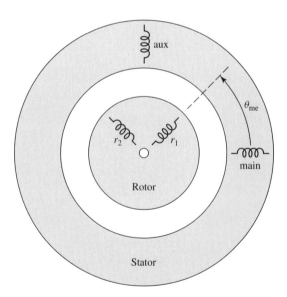

Figure 9.16 Schematic representation of a two-phase induction motor with an equivalent two-phase rotor.

L_{aux} = Self-inductance of the auxiliary winding

L_r = Self-inductance of the equivalent rotor windings

$\mathcal{L}_{main,r1}(\theta_{me})$ = Mutual inductance between the main winding and equivalent rotor winding 1

$\mathcal{L}_{main,r2}(\theta_{me})$ = Mutual inductance between the main winding and equivalent rotor winding 2

$\mathcal{L}_{aux,r1}(\theta_{me})$ = Mutual inductance between the auxiliary winding and rotor winding 1

$\mathcal{L}_{aux,r2}(\theta_{me})$ = Mutual inductance between the auxiliary winding and rotor winding 2

Assuming a sinusoidal distribution of air-gap flux, the mutual inductances between the main winding and the rotor will be of the form

$$\mathcal{L}_{main,r1}(\theta_{me}) = L_{main,r} \cos \theta_{me} \tag{9.26}$$

and

$$\mathcal{L}_{main,r2}(\theta_{me}) = -L_{main,r} \sin \theta_{me} \tag{9.27}$$

where $L_{main,r}$ is the amplitude of the mutual inductance.

The mutual inductances between the auxiliary winding will be of the same form with the exception that the auxiliary winding is displaced by 90 electrical degrees in space from the main winding. Hence we can write

$$\mathcal{L}_{aux,r1}(\theta_{me}) = L_{aux,r} \sin \theta_{me} \tag{9.28}$$

and

$$\mathcal{L}_{aux,r2}(\theta_{me}) = L_{aux,r} \cos \theta_{me} \tag{9.29}$$

Note that the auxiliary winding will typically have a different number of turns (and perhaps a different winding distribution) from that of the main winding. Thus, for modeling purposes, it is often convenient to write

$$L_{aux,r} = a\, L_{main,r} \tag{9.30}$$

where

$$a = \text{Turns ratio} = \frac{\text{Effective turns of auxiliary winding}}{\text{Effective turns of main winding}} \tag{9.31}$$

Similarly, if we write the self-inductance of the magnetizing branch as the sum of a leakage inductance $L_{main,l}$ and a magnetizing inductance L_m

$$L_{main} = L_{main,l} + L_m \tag{9.32}$$

then the self-inductance of the auxiliary winding can be written in the form

$$L_{aux} = L_{aux,l} + a^2 L_m \tag{9.33}$$

The voltage equations for this machine can be written in terms of the winding currents and flux linkages as

$$v_{main} = i_{main} R_{main} + \frac{d\lambda_{main}}{dt} \tag{9.34}$$

$$v_{aux} = i_{aux} R_{aux} + \frac{d\lambda_{aux}}{dt} \tag{9.35}$$

$$v_{r1} = 0 = i_{r1} R_r + \frac{d\lambda_{r1}}{dt} \tag{9.36}$$

$$v_{r2} = 0 = i_{r2} R_r + \frac{d\lambda_{r2}}{dt} \tag{9.37}$$

where R_{main}, R_{aux} and R_r are the resistances of the main, auxiliary, and rotor windings, respectively. Note that the rotor-winding voltages are set equal to zero because the rotor windings of an induction motor are internally shorted.

When modeling a split-phase induction motor (Section 9.2.1) the main and auxiliary windings are simply connected in parallel, and thus v_{main} and v_{aux} are both set equal to the single-phase supply voltage when the motor is started. Following the time that the auxiliary winding is disconnected, the auxiliary-winding current is zero, and the motor is represented by a reduced-order model which includes only the main winding and the two equivalent rotor windings.

When modeling the various capacitor motors of Section 9.2.2, the circuit equations must take into account the fact that, while the main winding is connected directly to the single-phase supply, a capacitor is connected between the supply and the auxiliary-winding terminals. Depending upon the type of motor being modeled, the auxiliary winding may or may not be switched out as the motor comes up to speed.

Finally, the techniques of Section 3.5 can be used to show that the electromagnetic torque of this motor can be written as

$$
\begin{aligned}
T_{mech} &= i_{main} i_{r1} \frac{d\mathcal{L}_{main,r1}(\theta_{me})}{d\theta_m} + i_{main} i_{r2} \frac{d\mathcal{L}_{main,r2}(\theta_{me})}{d\theta_m} \\
&\quad + i_{aux} i_{r1} \frac{d\mathcal{L}_{aux,r1}(\theta_{me})}{d\theta_m} + i_{aux} i_{r2} \frac{d\mathcal{L}_{aux,r2}(\theta_{me})}{d\theta_m} \\
&= \left(\frac{poles}{2}\right) [-L_{main,r} (i_{main} i_{r1} \sin\theta_{me} + i_{main} i_{r2} \cos\theta_{me}) \\
&\quad + L_{aux,r} (i_{aux} i_{r1} \cos\theta_{me} - i_{aux} i_{r2} \sin\theta_{me})]
\end{aligned} \tag{9.38}
$$

where $\theta_m = (2/poles)\theta_{me}$ is the rotor angle in radians.

Analogous to the development of the equivalent circuits derived in Chapter 6 for polyphase induction machines and earlier in this chapter for single-phase machines, the equations derived in this section can be further developed by assuming steady-state operation, with constant mechanical speed ω_{me}, corresponding to a slip s, and constant electrical supply frequency ω_e. Consistent with this assumption the rotor currents will be at frequencies $\omega_r = \omega_e - \omega_{me} = s\omega_e$ (produced by the stator positive-sequence field) and $\omega_r = \omega_e + \omega_{me} = (2-s)\omega_e$ (produced by the stator negative-sequence field).

After considerable algebraic manipulation, which includes using Eqs. 9.36 and 9.37 to eliminate the rotor currents, the main- and auxiliary-winding flux-linkage/current relationships of Eq. 9.25 can be written as phasor equations

$$\hat{\lambda}_{\text{main}} = \left[L_{\text{main}} - jL_{\text{main,r}}^2(\hat{K}^+ + \hat{K}^-) \right] \hat{I}_{\text{main}} + L_{\text{main,r}} L_{\text{aux,r}}(\hat{K}^+ - \hat{K}^-) \hat{I}_{\text{aux}}$$

(9.39)

and

$$\hat{\lambda}_{\text{aux}} = -L_{\text{main,r}} L_{\text{aux,r}}(\hat{K}^+ - \hat{K}^-) \hat{I}_{\text{main}} + \left[L_{\text{aux}} - jL_{\text{aux,r}}^2(\hat{K}^+ + \hat{K}^-) \right] \hat{I}_{\text{aux}} \quad (9.40)$$

where

$$\hat{K}^+ = \frac{s\omega_e}{2(R_r + js\omega_e L_r)}$$

(9.41)

and

$$\hat{K}^+ = \frac{(2-s)\omega_e}{2(R_r + j(2-s)\omega_e L_r)}$$

(9.42)

Similarly, the voltage equations, Eqs. 9.34 and 9.35 become

$$\hat{V}_{\text{main}} = \hat{I}_{\text{main}} R_{\text{main}} + j\omega_e \hat{\lambda}_{\text{main}}$$

(9.43)

$$\hat{V}_{\text{aux}} = \hat{I}_{\text{aux}} R_{\text{aux}} + j\omega_e \hat{\lambda}_{\text{aux}}$$

(9.44)

The rotor currents will each consist of positive- and negative-sequence components. The complex amplitudes of the positive sequence components (at frequency $s\omega_e$) are given by

$$\hat{I}_{r1}^+ = \frac{-js\omega_e[L_{\text{main,r}} \hat{I}_{\text{main}} + jL_{\text{aux,r}} \hat{I}_{\text{aux}}]}{2(R_r + js\omega_e L_r)}$$

(9.45)

and

$$\hat{I}_{r2}^+ = -j\hat{I}_{r1}^+$$

(9.46)

while the complex amplitudes of the negative sequence components (at frequency $(2-s)\omega_e$) are given by

$$\hat{I}_{r1}^- = \frac{-j(2-s)\omega_e[L_{\text{main,r}} \hat{I}_{\text{main}} - jL_{\text{aux,r}} \hat{I}_{\text{aux}}]}{2(R_r + j(2-s)\omega_e L_r)}$$

(9.47)

and

$$\hat{I}_{r2}^- = j\hat{I}_{r1}^-$$

(9.48)

Finally, again after careful algebraic manipulation, the time-averaged electromagnetic torque can be shown to be given by

$$<T_{\text{mech}}> = \left(\frac{\text{poles}}{2} \right) \text{Re} \left[(L_{\text{main,r}}^2 \hat{I}_{\text{main}} \hat{I}_{\text{main}}^* + L_{\text{aux,r}}^2 \hat{I}_{\text{aux}} \hat{I}_{\text{aux}}^*)(\hat{K}^+ - \hat{K}^-)^* \right.$$

$$\left. + jL_{\text{main,r}} L_{\text{aux,r}}(\hat{I}_{\text{main}}^* \hat{I}_{\text{aux}} - \hat{I}_{\text{main}} \hat{I}_{\text{aux}}^*)(\hat{K}^+ + \hat{K}^-)^* \right]$$

(9.49)

where the symbol Re[ ] again indicates the real part of a complex number and the superscript * indicates the complex conjugate. Note that Eq. 9.49 is derived based upon the assumption that the various currents are expressed as rms quantities.

EXAMPLE 9.4

Consider the case of a symmetrical two-phase motor such as is discussed in Section 9.4.1. In this case, Eqs. 9.25 through 9.37 simplify with equal self and mutual inductances and resistances for the two windings. Using the notation of Section 9.4.1, 'α' and 'β' replacing 'main' and 'aux', the flux-linkage/current relationships of Eq. 9.39 and 9.40 become

$$\hat{\lambda}_\alpha = \left[L_\alpha - jL_{\alpha,r}^2(\hat{K}^+ + \hat{K}^-)\right]\hat{I}_\alpha + L_{\alpha,r}^2(\hat{K}^+ - \hat{K}^-)\hat{I}_\beta$$

$$\hat{\lambda}_\beta = -L_{\alpha,r}^2(\hat{K}^+ - \hat{K}^-)\hat{I}_\alpha + \left[L_\alpha - jL_{\alpha,r}^2(\hat{K}^+ + \hat{K}^-)\right]\hat{I}_\beta$$

and the voltage equations (Eqs. 9.43 and 9.44) become

$$\hat{V}_\alpha = \hat{I}_\alpha R_\alpha + j\omega_e\hat{\lambda}_\alpha$$

$$\hat{V}_\beta = \hat{I}_\beta R_\alpha + j\omega_e\hat{\lambda}_\beta$$

Show that when operated from a positive sequence set of voltages such that $\hat{V}_\beta = -jV_\alpha$ the single-phase equivalent circuit is that of the forward-field (positive-sequence) equivalent circuit of Fig. 9.12a.

■ **Solution**

Substitution of the positive-sequence voltages in the above equations and solution for the impedance $Z_\alpha = \hat{V}_\alpha / \hat{I}_\alpha$ gives

$$Z_\alpha = R_\alpha + j\omega_e L_\alpha + \frac{(\omega_e L_{\alpha,r})^2}{(R_r/s + j\omega_e L_r)}$$

$$= R_\alpha + jX_\alpha + \frac{X_{\alpha,r}^2}{(R_r/s + jX_r)}$$

This equation can be rewritten as

$$Z_\alpha = R_\alpha + j(X_\alpha - X_{\alpha,r}) + \frac{jX_{\alpha,r}[j(X_r - X_{\alpha,r}) + R_r/s]}{(R_r/s + jX_r)}$$

Setting $R_\alpha \Rightarrow R_1$, $(X_\alpha - X_{\alpha,r}) \Rightarrow X_1$, $X_{\alpha,r} \Rightarrow X_m$, $(X_r - X_{\alpha,r}) \Rightarrow X_2$, and $R_r \Rightarrow R_2$, we see that this equation does indeed represent an equivalent circuit of the form of Fig. 9.12a.

Practice Problem 9.4

Analogous to the calculation of Example 9.4, show that when operated from a negative sequence set of voltages such that $\hat{V}_\beta = jV_\alpha$ the single-phase equivalent circuit is that of the backward-field (negative-sequence) equivalent circuit of Fig. 9.12b.

Solution

Under negative-sequence conditions, the impedance Z_α is equal to

$$Z_\alpha = R_\alpha + j\omega_e L_\alpha + \frac{(\omega_e L_{\alpha,r})^2}{(R_r/(2-s) + j\omega_e L_r)}$$

$$= R_\alpha + jX_\alpha + \frac{X_{\alpha,r}^2}{(R_r/(2-s) + jX_r)}$$

As in Example 9.4, this can be shown to correspond to an equivalent circuit of the form of Fig. 9.12b.

EXAMPLE 9.5

A two-pole, single-phase induction motor has the following parameters

$$L_{main} = 80.6 \text{ mH} \qquad R_{main} = 0.58 \ \Omega$$

$$L_{aux} = 196 \text{ mH} \qquad R_{aux} = 3.37 \ \Omega$$

$$L_r = 4.7 \ \mu\text{H} \qquad R_r = 37.6 \ \mu\Omega$$

$$L_{main,r} = 0.588 \text{ mH} \qquad L_{aux,r} = 0.909 \text{ mH}$$

It is operated from a single-phase, 230-V rms, 60-Hz source as a permanent-split-capacitor motor with a 35 μF capacitor connected in series with the auxiliary winding. In order to achieve the required phase shift of the auxiliary-winding current, the windings must be connected with the polarities shown in Fig. 9.17. The motor has a rotational losses of 40 W and 105 W of core loss.

Consider motor operation at 3500 r/min.

a. Find the main-winding, auxiliary-winding and source currents and the magnitude of the capacitor voltage.
b. Find the time-averaged electromagnetic torque and shaft output power.

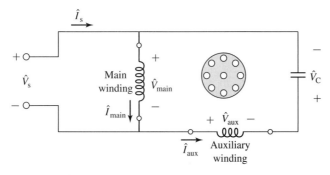

Figure 9.17 Permanent-split-capacitor induction-motor connections for Example 9.5.

c. Calculate the motor input power and its electrical efficiency. Note that since core loss isn't explicitly accounted for in the model derived in this section, you may simply consider it as an additional component of the input power.

d. Plot the motor time-averaged electromagnetic torque as a function of speed from standstill to synchronous speed.

■ **Solution**

MATLAB, with its ease of handling complex numbers, is ideal for the solution of this problem.

a. The main winding of this motor is directly connected to the single-phase source. Thus we directly set $\hat{V}_{\text{main}} = \hat{V}_{\text{s}}$. However, the auxiliary winding is connected to the single-phase source through a capacitor and its polarity is reversed. Thus we must write

$$\hat{V}_{\text{aux}} + \hat{V}_{\text{C}} = -\hat{V}_{\text{s}}$$

where the capacitor voltage is given by

$$\hat{V}_{\text{C}} = j \hat{I}_{\text{aux}} X_{\text{C}}$$

Here the capacitor impedance X_{C} is equal to

$$X_{\text{C}} = -\frac{1}{(\omega_{\text{e}} C)} = -\frac{1}{(120\pi \times 35 \times 10^{-6})} = -75.8 \ \Omega$$

Setting $\hat{V}_{\text{s}} = V_0 = 230$ V and substituting these expressions into Eqs. 9.43 and 9.44 and using Eqs. 9.39 and 9.40 then gives the following matrix equation for the main- and auxiliary-winding currents.

$$\begin{bmatrix} (R_{\text{main}} + j\omega_{\text{e}} \hat{A}_1) & j\omega_{\text{e}} \hat{A}_2 \\ -j\omega_{\text{e}} \hat{A}_2 & (R_{\text{aux}} + jX_{\text{c}} + j\omega_{\text{e}} \hat{A}_3) \end{bmatrix} \begin{bmatrix} \hat{I}_{\text{main}} \\ \hat{I}_{\text{aux}} \end{bmatrix} = \begin{bmatrix} V_0 \\ -V_0 \end{bmatrix}$$

where

$$\hat{A}_1 = L_{\text{main}} - jL_{\text{main,r}}^2 (\hat{K}^+ + \hat{K}^-)$$

$$\hat{A}_2 = L_{\text{main,r}} L_{\text{aux,r}} (\hat{K}^+ - \hat{K}^-)$$

and

$$\hat{A}_3 = L_{\text{aux}} - jL_{\text{aux,r}}^2 (\hat{K}^+ + \hat{K}^-)$$

The parameters $\hat{K}^+$ and $\hat{K}^-$ can be found from Eqs. 9.41 and 9.42 once the slip is found using Eq. 6.1

$$s = \frac{n_{\text{s}} - n}{n_{\text{s}}} == \frac{3600 - 3500}{3600} = 0.278$$

This matrix equation can be readily solved using MATLAB with the result

$$\hat{I}_{\text{main}} = 15.9\angle -37.6° \text{ A}$$

$$\hat{I}_{\text{aux}} = 5.20\angle -150.7° \text{ A}$$

and

$$\hat{I}_s = 18.5 \angle -22.7° \text{ A}$$

The magnitude of the capacitor voltage is

$$|\hat{V}_C| = |\hat{I}_{aux} X_C| = 374 \text{ V}$$

b. Using MATLAB the time-averaged electromagnetic torque can be found from Eq. 9.49 to be

$$<T_{mech}> = 9.74 \text{ N} \cdot \text{m}$$

The shaft power can then be found by subtracting the rotational losses P_{rot} from the air-gap power

$$
\begin{aligned}
P_{shaft} &= \omega_m <T_{mech}> - P_{rot} \\
&= \left(\frac{2}{poles}\right)(1-s)\omega_e(<T_{mech}>) - P_{rot} \\
&= 3532 \text{ W}
\end{aligned}
$$

c. The power input to the main winding can be found as

$$P_{main} = \text{Re}\left[V_0 \hat{I}_{main}^*\right] = 2893 \text{ W}$$

and that into the auxiliary winding, including the capacitor (which dissipates no power)

$$P_{aux} = \text{Re}\left[-V_0 \hat{I}_{aux}^*\right] = 1043 \text{ W}$$

The total input power, including the core loss power P_{core} is found as

$$P_{in} = P_{main} + P_{aux} + P_{core} = 4041 \text{ W}$$

Finally, the efficiency can be determined

$$\eta = \frac{P_{shaft}}{P_{in}} == 0.874 = 87.4\%$$

d. The plot of $<T_{mech}>$ versus speed generated by MATLAB is found in Fig. 9.18.

Here is the MATLAB script:

```
clc
clear

% Source parameters
V0 = 230;
omegae = 120*pi;

% Motor parameters
poles = 2;
Lmain = .0806;
Rmain = 0.58;
Laux = 0.196;
Raux = 3.37;
Lr = 4.7e-6;
```

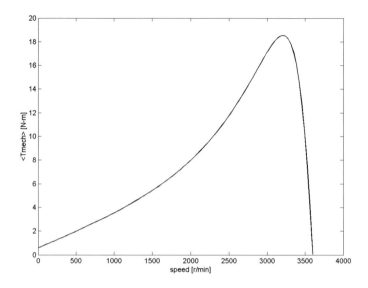

Figure 9.18 Time-averaged electromagnetic torque versus speed for the single-phase induction motor of Example 9.5.

```
Rr = 37.6e-6;
Lmainr = 5.88e-4;
Lauxr = 9.09e-4;
C = 35e-6;
Xc = -1/(omegae*C);
Prot = 40;
Pcore = 105;

% Run through program twice. If calcswitch = 1, then
% calculate at speed of 3500 r/min only. The second time
% program will produce the plot for part (d).

for calcswitch = 1:2

if calcswitch == 1
  mmax = 1;
else
   mmax = 101;
end

for m = 1:mmax

if calcswitch == 1
   speed(m)  = 3500;
else
   speed(m)  = 3599*(m-1)/100;
end
```

```
% Calculate the slip
ns = (2/poles)*3600;
s = (ns-speed(m))/ns;

% part (a)
% Calculate the various complex constants
Kplus = s*omegae/(2*(Rr + j*s*omegae*Lr));
Kminus = (2-s)*omegae/(2*(Rr + j*(2-s)*omegae*Lr));
A1 = Lmain - j*Lmainr^2*(Kplus+Kminus);
A2 = Lmainr*Lauxr*(Kplus-Kminus);
A3 = Laux - j*Lauxr^2*(Kplus+Kminus);

% Set up the matrix
M(1,1) = Rmain + j*omegae*A1;
M(1,2) = j*omegae*A2;
M(2,1) = -j*omegae*A2;
M(2,2) = Raux + j*Xc+ j*omegae*A3;

% Here is the voltage vector
V = [V0 ; -V0];

% Now find the current matrix

I = M\V;

Imain = I(1);
Iaux = I(2);
Is = Imain-Iaux;

magImain = abs(Imain);
angleImain = angle(Imain)*180/pi;
magIaux = abs(Iaux);
angleIaux = angle(Iaux)*180/pi;
magIs = abs(Is);
angleIs = angle(Is)*180/pi;

%Capacitor voltage
Vcap = Iaux*Xc;
magVcap = abs(Vcap);

% part (b)
Tmech1 = conj(Kplus-Kminus);
Tmech1 = Tmech1*(Lmainr^2*Imain*conj(Imain)+Lauxr^2*Iaux*conj(Iaux));
Tmech2 = j*Lmainr*Lauxr*conj(Kplus+Kminus);
Tmech2 = Tmech2*(conj(Imain)*Iaux-Imain*conj(Iaux));
Tmech(m) = (poles/2)*real(Tmech1+Tmech2);
Pshaft = (2/poles)*(1-s)*omegae*Tmech(m)-Prot;

%part (c)
Pmain = real(V0*conj(Imain));
Paux = real(-V0*conj(Iaux));
```

```
Pin = Pmain+Paux+Pcore;
eta = Pshaft/Pin;

if calcswitch == 1
  fprintf('part (a):')
  fprintf('\n  Imain = %g A at angle %g degrees',magImain,angleImain)
  fprintf('\n  Iaux = %g A at angle %g degrees',magIaux,angleIaux)
  fprintf('\n  Is = %g A at angle %g degrees',magIs,angleIs)
  fprintf('\n  Vcap = %g V\n',magVcap)
  fprintf('\npart (b):')
  fprintf('\n  Tmech = %g N-m',Tmech)
  fprintf('\n  Pshaft = %g W\n',Pshaft)
  fprintf('\npart (c):')
  fprintf('\n  Pmain = %g W',Pmain)
  fprintf('\n  Paux = %g W',Paux)
  fprintf('\n  Pin = %g W',Pin)
  fprintf('\n  eta = %g percent\n\n',100*eta)
else
  plot(speed,Tmech)
  xlabel('speed [r/min]')
  ylabel('<Tmech> [N-m]')
end

end %end of for m loop
end %end of for calcswitch loop
```

Practice Problem 9.5

(a) Calculate the efficiency of the single-phase induction motor of Example 9.5 operating at a speed of 3475 r/min. (b) Search over the range of capacitor values from 25 μF to 45 μF to find the capacitor value which will give the maximum efficiency at this speed and the corresponding efficiency.

Solution

    a. 86.4%

    b. 41.8 μF, 86.6%

9.5   SUMMARY

One theme of this chapter is a continuation of the induction-machine theory of Chapter 6 and its application to the single-phase induction motor. This theory is expanded by a step-by-step reasoning process from the simple revolving-field theory of the symmetrical polyphase induction motor. The basic concept is the resolution of the stator-mmf wave into two constant-amplitude traveling waves revolving around the air gap at synchronous speed in opposite directions. If the slip for the forward field is s, then that for the backward field is $(2 - s)$. Each of these component fields produces induction-motor action, just as in a symmetrical polyphase motor. From the viewpoint of the stator, the reflected effects of the rotor can be visualized and expressed quantitatively in terms of simple equivalent circuits. The ease with which the internal reactions can be accounted for in this manner is the essential reason for the usefulness of the double-revolving-field theory.

For a single-phase winding, the forward- and backward-component mmf waves are equal, and their amplitude is half the maximum value of the peak of the stationary pulsating mmf produced by the winding. The resolution of the stator mmf into its forward and backward components then leads to the physical concept of the single-phase motor described in Section 9.1 and finally to the quantitative theory developed in Section 9.3 and to the equivalent circuits of Fig. 9.11.

In most cases, single-phase induction motors are actually two-phase motors with unsymmetrical windings operated off of a single phase source. Thus to complete our understanding of single-phase induction motors, it is necessary to examine the performance of two-phase motors. Hence, the next step is the application of the double-revolving-field picture to a symmetrical two-phase motor with unbalanced applied voltages, as in Section 9.4.1. This investigation leads to the symmetrical-component concept, whereby an unbalanced two-phase system of currents or voltages can be resolved into the sum of two balanced two-phase component systems of opposite phase sequence. Resolution of the currents into symmetrical-component systems is equivalent to resolving the stator-mmf wave into its forward and backward components, and therefore the internal reactions of the rotor for each symmetrical-component system are the same as those which we have already investigated. A very similar reasoning process, not considered here, leads to the well-known three-phase symmetrical-component method for treating problems involving unbalanced operation of three-phase rotating machines. The ease with which the rotating machine can be analyzed in terms of revolving-field theory is the chief reason for the usefulness of the symmetrical-component method.

Finally, the chapter ends in Section 9.4.2 with the development of an analytical theory for the general case of a two-phase induction motor with unsymmetrical windings. This theory permits us to analyze the operation of single-phase motors running off both their main and auxiliary windings.

9.6 PROBLEMS

9.1 A 1-kW, 120-V, 60-Hz capacitor-start motor has the following parameters for the main and auxiliary windings (at starting):

$$Z_{main} = 4.82 + j7.25 \ \Omega \quad \text{main winding}$$

$$Z_{aux} = 7.95 + j9.21 \ \Omega \quad \text{auxiliary winding}$$

a. Find the magnitude and the phase angles of the currents in the two windings when rated voltage is applied to the motor under starting conditions.

b. Find the value of starting capacitance that will place the main- and auxiliary-winding currents in time quadrature at starting.

c. Repeat part (a) when the capacitance of part (b) is inserted in series with the auxiliary winding.

9.2 Repeat Problem 9.1 if the motor is operated from a 120-V, 50-Hz source.

9.3 Given the applied electrical frequency and the corresponding impedances Z_{main} and Z_{aux} of the main and auxiliary windings at starting, write a MATLAB script to calculate the value of the capacitance, which, when connected in series with the starting winding, will produce a starting winding current which will lead that of the main winding by 90°.

9.4 Repeat Example 9.2 for slip of 0.045.

9.5 A 500-W, four-pole, 115-V, 60-Hz single-phase induction motor has the following parameters (resistances and reactances in Ω/phase):

$$R_{1,\text{main}} = 1.68 \quad R_{2,\text{main}} = 2.96$$

$$X_{1,\text{main}} = 1.87 \quad X_{\text{m,main}} = 60.6 \quad X_{2,\text{main}} = 1.72$$

$$\text{Core loss} = 38 \text{ W} \quad \text{Friction and windage} = 11.8 \text{ W}$$

Find the speed, stator current, torque, power output, and efficiency when the motor is operating at rated voltage and a slip of 4.2 percent.

9.6 Write a MATLAB script to produce plots of the speed and efficiency of the single-phase motor of Problem 9.5 as a function of output power over the range $0 \le P_{\text{out}} \le 500$ W.

9.7 At standstill the rms currents in the main and auxiliary windings of a four-pole, capacitor-start induction motor are $I_{\text{main}} = 20.7$ A and $I_{\text{aux}} = 11.1$ A respectively. The auxiliary-winding current leads the main-winding current by 53°. The effective turns per pole (i.e., the number of turns corrected for the effects of winding distribution) are $N_{\text{main}} = 42$ and $N_{\text{aux}} = 68$. The windings are in space quadrature.

a. Determine the peak amplitudes of the forward and backward stator-mmf waves.

b. Suppose it were possible to adjust the magnitude and phase of the auxiliary-winding current. What magnitude and phase would produce a purely forward mmf wave?

9.8 Derive an expression in terms of $Q_{2,\text{main}}$ for the nonzero speed of a single-phase induction motor at which the internal torque is zero. (See Example 9.2.)

9.9 The equivalent-circuit parameters of an 8-kW, 230-V, 60-Hz, four-pole, two-phase, squirrel-cage induction motor in ohms per phase are

$$R_1 = 0.253 \quad X_1 = 1.14 \quad X_{\text{m}} = 32.7 \quad R_2 = 0.446 \quad X_2 = 1.30$$

This motor is operated from an unbalanced two-phase, 60-Hz source whose phase voltages are, respectively, 223 and 190 V, the smaller voltage leading the larger by 73°. For a slip of 0.045, find

a. the phase currents in each of the windings and

b. the internal mechanical power.

9.10 Consider the two-phase motor of Example 9.3.

    a. Find the starting torque for the conditions specified in the example.

    b. Compare the result of part (a) with the starting torque which the motor would produce if 220-V, balanced two-phase voltages were applied to the motor.

    c. Show that if the stator voltages $\hat{V}_\alpha$ and $\hat{V}_{\text{beta}}$ of a two-phase induction motor are in time quadrature but unequal in magnitude, the starting torque is the same as that developed when balanced two-phase voltages of magnitude $\sqrt{V_\alpha V_\beta}$ are applied.

9.11 The induction motor of Problem 9.9 is supplied from an unbalanced two-phase source by a four-wire feeder having an impedance $Z = 0.32 + j1.5\ \Omega/\text{phase}$. The source voltages can be expressed as

$$\hat{V}_\alpha = 235\angle 0° \quad \hat{V}_\alpha = 212\angle 78°$$

For a slip of 5 percent, show that the induction-motor terminal voltages correspond more nearly to a balanced two-phase set than do those of the source.

9.12 The equivalent-circuit parameters in ohms per phase referred to the stator for a two-phase, 1.0 kW, 220-V, four-pole, 60-Hz, squirrel-cage induction motor are given below. The no-load rotational loss is 65 W.

$$R_1 = 0.78 \quad R_2 = 4.2 \quad X_1 = X_2 = 5.3 \quad X_\text{m} = 93$$

    a. The voltage applied to phase α is $220\angle 0°$ V and that applied to phase β is $220\angle 65°$ V. Find the net air-gap torque at a slip $s = 0.035$.

    b. What is the starting torque with the applied voltages of part (a)?

    c. The applied voltages are readjusted so that $\hat{V}_\alpha = 220\angle 65°$ V and $\hat{V}_\beta = 220\angle 90°$ V. Full load on the machine occurs at $s = 0.048$. At what slip does maximum internal torque occur? What is the value of the maximum torque?

    d. While the motor is running as in part (c), phase β is open-circuited. What is the power output of the machine at a slip $s = 0.04$?

    e. What voltage appears across the open phase-β terminals under the conditions of part (d)?

9.13 A 120-V, 60-Hz, capacitor-run, two-pole, single-phase induction motor has the following parameters:

$$L_\text{main} = 47.2\ \text{mH} \qquad R_\text{main} = 0.38\ \Omega$$

$$L_\text{aux} = 102\ \text{mH} \qquad R_\text{aux} = 1.78\ \Omega$$

$$L_\text{r} = 2.35\ \mu\text{H} \qquad R_\text{r} = 17.2\ \mu\Omega$$

$$L_\text{main,r} = 0.342\ \text{mH} \qquad L_\text{aux,r} = 0.530\ \text{mH}$$

You may assume that the motor has 48 W of core loss and 23 W of rotational losses. The motor windings are connected with the polarity shown in Fig. 9.17 with a 40 μF run capacitor.

     a.   Calculate the motor starting torque.

With the motor operating at a speed of 3490 r/min, calculate

     b.   the main and auxiliary-winding currents,

     c.   the total line current and the motor power factor,

     d.   the output power and

     e.   the electrical input power and the efficiency.

Note that this problem is most easily solved using MATLAB.

9.14   Consider the single-phase motor of Problem 9.13. Write a MATLAB script to search over the range of capacitor values from 25 μF to 75 μF to find the value which will maximize the motor efficiency at a motor speed of 3490 r/min. What is the corresponding maximum efficiency?

9.15   In order to raise the starting torque, the single-phase induction motor of Problem 9.13 is to be converted to a capacitor-start, capacitor-run motor. Write a MATLAB script to find the minimum value of starting capacitance required to raise the starting torque to 0.5 N · m.

9.16   Consider the single-phase induction motor of Example 9.5 operating over the speed range 3350 r/min to 3580 r/min.

     a.   Use MATLAB to plot the output power over the given speed range.

     b.   Plot the efficiency of the motor over this speed range.

     c.   On the same plot as that of part (b), plot the motor efficiency if the run capacitor is increased to 45 μF.

CHAPTER 10

Introduction to Power Electronics

U
ntil the last few decades of the twentieth century, ac machines tended to be employed primarily as single-speed devices. Typically they were operated from fixed-frequency sources (in most cases this was the 50- or 60-Hz power grid). In the case of motors, control of motor speed requires a variable-frequency source, and such sources were not readily available. Thus, applications requiring variable speed were serviced by dc machines, which can provide highly flexible speed control, although at some cost since they are more complex, more expensive, and require more maintenance than their ac counterparts.

The availability of solid-state power switches changed this picture immensely. It is now possible to build power electronics capable of supplying the variable-voltage/current, variable-frequency drive required to achieve variable-speed performance from ac machines. Ac machines have now replaced dc machines in many traditional applications, and a wide range of new applications have been developed.

As is the case with electromechanics and electric machinery, power electronics is a discipline which can be mastered only through significant study. Many books have been written on this subject, a few of which are listed in the bibliography at the end of this chapter. It is clear that a single chapter in a book on electric machinery cannot begin to do justice to this topic. Thus our objectives here are limited. Our goal is to provide an overview of power electronics and to show how the basic building blocks can be assembled into drive systems for ac and dc machines. We will not focus much attention on the detailed characteristics of particular devices or on the many details required to design practical drive systems. In Chapter 11, we will build on the discussion of this chapter to examine the characteristics of some common drive systems.

10.1 POWER SWITCHES

Common to all power-electronic systems are switching devices. Ideally, these devices control current much like valves control the flow of fluids: turn them "ON," and they present no resistances to the flow of current; turn them "OFF," and no current flow is possible. Of course, practical switches are not ideal, and their specific characteristics significantly affect their applicability in any given situation. Fortunately, the essential performance of most power-electronic circuits can be understood assuming the switches to be ideal. This is the approach which we will adopt in this book. In this section we will briefly discuss some of the common switching devices and present simplified, idealized models for them.

10.1.1 Diodes

Diodes constitute the simplest of power switches. The general form of the v-i characteristics of a diode is shown in Fig. 10.1.

The essential features of a diode are captured in the idealized v-i characteristic of Fig. 10.2a. The symbol used to represent a diode is shown in Fig. 10.2b along with the reference directions for the current i and voltage v. Based upon terminology developed when rectifier diodes were electron tubes, diode current flows into the *anode* and flows out of the *cathode*.

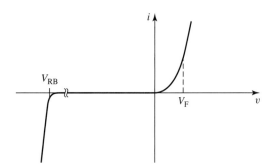

Figure 10.1 *v-i* characteristic of a diode.

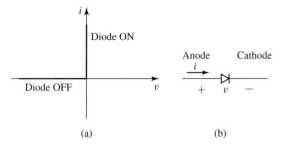

Figure 10.2 (a) *v-i* characteristic of an ideal diode.
(b) Diode symbol.

We can see that the ideal diode blocks current flow when the voltage is negative ($i = 0$ for $v < 0$) and passes positive current without voltage drop ($v = 0$ for $i \geq 0$). We will refer to the negative-voltage region as the diode's OFF state and the positive-current region as the diode's ON state. Comparison with the v-i characteristic shows that a practical diode varies from an ideal diode in that:

■ There is a finite *forward voltage drop,* labeled V_F in Fig. 10.1, for positive current flow. For low-power devices, this voltage range is typically on the order of 0.6–0.7 V while for high-power devices it can exceed 3 V.

■ Corresponding to this voltage drop is a power dissipation. Practical diodes have a maximum power dissipation (and a corresponding maximum current) which must not be exceeded.

■ A practical diode is limited in the negative voltage it can withstand. Known as the *reverse-breakdown voltage* and labeled V_{RB} in Fig. 10.1, this is the maximum reverse voltage that can be applied to the diode before it starts to conduct reverse current.

The diode is the simplest power switch in that it cannot be controlled; it simply turns ON when positive current begins to flow and turns OFF when the current attempts to reverse. In spite of this simple behavior, it is used in a wide variety of applications, the most common of which is as a rectifier to convert ac to dc.

The basic performance of a diode can be illustrated by the simple example shown in Example 10.1.

EXAMPLE 10.1

Consider the *half-wave rectifier* circuit of Fig. 10.3a in which a resistor R is supplied by a voltage source $v_s(t) = V_0 \sin \omega t$ through a diode. Assume the diode to be ideal. (*a*) Find the resistor voltage $v_R(t)$ and current $i_R(t)$. (*b*) Find the dc average resistor voltage V_{dc} and current I_{dc}.

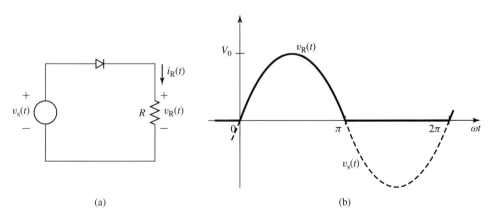

(a)   (b)

Figure 10.3 (a) Half-wave rectifier circuit for Example 10.1. (b) Resistor voltage.

■ Solution

a. This is a nonlinear problem in that it is not possible to write an analytic expression for the v-i characteristic of the ideal diode. However, it can readily be solved using the *method-of-assumed-states* in which, for any given value of the source voltage, the diode is alternately assumed to be ON (a short-circuit) or OFF (an open-circuit) and the current is found. One of the two solutions will violate the v-i characteristic of the diode (i.e., there will be negative current flow through the short-circuit or positive voltage across the open-circuit) and must be discarded; the remaining solution will be the correct one.

Following the above procedure, we find that the solution is given by

$$v_R(t) = \begin{cases} v_s(t) = V_0 \sin \omega t & v_s(t) \geq 0 \\ 0 & v_s(t) < 0 \end{cases}$$

This voltage is plotted in Fig. 10.3b. The current is identical in form and is found simply as $i_R(t) = v_R(t)/R$. The terminology *half-wave rectification* is applied to this system because voltage is applied to the resistor during only the half cycle for which the supply voltage waveform is positive.

b. The dc or average value of the voltage waveform is equal to

$$V_{dc} = \frac{\omega}{\pi} \int_0^{\frac{\pi}{\omega}} V_0 \sin (\omega t) \, dt = \frac{V_0}{\pi}$$

and hence the dc current through the resistor is equal to

$$I_{dc} = \frac{V_0}{\pi R}$$

Practice Problem 10.1

Calculate the average voltage across the resistor of Fig. 10.3 if the sinusoidal voltage source of Example 10.1 is replaced by a source of the same frequency but which produces a square wave of zero average value and peak-peak amplitude $2V_0$.

Solution

$$V_{dc} = \frac{V_0}{2}$$

10.1.2 Silicon Controlled Rectifiers and TRIACs

The characteristics of a *silicon controlled rectifier,* or *SCR,* also referred to as a *thyristor,* are similar to those of a diode. However, in addition to an anode and a cathode, the SCR has a third terminal known as the *gate*. Figure 10.4 shows the form of the v-i characteristics of a typical SCR.

As is the case with a diode, the SCR will turn ON only if the anode is positive with respect to the cathode. Unlike a diode, the SCR also requires a pulse of current i_G into the gate to turn ON. Note however that once the SCR turns ON, the gate signal

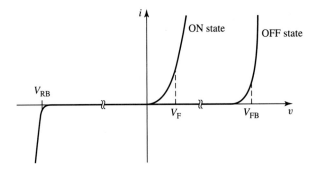

Figure 10.4 *v-i* characteristic of an SCR.

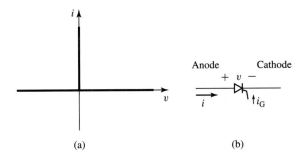

Figure 10.5 (a) Idealized SCR *v-i* characteristic.
(b) SCR symbol.

can be removed and the SCR will remain ON until the SCR current drops below a small value referred to as the *holding current,* at which point it will turn OFF just as a diode does.

As can be seen from Fig. 10.4, the ON-state characteristic of an SCR is similar to that of a diode, with a forward voltage drop V_F and a reverse-breakdown voltage V_{RB}. When the SCR is OFF, it does not conduct current over its normal operating range of positive voltage. However it will conduct if this voltage exceeds a characteristic voltage, labeled V_{FB} in the figure and known as the *forward-breakdown voltage.* As is the case for a diode, a practical SCR is limited in its current-carrying capability.

For our purposes, we will simplify these characteristics and assume the SCR to have the idealized characteristics of Fig. 10.5a. Our idealized SCR appears as an open-circuit when it is OFF and a short-circuit when it is ON. It also has a holding current of zero; i.e., it will remain ON until the current drops to zero and attempts to go negative. The symbol used to represent an SCR is shown in Fig. 10.5b.

Care must be taken in the design of gate-drive circuitry to insure that an SCR turns on properly; e.g., the gate pulse must inject enough charge to fully turn on the SCR, and so forth. Similarly, an additional circuit, typically referred to as a *snubber* circuit, may be required to protect an SCR from being turned on inadvertently, such as might occur if the rate of rise of the anode-to-cathode voltage is excessive. Although these details must be properly accounted for to achieve successful SCR performance in practical circuits, they are not essential for the present discussion.

The basic performance of an SCR can be understood from the following example.

EXAMPLE 10.2

Consider the *half-wave rectifier* circuit of Fig. 10.6 in which a resistor R is supplied by a voltage source $v_s(t) = V_0 \sin \omega t$ through an SCR. Note that this is identical to the circuit of Example 10.1, with the exception that the diode has been replaced by an SCR.

Assume that a pulse of gate current is applied to the SCR at time t_0 $(0 \leq t_0 < \pi/\omega)$ following each zero-crossing of the source voltage, as shown in Fig. 10.7a. It is common to describe this *firing-delay time* in terms of a *firing-delay angle*, $\alpha_0 \equiv \omega t_0$. Find the resistor voltage $v_R(t)$ as a function of α_0. Assume the SCR to be ideal and that the gate pulses supply sufficient charge to properly turn ON the SCR.

■ Solution

The solution follows that of Example 10.1 with the exception that, independent of the polarity of the voltage across it, once the SCR turns OFF, it will remain OFF until both the SCR voltage becomes positive and a pulse of gate current is applied. Once a gate pulse has been applied, the method-of-assumed-states can be used to solve for the state of the SCR.

Following the above procedure, we find the solution is given by

$$v_R(t) = \begin{cases} 0 & v_s(t) \geq 0 \quad \text{(prior to the gate pulse)} \\ v_s(t) = V_0 \sin \omega t & v_s(t) \geq 0 \quad \text{(following the gate pulse)} \\ 0 & v_s(t) < 0 \end{cases}$$

This voltage is plotted in Fig. 10.7b. Note that this system produces a half-wave rectified voltage similar to that of the diode system of Example 10.1. However, in this case, the dc value of the rectified voltage can be controlled by controlling the timing of the gate pulse. Specifically, it is given by

$$V_{dc} = \frac{V_0}{2\pi}(1 + \cos \alpha_0)$$

Note that when there is no delay in firing the SCR $(\alpha_0 = 0)$, this system produces a dc voltage of V_0/π, equal to that of the diode rectifier system of Example 10.1. However, as the gate pulse of the SCR is delayed (i.e., by increasing α_0), the dc voltage can be reduced. In fact, by delaying the gate pulse a full half cycle $(\alpha_0 = \pi)$ the dc voltage can be reduced to

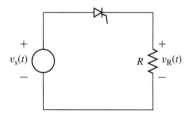

Figure 10.6 Half-wave SCR rectifier circuit for Example 10.2.

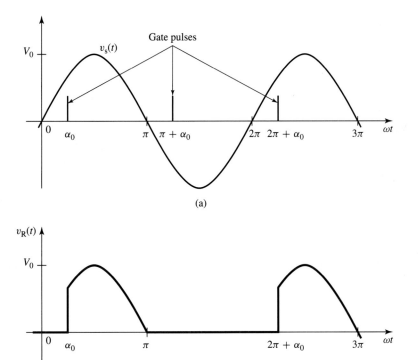

Figure 10.7 (a) Gate pulses for Example 10.2. (b) Resistor voltage.

zero. This system is known as a *phase-controlled rectifier* because the dc output voltage can be varied by controlling the phase angle of the gate pulse relative to the zero crossing of the source voltage.

Practice Problem 10.2

Calculate the resistor average voltage as a function of the delay angle α_0 if the sinusoidal source of Example 10.2 is replaced by a source of the same frequency, but which produces a square wave of zero average value and peak-peak amplitude $2V_0$.

Solution

$$V_{dc} = \frac{V_0}{2}\left(1 - \frac{\alpha_0}{\pi}\right)$$

Example 10.2 shows that the SCR provides a significant advantage over the diode in systems where voltage control is desired. However, this advantage comes at the additional expense of the SCR as well as the circuitry required to produce the gate pulses used to fire the SCR.

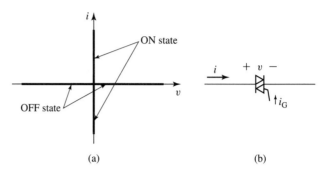

Figure 10.8 (a) Idealized TRIAC *v-i* characteristic.
(b) TRIAC symbol.

Another phase-controlled device is the *TRIAC,* which behaves much like two back-to-back SCRs sharing a common gate. The idealized *v-i* characteristic of a TRIAC is shown in Fig. 10.8a and its symbol in Fig. 10.8b. As with an SCR, TRIACs can be turned ON by the application of a pulse of current at their gate. Unlike an SCR, provided the current pulses inject sufficient charge, both positive and negative gate current pulses can be used to turn ON a TRIAC.

The use of a TRIAC is illustrated in the following example.

EXAMPLE 10.3

Consider the circuit of Fig. 10.9 in which the SCR of Example 10.2 has been replaced by a TRIAC.

Assume again that a short gate pulse is applied to the SCR at a delay angle α_0 ($0 \le \alpha_0 < \pi$) following each zero-crossing of the source voltage, as shown in Fig. 10.10a. Find the resistor voltage $v_R(t)$ and its rms value $V_{R,\text{rms}}$ as a function of α_0. Assume the TRIAC to be ideal and that the gate pulses inject sufficient charge to properly turn it ON.

■ **Solution**

The solution to this example is similar to that of Example 10.2 with the exception that the TRIAC, which will permit current to flow in both directions, turns on each half cycle of the source-voltage waveform.

$$v_R(t) = \begin{cases} 0 & \text{(prior to the gate pulse)} \\ v_s(t) = V_0 \sin \omega t & \text{(following the gate pulse)} \end{cases}$$

Unlike the rectification of Example 10.2, in this case the resistor voltage, shown in Fig. 10.10b, has no dc component. However, its rms value varies with α_0:

$$V_{R,\text{rms}} = V_0 \sqrt{\left(\frac{\omega}{\pi} \int_{\frac{\alpha_0}{\omega}}^{\frac{\pi}{\omega}} \sin^2 (\omega t) \, dt \right)}$$

$$= V_0 \sqrt{\left(\frac{1}{2} - \frac{\alpha_0}{2\pi} + \frac{1}{4\pi} \sin (2\alpha_0) \right)}$$

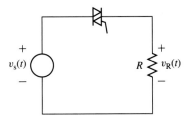

Figure 10.9 Circuit for Example 10.3.

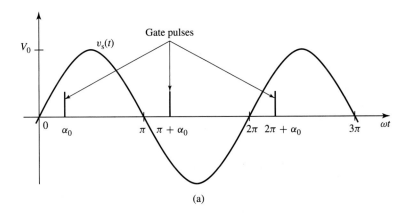

(a)

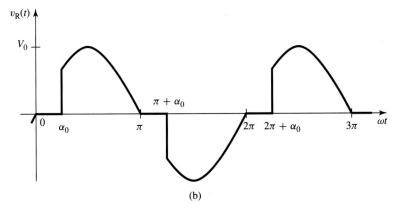

(b)

Figure 10.10 (a) Gate pulses for Example 10.3. (b) Resistor voltage.

Notice that when $\alpha_0 = 0$, the TRIAC is ON all the time and it appears that the resistor is connected directly to the voltage source. In this case, $V_{R,rms} = V_0/\sqrt{2}$ as expected. As α_0 is increased to π, the rms voltage decreases towards zero.

This simple type of controller can be applied to an electric light bulb (in which case it serves as *light dimmer*) as well as to a resistive heater. It is also used to vary the speed of a universal motor and finds widespread application as a speed-controller in small ac hand tools, such as hand drills, as well as in small appliances, such as electric mixers, where continuous speed variation is desired.

Find the rms resistor voltage for the system of Example 10.3 if the sinusoidal source has been replaced by a source of the same frequency but which produces a square wave of zero average value and peak-peak amplitude $2V_0$.

Solution

$$V_{R,rms} = V_0 \sqrt{\left(1 - \frac{\alpha_0}{\pi}\right)}$$

10.1.3 Transistors

For power-electronic circuits where control of voltages and currents is required, power transistors have become a common choice for the controllable switch. Although a number of types are available, we will consider only two: the *metal-oxide-semiconductor field effect transistor (MOSFET)* and the *insulated-gate bipolar transistor (IGBT)*.

MOSFETs and IGBTs are both three-terminal devices. Figure 10.11a shows the symbols for n- and p-channel MOSFETs, while Fig. 10.11b shows the symbol for n- and p-channel IGBTs. In the case of the MOSFET, the three terminals are referred to as the *source, drain,* and *gate,* while in the case of the IGBT the corresponding

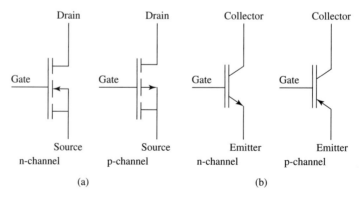

Figure 10.11   (a) Symbols for n- and p-channel MOSFETs. (b) Symbols for n- and p-channel IGBTs.

terminals are the *emitter, collector,* and *gate.* For the MOSFET, the control signal is the gate-source voltage, v_{GS}. For the IGBT, it is the gate-emitter voltage, v_{GE}. In both the MOSFET and the IGBT, the gate electrode is capacitively coupled to the remainder of the device and appears as an open circuit at dc, drawing no current, and drawing only a small capacitive current under ac operation.

Figure 10.12a shows the *v-i* characteristic of a typical n-channel MOSFET. The characteristic of the corresponding p-channel device looks the same, with the exception that the signs of the voltages and the currents are reversed. Thus, in an n-channel device, current flows from the drain to the source when the drain-source and

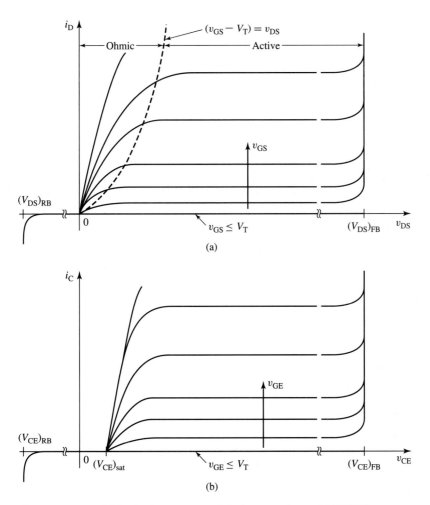

Figure 10.12 (a) Typical *v-i* characteristic for an n-channel MOSFET. (b) Typical *v-i* characteristic for an n-channel IGBT.

gate-source voltages are positive, while in a p-channel device current flows from the source to the drain when the drain-source and gate-source voltages are negative.

Note the following features of the MOSFET and IGBT characteristics:

■ In the case of the MOSFET, for positive drain-source voltage v_{DS}, no drain current will flow for values of v_{GS} less than a *threshold voltage* which we will refer to by the symbol V_T. Once v_{GS} exceeds V_T, the drain current i_D increases as v_{GS} is increased.

  In the case of the IGBT, for positive collector-emitter voltage v_{CE}, no collector current will flow for values of v_{GE} less than a threshold voltage V_T. Once v_{GE} exceeds V_T, the collector current i_C increases as v_{GE} is increased.

■ In the case of the MOSFET, no drain current flows for negative drain-source voltage.

  In the case of the IGBT, no collector current flows for negative collector-emitter voltage.

■ Finally, the MOSFET will fail if the drain-source voltage exceeds its breakdown limits; in Fig. 10.12a, the forward breakdown voltage is indicated by the symbol $(V_{DS})_{FB}$ while the reverse breakdown voltage is indicated by the symbol $(V_{DS})_{RB}$.

  Similarly, the IGBT will fail if the collector-emitter voltage exceeds its breakdown values; in Fig. 10.12b, the forward breakdown voltage is indicated by the symbol $(V_{CE})_{FB}$ while the reverse breakdown voltage is indicated by the symbol $(V_{CE})_{RB}$.

■ Although not shown in the figure, a MOSFET will fail due to excessive gate-source voltage as well as excessive drain current which leads to excessive power dissipation in the device. Similarly an IGBT will fail due to excessive gate-emitter voltage and excessive collector current.

Note that for small values of v_{CE}, the IGBT voltage approaches a constant value, independent of the drain current. This *saturation voltage,* labeled $(V_{CE})_{sat}$ in the figure, is on the order of a volt or less in small devices and a few volts in high-power devices. Correspondingly, in the MOSFET, for small values of v_{DS}, v_{DS} is proportional to the drain current and the MOSFET behaves as a small resistance whose value decreases with increasing v_{GS}.

Fortunately, for our purposes, the details of these characteristics are not important. As we will see in the following example, with a sufficient large gate signal, the voltage drop across both the MOSFET and the IGBT can be made quite small. In this case, these devices can be modeled as a short circuit between the drain and the source in the case of the MOSFET and between the collector and the emitter in the case of the IGBT. Note, however, these "switches" when closed carry only unidirectional current, and hence we will model them as a switch in series with an ideal diode. This *ideal-switch model* is shown in Fig. 10.13a.

In many cases, these devices are commonly protected by reverse-biased protection diodes connected between the drain and the source (in the case of a MOSFET) or between the collector and emitter (in the case of an IGBT). These protection

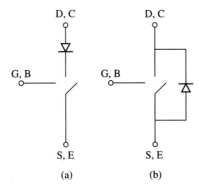

Figure 10.13 (a) Ideal-switch model for a MOSFET or an IGBT showing the series ideal diode which represents the unidirectional-current device characteristic. (b) Ideal-switch model for devices which include a reverse-biased protection diode. The symbols G, D, and S apply to the MOSFET while the symbols B, C, and E apply to the IBGT.

devices are often included as integral components within the device package. If these protection diodes are included, there is actually no need to include the series diode, in which case the model can be reduced to that of Fig. 10.13b.

EXAMPLE 10.4

Consider the circuit of Fig. 10.14a. Here we see an IGBT which is to be used to control the current through the resistor R as supplied from a dc source V_0. Assume that the IGBT characteristics are those of Fig. 10.12b and that V_0 is significantly greater than the saturation voltage. Show a graphical procedure that can be used to find v_{CE} as a function of v_{GE}.

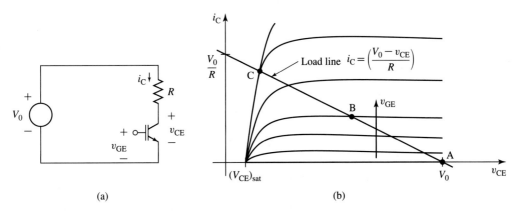

Figure 10.14 (a) Circuit for Example 10.4. (b) IGBT characteristic showing load line and operating point.

■ Solution

Writing KVL for the circuit of Fig. 10.14a gives

$$V_0 = i_C R + v_{CE}$$

Solving for i_C gives

$$i_C = \frac{(V_0 - v_{CE})}{R}$$

Note that this linear relationship, referred to as the *load line,* represents a constraint imposed by the external circuit on the relationship between the IGBT terminal variables i_C and v_{CE}. The corresponding constraint imposed by the IGBT itself is given by the v-i characteristic of Fig. 10.12b.

The operating point of the circuit is that point at which both these constraints are simultaneously satisfied. It can be found most easily by plotting the load line on the v-i relationship of the IGBT. This is done in Fig. 10.14b. The operating point is then found from the intersection of the load line with the v-i characteristic of the IGBT.

Consider the operating point labeled A in Fig. 10.14b. This is the operating point corresponding to values of v_{GE} less than or equal to the threshold voltage V_T. Under these conditions, the IGBT is OFF, there is no collector current, and hence $v_{CE} = V_0$. As v_{GE} is increased past V_T, collector current begins to flow, the operating point begins to climb up the load line, and v_{CE} decreases; the operating point labeled B is a typical example.

Note however that as v_{GE} is further increased, the operating point approaches that portion of the IGBT characteristic for which the curves crowd together (see the operating point labeled C in Fig. 10.14b). Once this point is reached, any further increase in v_{GE} will result in only a minimal decrease in v_{CE}. Under this condition, the voltage across the IGBT is approximately equal to the saturation voltage $(V_{CE})_{sat}$.

If the IGBT of this example were to be replaced by a MOSFET the result would be similar. As the gate-source voltage v_{GS} is increased, a point is reached where the voltage drop across approaches a small constant value. This can be seen by plotting the load line on the MOSFET characteristic of Fig. 10.12a.

The load line intersects the vertical axis at a collector current of $i_C = V_0/R$. Note that the larger the resistance, the lower this intersection and hence the smaller the value of v_{GE} required to saturate the transistor. Thus, in systems where the transistor is to be used as a switch, it is necessary to insure that the device is capable of carrying the required current and that the gate-drive circuit is capable of supplying sufficient drive to the gate.

Example 10.4 shows that when a sufficiently large gate voltage is applied, the voltage drop across a power transistor can be reduced to a small value. Under these conditions, the IGBT will look like a constant voltage while the MOSFET will appear as a small resistance. In either case, the voltage drop will be small, and it is sufficient to approximate it as a closed switch (i.e., the transistor will be ON). When the gate drive is removed (i.e., reduced below V_T), the switch will open and the transistor will turn OFF.

10.2 RECTIFICATION: CONVERSION OF AC TO DC

The power input to many motor-drive systems comes from a constant-voltage, constant-frequency source (e.g., a 50- or 60-Hz power system), while the output must provide variable-voltage and/or variable-frequency power to the motor. Typically such systems convert power in two stages: the input ac is first *rectified* to dc, and the dc is then converted to the desired ac output waveform. We will thus begin with a discussion of rectifier circuits. We will then discuss inverters, which convert dc to ac, in Section 10.3.

10.2.1 Single-Phase, Full-Wave Diode Bridge

Example 10.1 illustrates a half-wave rectifier circuit. Such rectification is typically used only in small, low-cost, low-power applications. Full-wave rectifiers are much more common. Consider the *full-wave rectifier* circuit of Fig. 10.15a. Here the resistor R is supplied from a voltage source $v_s(t) = V_0 \sin \omega t$ through four diodes connected in a *full-wave bridge* configuration.

If we assume the diodes to be ideal, we can use the method-of-assumed states to show that the allowable diode states are:

Diodes D1 and D3 ON, diodes D2 and D4 OFF for $v_s(t) > 0$
Diodes D2 and D4 ON, diodes D1 and D3 OFF for $v_s(t) < 0$

The resistor voltage, plotted in Fig. 10.15b, is then given by

$$v_R(t) = \begin{cases} v_s(t) = V_0 \sin \omega t & v_s(t) \geq 0 \\ -v_s(t) = -V_0 \sin \omega t & v_s(t) < 0 \end{cases} \qquad (10.1)$$

Now notice that the resistor voltage is positive for both polarities of the source voltage, hence the terminology *full-wave rectification*. The dc or average value of this waveform can be seen to be twice that of the half-wave rectified waveform of Example 10.1.

$$V_{dc} = \left(\frac{2}{\pi}\right) V_0 \qquad (10.2)$$

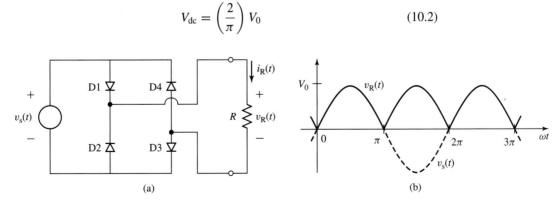

(a) (b)

Figure 10.15 (a) Full-wave bridge rectifier. (b) Resistor voltage.

The rectified waveforms of Figs. 10.3b and 10.15b are clearly not the sort of "dc" waveforms that are considered desirable for most applications. Rather, to be most useful, the rectified dc should be relatively constant and ripple free. Such a waveform can be achieved using a filter capacitor, as illustrated in Example 10.5.

EXAMPLE 10.5

As shown in Fig. 10.16, a filter capacitor has been added in parallel with the load resistor in the full-wave rectifier system of Fig. 10.15. For the purposes of this example, assume that $v_s(t) = V_0 \sin \omega t$ with $V_0 = \sqrt{2} \, (120)$ V, $\omega = (2\pi)60 \approx 377$ rad/sec and that $R = 10 \, \Omega$ and $C = 10^4 \, \mu$F. Plot the resistor voltage, $v_R(t)$, current, $i_R(t)$, and the total bridge current, $i_B(t)$.

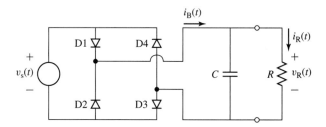

Figure 10.16 Full-wave bridge rectifier with capacitive filter for Example 10.5.

■ **Solution**

The addition of the filter capacitor will tend to maintain the resistor voltage $v_R(t)$ as the source voltage drops. The diodes will remain ON as long as the bridge output current remains positive and will switch OFF when this current starts to reverse.

This example can be readily solved using MATLAB.[†] Figure 10.17a shows the resistor voltage $v_R(t)$ plotted along with the rectified source voltage. During the time that the bridge is ON, i.e., one pair of diodes is conducting, the resistor voltage is equal to the rectified source voltage. When the bridge is OFF, the resistor voltage decays exponentially.

Notice that because the capacitor is relatively large (the RC time constant is 100 msec as compared to the period of the rectified source voltage, which is slightly over 8.3 msec) the diodes conduct only for a short amount of time around the peak of the rectified-source-voltage waveform. This can be readily seen from the expanded plots of the resistor current and the bridge current in Fig. 10.17b. Although the resistor current remains continuous and relatively constant, varying between 15.8 and 17 A, the bridge output current consists essentially of a

---

[†] MATLAB is a registered trademark of The MathWorks, Inc.

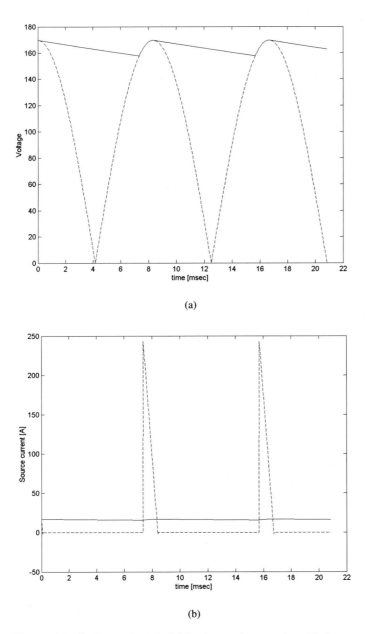

(a)

(b)

Figure 10.17 Example 10.5. (a) Resistor voltage and rectified source voltage (dashed). (b) Resistor current and total bridge current (dashed).

current pulse which flows for less than 0.9 msec near the peak of the rectified voltage waveform and has a peak value of 250 A. It should be pointed out that the peak current in a practical circuit will be smaller than 250 A, being limited by circuit impedances, diode drops, and so on.

Using MATLAB, it is possible to calculate the rms value of the resistor current to be 16.4 A while that of the bridge current is 51.8 A. We see therefore that the bridge diodes in such a system must be rated for rms currents significantly in excess of that of the load. The data sheets for power-supply diodes typically indicate their rms current ratings, specifically with these sorts of applications in mind. Such peaked supply currents are characteristic of rectifier circuits with capacitive loads and can significantly affect the voltage waveforms on ac power systems when they become a significant fraction of the overall system load.

The *ripple voltage* in the resistor voltage is defined as the difference between its maximum and minimum values. In this example, the maximum value is equal to the peak value of the source voltage, or 169.7 V. The minimum value can be found from the MATLAB solution to be 157.8 V. Thus the ripple voltage is 11.9 V. Clearly the ripple voltage can be decreased by increasing the value of the filter capacitor. Note however that this comes at the expense of increased cost as well as shorter current pulses and higher rms current through the rectifier diodes.

Here is the MATLAB script for Example 10.5.

```
clc
clear

%parameters
omega = 2*pi*60;
R = 10;
C = 0.01;
V0 = 120*sqrt(2);
tau = R*C;
Nmax = 800;

% diode = 1 when rectifier bridge is conducting
diode = 1;

%Here is the loop that does the work.
for n = 1:Nmax+1
    t(n) = (2.5*pi/omega)*(n-1)/Nmax; %time
    vs(n) = V0*cos(omega*t(n)); %source voltage
    vrect(n) = abs(vs(n)); %full-wave rectified source voltage

%Calculations if the rectifier bridge is ON
if diode == 1

%If the bridge is ON, the resistor voltage is equal to the rectified
%source voltage.
vR(n) = vrect(n);
%Check the total current out of rectifiers
if (omega*t(n)) <= pi/2.
iB(n) = vR(n)/R - V0*C*omega*sin(omega*t(n));
elseif (omega*t(n)) <= 3.*pi/2.
```

```
iB(n) = vR(n)/R + V0*C*omega*sin(omega*t(n));
else
iB(n) = vR(n)/R - V0*C*omega*sin(omega*t(n));
end

%If the current tries to go negative, the diodes will switch OFF
if iB(n) < 0;
diode = 0;
toff = t(n);
Voff = vrect(n);
end
   else
%When the diodes are off, the resistor/capacitor voltage decays
%exponentially.
vR(n) = Voff*exp(-(t(n)-toff)/tau);
iB(n) = 0;
if (vrect(n) - vR(n)) > 0;
diode = 1;
end
   end
end

%Calculate the resistor current
iR = vR/R;

%Now plot vR as a solid line and vrect as a dashed line
plot(1000*t,vR)
xlabel('time [msec]')
ylabel('Voltage [V]')
axis ([0 22 0 180])
hold
plot(1000*t,vrect,'--')
hold

fprintf('\nHit any key to continue\n')
pause

%Now plot iR as a solid line and iL as a dashed line
plot(1000*t,iR)
xlabel('time [msec]')
ylabel('Source current [A]')
axis ([0 22 -50 250])
hold
plot(1000*t,iB,'--')
hold
```

Use MATLAB to calculate the ripple voltage and rms diode current for the system of Example 10.5 for (*a*) $C = 5 \times 10^4 \ \mu F$ and (*b*) $C = 5 \times 10^3 \ \mu F$. (Hint: Note that the rms current must be calculated over an integral number of cycles of the current waveform).

Solution

a. 2.64 V and 79.6 A rms
b. 21.6 V and 42.8 A rms

In Example 10.5 we have seen that a capacitor can significantly decrease the ripple voltage across a resistive load. However, this comes at the cost of large bridge current pulses since the current must be delivered to the capacitor in the short time period during which the rectified source voltage is near its peak value.

Figure 10.18 shows the addition of an inductor L at the output of the bridge, in series with the filter capacitor and its load. If the impedance of the inductor is chosen to be large compared to that of the capacitor/load combination at the frequency of the rectified source voltage, very little of the ac component of the rectified source voltage will appear across the capacitor, and thus the resultant L-C filter will produce low voltage ripple while drawing a relatively constant current from the diode bridge.

We have seen how the addition of a filter capacitor across a dc load can significantly reduce the ripple voltage seen by the load. In fact, the addition of significant capacitance can "stiffen" the rectified voltage to the point that it appears as a constant-voltage dc source to a load. In an analogous fashion, an inductor in series with a load will reduce the current ripple out of a rectifier. Under these conditions, the rectified source will appear as a constant-current dc source to a load.

The combination of a rectifier and an inductor at the output to supply a constant dc current to a load is of significant importance in power-electronic applications. It can be used, for example, as the front end of a current-source inverter which can be used to supply ac current waveforms to a load. We will investigate the behavior of such rectifier systems in the next section.

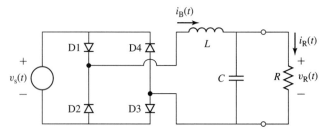

Figure 10.18 Full-wave bridge rectifier with an L-C filter supplying a resistive load.

10.2.2 Single-Phase Rectifier with Inductive Load

In this section we will examine the performance of a single-phase rectifier driving an inductive load. This situation covers both the case where the inductor is included as part of the rectifier system as a filter to smooth out current pulses as well as the case where the load itself is primarily inductive.

Let us examine first the half-wave rectifier circuit of Fig. 10.19. Here, the load consists of an inductor L in series with a resistor R. The source voltage is equal to $v_s(t) = V_0 \cos \omega t$.

Consider first the case where L is small ($\omega L \ll R$). In this case, the load looks essentially resistive and the load current $i_L(t)$ will vary only slightly from the current for a purely resistive load as seen in Example 10.1. This current, obtained from a detailed analytical solution, is plotted in Fig. 10.20a, along with the current for a purely resistive load.

Note that the effect of the inductance is to decrease both the initial rate of rise of the current and the peak current. More significantly, the diode *conduction angle* increases; current flows for longer than the half-period that is the case for a purely resistive load. As can be seen in Fig. 10.20a, this effect increases as the inductance is increased; current flows for a greater fraction of the cycle, and the peak current as well as the current ripple is reduced.

Figure 10.20b, which shows the inductor voltage, illustrates an important point that applies to all situations in which an inductor is subjected to steady-state, periodic conditions: *the time-averaged voltage across the inductor must equal zero.* This can be readily seen from the basic *v-i* relationship for an inductor

$$v = L\frac{di}{dt} \tag{10.3}$$

If we consider the operation of an inductor over a period of the excitation frequency and recognize that, under steady-state conditions, the change in the inductor current over that period must equal zero (i.e., it must have the same value at time t at the beginning of the period as it does one period later at time $t + T$), then we can write

$$i(t + T) - i(T) = 0 = \frac{1}{L}\int_t^{t+T} v \, dt \tag{10.4}$$

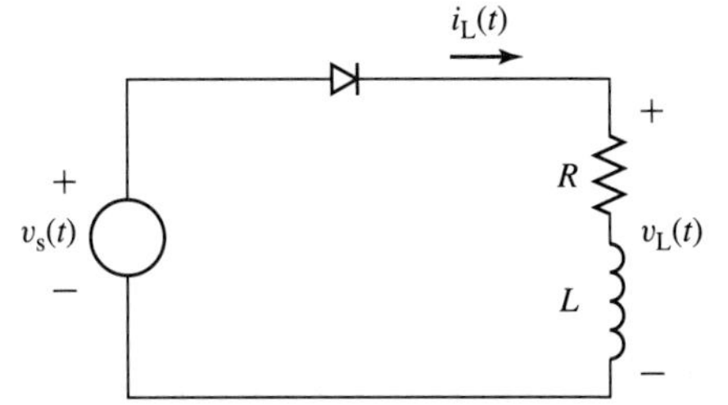

Figure 10.19 Half-wave rectifier with an inductive load.

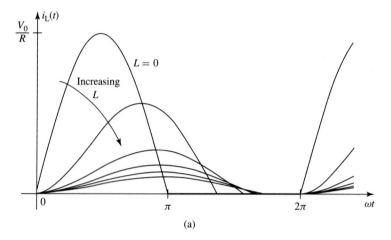

(a)

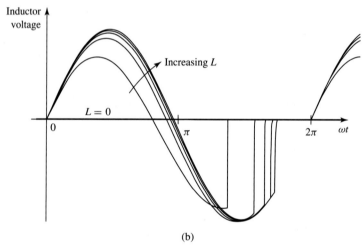

(b)

Figure 10.20   Effect of increasing the series inductance in the circuit of Fig. 10.19 on (a) the load current and (b) the inductor voltage.

from which we can see that the net volt-seconds (and correspondingly the average voltage) across the inductor during a cycle must equal zero

$$\int_t^{t+T} v\, dt = 0 \tag{10.5}$$

For this half-wave rectifier, note that as the inductance increases both the ripple current and the dc current will decrease. In fact, for large inductance ($\omega L \gg R$) the dc load current will tend towards zero. This can be easily seen by the following argument:

> As the inductance increases, the conduction angle of the diode will increase from 180° and approach 360° for large values of L.

> In the limit of a 360° conduction angle, the diode can be replaced by a continuous short circuit, in which case the circuit reduces to the ac voltage source connected directly across the series combination of the resistor and the inductor.

> Under this situation, no dc current will flow since the source is purely ac. In addition, since the impedance $Z = R + j\omega L$ becomes large with large L, the ac (ripple) current will also tend to zero.

Figure 10.21a shows a simple modification which can be made to the half-wave rectifier circuit of Fig. 10.19. The *free-wheeling diode* D2 serves as an alternate path for inductor current.

To understand the behavior of this circuit, consider the condition when the source voltage is positive, and the rectifier diode D1 is ON. The equivalent circuit for this operating condition is shown in Fig. 10.21b. Note that under this condition, the voltage across diode D2 is equal to the negative of the source voltage and diode D2 is turned OFF.

This operating condition will remain in effect as long as the source voltage is positive. However, as soon as the source voltage begins to go negative, the voltage across diode D2 will begin to go positive and it will turn ON. Since the load is

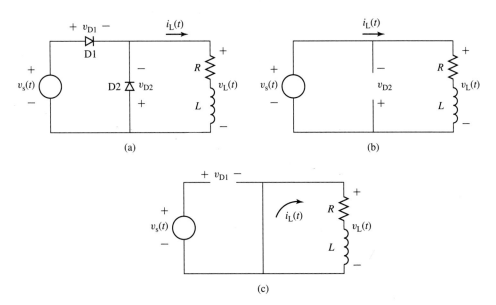

Figure 10.21 (a) Half-wave rectifier with an inductive load and a free-wheeling diode. (b) Equivalent circuit when $v_s(t) > 0$ and diode D1 is conducting. (c) Equivalent circuit when $v_s(t) < 0$ and the free-wheeling diode D2 is conducting.

inductive, a positive load current will be flowing at this time, and that load current will immediately transfer to the short circuit corresponding to diode D2. At the same time, the current through diode D1 will immediately drop to zero, diode D1 will be reverse biased by the source voltage, and it will turn OFF. This operating condition is shown in Fig. 10.21c. Thus, the diodes in this circuit alternately switch ON and OFF each half cycle: D1 is ON when $v_s(t)$ is positive, and D2 is ON when it is negative.

Based upon this discussion, we see that the voltage $v_L(t)$ across the load (equal to the negative of the voltage across diode D2) is a half-wave rectified version of $v_s(t)$ as seen in Fig. 10.22a. As shown in Example 10.1, the average of this voltage is $V_{dc} = V_0/\pi$. Furthermore, the average of the steady-state voltage across the inductor must equal zero, and hence the average of the voltage $v_L(t)$ will appear across the resistor. Thus the dc load current will equal $V_0/(\pi R)$. This value is independent of the inductor value and hence does not approach zero as the inductance is increased.

Figure 10.22b shows the diode and load currents for a relatively small value of inductance $(\omega L < R)$, and Fig. 10.22c shows these same currents for a large inductance $\omega L \gg R$. In each case we see the load current, which must be continuous due to the presence of the inductor, instantaneously switching between the diodes depending on the polarity of the source voltage. We also see that during the time diode D1 is ON, the load current increases due to the application of the sinusoidal source voltage, while during the time diode D2 is ON, the load current simply decays with the L/R time constant of the load itself.

As expected, in each case the average current through the load is equal to $V_0/(\pi R)$. In fact, the presence of a large inductor can be seen to reduce the ripple current to the point that the load current is essentially a dc current equal to this value.

Let us now consider the case where the half-wave bridge of Fig. 10.19 is replaced by a full-wave bridge as in Fig. 10.23a. In this circuit, the voltage applied to the load is the full-wave-rectified source voltage as shown in Fig. 10.15 and the average (dc) voltage applied to the load will equal $2V_0/\pi$. Here again, the presence of the inductor will tend to reduce the ac ripple. Figure 10.24, again obtained from a detailed analytical solution, shows the current through the resistor as the inductance is increased.

If we assume a large inductor $(\omega L \gg R)$, the load current will be relatively ripple free and constant. It is therefore common practice to analyze the performance of this circuit by replacing the inductor by a dc current source I_{dc} as shown in Fig. 10.23b, where $I_{dc} = 2V_0/(\pi R)$. This is a commonly used technique in the analysis of power-electronic circuits which greatly simplifies their analysis.

Under this assumption, we can easily show that the diode and source currents of this circuit are given by waveforms of Fig. 10.25. Figure 10.25a shows the current through one pair of diodes (e.g., diodes D1 and D3), and Fig. 10.25b shows the source current $i_s(t)$. The essentially constant load current I_{dc} flows through each pair of diodes for one half cycle and appears as a square wave of amplitude I_{dc} at the source.

In a fashion similar to that which we saw in the half-wave rectifier circuit with the free-wheeling diode, here a pair of diodes (e.g., diodes D1 and D3) are carrying current when the source voltage reverses, turning ON the other pair of diodes and switching OFF the pair that were previously conducting. In this fashion, the load current remains continuous and simply switches between the diode pairs.

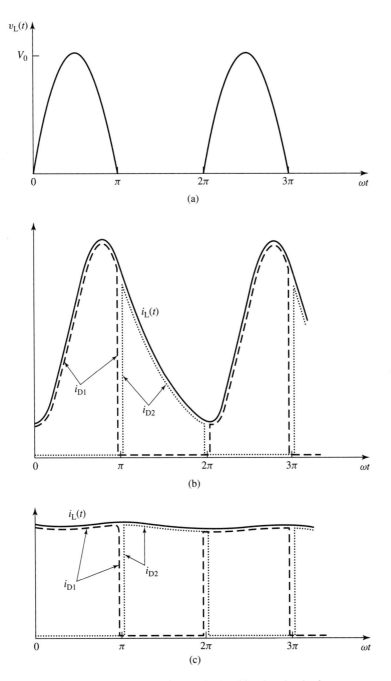

Figure 10.22 (a) Voltage applied to the load by the circuit of
Fig. 10.21. (b) Load and diode currents for small *L*. (c) Load and diode
currents for large *L*.

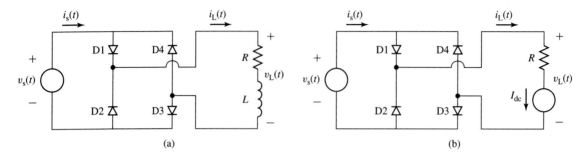

Figure 10.23   (a) Full-wave rectifier with an inductive load. (b) Full-wave rectifier with the inductor replaced by a dc current source.

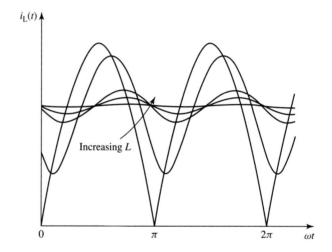

Figure 10.24   Effect of increasing the series inductance in the circuit of Fig. 10.23a on the load current.

10.2.3   Effects of Commutating Inductance

Our analysis and the current waveforms of Fig. 10.25 show that the current commutes instantaneously from one diode pair to the next. In practical circuits, due to the presence of source inductance, *commutation* of the current between the diode pairs does not occur instantaneously. The effect of source inductance, typically referred to as *commutating inductance,* will be examined by studying the circuit of Fig. 10.26 in which a source inductance L_s has been added in series with the voltage source in the full-wave rectifier circuit of Fig. 10.23b. We have again assumed that the load time constant is large ($\omega L/R \gg 1$) and have replaced the inductor with a dc current source I_{dc}.

Figure 10.27a shows the situation which occurs when diodes D2 and D4 are ON and carrying current I_{dc} and when $v_s < 0$. Commutation begins when v_s reaches zero and begins to go positive, turning ON diodes D1 and D3. Note that *because the current in the source inductance L_s cannot change instantaneously,* the circuit condition at

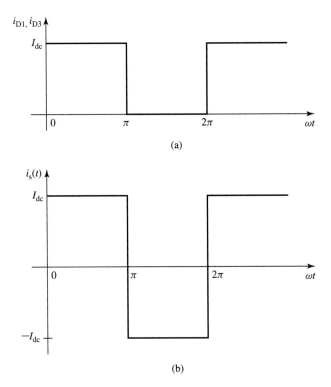

(a)

(b)

Figure 10.25 (a) Current through diodes D1 and D3 and (b) source current for the circuit of Fig. 10.23, both in the limit of large inductance.

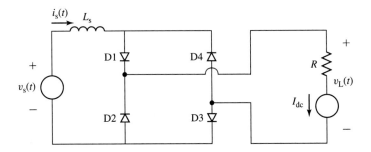

Figure 10.26 Full-wave bridge rectifier with source inductance. Dc load current assumed.

this time is described by Fig. 10.27b: the current through L_s is equal to $-I_{dc}$, the current through diodes D2 and D4 is equal to I_{dc}, while the current through diodes D1 and D3 is zero.

Under this condition with all four diodes ON, the source voltage $v_s(t)$ appears directly across the source inductance L_s. Noting that commutation starts at the time

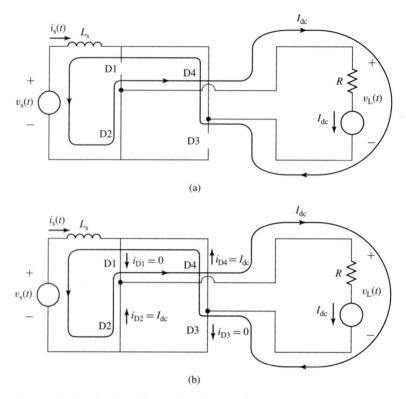

Figure 10.27   (a) Condition of the full-wave circuit of Fig. 10.26 immediately before diodes D1 and D3 turn ON. (b) Condition immediately after diodes D1 and D3 turn ON.

when $v_s(t) = 0$, the current through L_s can be written as

$$i_s(t) = -I_{dc} + \frac{1}{L_s} \int_0^t v_s(t)\, dt$$

$$= -I_{dc} + \left(\frac{V_0}{\omega L_s}\right) (1 - \cos \omega t) \tag{10.6}$$

Noting that $i_s = i_{D1} - i_{D4}$, that $i_{D1} + i_{D2} = I_{dc}$ and, that by symmetry, $i_{D4} = i_{D2}$, we can write that

$$i_{D2} = \frac{I_{dc} - i_s(t)}{2} \tag{10.7}$$

Diode D2 (and similarly diode D4) will turn OFF when i_{D2} reaches zero, which will occur when $i_s(t) = I_{dc}$. In other words, commutation will be completed at time t_c, when the current through L_s has completely reversed polarity and when all of the load current is flowing through diodes D1 and D3.

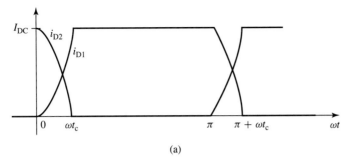

(a)

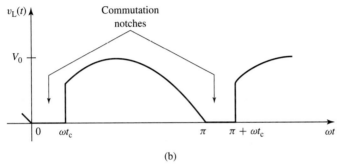

(b)

Figure 10.28 (a) Currents through diodes D1 and D2 showing the finite commutation interval. (b) Load voltage showing the commutation notches due to the finite commutation time.

Setting $i_s(t_c) = I_{dc}$ and solving Eq. 10.6 gives an expression for the *commutation interval* t_c as a function of I_{dc}

$$t_c = \frac{1}{\omega} \cos^{-1}\left[1 - \left(\frac{2I_{dc}\omega L_s}{V_0}\right)\right] \tag{10.8}$$

Figure 10.28a shows the currents through diodes D1 and D2 as the current commutates between them. The finite commutation time t_c can clearly be seen. There is a second effect of commutation which can be clearly seen in Fig. 10.28b which shows the rectified load voltage $v_L(t)$. Note that during the time of commutation, with all the diodes on, the rectified load voltage is zero. These intervals of zero voltage on the rectified voltage waveform are known as *commutation notches*.

Comparing the ideal full-wave rectified voltage of Fig. 10.15b to the waveform of Fig. 10.28b, we see that the effect of the commutation notches is to reduce the dc output of the rectifier. Specifically, the dc voltage in this case is given by

$$V_{dc} = \left(\frac{\omega}{\pi}\right) \int_{t_c}^{\frac{\pi}{\omega}} V_0 \sin \omega t \, dt$$

$$= \frac{V_0}{\pi} (1 + \cos \omega t_c) \tag{10.9}$$

where t_c is the commutation interval as calculated by Eq. 10.8.

Finally, the dc load current can be calculated as function of t_c

$$I_{dc} = \frac{V_{dc}}{R} = \frac{V_0}{\pi R}(1 + \cos \omega t_c) \qquad (10.10)$$

Substituting Eq. 10.8 into Eq. 10.10 gives a closed-form solution for I_{dc}

$$I_{dc} = \frac{2V_0}{\pi R + 2\omega L_s} \qquad (10.11)$$

and hence

$$V_{dc} = I_{dc}R = \frac{2V_0}{\pi + \frac{2\omega L_s}{R}} \qquad (10.12)$$

We have seen that commutating inductance (which is to a great extent unavoidable in practical circuits) gives rise to a finite commutation time and produces commutation notches in the rectified-voltage waveform which reduces the dc voltage applied to the load.

EXAMPLE 10.6

Consider a full-wave rectifier driving an inductive load as shown in Fig. 10.29. For a 60-Hz, 230-V rms source voltage, $R = 5.6\ \Omega$ and large L ($\omega L \gg R$), plot the dc current through the load I_{dc} and the commutation time t_c as the source inductance L_s varies from 1 to 100 mH.

■ Solution

The solution can be obtained by substitution into Eqs. 10.8 and 10.11. This is easily done using MATLAB, and the plots of I_{dc} and t_c are shown in Figs. 10.30a and b respectively. Note that the maximum achievable dc current, corresponding to $L_s = 0$, is equal to $2V_0/(\pi R) = 37$ A. Thus, commutating inductances on the order of 1 mH can be seen to have little effect on the performance of the rectifier and can be ignored. On the other hand, a commutating inductance of 100 mH can be seen to reduce the dc current to approximately 7 A, significantly reducing the capability of the rectifier circuit.

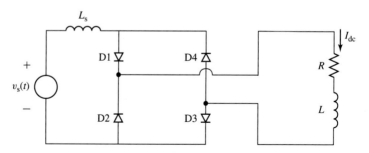

Figure 10.29   Full-wave bridge rectifier with source inductance for Example 10.6.

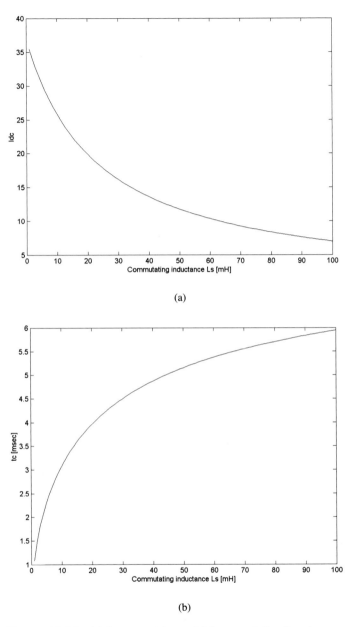

(a)

(b)

Figure 10.30   (a) Dc current I_{dc} and (b) commutation time t_c for Example 10.6.

Here is the MATLAB script for Example 10.6.

```
clc
clear

%parameters
omega = 2*pi*60;
R = 5.6;
V0 = 230*sqrt(2);

for n = 1:100
    Ls(n) = n*1e-3;
    Idc(n) = 2*V0/(pi*R + 2*omega*Ls(n));
    tc(n) = (1/omega)*acos(1-(2*Idc(n)*omega*Ls(n))/V0);
end

plot(Ls*1000,Idc)
xlabel('Commutating inductance Ls [mH]')
ylabel('Idc')

fprintf('\nHit any key to continue\n')
pause

plot(Ls*1000,tc*1000)
xlabel('Commutating inductance Ls [mH]')
ylabel('tc [msec]')
```

Practice Problem 10.5

Calculate the commutating inductance and the corresponding commutation time for the circuit of Example 10.6 if the dc load current is observed to be 29.7 A.

Solution

$L_s = 5.7$ mH and $t_c = 2.4$ msec

10.2.4 Single-Phase, Full-Wave, Phase-Controlled Bridge

Figure 10.31 shows a full-wave bridge in which the diodes of Fig. 10.15 have been replaced by SCRs. We will assume that the load inductance L is sufficiently large that the load current is essentially constant at a dc value I_{dc}. We will also ignore any effects of commutating inductance, although they clearly would play the same role in a phase-controlled rectifier system as they do in a diode-rectifier system.

Figure 10.32 shows the source voltage and the timing of the SCR gate pulses under a typical operating condition for this circuit. Here we see that the firing pulses are delayed by an angle α_d after the zero-crossing of the source-voltage waveform, with the firing pulses for SCRs T1 and T3 occuring after the positive-going transition of $v_s(t)$ and those for SCRs T2 and T4 occuring after the negative-going transition.

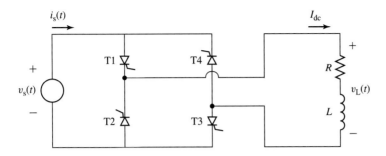

Figure 10.31 Full-wave, phase-controlled SCR bridge.

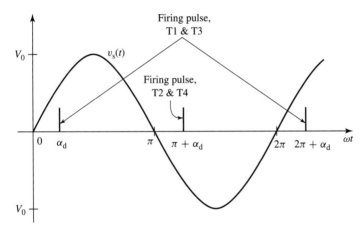

Figure 10.32 Source voltage and firing pulses for the phase-controlled SCR bridge of Fig. 10.31.

Figure 10.33a shows the current through SCRs T1 and T3. Note that these SCRs do not turn ON until they receive firing pulses at angle α_d after they are forward biased following positive-going zero crossing of the source voltage. Furthermore, note that SCRs T2 and T4 do not turn ON following the next zero crossing of the source voltage. Hence, SCRs T1 and T3 remain ON, carrying current until SCRs T2 and T4 are turned ON by gate pulses. Rather, T2 and T4 turn ON only after they receive their respective gate pulses (for example, at angle $\pi + \alpha_d$ in Fig. 10.33). This is an example of *forced commutation,* in that one pair of SCRs does not naturally commutate OFF but rather must be forcibly commutated when the other pair is turned ON.

Figure 10.33b shows the resultant load voltage $v_L(t)$. We see that the load voltage now has a negative component, which will increase as the firing-delay angle α_d is increased. The dc value of this waveform is equal to

$$V_{dc} = \left(\frac{2V_0}{\pi}\right) \cos \alpha_d \quad (0 \leq \alpha_d \leq \pi) \qquad (10.13)$$

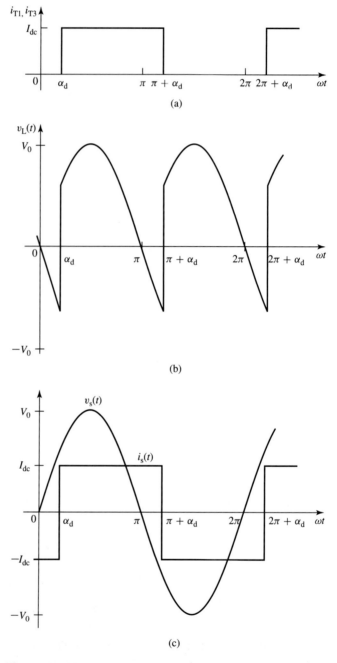

Figure 10.33 Waveforms for the phase-controlled SCR bridge of Fig. 10.31. (a) Current through SCRs T1 and T3. (b) Load voltage. (c) Source voltage and current.

from which the firing-delay angle corresponding to a given value of dc voltage can be seen to be

$$\alpha_d = \cos^{-1}\left(\frac{\pi V_{dc}}{2V_0}\right) \tag{10.14}$$

From Eq. 10.13 we see that the dc voltage applied to the load can vary from $2V_0/\pi$ to $-2V_0/\pi$. This is a rather surprising result in that it is hard to understand how a rectifier bridge can supply negative voltage. However, in this case, it is necessary to recognize that this result applies to an inductive load which maintains positive current flow through the SCRs in spite of the reversal of polarity of the source voltage. If the load were purely resistive, the current through the conducting SCRs would go to zero as the source-voltage reversed polarity, and they would simply turn OFF; no load current would flow until the next pair of SCRs is turned ON.

Figure 10.33c shows the source voltage and current waveforms for the phase-controlled SCR-bridge. We see that the square-wave source current is out of phase with the source voltage. Its fundamental-harmonic is given by

$$i_{s,1}(t) = \left(\frac{4}{\pi}\right) I_{dc} \sin(\omega t - \alpha_d) \tag{10.15}$$

and thus the real power supplied to the load is given by

$$P = V_{dc}I_{dc} = \frac{2}{\pi}V_0 I_{dc}\cos\alpha_d \tag{10.16}$$

and the reactive power supplied is

$$Q = -\frac{2}{\pi}V_0 I_{dc}\sin\alpha_d \tag{10.17}$$

Under steady-state operation at a load current I_{dc}, $V_{dc} = I_{dc}R$ and the steady-state firing-delay angle can be found from Eq. 10.14 to be $\alpha_{ss} = \cos^{-1}\left(\frac{\pi I_{dc}R}{2V_0}\right)$. Under this condition, the real power simply supplies the losses in the resistor and hence $P = I_{dc}^2 R$. It may seem strange to be supplying reactive power to a "dc" load. However, careful analysis will show that this reactive power supplies the energy associated with the small but finite ripple current through the inductor.

If the delay angle is suddenly reduced ($\alpha_d < \alpha_{ss}$), the dc voltage applied to the load will increase (see Eq. 10.13) as will the power supplied to the load (see Eq. 10.16). As a result, I_{dc} will begin to increase and the increased power will increase the energy storage in the inductor. Similarly, if the delay time is suddenly increased ($t_d > t_{d0}$), V_{dc} will decrease (it may even go negative) and the power into the load will decrease, corresponding to a decrease in I_{dc} and a decrease in the energy storage in the inductor.

Note that if $\alpha_d > \pi/2$, V_{dc} will be negative, a condition which will continue until I_{dc} reaches zero at which time the SCR bridge will turn OFF. Under this condition, the real power P will also be negative. Under this condition, power is being supplied from the load to the source and the system is said to be *regenerating*.

EXAMPLE 10.7

A small superconducting magnet has an inductance $L = 1.2$ H. Although the resistance of the magnet itself is essentially zero, the resistance of the external leads is 12.5 mΩ. Current is supplied to the magnet from a 60-Hz, 15-V peak, single-phase source through a phase-controlled SCR bridge as in Fig. 10.31.

a. The magnet is initially operating in the steady state at a dc current of 35 A. Calculate the dc voltage applied to the magnet, the power supplied to the magnet, and the delay angle α_d in msec. Plot the magnet voltage $v_L(t)$.

b. In order to quickly discharge the magnet, the delay angle is suddenly increased to $\alpha_d = 0.9\pi = 162°$. Plot the corresponding magnet voltage. Calculate the time required to discharge the magnet and the maximum power regenerated to the source.

■ **Solution**

The example is most easily solved using MATLAB, which can easily produce the required plots.

a. Under this steady-state condition, $V_{dc} = I_{dc}R = 35 \times 0.0125 = 0.438$ V. The power supplied to the magnet is equal to $P = V_{dc}I_{dc} = 0.438 \times 35 = 15.3$ W, all of which is going into supplying losses in the lead resistance. The delay angle can be found from Eq. 10.14.

$$\alpha_d = \cos^{-1}\left(\frac{\pi R I_{dc}}{2V_0}\right) = \cos^{-1}\left(\frac{\pi \times 0.0125 \times 35}{2 \times 15}\right)$$

$$= 1.52 \text{ rad} = 87.4°$$

A plot of $v_L(t)$ for this condition is given in Fig. 10.34a.

b. For a delay angle of 0.9π, the dc load voltage will be

$$V_{dc} = \left(\frac{2V_0}{\pi}\right)\cos\alpha_d = \left(\frac{2 \times 15}{\pi}\right)\cos(0.9\pi) = -9.1 \text{ V}$$

A plot of $v_L(t)$ for this condition is given in Fig. 10.34b.

The magnet current i_m can be calculated from the differential equation

$$V_{dc} = i_m R + L\frac{di_m}{dt}$$

subject to the initial condition that $i_m(0) = 35$ A. Thus

$$i_m(t) = \frac{V_{dc}}{R} + \left(i_m(0) - \frac{V_{dc}}{R}\right)e^{-\left(\frac{R}{L}\right)t}$$

From this equation we find that the magnet current will reach zero at time $t = 4.5$ seconds, at which time the bridge will shut off. The power regenerated to the source will be given by $-V_{dc}i_m(t)$. It has a maximum value of $9.1 \times 35 = 318$ W at time $t = 0$.

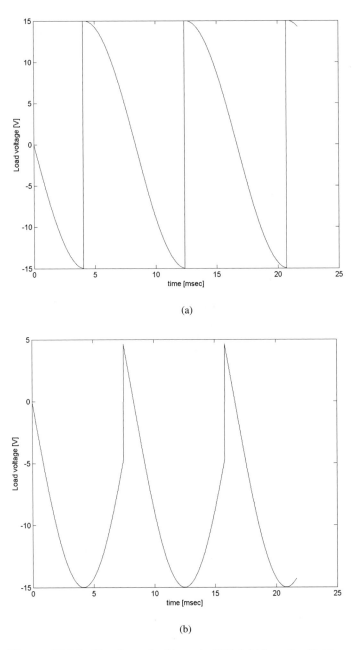

(a)

(b)

Figure 10.34 Waveforms for Example 10.7. (a) Magnet-voltage for $\alpha_d = 87.4°$, $V_{dc} = 0.438$ V. (b) Magnet-voltage for $\alpha_d = 162°$, $V_{dc} = -9.1$ V.

Here is the MATLAB script for Example 10.7.

```
clc
clear

% system parameters
R = 12.5e-3;
L = 1.2;
V0 = 15;
omega = 120*pi;

% part (a)

% dc current
Idc = 35;

% dc voltage
Vdc_a = R*Idc;

% Power
P = Vdc_a*Idc;

%Calculate the delay angle
alpha_da = acos(pi*R*Idc/(2*V0));

%Now calculate the load voltage
for n = 1:1300
   theta(n) = 2*pi*(n-1)/1000;
   t(n) = theta(n)/omega;
   vs(n) = V0*sin(theta(n));

   if theta(n) < alpha_da
vL(n) = -vs(n);
   elseif (theta(n) < pi + alpha_da)
vL(n) = vs(n);
   elseif theta(n) < 2*pi + alpha_da
vL(n) = -vs(n);
   elseif theta(n) < 3*pi + alpha_da
vL(n) = vs(n);
   elseif theta(n) < 4*pi + alpha_da
vL(n) = -vs(n);
   else
vL(n) = vs(n);
   end
end

plot(1000*t,vL)
xlabel('time [msec]')
ylabel('Load voltage [V]')

pause

% part (b)
```

```
% delay angle
alpha_db = 0.9*pi;

% Find the new dc voltage
Vdc_b = (2*V0/pi)*cos(alpha_db);

% Time constants
tau = L/R;

% Initial current
im0 = Idc;

% Calculate the time at which the current reaches zero
tzero = -tau*log((-Vdc_b/R)/(im0-Vdc_b/R));

% Now plot the load voltage
for n = 1:1300
    theta(n) = 2*pi*(n-1)/1000;
    t(n) = theta(n)/omega;
    vs(n) = V0*sin(theta(n));

    if theta(n) < alpha_db
vL(n) = -vs(n);
    elseif (theta(n) < pi + alpha_db)
vL(n) = vs(n);
    elseif theta(n) < 2*pi + alpha_db
vL(n) = -vs(n);
    elseif theta(n) < 3*pi + alpha_db
vL(n) = vs(n);
    elseif theta(n) < 4*pi + alpha_db
vL(n) = -vs(n);
    else
vL(n) = vs(n);
    end
end

plot(1000*t,vL)
xlabel('time [msec]')
ylabel('Load voltage [V]')

fprintf('part (a):')
fprintf('\n Vdc_a = %g [mV]',1000*Vdc_a)
fprintf('\n Power = %g [W]',P);
fprintf('\n alpha_d = %g [rad] = %g [degrees]',alpha_da,180*alpha_da/pi)
fprintf('\npart (b):')
fprintf('\n alpha_d = %g [rad] = %g [degrees]',alpha_db,180*alpha_db/pi)
fprintf('\n Vdc_b = %g [V]',Vdc_b)
fprintf('\n Current will reach zero at %g [sec]',tzero)
fprintf('\n')
```

The field winding of a small synchronous generator has a resistance of 0.3 Ω and an inductance of 250 mH. It is fed from a 24-V peak, single-phase 60-Hz source through a full-wave phase-controlled SCR bridge. (*a*) Calculate the dc voltage required to achieve a dc current of 18 A through the field winding and the corresponding firing-delay angle. (*b*) Calculate the field current corresponding to a delay angle of 45°.

Solution

a. 5.4 V, 69°
b. 36.0 A

10.2.5 Inductive Load with a Series DC Source

As we have seen in Chapter 9, dc motors can be modeled as dc voltage sources in series with an inductor and a resistor. Thus, it would be useful to briefly investigate the case of a dc voltage source in series with an inductive load.

Let us examine the full-wave, phase-controlled SCR rectifier system of Fig. 10.35. Here we have added a dc source E_{dc} in series with the load. Again assuming that $\omega L \gg R$ so that the load current is essentially dc, we see that the load voltage $v_L(t)$ depends solely on the timing of the SCR gate pulses and hence is unchanged by the presence of the dc voltage source E_{dc}. Thus the dc value of $v_L(t)$ is given by Eq. 10.13 as before.

In the steady-state, the dc current through the load can be found from the net dc voltage across the resistor as

$$I_{dc} = \frac{V_{dc} - E_{dc}}{R} \quad (V_{dc} \geq E_{dc}) \tag{10.18}$$

where V_{dc} is found from Eq. 10.13. Under transient conditions, it is the difference voltage, $V_{dc} - E_{dc}$, that drives a change in the dc current through the series R-L combination, in a fashion similar to that illustrated in Example 10.7.

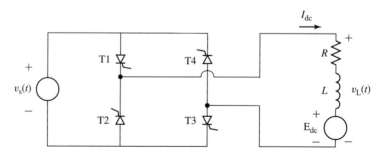

Figure 10.35 Full-wave, phase-controlled SCR bridge with an inductive load including a dc voltage source.

EXAMPLE 10.8

A small permanent-magnet dc motor is to be operated from a phase-controlled bridge. The 60-Hz ac waveform has an rms voltage of 35 volts. The dc motor has an armature resistance of 3.5 Ω and an armature inductance of 17.5 mH. It achieves a no-load speed of 8000 r/min at an armature voltage of 50 V.

Calculate the no-load speed in r/min of the motor as a function of the firing delay angle α_d.

■ Solution

In Section 7.7 we see that the equivalent circuit for a permanent-magnet dc motor consists of a dc source (proportional to motor speed) in series with an inductance and a resistance. Thus, the equivalent circuit of Fig. 10.35 applies directly to the situation of this problem.

From Eq. 7.26, the generated voltage from the dc motor (E_{dc} in Fig. 10.35) is proportional to the speed of the dc motor. Thus,

$$n = \left(\frac{8000}{50}\right) E_{dc} = 160 E_{dc} \text{ r/min}$$

Under steady state operation, the dc voltage drop across the armature inductance will be zero. In addition, at no load, the armature current will be sufficiently small that the voltage drop across the armature resistance can be neglected. Thus, setting $E_{dc} = V_{dc}$ and substituting the expression for V_{dc} from Eq. 10.13 give,

$$E_{dc} = V_{dc} = \left(\frac{2V_0}{\pi}\right) \cos \alpha_d$$

$$= \left(\frac{2\sqrt{2}\,35}{\pi}\right) \cos \alpha_d = 31.5 \cos \alpha_d$$

Note that because the bridge can only supply positive current to the dc motor (and hence, in the steady-state, the dc voltage must be positive), this expression is valid only for $0 \le \alpha_d \le \pi/2$.

Finally, substituting the expression for the speed n in terms of E_{dc} gives

$$n = 160 \times (31.5 \cos \alpha_d) = 5040 \cos \alpha_d \text{ r/min} \quad (0 \le \alpha_d \le \pi/2)$$

The dc-motor of Example 10.8 is observed to be operating at a speed of 3530 r/min and drawing a dc current of 1.75 ampere. Calculate the corresponding firing delay angle α_d.

Solution

$$\alpha_d = 0.15\pi \text{ rad} = 27°$$

10.2.6 Three-Phase Bridges

Although many systems with ratings ranging up to five or more kilowatts run off single-phase power, most large systems are supplied from three-phase sources. In general, all of the issues which we have discussed with regard to single-phase,

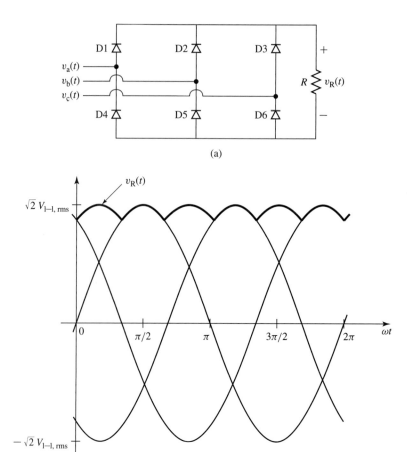

Figure 10.36   (a) Three-phase, six-pulse diode bridge with resistive load.
(b) Line-to-line voltages and resistor voltage.

full-wave bridges apply directly to situations with three-phase bridges. As a result, we will discuss three-phase bridges only briefly.

Figure 10.36a shows a system in which a resistor R is supplied from a three-phase source through a *three-phase, six-pulse diode bridge*. Figure 10.36b shows the three-phase line-to-line voltages (peak value $\sqrt{2}V_{l-l,\mathrm{rms}}$ where $V_{l-l,\mathrm{rms}}$ is the rms value of the line-to-line voltage) and the resistor voltage $v_R(t)$, found using the method-of-assumed-states and assuming that the diodes are ideal.

Note that v_R has six pulses per cycle. Unlike the single-phase, full-wave bridge of Fig. 10.15a, the resistor voltage does not go to zero. Rather, the three-phase diode bridge produces an output voltage equal to the instantaneous maximum of the absolute value of the three line-to-line voltages. The dc average of this voltage (which can be obtained by integrating over 1/6 of a cycle) is given by

$$
V_{dc} = \frac{3\omega}{\pi} \int_0^{\frac{\pi}{3\omega}} -v_{bc}(t) \, dt
$$

$$
= -\frac{3\omega}{\pi} \int_0^{\frac{\pi}{3\omega}} \sqrt{2} V_{l-l,rms} \sin\left(\omega t - \frac{2\pi}{3}\right) dt
$$

$$
= \left(\frac{3\sqrt{2}}{\pi}\right) V_{l-l,rms} \tag{10.19}
$$

where $V_{l-l,rms}$ is the rms value of the line-to-line voltage.

Table 10.1 shows the diode-switching sequence for the three-phase bridge of Fig. 10.36a corresponding to a single period of the three-phase voltage of waveforms of Fig. 10.36b. Note that only two diodes are on at any given time and that each diode is on for 1/3 of a cycle (120°).

Analogous to the single-phase, full-wave, phase-controlled SCR bridge of Figs. 10.31 and 10.35, Fig. 10.37 shows a three-phase, phase-controlled SCR bridge. Assuming continuous load current, corresponding for example to the condition $\omega L \gg R$, in which case the load current will be essentially a constant dc current I_{dc}, this bridge is capable of applying a negative voltage to the load and of regenerating power in a fashion directly analogous to the single-phase, full-wave, phase-controlled SCR bridge which we discussed in Section 10.2.4.

Table 10.1 Diode conduction times for the three-phase diode bridge of Fig. 10.36a.

α_d	$0-\pi/3$	$\pi/3-2\pi/3$	$2\pi/3-\pi$	$\pi-4\pi/3$	$4\pi/3-5\pi/3$	$5\pi/3-2\pi$
D1	OFF	ON	ON	OFF	OFF	OFF
D2	OFF	OFF	OFF	ON	ON	OFF
D3	ON	OFF	OFF	OFF	OFF	ON
D4	OFF	OFF	OFF	OFF	ON	ON
D5	ON	ON	OFF	OFF	OFF	OFF
D6	OFF	OFF	ON	ON	OFF	OFF

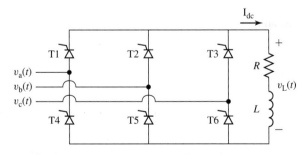

Figure 10.37 Three-phase, phase-controlled SCR bridge circuit with an inductive load.

It is a relatively straightforward matter to show that maximum output voltage of this bridge configuration will occur when the SCRs are turned ON at the times when the diodes in a diode bridge would naturally turn ON. These times can be found from Table 10.1. For example, we see that SCR T5 must be turned ON at angle $\alpha_d = 0$ (i.e., at the positive zero crossing of $v_{ab}(t)$). Similarly, SCR T1 must be turned ON at time $\alpha_d = \pi/3$, and so on.

Thus, one possible scheme for generating the SCR gate pulses is to use the positive-going zero crossing of $v_{ab}(t)$ as a reference from which to synchronize a pulse train running at six times the fundamental frequency (i.e., there will be six uniformly-spaced pulses in each cycle of the applied voltage). SCR T5 would be fired first, followed by SCRs T1, T6, T2, T4, and T3 in that order, each separated by 60° in phase delay.

If the firing pulses are timed to begin immediately following the zero crossing of $v_{ab}(t)$ the load voltage waveform $v_L(t)$ will be that of Fig. 10.36b. If the firing pulses are delayed by an angle α_d, then the load-voltage waveforms will appear as in Fig. 10.38a (for $\alpha_d = 0.1\pi$) and Fig. 10.38b (for $\alpha_d = 0.9\pi$).

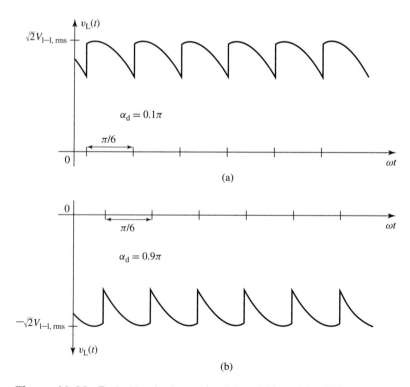

Figure 10.38 Typical load voltages for delayed firing of the SCRs in the three-phase, phase-controlled rectifier of Fig. 10.37; (a) $\alpha_d = 0.1\pi$, (b) $\alpha_d = 0.9\pi$.

The dc average of the output voltage of the phase-controlled bridge can be found as

$$V_{dc} = \frac{3\omega}{\pi} \int_{\frac{\alpha_d}{\omega}}^{\frac{\alpha_d+\pi/3}{\omega}} -v_{bc}(t)\, dt$$

$$= -\frac{3\omega}{\pi} \int_{\frac{\alpha_d}{\omega}}^{\frac{\alpha_d+\pi/3}{\omega}} \sqrt{2}\, V_{l-l,rms} \sin\left(\omega t - \frac{2\pi}{3}\right) dt$$

$$= \left(\frac{3\sqrt{2}}{\pi}\right) V_{l-l,rms} \cos\alpha_d \quad (0 \le \alpha_d \le \pi) \tag{10.20}$$

where $V_{l-l,rms}$ is the rms value of the line-to-line voltage.

EXAMPLE 10.9

A large magnet with an inductance of 14.7 H and resistance of 68 Ω is to be supplied from a 60-Hz, 460-V, three-phase source through a phase-controlled SCR bridge as in Fig. 10.37. Calculate (a) the maximum dc voltage $V_{dc,max}$ and current $I_{dc,max}$ which can be supplied from this source and (b) the delay angle α_d required to achieve a magnet current of 2.5 A.

■ **Solution**

a. From Eq. 10.20, the maximum voltage (corresponding to $\alpha_d = 0$) is equal to

$$V_{dc,max} = \left(\frac{3\sqrt{2}}{\pi}\right) V_{l-l,rms} = \left(\frac{3\sqrt{2}}{\pi}\right) 460 = 621 \text{ V}$$

and $I_{dc,max} = V_{dc,max}/R = 9.1$ A

b. The delay angle for a current of 2.5 A, corresponding to $V_{dc} = I_{dc}R = 170$ V, can be found from inverting Eq. 10.20 as

$$\alpha_d = \cos^{-1}\left[\left(\frac{\pi}{3\sqrt{3}}\right)\left(\frac{V_{dc}}{V_{l-l,rms}}\right)\right] = 1.29 \text{ rad} = 74.1°$$

Practice Problem 10.8

Repeat Example 10.9 for the case in which the 60-Hz source is replaced by a 50-Hz, 400-V, three-phase source.

Solution

a. $V_{dc,max} = 540$ V, $I_{dc,max} = 7.94$ V
b. $\alpha_d = 1.25$ rad $= 71.6°$

The derivations for three-phase bridges presented here have ignored issues such as the effect of commutating inductance, which we considered during our examination of single-phase rectifiers. Although the limited scope of our presentation does not permit us to specifically discuss them here, the effects in three-phase rectifiers are similar to those for single-phase systems and must be considered in the design and analysis of practical three-phase rectifier systems.

10.3 INVERSION: CONVERSION OF DC TO AC

In Section 10.2 we discussed various rectifier configurations that can be used to convert ac to dc. In this section, we will discuss some circuit configurations, referred to as *inverters*, which can be used to convert dc to the variable-frequency, variable-voltage power required for many motor-drive applications. Many such configurations and techniques are available, and we will not attempt to discuss them all. Rather, consistent with the aims of this chapter, we will review some of the common inverter configurations and highlight their basic features and characteristics.

For the purposes of this discussion, we will assume the inverter is preceded by a "stiff" dc source. For example, in Section 10.2, we saw how an LC filter can be used to produce a relatively constant dc output voltage from a rectifier. Thus, as shown in Fig. 10.39a, for our study of inverters we will represent such rectifier systems by a constant dc voltage source V_0, known as the *dc bus voltage* at the inverter input. We will refer to such a system, with a constant-dc input voltage, as a *voltage-source inverter*.

Similarly, we saw that a "large" inductor in series with the rectifier output produces a relatively constant dc current, known as the *dc link current*. We will therefore

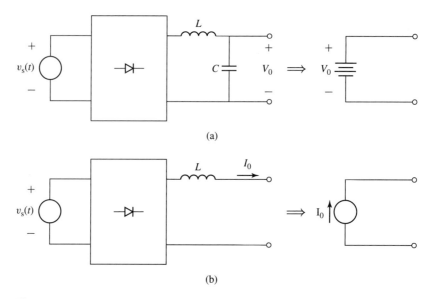

(a)

(b)

Figure 10.39 Inverter-input representations. (a) Voltage source. (b) Current source.

represent such a rectifier system by a current source I_0 at the inverter input. We will refer to this type of inverter as a *current-source inverter*.

Note that, as we have seen in Section 10.2, the values of these dc sources can be varied by appropriate controls applied to the rectifier stage, such as the timing of gate pulses to SCRs in the rectifier bridge. Control of the magnitude of these sources in conjunction with controls applied to the inverter stage provides the flexibility required to produce a wide variety of output waveforms for various motor-drive applications.

10.3.1 Single-Phase, H-bridge Step-Waveform Inverters

Figure 10.40a shows a single-phase inverter configuration in which a load (consisting here of a series RL combination) is fed from a dc voltage source V_0 through a set of four IGBTs in what is referred to as an *H-bridge* configuration. MOSFETs or other switching devices are equally applicable to this configuration. As we discussed in Section 10.1.3, the IGBTs in this system are used simply as switches. Because the IGBTs in this H-bridge include protection diodes, we can analyze the performance of this circuit by replacing the IGBTs by the ideal-switch model of Fig. 10.13b as shown in Fig. 10.40b.

For our analysis of this inverter, we will assume that the switching times of this inverter (i.e., the length of time the switches remain in a constant state) are much longer than the load time constant L/R. Hence, on the time scale of interest, the load current will simply be equal to V_L/R, with V_L being determined by the state of the switches.

Let us begin our investigation of this inverter configuration assuming that switches S1 and S3 are ON and that i_L is positive, as shown in Fig. 10.41a. Under this condition the load voltage is equal to V_0 and the load current is thus equal to V_0/R.

Let us next assume that switch S1 is turned OFF, while S3 remains ON. This will cause the load current, which cannot change instantaneously due to the presence of the inductor, to commutate from switch S1 to diode D2, as shown in Fig. 10.41b. Note that under this condition, the load voltage is zero and hence there will be zero load current. Note also that this same condition could have been reached by turning switch S3 OFF with S1 remaining ON.

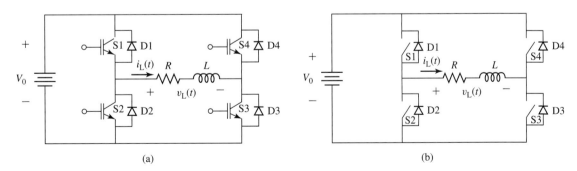

(a)   (b)

Figure 10.40 Single-phase H-bridge inverter configuration. (a) Typical configuration using IGBTs. (b) Generic configuration using ideal switches.

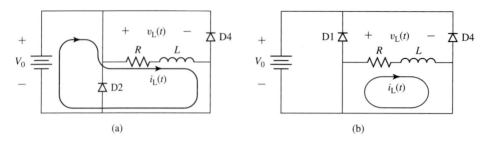

Figure 10.41   Analysis of the H-bridge inverter of Fig. 10.40b. (a) Switches S1 and S3 ON. (b) Switch S3 ON.

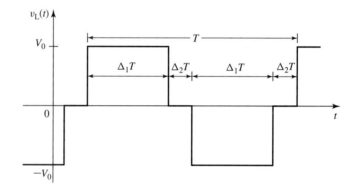

Figure 10.42   Typical output-voltage waveform for the H-bridge of Fig. 10.40.

At this point, it is possible to reverse the load voltage and current by turning ON switches S2 and S4, in which case $V_L = -V_0$ and $i_L = -V_0/R$. Finally, the current can be again brought to zero by turning OFF either switch S2 or switch S4. At this point, one cycle of an applied load-voltage waveform of the form of Fig. 10.42 has been completed.

A typical waveform produced by the switching sequence described above is shown in Fig. 10.42, with an ON time of $\Delta_1 T$ and OFF time of $\Delta_2 T$ ($\Delta_2 = 0.5 - \Delta_1$) for both the positive and negative portions of the cycle. Such a waveform consists of a fundamental ac component of frequency $f_0 = 1/T$, where T is the period of the switching sequence, and components at odd-harmonics frequencies of that fundamental.

The waveform of Fig. 10.42 can be considered a simple one-step approximation to a sinusoidal waveform. Fourier analysis can be used to show that it has a fundamental component of peak amplitude

$$V_{L,1} = \left(\frac{4}{\pi}\right) V_0 \sin(\Delta_1 \pi) \tag{10.21}$$

and n'th-harmonic components $(n = 3, 5, 7, \ldots)$ of peak amplitude

$$V_{L,n} = \left(\frac{4}{n\pi}\right) V_0 \sin(n \, \Delta_1 \pi) \tag{10.22}$$

Although this stepped waveform appears to be a rather crude approximation to a sinusoid, it clearly contains a significant fundamental component. In many applications it is perfectly adequate as the output voltage of a motor-drive. For example, three-such waveforms, separated by $120°$ in time phase, could be used to drive a three-phase motor. The fundamental components would combine to produce a rotating flux wave as discussed in Chapter 4. In some motor-drive systems, LC filters, consisting of shunt capacitors operating in conjunction with the motor phase inductances, are used to reduce the harmonic voltages applied to the motor phase windings.

In general, the higher-order harmonics, whose amplitudes vary inversely with their harmonic number, as seen from Eq. 10.22, will produce additional core loss in the stator as well as dissipation in the rotor. Provided that these additional losses are acceptable both from the point of view of motor heating as well as motor efficiency, a drive based upon this switching scheme will be quite adequate for many applications.

EXAMPLE 10.10

A three-phase, H-bridge, voltage-source, step-waveform inverter will be built from three H-bridge inverter stages of the type shown in Fig. 10.40b. Each phase will be identical, with the exception that the switching pattern of each phase will be displaced by $1/3$ of a period in time phase. This system will be used to drive a three-phase, four-pole motor with $N_{ph} = 34$ turns per phase and winding factor $k_w = 0.94$. The motor is Y-connected, and the inverters are each connected phase-to-neutral.

For a dc supply voltage of 125 V, a switching period T of 20 msec and with $\Delta_1 = 0.44$, calculate (a) the frequency and synchronous speed in rpm of the resultant air-gap flux wave and (b) the rms amplitude of the line-to-line voltage applied to the motor.

■ **Solution**

a. The frequency f_e of the fundamental component of the drive voltage will equal $f_e = 1/T = 50$ Hz. From Eq. 4.41 this will produce an air-gap flux wave which rotates at

$$n_s = \left(\frac{120}{\text{poles}}\right) f_e = \left(\frac{120}{4}\right) 50 = 1500 \text{ r/min}$$

b. The peak of the fundamental component of the applied line-to-neutral voltage can be found from Eq. 10.21.

$$V_{a,\text{peak}} = \left(\frac{4}{\pi}\right) V_0 \sin(\Delta_1 \pi) = \left(\frac{4}{\pi}\right) 125 \sin(0.44 \, \pi) = 156 \text{ V}$$

The resultant rms, line-to-line voltage is thus given by

$$V_{l-l,\text{rms}} = \sqrt{\frac{3}{2}} V_{a,\text{peak}} = 191 \text{ V}$$

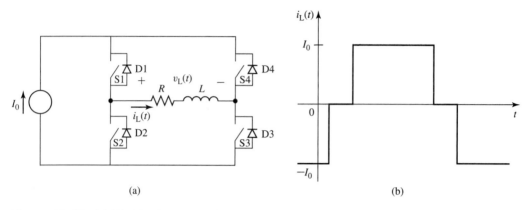

Figure 10.43 (a) H-bridge inverter configuration fed by a current source. (b) Typical stepped load-current waveform.

Practice Problem 10.9

For the three-phase inverter system of Example 10.10, (*a*) find the ON-time fraction Δ_1 for which the 5'th-harmonic component of the applied voltage will be zero. (*b*) Calculate the corresponding peak amplitude of the fundamental-component of the line-to-neutral voltage.

Solution

a. 0.2

b. 93 V

Figure 10.43a shows an H-bridge current-source inverter. This inverter configuration is analogous to the voltage-source configuration of Fig. 10.40. In fact, the discussion of the voltage-source inverter applies directly to the current-source configuration with the exception that the switches control the load current instead of the load voltage. Thus, again assuming that the load time constant (L/R) is much shorter than the switching time, a typical load current waveform would be similar to that shown in Fig. 10.43b.

EXAMPLE 10.11

Determine a switching sequence for the inverter of Fig. 10.43a that will produce the stepped waveform of Fig. 10.43b.

■ Solution

Table 10.2 shows one such switching sequence, starting at time $t = 0$ at which point the load current $i_L(t) = -I_0$.

Note that zero load current is achieved by turning on two switches so as to bypass the load and to directly short the current source. When this is done, the load current will quickly decay to zero, flowing through one of the switches and one of the reverse-polarity diodes. In general, one would not apply such a direct short across the voltage source in a voltage-source inverter

Table 10.2 Switching sequence used to produce the load current waveform of Fig. 10.43b.

$i_L(t)$	S1	S2	S3	S4
$-I_0$	OFF	ON	OFF	ON
0	ON	ON	OFF	OFF
I_0	ON	OFF	ON	OFF
0	OFF	OFF	ON	ON
$-I_0$	OFF	ON	OFF	ON

because the resultant current would most likely greatly exceed the ratings of the switches. However, in the case of a current-source inverter, the switch current cannot exceed that of the current source, and hence the direct short can be (and, in fact, must be) maintained for as long as it is desired to maintain zero load current.

EXAMPLE 10.12

Consider the current-source inverter of Fig. 10.44a. Here the load consists of a sinusoidal voltage source $V_a \cos \omega t$. Assume the inverter switches are controlled such that the load current is a square-wave, also at frequency $f = \omega/(2\pi)$, as shown in Fig. 10.44b. Calculate the time-average power delivered to the load as a function of the delay angle α_d as defined in Fig. 10.44b.

■ **Solution**

Because the load voltage is sinusoidal, time-average power will only be produced by the fundamental component of the load current. By analogy to Eq. 10.21, with I_0 replacing V_0 and with $\Delta_1 = 0.5$, the amplitude of the fundamental-component of the load current is

$$I_{L,1} = \left(\frac{4}{\pi}\right) I_0$$

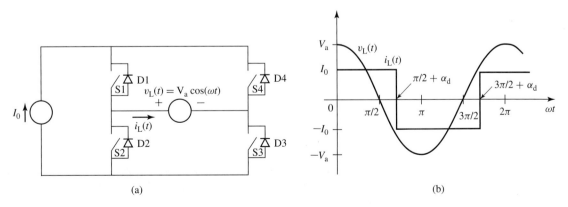

(a) (b)

Figure 10.44 (a) Current-source inverter for Example 10.12. (b) Load-current waveform.

and therefore the fundamental component of the load current is equal to

$$i_{L,1}(t) = I_{L,1} \cos(\omega t - \alpha_d) = \left(\frac{4}{\pi}\right) I_0 \cos(\omega t - \alpha_d)$$

The complex amplitude of the load voltage is thus given by $\hat{V}_L = V_a$ and that of the load current is $\hat{I}_L = I_{L,1} e^{-j\alpha_d}$. Thus the time average power is equal to

$$P = \frac{1}{2}\text{Re}[\hat{I}_L \hat{V}_L^*] = \left(\frac{2}{\pi}\right) V_a I_0 \cos \alpha_d$$

By varying the firing-delay angle α_d, the power transferred from the source to the load can be varied. In fact, as α_d is varied over the range $0 \le \alpha_d \le \pi$, the power will vary over the range

$$\left(\frac{2}{\pi}\right) V_a I_0 \ge P \ge -\left(\frac{2}{\pi}\right) V_a I_0$$

Note that this inverter can regenerate; i.e., for $\pi/2 < \alpha_d \le \pi$, $P < 0$ and hence power will flow from the load back into the inverter.

Practice Problem 10.10

The inverter of Example 10.12 is operated with a fixed delay angle $\alpha_d = 0$ but with a variable ON-time fraction Δ_1. Find an expression for the time-average power delivered to the load as a function of Δ_1.

Solution

$$P = \left(\frac{2}{\pi}\right) V_a I_0 \sin(\Delta_1 \pi)$$

10.3.2   Pulse-Width-Modulated Voltage-Source Inverters

Let us again consider the H-bridge configuration of Fig. 10.40b, repeated again in Fig. 10.45. Again, an RL load is fed from a voltage source through the H-bridge. However, in this case, let us assume that the switching time of the inverter is much shorter than the load time constant (L/R).

Consider a typical operating condition as shown in Fig. 10.46. Under this condition, the switches are operated with a period T and a *duty cycle* D ($0 \le D \le 1$). As can be seen from Fig. 10.46a, for a time DT switches S1 and S3 are ON, and the load voltage is V_0. This is followed by a time $(1 - D)T$ during which switches S1 and S3 are OFF, and the current is transferred to diodes D2 and D4, setting the load voltage equal to $-V_0$. The duty cycle D is thus a fraction of the total period, in this case the fraction of the period during which the load voltage is V_0.

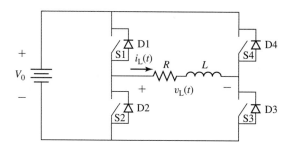

Figure 10.45 Single-phase H-bridge inverter configuration.

Note that although switches S2 and S4 would normally be turned ON after switches S1 and S3 are turned OFF (but not before they are turned OFF, to avoid a direct short across the voltage source) because they are actually semiconductor devices, they will not carry any current unless the load current goes negative. Rather, the current will flow through the protection diodes D1 and D3. Alternatively, if the load current is negative, then the current will be controlled by the operation of switches S2 and S4 in conjunction with diodes D1 and D3. Under this condition, switches S1 and S3 will not carry current.

This type of control is referred to as *pulse-width modulation,* or *PWM,* because it is implemented by varying the width of the voltage pulses applied to the load. As can be seen from Fig. 10.46a, the average voltage applied to the load is equal to

$$(v_L)_{avg} = (2D - 1)V_0 \tag{10.23}$$

As we will now show, varying the duty cycle under PWM control can produce a continuously varying load current.

A typical load-current waveform is shown in Fig. 10.46b. In the steady state, the average current through the inductor will be constant and hence voltage across the inductor must equal zero. Thus, the average load current will equal the average

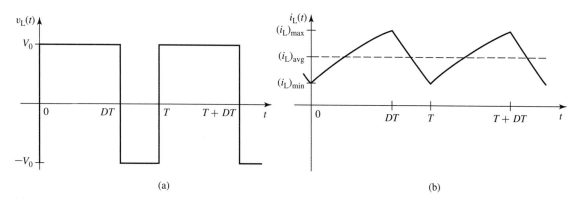

Figure 10.46 Typical (a) voltage and (b) current waveform under PWM operation.

voltage divided by the resistance or

$$(i_L)_{avg} = \frac{(v_L)_{avg}}{R} = \frac{[2D - 1]V_0}{R} \tag{10.24}$$

Thus we see that by varying the duty cycle D over the range of 0 to 1, we can vary the load current over the range $-V_0/R \leq (i_L)_{avg} \leq V_0/R$.

Because the current waveform is periodic, the maximum and minimum current, and hence the current ripple, can be easily calculated. Assigning time $t = 0$ to the time switches S1 and S3 are first turned ON and the load current is minimum, the current during this time period will be given by

$$i_L(t) = \frac{V_0}{R} + \left((i_L)_{min} - \frac{V_0}{R} \right) e^{-\frac{t}{\tau}} \quad (0 \leq t \leq DT) \tag{10.25}$$

where $\tau = L/R$. The maximum load current $(i_L)_{max}$ is reached at time DT

$$(i_L)_{max} = \frac{V_0}{R} + \left((i_L)_{min} - \frac{V_0}{R} \right) e^{-\frac{DT}{\tau}} \tag{10.26}$$

After switches S1 and S3 are turned OFF, the load voltage is $-V_0$ and the current is given by

$$i_L(t) = -\frac{V_0}{R} + \left((i_L)_{max} + \frac{V_0}{R} \right) e^{-\frac{(t - DT)}{\tau}} \quad (DT < t \leq T) \tag{10.27}$$

Because the current is periodic with period T, $i_L(t)$ again will be equal to $(i_L)_{min}$ at time T. Thus

$$(i_L)_{min} = -\frac{V_0}{R} + \left((i_L)_{max} + \frac{V_0}{R} \right) e^{-\frac{(T - DT)}{\tau}} \tag{10.28}$$

Solving Eqs. 10.26 and 10.28 gives

$$(i_L)_{min} = -\left(\frac{V_0}{R} \right) \frac{\left[1 - 2e^{\frac{-T(1 - D)}{\tau}} + e^{-\frac{T}{\tau}} \right]}{\left(1 - e^{-\frac{T}{\tau}} \right)} \tag{10.29}$$

and

$$(i_L)_{max} = \left(\frac{V_0}{R} \right) \frac{\left[1 - 2e^{\frac{-DT}{\tau}} + e^{-\frac{T}{\tau}} \right]}{\left(1 - e^{-\frac{T}{\tau}} \right)} \tag{10.30}$$

The current ripple Δi_L can be calculated as the difference between the maximum and minimum current.

$$\Delta i_L = (i_L)_{max} - (i_L)_{min} \tag{10.31}$$

In the limit that $T \ll \tau$, this can be written as

$$\Delta i_L \approx \left(\frac{2V_0}{R} \right) \left(\frac{T}{\tau} \right) D(1 - D) \tag{10.32}$$

EXAMPLE 10.13

A PWM inverter such as that of Fig. 10.45 is operating from a dc voltage of 48 V and driving a load with $L = 320$ mH and $R = 3.7\ \Omega$. For a switching frequency of 1 kHz, calculate the average load current, the minimum and maximum current, and the current ripple for a duty cycle $D = 0.8$.

■ **Solution**

From Eq. 10.24, the average load current will equal

$$(i_L)_{avg} = \frac{[2D - 1]V_0}{R} = \frac{0.6 \times 48}{3.7} = 7.78\ A$$

For a frequency of 1 kHz, the period $T = 1$ msec. The time constant $\tau = L/R = 86.5$ msec. $(i_L)_{min}$ and $(i_L)_{max}$ can then be found from Eqs. 10.29 and 10.30 to be $(i_L)_{min} = 7.76$ A and $(i_L)_{max} = 7.81$ A. Thus the current ripple, calculated from Eq. 10.31, is 0.05 A, which is equal to 0.6 percent of the average load current. Alternatively, using the fact that $T/\tau = 0.012 \ll 1$, the current ripple could have been calculated directly from Eq. 10.32

$$\Delta i_L = \left(\frac{2V_0}{R}\right)\left(\frac{T}{\tau}\right)D(1 - D) = \left(\frac{2 \times 48}{3.7}\right)\left(\frac{1}{86.5}\right) \times 0.8 \times 0.2 = 0.05\ A$$

Calculate (a) the average current and (b) the current ripple for the PWM inverter of Example 10.13 if the switching frequency is reduced to 250 Hz.

Solution

    a. 7.78 A (unchanged from Example 10.13)
    b. 0.19 A

Now let us consider the situation for which the duty cycle is varied with time, i.e., $D = D(t)$. If $D(t)$ varies slowly in comparison to the period T of the switching frequency, from Eq. 10.23, the average load voltage will be equal to

$$(v_L)_{avg} = [2D(t) - 1]V_0 \tag{10.33}$$

and the average load current will be

$$(i_L)_{avg} = \frac{[2D(t) - 1]V_0}{R} \tag{10.34}$$

Figure 10.47a illustrates a method for producing the variable duty cycle for this system. Here we see a saw-tooth waveform which varies between -1 and 1. Also shown is a reference waveform $W_{ref}(t)$ which is constrained to lie within the range -1 and 1. The switches will be controlled in pairs. During the time that $W_{ref}(t)$ is greater than the saw-tooth waveform, switches S1 and S3 will be ON and the load voltage will be V_0. Similarly, when $W_{ref}(t)$ is less than the saw-tooth waveform, switches S2

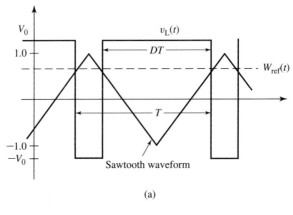

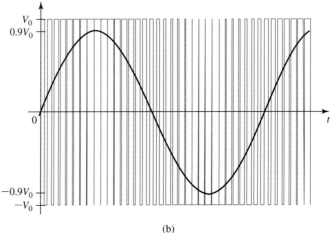

Figure 10.47 (a) Method for producing a variable duty cycle from a reference waveform $W_{ref}(t)$. (b) Load voltage and average load voltage for $W_{ref}(t) = 0.9 \sin \omega t$.

and S4 will be ON and the load voltage will be $-V_0$. Thus,

$$D(t) = \frac{(1 + W_{ref}(t))}{2} \tag{10.35}$$

and thus

$$(v_L)_{avg} = \left[2 \left(\frac{1 + W_{ref}(t)}{2} \right) - 1 \right] V_0 = W_{ref}(t) V_0 \tag{10.36}$$

Figure 10.47b shows the load voltage $v_L(t)$ for a sinusoidal reference waveform $W_{ref}(t) = 0.9 \sin \omega t$. The average voltage across the load, $(v_L)_{avg} = W_{ref}(t) V_0$, is also shown.

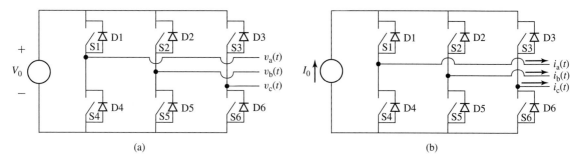

Figure 10.48 Three-phase inverter configurations. (a) Voltage-source. (b) Current-source.

Note that the H-bridge inverter configuration of Fig. 10.43 can be used to produce a PWM current-source inverter. In a fashion directly analogous to the derivation of Eq. 10.36, one can show that such an inverter would produce an average current of the form

$$(i_{\mathrm{L}})_{\mathrm{avg}} = W_{\mathrm{ref}}(t) I_0 \tag{10.37}$$

where I_0 is the magnitude of the dc link current feeding the H-bridge. Note, however, that the sudden current swings between I_0 and $-I_0$ associated with such an inverter will produce large voltages should the load have any inductive component. As a result, practical inverters of this type require large capacitive filters to absorb the harmonic components of the PWM current and to protect the load against damage due to voltage-induced insulation failure.

10.3.3 Three-Phase Inverters

Although the single-phase motor drives of Section 10.3.2 demonstrate the important characteristics of inverters, most variable-frequency drives are three phase. Figures 10.48a and 10.48b show the basic configuration of three-phase motor inverters (voltage- and current-source respectively). Here we have shown the switches as ideal switches, recognizing that in a practical implementation, bidirectional capability will be achieved by a combination of a semiconductor switching device, such as an IGBT and a MOSFET, and a reverse-polarity diode.

These configurations can be used to produce both stepped waveforms (either voltage-source or current-source) as well as pulse-width-modulated waveforms. This will be illustrated in the following example.

EXAMPLE 10.14

The three-phase current-source inverter configuration of Fig. 10.48b is to be used to produce a three-phase stepped current waveform of the form shown in Fig 10.49. (*a*) Determine the switch sequence over the period $0 \le t \le T$ and (*b*) calculate the fundamental, third, fifth, and seventh harmonics of the phase-*a* current waveform.

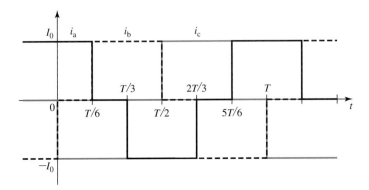

Figure 10.49 Three-phase stepped current waveform for Example 10.14.

■ **Solution**

a. By observing that switch S1 is ON when the phase-a current is positive, switch S4 is ON when it is negative, and so forth, the following table of switching operations can be produced.

t	0–$(T/6)$	$(T/6)$–$(T/3)$	$(T/3)$–$(T/2)$	$(T/2)$–$(2T/3)$	$(2T/3)$–$(5T/6)$	$(5T/6)$–T
S1	ON	OFF	OFF	OFF	OFF	ON
S2	OFF	ON	ON	OFF	OFF	OFF
S3	OFF	OFF	OFF	ON	ON	OFF
S4	OFF	OFF	ON	ON	OFF	OFF
S5	OFF	OFF	OFF	OFF	ON	ON
S6	ON	ON	OFF	OFF	OFF	OFF

b. The amplitudes of the harmonic components of the phase current can be determined from Eqs. 10.21 and 10.22 by setting $\Delta_1 = 1/3$. Thus,

$$I_{a,1} = \left(\frac{2\sqrt{3}}{\pi}\right) I_0 \qquad I_{a,3} = 0$$

$$I_{a,5} = -\left(\frac{2\sqrt{3}}{5\pi}\right) I_0 \qquad I_{a,7} = \left(\frac{2\sqrt{3}}{7\pi}\right) I_0$$

10.4 SUMMARY

The goal of this chapter is relatively modest. Our focus has been to introduce some basic principles of power electronics and to illustrate how they can be applied to the design of various power-conditioning circuits that are commonly found in motor drives. Although the discussion in this chapter is neither complete nor extensive, it is intended to provide the background required to support the various discussions of motor control which are presented in this book.

We began with a brief overview of a few of the available solid-state switching devices: diodes, SCRs, IGBTs and MOSFETs, and so on. We showed that, for the purposes of a preliminary analysis, it is quite sufficient to represent these devices as ideal switches. To emphasize the fact that they typically can pass only unidirectional current, we included ideal diodes in series with these switches. The simplest of these devices is the diode, which has only two terminals and is turned ON and OFF simply by the conditions of the external circuit. The remainder have a third terminal which can be used to turn the device ON and, in the case of transistors such as MOSFETS and IGBTs, OFF again.

A typical variable-frequency, variable-voltage motor-drive system can be considered to consist of three sections. The input section rectifies the power-frequency, fixed-voltage ac input and produces a dc voltage or current. The middle section filters the rectifier output, producing a relatively constant dc current or voltage, depending upon the type of drive under consideration. The output inverter section converts the dc to variable-frequency, variable-voltage ac voltages or currents which can be applied to the terminals of a motor.

The simplest inverters we investigated produce stepped voltage or current waveforms whose amplitude is equal to that of the dc source and whose frequency can be controlled by the timing of the inverter switches. To produce a variable-amplitude output waveform, it is necessary to apply additional control to the rectifier stage to vary the amplitude of the dc bus voltage or link current supplied to the inverter.

We also discussed pulse-width-modulated voltage-source inverters. In this type of inverter, the voltage to the load is switched between V_0 and $-V_0$ such that the average load voltage is determined by the duty cycle of the switching waveform. Loads whose time constant is long compared to the switching time of the inverter will act as filters, and the load current will then be determined by the average load voltage. Pulse-width modulated current-source inverters were also discussed briefly.

The reader should approach the presentation here with great caution. It is important to recognize that a complete treatment of power electronics and motor drives is typically the topic of a multiple-course sequence of study. Although the basic principles discussed here apply to a wide range of motor drives, there are many details which must be included in the design of practical motor drives. Drive circuitry to turn ON the "switches" (gate drives for SCRs, MOSFETs, IGBTs, etc.) must be carefully designed to provide sufficient drive to fully turn on the devices and to provide the proper switching sequences. The typical inverter includes a controller and a protection system which is quite elaborate. Typically, the design of a specific drive is dominated by the current and voltage ratings of available switches devices. This is especially true in the case of high-power drive systems in which switches must be connected in series and/or parallel to achieve the desired power rating. The reader is referred to references in the bibliography for a much more complete discussion of power electronics and inverter systems than has been presented here.

Motor drives based upon the configurations discussed here can be used to control motor speed and motor torque. In the case of ac machines, the application of power-electronic based motor drives has resulted in performance that was previously available only with dc machines and has led to widespread use of these machines in most applications.

10.5 BIBLIOGRAPHY

This chapter is intended to serve as an introduction to the discipline of power electronics. For readers who wish to study this topic in more depth, this bibliography lists a few of the many textbooks which have been written on this subject.

Bird, B. M., K. G. King, and D. A. G. Pedder, *An Introduction to Power Electronics, 2/e*. New York: John Wiley & Sons, 1993.

Dewan, S. B., and A. Straughen, *Power Semiconductor Circuits*. New York: John Wiley & Sons, 1975.

Hart, D. W., *Introduction to Power Electronics*. Englewood Cliffs, New Jersey: Prentice-Hall, 1998.

Kassakian, J. G., M. F. Schlecht, and G. C. Verghese, *Principles of Power Electronics*. Reading, Massachusetts: Addison-Wesley, 1991.

Mohan, N., T. M. Undeland, and W. P. Robbins, *Power Electronics: Converters, Applications, and Design, 3/e*. New York: John Wiley & Sons, 2002.

Rahsid, M. H., *Power Electronics: Circuits, Devices and Applications, 2/e*. Englewood Cliffs, New Jersey: Prentice-Hall, 1993.

Subrahmanyam, V., *Electric Drives: Concepts and Applications*. New York: McGraw-Hill, 1996.

Thorborg, K., *Power Electronics*. Englewood Cliffs, New Jersey: Prentice Hall International (U.K.) Ltd, 1988.

10.6 PROBLEMS

10.1 Consider the half-wave rectifier circuit of Fig. 10.3a. The circuit is driven by a triangular voltage source $v_s(t)$ of amplitude $V_0 = 9$ V as shown in Fig. 10.50. Assuming the diode to be ideal and for a resistor $R = 1.5$ kΩ:

  a. Plot the resistor voltage $v_R(t)$.

  b. Calculate the rms value of the resistor voltage.

  c. Calculate the time-averaged power dissipation in the resistor.

10.2 Repeat Problem 10.1 assuming the diode to have a fixed 0.6 V voltage drop when it is ON but to be otherwise ideal. In addition, calculate the time-averaged power dissipation in the diode.

10.3 Consider the half-wave SCR rectifier circuit of Fig. 10.6 supplied from the triangular voltage source of Fig. 10.50. Assuming the SCR to be ideal, calculate the rms resistor voltage as a function of the firing-delay time t_d $(0 \le t_d \le T/2)$.

10.4 Consider the rectifier system of Example 10.5. Write a MATLAB script to plot the ripple voltage as a function of filter capacitance as the filter

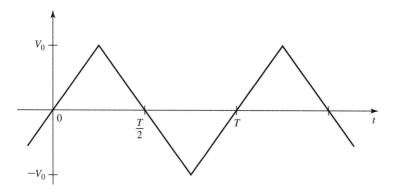

Figure 10.50 Triangular voltage waveform.

capacitance is varied over the range $3000 \ \mu\text{F} \le C \le 10^5 \ \mu\text{F}$. Assume the diode to be ideal. Use a log-scale for the capacitance.

10.5 Consider the full-wave rectifier system of Fig. 10.16 with $R = 500 \ \Omega$ and $C = 200 \ \mu\text{F}$. Assume each diode to have a constant voltage drop of 0.7 V when it is ON but to be otherwise ideal. For a 220 V rms, 50 Hz sinusoidal source, write a MATLAB script to calculate

a. the peak voltage across the load resistor.

b. the magnitude of the ripple voltage.

c. the time-averaged power supplied to the load resistor.

d. the time-averaged power dissipation in the diode bridge.

10.6 Consider the half-wave rectifier system of Fig. 10.51. The voltage source is $v_s(t) = V_0 \sin \omega t$ where $V_0 = 15$ V, and the frequency is 100 Hz. For $L = 1$ mH and $R = 1 \ \Omega$, plot the inductor current $i_L(t)$ for the first 1-1/2 cycles of the applied waveform assuming the switch closes at time $t = 0$.

10.7 Repeat Problem 10.6, using MATLAB to plot the inductor current for the first 10 cycles following the switch closing at time $t = 0$. (Hint: This problem can be easily solved, using simple Euler integration to solve for the current.)

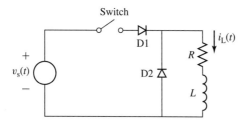

Figure 10.51 Half-wave rectifier system for Problem 10.6.

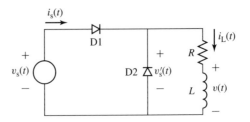

Figure 10.52 Half-wave rectifier system for Problem 10.8.

10.8 Consider the half-wave rectifier system of Fig. 10.52 as L becomes sufficiently large such that $\omega(L/R) \gg 1$, where ω is the supply frequency. In this case, the inductor current will be essentially constant. For $R = 5\,\Omega$ and $v_s(t) = V_0 \sin \omega t$ where $V_0 = 45$ V and $\omega = 100\pi$ rad/sec. Assume the diodes to be ideal.

a. Calculate the average (dc) value V_{dc} of the voltage $v_s'(t)$ across the series resistor/inductor combination.

b. Using the fact that, in the steady state, there will be zero average voltage across the inductor, calculate the dc inductor current I_{dc}.

c. Plot the instantaneous inductor voltage $v(t)$ over one cycle of the supply voltage.

d. Plot the instantaneous source current $i_s(t)$.

10.9 Consider the half-wave, phase-controlled rectifier system of Fig. 10.53. This is essentially the same circuit as that of Problem 10.8 with the exception that diode D1 of Fig. 10.52 has been replaced by an SCR, which you can consider to be ideal. Let $R = 5\,\Omega$ and $v_s(t) = V_0 \sin \omega t$, where $V_0 = 45$ V and $\omega = 100\pi$ rad/sec. Assume that the inductor L is sufficiently large such that $\omega(L/R) \gg 1$ and that the SCR is triggered ON at time t_d ($0 \leq t_d \leq \pi/\omega$).

a. Find an expression for the average (dc) value V_{dc} of the voltage $v_s'(t)$ across the series resistor/inductor combination as a function of the delay time t_d.

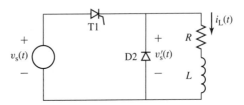

Figure 10.53 Half-wave, phase-controlled rectifier system for Problem 10.9.

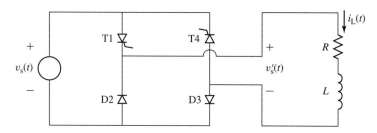

Figure 10.54 Full-wave, phase-controlled rectifier system for Problem 10.10.

    b. Using the fact that, in the steady state, there will be zero average voltage across the inductor, find an expression for the dc inductor current I_{dc}, again as a function of the delay time t_d.

    c. Plot I_{dc} as a function of t_d for $(0 \leq t_d \leq \pi/\omega)$.

10.10 The half-wave, phase-controlled rectifier system of Problem 10.9 and Fig. 10.53 is to be replaced by the full-wave, phase-controlled system of Fig. 10.54. SCR T1 will be triggered ON at time t_d $(0 \leq t_d \leq \pi/\omega)$, and SCR T4 will be triggered on exactly one half cycle later.

    a. Find an expression for the average (dc) value V_{dc} of the voltage $v_s'(t)$ across the series resistor/inductor combination as a function of the delay time t_d.

    b. Using the fact that, in the steady state, there will be zero average voltage across the inductor, find an expression for the dc inductor current I_{dc}, again as a function of the delay time t_d.

    c. Plot I_{dc} as a function of t_d for $(0 \leq t_d \leq \pi/\omega)$.

    d. Plot the source current $i_s(t)$ for one cycle of the source voltage for $t_d = 3$ msec.

10.11 The full-wave, phase-controlled rectifier of Fig. 10.55 is supplying a highly inductive load such that the load current can be assumed to be purely dc, as represented by the current source I_{dc} in the figure. The source voltage is a

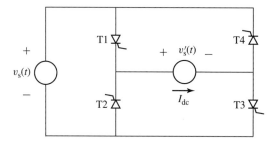

Figure 10.55 Full-wave, phase-controlled rectifier for Problem 10.11.

sinusoid, $v_s(t) = V_0 \sin \omega t$. As shown in Fig. 10.31, SCRs T1 and T3 are triggered together at delay angle α_d ($0 \le \alpha_d \le \pi$), and SCRs T2 and T4 are triggered exactly one-half cycle later.

a. For $\alpha_d = \pi/4$:
    (i)   Sketch the load voltage $v_s'(t)$.
    (ii) Calculate the average (dc) value V_{dc} of $v_s'(t)$.
    (iii) Calculate the time-averaged power supplied to the load.
b. Repeat part (a) for $\alpha_d = 3\pi/4$.

10.12 A full-wave diode rectifier is fed from a 50-Hz, 220-V rms source whose series inductance is 12 mH. It drives a load with a resistance 8.4 Ω which is sufficiently inductive that the load current can be considered to be essentially dc.

a. Calculate the dc load current I_{dc} and the commutation time t_c.
b. Compare the dc current of part (a) with the dc current which would result if the commutating inductance could be eliminated from the system.

10.13 A 1-kW, 85-V, permanent-magnet dc motor is to be driven from a full-wave, phase-controlled bridge such as is shown in Fig. 10.56. When operating at its rated voltage, the dc-motor has a no-load speed of 1725 r/min and an armature resistance $R_a = 0.82$ Ω. A large inductor ($L = 580$ mH) with resistance $R_L = 0.39$ Ω has been inserted in series with the output of the rectifier bridge to reduce the ripple current applied to the motor. The source voltage is a 115-V rms, 60-Hz sinusoid.

With the motor operating at a speed of 1650 r/min, the motor current is measured to be 7.6 A.

a. Calculate the motor input power.
b. Calculate the firing delay angle α_d of the SCR bridge.

10.14 Consider the dc-motor drive system of Problem 10.13. To limit the starting current of the dc motor to twice its rated value, a controller will be used to adjust the initial firing-delay angle of the SCR bridge. Calculate the required firing-delay angle α_d.

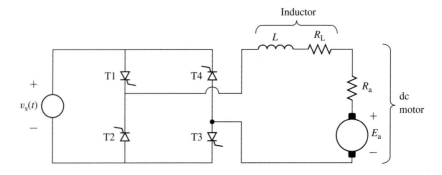

Figure 10.56 Dc motor driven from a full-wave, phase-controlled rectifier. Problem 10.13.

10.15 A three-phase diode bridge is supplied by a three-phase autotransformer such that the line-to-line input voltage to the bridge can be varied from zero to 230 V. The output of the bridge is connected to the shunt field winding of a dc motor. The resistance of this winding is 158 Ω. The autotransformer is adjusted to produce a field current of 1.75 A. Calculate the rms output voltage of the autotransformer.

10.16 A dc-motor shunt field winding of resistance 210 Ω is to be supplied from a 220-V rms, 50-Hz, three-phase source through a three-phase, phase-controlled rectifier. Calculate the delay angle α_d which will result in a field current of 1.1 A.

10.17 A superconducting magnet has an inductance of 4.9 H, a resistance of 3.6 mΩ, and a rated operating current of 80 A. It will be supplied from a 15-V rms, three-phase source through a three-phase, phase-controlled bridge. It is desired to "charge" the magnet at a constant rate to achieve rated current in 25 seconds.

    a. Calculate the firing-delay angle α_d required to achieve this objective.

    b. Calculate the firing-delay angle required to maintain a constant current of 80 A.

10.18 A voltage-source H-bridge inverter is used to produce the stepped waveform $v(t)$ shown in Fig. 10.57. For $V_0 = 50$ V, $T = 10$ msec and $D = 0.3$:

    a. Using Fourier analysis, find the amplitude of the fundamental time-harmonic component of $v(t)$.

    b. Use the MATLAB 'fft()' function to find the amplitudes of the first 10 time harmonics of $v(t)$.

10.19 Consider the stepped voltage waveform of Problem 10.18 and Fig. 10.57.

    a. Using Fourier analysis, find the value of D ($0 \le D \le 0.5$) such that the amplitude of the third-harmonic component of the voltage waveform is zero.

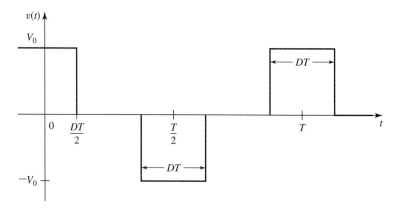

Figure 10.57 Stepped voltage waveform for Problem 10.18.

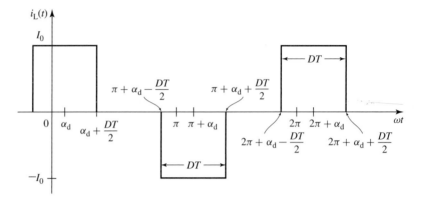

Figure 10.58   Stepped current waveform for Problem 10.20.

   b. Use the MATLAB 'fft()' function to find the amplitudes of the first 10
       time harmonics of the resulting waveform.

10.20 Consider Example 10.12 in which a current-source inverter is driving a load
   consisting of a sinusoidal voltage. The inverter is controlled to produce the
   stepped current waveform shown in Fig. 10.58.

   a. Create a table showing the switching sequence required to produce the
       specified waveform and the time period during which each switch is
       either ON or OFF.

   b. Express the fundamental component of the current waveform in the form

$$i_1(t) = I_1 \cos(\omega t + \phi_1)$$

   where I_1 and ϕ_1 are functions of I_0, D and the delay angle α_d.

   c. Derive an expression for the time-averaged power delivered to the
       voltage source $v_L(t) = V_a \cos \omega t$.

10.21 A PWM inverter such as that of Fig. 10.45 is operating from a dc voltage of
   75 V and driving a load with $L = 53$ mH and $R = 1.7\ \Omega$. For a switching
   frequency of 1500 Hz, calculate the average load current, the minimum and
   maximum current, and the current ripple for a duty cycle $D = 0.7$.

CHAPTER 11

Speed and Torque Control

The objective of this chapter is to discuss various techniques for the control of electric machines. Since an in-depth discussion of this topic is both too extensive for a single chapter and beyond the scope of this book, the presentation here will necessarily be introductory in nature. We will present basic techniques for speed and torque control and will illustrate typical configurations of drive electronics that are used to implement the control algorithms. This chapter will build upon the discussion of power electronics in Chapter 10.

Note that the discussion of this chapter is limited to steady-state operation. The steady-state picture presented here is quite adequate for a wide variety of electric-machine applications. However, the reader is cautioned that system dynamics can play a critical role in some applications, with concerns ranging from speed of response to overall system stability. Although the techniques presented here form the basis for dynamic analyses, the constraints of an introductory textbook are such that a more extensive discussion, including transient and dynamic behavior, is not possible.

In the discussion of torque control for synchronous and induction machines, the techniques of field-oriented or vector control are introduced and the analogy is made with torque control in dc motors. This material is somewhat more sophisticated mathematically than the speed-control discussion and requires application of the dq0 transformations developed in Appendix C. The chapter is written such that this material can be omitted at the discretion of the instructor without detracting from the discussion of speed control.

11.1 CONTROL OF DC MOTORS

Before the widespread application of power-electronic drives to control ac machines, dc motors were by far the machines of choice in applications requiring flexibility of control. Although in recent years ac drives have become quite common, the ease of control of dc machines insure their continued use in many applications.

11.1.1 Speed Control

The three most common speed-control methods for dc motors are adjustment of the flux, usually by means of field-current control, adjustment of the resistance associated with the armature circuit, and adjustment of the armature terminal voltage.

Field-Current Control   In part because it involves control at a relatively low power level (the power into the field winding is typically a small fraction of the power into the armature of a dc machine), *field-current control* is frequently used to control the speed of a dc motor with separately excited or shunt field windings. The equivalent circuit for a separately excited dc machine is found in Fig. 7.4a and is repeated in Fig. 11.1. The method is, of course, also applicable to compound motors. The shunt field current can be adjusted by means of a variable resistance in series with the shunt field. Alternatively, the field current can be supplied by power-electronic circuits which can be used to rapidly change the field current in response to a wide variety of control signals.

Figure 11.2a shows in schematic form a switching scheme for pulse-width modulation of the field voltage. This system closely resembles the pulse-width modulation system discussed in Section 10.3.2. It consists of a rectifier which rectifies the ac input voltage, a dc-link capacitor which filters the rectified voltage, producing a dc voltage V_{dc}, and a pulse-width modulator.

In this system, because only a unidirectional field current is required, the pulse-width modulator consists of a single switch and a free-wheeling diode rather than the more complex four-switch arrangement of Fig. 10.45. Assuming both the switch and diode to be ideal, the average voltage across the field winding will be equal to

$$V_f = D V_{dc} \tag{11.1}$$

where D is the duty cycle of the switching waveform; i.e., D is the fraction of time that the switch S is on.

Figure 11.2b shows the resultant field current. Because in the steady-state the average voltage across the inductor must equal zero, the average field current I_f will thus be equal to

$$I_f = \frac{V_f}{R_f} = D\left(\frac{V_{dc}}{R_f}\right) \tag{11.2}$$

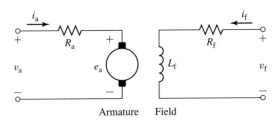

Armature   Field

Figure 11.1 Equivalent circuit for a separately excited dc motor.

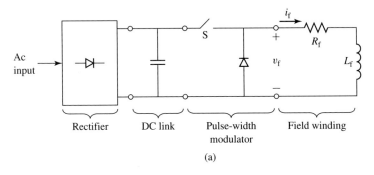

(a)

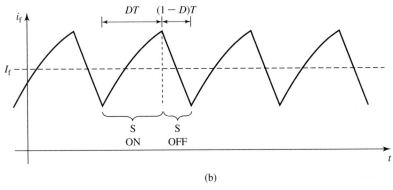

(b)

Figure 11.2 (a) Pulse-width modulation system for a dc-machine field winding. (b) Field-current waveform.

Thus, the field current can be controlled simply by controlling the duty cycle of the pulse-width modulator. If the field-winding time constant L_f/R_f is long compared to the switching time, the ripple current Δi_f will be small compared to the average current I_f.

EXAMPLE 11.1

A 25-kW, 3600 r/min, 240-V dc motor has an armature resistance of 47 mΩ and a shunt-field with a resistance of 187 Ω and an inductance of 4.2 H. Calculate (a) the average field current I_f and (b) the magnitude of the current ripple Δi_f when the field winding is supplied from a 240 V dc source by pulse-width modulation with a duty cycle $D = 0.75$ and a switching period of 1 msec.

■ **Solution**

a. The average field current is readily found from Eq. 11.2

$$I_f = D\left(\frac{V_{dc}}{R_f}\right) = 0.75\left(\frac{240}{187}\right) = 0.96 \text{ A}$$

b. The field time constant $\tau = L_f/R_f = 22.5$ msec is much longer that the switching period of 1 msec. Thus, the ripple current can be calculated using Eq. 10.32 as

$$
\Delta i_f = \left(\frac{2V_{dc}}{R_f}\right) \left(\frac{T}{\tau}\right) D(1 - D)
$$

$$
= \left(\frac{2 \times 240}{187}\right) \left(\frac{1}{22.5}\right) 0.75 \times (1 - 0.75)
$$

$$
= 21.4 \text{ mA}
$$

Practice Problem 11.1

The duty-cycle D of the dc-motor controller of Example 11.1 is suddenly switched from 0.75 to 1.0. Calculate (a) the resultant steady-state field current and (b) the time constant for the change from the initial value of 1.08 A to the new final value.

Solution

   a. 1.28 A
   b. 22.5 msec

To examine the effect of field-current control, let us begin with the case of a dc motor driving a load of constant torque T_{load}. From Eqs. 7.9 and 7.14, the generated voltage of a dc motor can be written as

$$
E_a = K_f I_f \omega_m \tag{11.3}
$$

where I_f is the average field current, ω_m is the angular velocity in rad/sec, and $K_f = K_a \mathcal{P}_d N_f$ is a geometric constant which depends upon the dimensions of the motor, the properties of the magnetic material used to construct the motor, as well as the number of turns in the field winding. Note that strictly speaking, K_f is not constant since it is proportional to the direct-axis permeance, which typically varies as the flux-level in the motor increases to the point that the effects of magnetic saturation become significant.

The electromagnetic torque is given by Eq. 7.16 as

$$
T_{mech} = \frac{E_a I_a}{\omega_m} = K_f I_f I_a \tag{11.4}
$$

and the armature current can be seen from the equivalent circuit of Fig. 11.1 to be given by

$$
I_a = \frac{(V_a - E_a)}{R_a} \tag{11.5}
$$

Setting the motor torque equal to T_{load}, Eqs. 11.3 through 11.5 can be solved for ω_{m}

$$\omega_{\text{m}} = \frac{(V_{\text{a}} - I_{\text{a}}R_{\text{a}})}{K_{\text{f}}I_{\text{f}}} = \frac{\left(V_{\text{a}} - \frac{T_{\text{load}}R_{\text{a}}}{K_{\text{f}}I_{\text{f}}}\right)}{K_{\text{f}}I_{\text{f}}} \qquad (11.6)$$

From Eq. 11.6, recognizing that the armature resistance voltage drop $I_{\text{a}}R_{\text{a}}$ is generally quite small in comparison to the armature voltage V_{a}, we see that for a given load torque, the motor speed will increase with decreasing field current and decrease as the field current is increased. The lowest speed obtainable is that corresponding to maximum field current (the field current is limited by heating considerations); the highest speed is limited mechanically by the mechanical integrity of the rotor and electrically by the effects of armature reaction under weak-field conditions giving rise to poor commutation.

Armature current is typically limited by motor cooling capability. In many dc motors, cooling is aided by a shaft-driven fan whose cooling capacity is a function of motor speed. To examine in an approximate fashion the limitations on the allowable continuous motor output as the speed is changed, we will neglect the influence of changing ventilation and assume that the armature current I_{a} cannot exceed its rated value, in order to insure that the motor will not overheat. In addition, in our approximate argument we will neglect the effect of rotational losses (which of course also change with motor speed). Because the voltage drop across the armature resistance is relatively small, the speed voltage E_{a} will remain essentially constant at a value slightly below the applied armature voltage; any change in field current will be compensated for by a change in motor speed.

Thus under constant-terminal-voltage operation with varying field current, the $E_{\text{a}}I_{\text{a}}$ product, and hence the allowable motor output power, remain substantially constant as the speed is varied. A dc motor controlled in this fashion is referred to as a *constant-power drive*. Torque, however, varies directly with field flux and therefore has its highest allowable value at the highest field current and hence lowest speed. Field-current control is thus best suited to drives requiring increased torque at low speeds. When a motor so controlled is used with a load requiring constant torque over the speed range, the rating and size of the machine are determined by the product of the torque and the highest speed. Such a drive is inherently oversized at the lower speeds, which is the principal economic factor limiting the practical speed range of large motors.

EXAMPLE 11.2

With an armature terminal voltage of 240 V and with a shunt-field current of 0.34 A, the no-load speed of the dc motor of Example 11.1 is found to be 3600 r/min. In this example, the motor is assumed to be driving a load which varies with speed as

$$P_{\text{load}} = 22.4 \left(\frac{n}{3600}\right)^3 \text{ kW}$$

where n is the motor speed in r/min. A rheostat is to be installed in series with the shunt field to vary the speed. Assuming the armature terminal voltage to remain constant at 240 V, calculate the required resistance range if the speed is to be varied between 1800 and 3600 r/min. The effect of rotational losses can be ignored.

■ Solution

The load torque is equal to the load power divided by the motor speed ω_m expressed in rad/sec. First expressing the power in terms of ω_m

$$P_{\text{load}} = 22.4 \left(\frac{\omega_m}{120\pi} \right)^3 \text{ kW}$$

The load torque is then given by

$$T_{\text{load}} = \frac{P_{\text{load}}}{\omega_m} = 22.4 \left(\frac{\omega_m^2}{(120\pi)^3} \right) = 4.18 \times 10^{-4} \omega_m^2 \text{ N} \cdot \text{m}$$

Thus, at 1800 r/min, $\omega_m = 60\pi$ and $T_{\text{load}} = 14.9$ N·m. At 3600 r/min, $\omega_m = 120\pi$ and $T_{\text{load}} = 59.4$ N·m.

Before solving for I_f, we must find the value of K_f, which can be found from the no-load data. Specifically, we see that with a terminal voltage of 240-V and at a no-load speed of 3600 r/min ($\omega_m = 120\pi$), the corresponding field current is 0.34 A. Since under no-load conditions $E_a \approx V_a$, we can find K_f from Eq. 11.3 as

$$K_f = \frac{E_a}{I_f \omega_m} = \frac{240}{0.34 \times 120\pi} = 1.87 \text{ V/(A} \cdot \text{rad/sec)}$$

To find the required field current, we can solve Eq. 11.6 for I_f

$$I_f = \frac{V_a}{2K_f \omega_m} \left(1 \pm \sqrt{1 - \frac{4\omega_m T_{\text{load}} R_a}{V_a^2}} \right)$$

Recognizing that R_a is small and hence that $I_f \approx V_a/(K_f \omega_m)$ we see that the positive sign should be used and thus

$$I_f = \frac{V_a}{2K_f \omega_m} \left(1 + \sqrt{1 - \frac{4\omega_m T_{\text{load}} R_a}{V_a^2}} \right)$$

Once the field current has been found, the total field resistance can be found as

$$(R_f)_{\text{total}} = \frac{V_a}{I_f} = \frac{240}{I_f}$$

and the required added rheostat resistance can be found by subtracting the resistance of the shunt-field winding (187 Ω) from $(R_f)_{\text{total}}$.

This leads to the following table:

r/min	T_{load} [N · m]	I_f [A]	$(R_f)_{\text{total}}$ [Ω]	R_{rheostat} [Ω]
1800	14.9	0.678	354	167
3600	59.4	0.334	719	532

Thus, the rheostat must be able to cover the range from 166 Ω to 532 Ω.

The rheostat of Example 11.2 is to be replaced by a duty cycle controller operating from the 240-V dc supply. Calculate the duty-cycle range required to achieve operation over the speed range of 1800–3600 r/min as specified in Example 11.2.

Solution

$0.26 \leq D \leq 0.53$

Armature-Circuit Resistance Control   *Armature-circuit resistance control* provides a means of obtaining reduced speed by the insertion of external series resistance in the armature circuit. It can be used with series, shunt, and compound motors; for the last two types, the series resistor must be connected between the shunt field and the armature, not between the line and the motor. It is a common method of speed control for series motors and is generally analogous in action to wound-rotor-induction-motor control by the addition of external series rotor resistance.

Depending upon the value of the series armature resistance, the speed may vary significantly with load, since the speed depends on the voltage drop in this resistance and hence on the armature current demanded by the load. For example, a 1200-r/min shunt motor whose speed under load is reduced to 750 r/min by series armature resistance will return to almost 1200-r/min operation if the load is removed because the no-load current produces a voltage drop across the series resistance which is insignificant. The disadvantage of poor speed regulation may not be important in a series motor, which is used only where varying-speed service is required or can be tolerated.

A significant disadvantage of this method of speed control is that the power loss in the external resistor is large, especially when the speed is greatly reduced. In fact, for a constant-torque load, the power input to the motor plus resistor remains constant, while the power output to the load decreases in proportion to the speed. Operating costs are therefore comparatively high for lengthy operation at reduced speeds. Because of its low initial cost however, the series-resistance method (or the variation of it discussed in the next paragraph) will often be attractive economically for applications which require only short-time or intermittent speed reduction. Unlike field-current control, armature-resistance control results in a *constant-torque drive* because both the field-flux and, to a first approximation, the allowable armature current remain constant as speed changes.

A variation of this control scheme is given by the *shunted-armature method*, which may be applied to a series motor, as in Fig. 11.3a, or a shunt motor, as in Fig. 11.3b. In effect, resistors R_1 and R_2 act as a voltage divider applying a reduced voltage to the armature. Greater flexibility is possible because two resistors can now be adjusted to provide the desired performance. For series motors, the no-load speed can be adjusted to a finite, reasonable value, and the scheme is therefore applicable to the production of slow speeds at light loads. For shunt motors, the speed regulation in the low-speed range is appreciably improved because the no-load speed is definitely lower than the value with no controlling resistors.

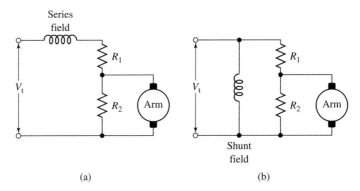

Series
field

R_1

V_t

R_2 Arm

(a)

R_1

V_t

R_2 Arm

Shunt
field

(b)

Figure 11.3 Shunted-armature method of speed control applied to (a) a series motor and (b) a shunt motor.

Armature-Terminal Voltage Control *Armature-terminal voltage control* can be readily accomplished with the use of power-electronic systems such as those discussed in Chapter 10. Figure 11.4 shows in somewhat schematic form three possible configurations. In Fig. 11.4a, a phase-controlled rectifier in combination with a dc-link filter capacitor can be used to produce a variable dc-link voltage which can be applied directly to the armature terminals of the dc motor.

In Fig. 11.4b, a constant dc-link voltage is produced by a diode rectifier in combination with a dc-link filter capacitor. The armature terminal voltage is then varied by a pulse-width modulation scheme in which switch S is alternately opened and closed. When switch S is closed, the armature voltage is equal to the dc-link voltage V_{dc}, and when the switch is opened, current transfers to the freewheeling diode, essentially setting the armature voltage to zero. Thus the average armature voltage under this condition is equal to

$$V_a = DV_{dc} \tag{11.7}$$

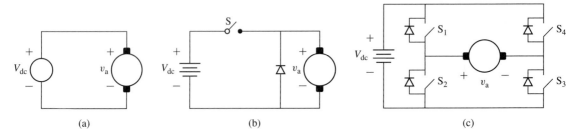

(a)

(b)

(c)

Figure 11.4 Three typical configurations for armature-voltage control. (a) Variable dc-link voltage (produced by a phase-controlled rectifier) applied directly to the dc-motor armature terminals. (b) Constant dc-link voltage with single-polarity pulse-width modulation. (c) Constant dc-link voltage with a full H-bridge.

where

V_a = average armature voltage (V)

V_{dc} = dc-link voltage (V)

D = PWM duty cycle (fraction of time that switch S is closed)

Figure 11.4c shows an H-bridge configuration as is discussed in the context of inverters in Section 10.3.3. Note that if switch S3 is held closed while switch S4 remains open, this configuration reduces to that of Fig. 11.4b. However, the H-bridge configuration is more flexible because it can produce both positive- and negative-polarity armature voltage. For example, with switches S1 and S3 closed, the armature voltage is equal to V_{dc} while with switches S2 and S4 closed, the armature voltage is equal to $-V_{dc}$. Clearly, using such an H-bridge configuration in combination with an appropriate choice of control signals to the switches allows this PWM system to achieve any desired armature voltage in the range $-V_{dc} \leq V_a \leq V_{dc}$.

Armature-voltage control takes advantage of the fact that, because the voltage drop across the armature resistance is relatively small, a change in the armature terminal voltage of a shunt motor is accompanied in the steady state by a substantially equal change in the speed voltage. With constant shunt field current and hence field flux, this change in speed voltage must be accompanied by a proportional change in motor speed. Thus, motor speed can be controlled directly by means of the armature terminal voltage.

EXAMPLE 11.3

A 500-V, 100-hp, 2500 r/min, separately excited dc motor has the following parameters:

Field resistance:	$R_f = 109\ \Omega$
Rated field voltage:	$V_{f0} = 300$ V
Armature resistance:	$R_a = 0.084\ \Omega$
Geometric constant:	$K_f = 0.694$ V/(A · rad/sec)

Assuming the field voltage to be held constant at 300 V, use MATLAB[†] to plot the motor speed as a function of armature voltage with the motor operating under no-load and also under rated full-load torque as the armature voltage is varied from 250 V to 500 V.

■ **Solution**

From Eq. 11.4

$$I_a = \frac{T_{mech}}{K_f I_f}$$

and from Eq. 11.5

$$I_a = \frac{V_a - E_a}{R_a} = \frac{V_a - K_f I_f \omega_m}{R_a}$$

[†] MATLAB is a registered trademark of The MathWorks, Inc.

Hence we can solve for ω_m

$$\omega_m = \frac{V_a - \left(\frac{T_{mech}R_a}{K_f I_f}\right)}{K_f I_f}$$

and the speed in r/min as

$$n = \left(\frac{30}{\pi}\right)\omega_m$$

Finally, the field current is

$$I_f = \frac{V_f}{R_f} = \frac{300}{109} = 2.75 \text{ A}$$

and the rated full-load torque is given by

$$T_{rated} = \frac{P_{rated}}{(\omega_m)_{rated}} = \frac{100 \times 746}{2500 \times \left(\frac{\pi}{30}\right)} = 285 \text{ N} \cdot \text{m}$$

Figure 11.5 is the desired plot. Notice that the speed drops approximately 63 r/min as the torque is increased from zero to full-load, independent of the armature voltage and machine speed.

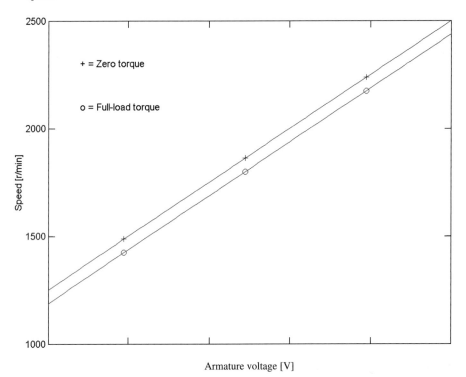

Figure 11.5   A plot of speed versus armature voltage for the dc motor of Example 11.3.

Here is the MATLAB script:

```
clc
clear

% Motor parameters
Rf = 109;
Ra = 0.084;
Kf = 0.694;

% Constant field voltage
Vf = 300;

% Resulting field current
If = Vf/Rf;

% Rated speed in rad/sec
omegarated = 2500*(pi/30);

% Rated power in Watts
Prated = 100*746;

% Rated torque in N-m
Trated = Prated/omegarated;

% Vary the armature voltage from 250 to 500 V
% and calculate speed.
for n=1:101
  Va(n) = 250 * (1 + (n-1)/100);

  % Zero torque
  T = 0;
  omega = (Va(n)- T*Ra/(Kf*If))/(Kf*If);
  NoLoadRPM(n) = omega*30/pi;

  % Full-load torque
  T = Trated;
  omega = (Va(n)- T*Ra/(Kf*If))/(Kf*If);
  FullLoadRPM(n) = omega*30/pi;
end

plot(Va,NoLoadRPM)
hold
plot(Va(20),NoLoadRPM(20),'+')
plot(Va(50),NoLoadRPM(50),'+')
plot(Va(80),NoLoadRPM(80),'+')
plot(Va,FullLoadRPM)
plot(Va(20),FullLoadRPM(20),'o')
plot(Va(50),FullLoadRPM(50),'o')
plot(Va(80),FullLoadRPM(80),'o')
hold

xlabel('Armature voltage [V]')
```

```
ylabel('Speed [r/min]')
text(270,2300,'+ = Zero torque')
text(270,2100,'o = Full-load torque')
```

Practice Problem 11.3

Calculate the change in armature voltage required to maintain the motor of Example 11.3 at a speed of 2000 r/min as the load is changed from zero to full-load torque.

Solution

12.5 V

Frequently the control of motor voltage is combined with field-current control in order to achieve the widest possible speed range. With such dual control, base speed can be defined as the normal-armature-voltage, full-field speed of the motor. Speeds above base speed are obtained by reducing the field current; speeds below base speed are obtained by armature-voltage control. As discussed in connection with field-current control, the range above base speed is that of a constant-power drive. The range below base speed is that of a constant-torque drive because, as in armature-resistance control, the flux and the allowable armature current remain approximately constant. The overall output limitations are therefore as shown in Fig. 11.6a for approximate allowable torque and in Fig. 11.6b for approximate allowable power. The constant-torque characteristic is well suited to many applications in the machine-tool industry, where many loads consist largely of overcoming the friction of moving parts and hence have essentially constant torque requirements.

The speed regulation and the limitations on the speed range above base speed are those already presented with reference to field-current control; the maximum speed thus does not ordinarily exceed four times base speed and preferably not twice base speed. For conventional machines, the lower limit for reliable and stable operation is

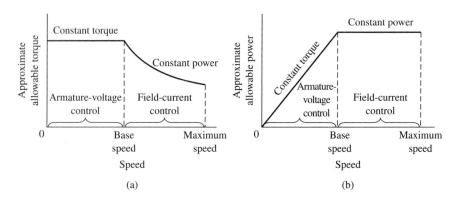

Figure 11.6 (a) Torque and (b) power limitations of combined armature-voltage and field-current methods of speed control.

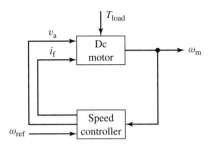

Figure 11.7 Block diagram for a speed-control system for a separately excited or shunt-connected dc motor.

about one-tenth of base speed, corresponding to a total maximum-to-minimum range not exceeding 40:1.

With armature reaction ignored, the decrease in speed from no-load to full-load torque is caused entirely by the full-load armature-resistance voltage drop in the dc generator and motor. This full-load armature-resistance voltage drop is constant over the voltage-control range, since full-load torque and hence full-load current are usually regarded as constant in that range. When measured in r/min, therefore, the speed decrease from no-load to full-load torque is a constant, independent of the no-load speed, as we saw in Example 11.3. The torque-speed curves accordingly are closely approximated by a series of parallel straight lines for the various motor-field adjustments. Note that a speed decrease of, say, 40 r/min from a no-load speed of 1200 r/min is often of little importance; a decrease of 40 r/min from a no-load speed of 120 r/min, however, may at times be of critical importance and require corrective steps in the layout of the system.

Figure 11.7 shows a block diagram of a feedback-control system that can be used to regulate the speed of a separately excited or shunt-connected dc motor. The inputs to the dc-motor block include the armature voltage and the field current as well as the load torque T_{load}. The resultant motor speed ω_m is fed back to a controller block which represents both the control logic and power electronics and which controls the armature voltage and field current applied to the dc motor, based upon a reference speed signal ω_{ref}. Depending upon the design of the controller, with such a scheme it is possible to control the steady-state motor speed to a high degree of accuracy independent of the variations in the load torque.

EXAMPLE 11.4

Figure 11.8 shows the block diagram for a simple speed control system to be applied to the dc motor of Example 11.3. In this controller, the field voltage is held constant (not shown) at its rated value of 300 V. Thus, the control is applied only to the armature voltage and takes the form

$$V_a = V_{a0} + G(\omega_{ref} - \omega_m)$$

where V_{a0} is the armature voltage when $\omega_m = \omega_{ref}$ and G is a multiplicative constant.

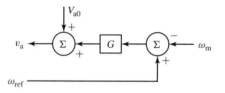

Figure 11.8 Simple dc-motor speed
controller for Example 11.4.

With the reference speed set to 2000 r/min ($\omega_{ref} = 2000 \times \pi/30$), calculate V_{a0} and G so
that the motor speed is 2000 r/min at no load and drops only by 25 r/min when the torque is
increased to its rated full-load value.

■ Solution

As was found in Example 11.3, the field current under this condition will be 2.75 A. At no
load, 2000 r/min,

$$V_a \approx E_a = K_f I_f \omega_m = 0.694 \times 2.75 \times 2000 \left(\frac{\pi}{30}\right) = 400 \text{ V}$$

and thus $V_{a0} = 400$ V.

The full load torque was found in Example 11.3 to be $T_{rated} = 285$ N·m and thus the
armature current required to achieve rated full-load torque can be found from Eq. 11.4

$$I_a = \frac{T_{rated}}{K_f I_f} = \frac{285}{0.694 \times 2.75} = 149 \text{ A}$$

At a speed of 1975 r/min, E_a will be given by

$$E_a = K_f I_f \omega_m = 0.694 \times 2.75 \times 1975 \left(\frac{\pi}{30}\right) = 395 \text{ V}$$

and thus

$$V_a = E_a + I_a R_a = 395 + 149 \times 0.084 = 408 \text{ V}$$

Solving for G gives

$$G = \frac{V_a - V_{a0}}{\omega_{ref} - \omega_m} = \frac{408 - 400}{(2000 - 1975)\left(\frac{\pi}{30}\right)} = 3.06 \text{ V} \cdot \text{sec/rad}$$

Practice Problem 11.4

If the load torque in Example 11.4 is equal to half of the rated full-load torque, calculate (*a*) the
speed of the motor and (*b*) the corresponding load power.

Solution

    a. 1988 r/min
    b. 29.6 kW

In the case of permanent-magnet dc motors, the field flux is, of course, fixed by the permanent magnet (with the possible exception of any effects of temperature changes on the magnet properties as the motor heats up). From Eqs. 11.3 and 11.4, we see that the voltage generated voltage can be written in the form

$$E_a = K_m \omega_m \qquad (11.8)$$

and that the electromagnetic torque can be written as

$$T_{mech} = K_m I_a \qquad (11.9)$$

Comparison of Eqs. 11.8 and 11.9 with Eqs. 11.3 and 11.4 show that the analysis of a permanent-magnet dc motor is identical to that of a shunt or separately excited dc motor with the exception that the torque-constant K_m must be substituted for the term $K_f I_f$.

EXAMPLE 11.5

The permanent-magnet dc motor of Example 7.9 has an armature resistance of 1.03 Ω and a torque constant $K_m = 0.22$ V/(rad/sec). Assume the motor to be driving a constant power load of 800 W (including rotational losses), and calculate the motor speed as the armature voltage is varied from 40 to 50 V.

■ **Solution**
The motor power output (including rotational losses) is given by the product $E_a I_a$ and thus we can write

$$P_{load} = E_a I_a = K_m \omega_m I_a$$

Solving for ω_m gives

$$\omega_m = \frac{P_{load}}{K_m I_a}$$

The armature current can be written as

$$I_a = \frac{(V_a - E_a)}{R_a} = \frac{(V_a - K_m \omega_m)}{R_a}$$

These two equations can be combined to give an equation for ω_m of the form

$$\omega_m^2 - \left(\frac{V_a}{K_m}\right) \omega_m + \frac{P_{load} R_a}{K_m^2} = 0$$

from which we can find

$$\omega_m = \frac{V_a}{2K_m} \left[1 \pm \sqrt{1 - \frac{4 P_{load} R_a}{V_a^2}} \right]$$

Recognizing that, if the voltage drop across the armature resistance is small, $V_a \approx E_a = K_m \omega_m$, we pick the positive sign and thus

$$\omega_m = \frac{V_a}{2K_m} \left[1 + \sqrt{1 - \frac{4 P_{load} R_a}{V_a^2}} \right]$$

Substituting values, we find that for $V_a = 40$ V, $\omega_m = 169.2$ rad/sec (1616 r/min) and for $V_a = 50$ V, $\omega_m = 217.5$ rad/sec (2077 r/min).

Calculate the speed variation (in r/min) of the permanent-magnet dc motor of Example 11.5 if the armature voltage is held constant at 50 V and the load power varies from 100 W to 500 W.

Solution

2077 r/min to 1540 r/min

11.1.2 Torque Control

As we have seen, the electromagnetic torque of a dc motor is proportional to the armature current I_a and is given by

$$T_{mech} = K_f I_f I_a \tag{11.10}$$

in the case of a separately excited or shunt motor and

$$T_{mech} = K_m I_a \tag{11.11}$$

in the case of a permanent-magnet motor.

From these equations we see that torque can be controlled directly by controlling the armature current. Fig. 11.9 shows three possible configurations. In Fig. 11.9a, a phase-controlled rectifier, in combination with a dc-link filter inductor, can be used to create a variable dc-link current which can be applied directly to the armature terminals of the dc motor.

In Fig. 11.9b, a constant dc-link current is produced by a diode rectifier. The armature terminal voltage is then varied by a pulse-width modulation scheme in which switch S is alternately opened and closed. When switch S is opened, the current I_{dc} flows into the dc-motor armature while when switch S is closed, the armature is

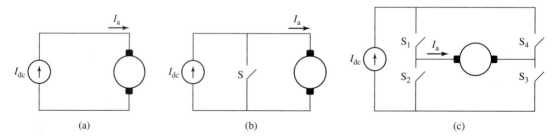

Figure 11.9 Three typical configurations for armature-current control. (a) Variable dc-link current (produced by a phase-controlled rectifier) applied directly to the dc-motor armature terminals. (b) Constant dc-link current with single-polarity pulse-width modulation. (c) Constant dc-link current with a full H-bridge.

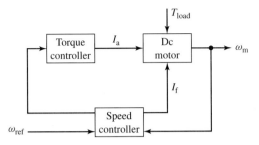

Figure 11.10 Block diagram of a dc-motor speed-control system using direct-control of motor torque.

shorted and I_a decays. Thus, the duty cycle of switch S will control the average current into the armature.

Finally, Fig 11.9c shows an H-bridge configuration as is discussed in the context of inverters in Section 10.3.2. Appropriate control of the four switches S1 through S4 allows this PWM system to achieve any desired armature average current in the range $-I_{dc} \leq I_a \leq I_{dc}$.

Note that in each of the PWM configurations of Fig. 11.9b and c, rapid changes in instantaneous current through the dc machine armature can give rise to large voltage spikes, which can damage the machine insulation as well as give rise to flashover and voltage breakdown of the commutator. In order to eliminate these effects, a practical system must include some sort of filter across the armature terminals (such as a large capacitor) to limit the voltage rise and to provide a low-impedance path for the high-frequency components of the drive current.

Figure 11.10 shows a typical configuration in which the torque control is surrounded by a speed-feedback loop. This looks similar to the speed control of Fig. 11.7. However, instead of controlling the armature voltage, in this case the output of the speed controller is a torque reference signal T_{ref} which in turn serves as the input to the torque controller. One advantage of such a system is that it automatically limits the dc-motor armature current to acceptable levels under all operating conditions, as is shown in Example 11.6.

EXAMPLE 11.6

Consider the 100-hp dc motor of Examples 11.3 and 11.4 to be driving a load whose torque varies linearly with speed such that it equals rated full-load torque (285 N·m) at a speed of 2500 r/min. We will assume the combined moment of inertia of the motor and load to equal 0.92 kg·m². The field voltage is to be held constant at 300 V.

a. Calculate the armature voltage and current required to achieve speeds of 2000 and 2500 r/min.

b. Assume that the motor is operated from an armature-voltage controller and that the armature voltage is suddenly switched from its 2000 r/min to its 2500 r/min value. Calculate the resultant motor speed and armature current as a function of time.

c. Assume that the motor is operated from an armature-current controller and that the armature current is suddenly switched from its 2000 r/min to its 2500 r/min value. Calculate the resultant motor speed as a function of time.

■ **Solution**

a. Neglecting any rotational losses, the armature current can be found from Eq. 11.4 by setting $T_{mech} = T_{load}$

$$I_a = \frac{T_{load}}{K_f I_f}$$

Substituting

$$T_{load} = \left(\frac{n}{n_f}\right) T_{fl}$$

where n is the motor speed in r/min, $n_f = 2500$ r/min and $T_{fl} = 285$ N·m gives

$$I_a = \frac{n T_{fl}}{n_f K_f I_f}$$

Solving for $V_a = E_a + I_a R_a$ then allows us to complete the following table:

r/min	ω_m [rad/sec]	V_a [V]	I_a [A]	T_{load} [N · m]
2000	209	410	119	228
2500	262	513	149	285

b. The dynamic equation governing the speed of the motor is

$$J \frac{d\omega_m}{dt} = T_{mech} - T_{load}$$

Substituting $\omega_m = (\pi/30)n$ and $\omega_r = (\pi/30)n_f$ we can write

$$T_{load} = \left(\frac{T_{fl}}{\omega_f}\right) \omega_m$$

Under armature-voltage control,

$$T_{mech} = K_f I_f I_a = K_f I_f \left(\frac{V_a - E_a}{R_a}\right)$$

$$= K_f I_f \left(\frac{V_a - K_f I_f \omega_m}{R_a}\right)$$

and thus the governing differential equation is

$$J \frac{d\omega_m}{dt} = K_f I_f \left(\frac{V_a - K_f I_f \omega_m}{R_a}\right) - \left(\frac{T_{fl}}{\omega_f}\right) \omega_m$$

or

$$\frac{d\omega_m}{dt} + \frac{1}{J}\left(\frac{T_{fl}}{\omega_f} + \frac{(K_f I_f)^2}{R_a}\right) \omega_m - \frac{K_f I_f V_a}{J R_a}$$

$$= \frac{d\omega_m}{dt} + 48.4\,\omega_m - 24.7 V_a = 0$$

From this differential equation, we see that with the motor initially at $\omega_m = \omega_i = 209$ rad/sec, if the armature voltage V_a is suddenly switched from $V_i = 413$ V to $V_f = 513$ V, the speed will rise exponentially to $\omega_m = \omega_f = 262$ rad/sec as

$$\omega_m = \omega_f + (\omega_i - \omega_f)e^{-t/\tau}$$

$$= 262 - 53e^{-t/\tau} \text{ rad/sec}$$

where $\tau = 1/48.4 = 20.7$ msec. Expressed in terms of r/min

$$n = 2500 - 50e^{-t/\tau} \text{ r/min}$$

The armature current will decrease exponentially with the same 20.7 msec time constant from an initial value of $(V_f - V_i)/R_a = 1190$ A to its final value of 149 A. Thus,

$$I_a = 149 + 1041e^{-t/\tau} \text{ A}$$

Notice that it is unlikely that the supply to the dc motor can supply this large initial current (eight times the rated full-load armature current) and, in addition, the high current and corresponding high torque could potentially cause damage to the dc motor commutator, brushes, and armature winding. Hence, as a practical matter, a practical controller would undoubtedly limit the rate of change of the armature voltage to avoid such sudden steps in voltage, with the result that the speed change would not occur as rapidly as calculated here.

c. The dynamic equation governing the speed of the motor remains the same as that in part (b) as does the equation for the load torque. However, in this case, because the motor is being operated from a current controller, the electromagnetic torque will remain constant at $T_{mech} = T_f = 285$ N·m after the current is switched from its initial value of 119 A to its final value of 149 A.

Thus

$$J\frac{d\omega_m}{dt} = T_{mech} - T_{load} = T_f - \left(\frac{T_f}{\omega_f}\right)\omega_m$$

or

$$\frac{d\omega_m}{dt} + \left(\frac{T_{fl}}{J\omega_f}\right)\omega_m - \frac{T_f}{J}$$

$$= \frac{d\omega_m}{dt} + 1.18\omega_m - 310 = 0$$

In this case, the speed will rise exponentially to $\omega_m = \omega_f = 262$ rad/sec as

$$\omega_m = \omega_f + (\omega_i - \omega_f)e^{-t/\tau}$$

$$= 262 - 53e^{-t/\tau} \text{ rad/sec}$$

where now the time constant $\tau = 1/1.18 = 845$ msec.

Clearly, the change in motor speed under the current controller is much slower. However, at no point during this transient do either the motor current or the motor torque exceed their rated value. In addition, should faster response be desired, the armature current (and hence motor torque) could be set temporarily to a fixed value higher than the rated value (e.g., two or three times rated as compared to the factor of 8 found in part (b)), thus limiting the potential for damage to the motor.

Practice Problem 11.6

Consider the dc motor/load combination of Example 11.6 operating under current (torque) control to be operating in the steady-state at a speed of 2000 r/min at an armature current of 119 A. If the armature current is suddenly switched to 250 A, calculate the time required for the motor to reach a speed of 2500 r/min.

Solution

0.22 sec

11.2 CONTROL OF SYNCHRONOUS MOTORS

11.2.1 Speed Control

As discussed in Chapters 4 and 5, synchronous motors are essentially constant-speed machines, with their speed being determined by the frequency of the armature currents as described by Eqs. 4.40 and 4.41. Specifically, Eq. 4.40 shows that the synchronous angular velocity is proportional to the electrical frequency of the applied armature voltage and inversely proportional to the number of poles in the machine

$$\omega_s = \left(\frac{2}{\text{poles}}\right)\omega_e \tag{11.12}$$

where

ω_s = synchronous spatial angular velocity of the air-gap mmf wave [rad/sec]

$\omega_e = 2\pi f_e$ = angular frequency of the applied electrical excitation [rad/sec]

f_e = applied electrical frequency [Hz]

Clearly, the simplest means of synchronous motor control is speed control via control of the frequency of the applied armature voltage, driving the motor by a polyphase voltage-source inverter such as the three-phase inverter shown in Fig. 11.11. As is discussed in Section 10.3.3, this inverter can either be used to supply stepped ac voltage waveforms of amplitude V_{dc} or the switches can be controlled to produce pulse-width-modulated ac voltage waveforms of variable amplitude. The dc-link voltage V_{dc} can itself be varied, for example, through the use of a phase-controlled rectifier.

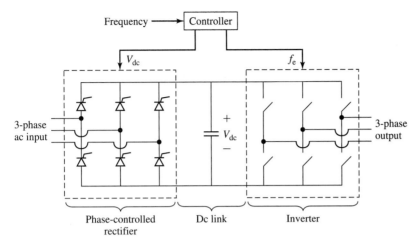

Figure 11.11 Three-phase voltage-source inverter.

The frequency of the inverter output waveforms can of course be varied by controlling the switching frequency of the inverter switches. For ac-machine applications, coupled with this frequency control must be control of the amplitude of the applied voltage, as we will now see.

From Faraday's Law, we know that the air-gap component of the armature voltage in an ac machine is proportional to the peak flux density in the machine and the electrical frequency. Thus, if we neglect the voltage drop across the armature resistance and leakage reactance, we can write

$$V_a = \left(\frac{f_e}{f_{rated}} \right) \left(\frac{B_{peak}}{B_{rated}} \right) V_{rated} \tag{11.13}$$

where V_a is the amplitude of the armature voltage, f_e is the operating frequency, and B_{peak} is the peak air-gap flux density. V_{rated}, f_{rated}, and B_{rated} are the corresponding rated-operating-point values.

Consider a situation in which the frequency of the armature voltage is varied while its amplitude is maintained at its rated value ($V_a = V_{rated}$). Under these conditions, from Eq. 11.13 we see that

$$B_{peak} = \left(\frac{f_{rated}}{f_e} \right) B_{rated} \tag{11.14}$$

Equation 11.14 clearly demonstrates the problem with constant-voltage, variable-frequency operation. Specifically, for a given armature voltage, the machine flux density is inversely proportional to frequency and thus as the frequency is reduced, the flux density will increase. Thus for a typical machine which operates in saturation at rated voltage and frequency, any reduction in frequency will further increase the flux density in the machine. In fact, a significant drop in frequency will increase the flux density to the point of potential machine damage due both to increased core loss and to the increased machine currents required to support the higher flux density.

As a result, for frequencies less than or equal to rated frequency, it is typical to operate a machine at constant flux density. From Eq. 11.13, with $B_{peak} = B_{rated}$

$$V_a = \left(\frac{f_e}{f_{rated}} \right) V_{rated} \tag{11.15}$$

which can be rewritten as

$$\frac{V_a}{f_e} = \frac{V_{rated}}{f_{rated}} \tag{11.16}$$

From Eq. 11.16, we see that constant-flux operation can be achieved by maintaining a constant ratio of armature voltage to frequency. This is referred to as *constant-volts-per-hertz* (constant V/Hz) operation. It is typically maintained from rated frequency down to the low frequency at which the armature resistance voltage drop becomes a significant component of the applied voltage.

Similarly, we see from Eq. 11.13 that if the machine is operated at frequencies in excess of rated frequency with the voltage at its rated value, the air-gap flux density will drop below its rated value. Thus, in order to maintain the flux density at its rated value, it would be necessary to increase the terminal voltage for frequencies in excess of rated frequency. In order to avoid insulation damage, it is common to maintain the machine terminal voltage at its rated value for frequencies in excess of rated frequency.

The machine terminal current is typically limited by thermal constraints. Thus, provided the machine cooling is not affected by rotor speed, the maximum permissible terminal current will remain constant at its rated value I_{rated}, independent of the applied frequency. As a result, for frequencies below rated frequency, with V_a proportional to f_e, the maximum machine power will be proportional to $f_e V_{rated} I_{rated}$. The maximum torque under these conditions can be found by dividing the power by the rotor speed ω_s, which is also proportional to f_e as can be seen from Eq. 11.12. Thus, we see that the maximum torque is proportional to $V_{rated} I_{rated}$, and hence it is constant at its rated-operating-point value.

Similarly, for frequencies in excess of rated frequency, the maximum power will be constant and equal to $V_{rated} I_{rated}$. The corresponding maximum torque will then vary inversely with machine speed as $V_{rated} I_{rated}/\omega_s$. The maximum operating speed for this operating regime will be determined either by the maximum frequency which can be supplied by the drive electronics or by the maximum speed at which the rotor can be operated without risk of damage due to mechanical concerns such as excessive centrifugal force or to the presence of a resonance in the shaft system.

Figure 11.12 shows a plot of maximum power and maximum torque versus speed for a synchronous motor under variable-frequency operation. The operating regime below rated frequency and speed is referred to as the *constant-torque regime* and that above rated speed is referred to as the *constant-power regime*.

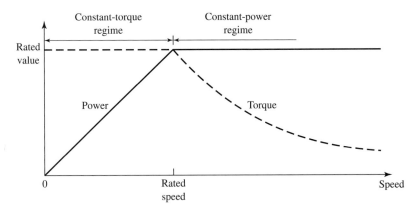

Figure 11.12 Variable-speed operating regimes for a synchronous motor.

EXAMPLE 11.7

The 45-kVA, 220-V, 60-Hz, six-pole, three-phase synchronous machine of Example 5.4 is to be operated as a motor and driven from a variable-frequency, three-phase voltage-source inverter which provides 220 V at 60 Hz and which maintains constant V/Hz as the frequency is reduced. The machine has a saturated synchronous reactance of 0.836 per unit and achieves rated open-circuit voltage at a field current of 2.84 A. For the purposes of this example, assume that the motor losses are negligible.

a. With the motor operating at 60 Hz, 220 V and at rated power, unity power factor, calculate (i) the motor speed in r/min and (ii) the motor field current.
b. If the inverter frequency is reduced to 50 Hz and the motor load adjusted to rated torque, calculate the (i) the resulting motor speed and (ii) and the motor field current required to again achieve unity power factor.

■ **Solution**

a. (i) The motor will operate at its synchronous speed which can be found from Eq. 4.41

$$n_s = \left(\frac{120}{\text{poles}}\right) f_e = \left(\frac{120}{6}\right) 60 = 1200 \text{ r/min}$$

(ii) As seen in Chapter 5, the field current can be determined from the generated voltage. For motor operation,

$$\hat{E}_{af} = \hat{V}_a - j X_s \hat{I}_a = 1.0 - j0.836 \times 1.0 = 1.30 \angle -39.9°$$

where V_a has been chosen as the reference phasor. Thus the field current is

$$I_f = 1.30 \times 2.84 = 3.69 \text{ A}$$

Note that we have chosen to solve for E_{af} in per unit. A solution in real units would have of course produced the same result.

b. (i) When the frequency is reduced from 60 Hz to 50 Hz, the motor speed will drop from 1200 r/min to 1000 r/min.

(*ii*) Let us again consider the equation for the generated voltage

$$\hat{E}_{af} = \hat{V}_a - jX_s\hat{I}_a$$

where here we will assume that the equation is written in real units.

As the inverter frequency is reduced from 60 Hz, the inverter voltage will drop proportionally since the inverter maintains constant V/Hz. Thus we can write

$$V_a = \left(\frac{\omega_m}{\omega_{m0}}\right) V_{a0}$$

where the subscript 0 is used to indicate a 60-Hz value as found in part (a). Reactance is also proportional to frequency and thus

$$X_s = \left(\frac{\omega_m}{\omega_{m0}}\right) X_{s0}$$

The generated voltage is proportional to both the motor speed (and hence the frequency) and the field current, and thus we can write

$$E_{af} = \left(\frac{\omega_m}{\omega_{m0}}\right)\left(\frac{I_f}{I_{f0}}\right) E_{af0}$$

Finally, if we recognize that, to operate at rated torque and unity power factor under this reduced frequency condition, the motor armature current will have to be equal to the value found in part (a), i.e., $I_a = I_{a0}$, we can write the generated voltage equation as

$$\left(\frac{\omega_m}{\omega_{m0}}\right)\left(\frac{I_f}{I_{f0}}\right)\hat{E}_{af0} = \left(\frac{\omega_m}{\omega_{m0}}\right)\hat{V}_{a0} - j\left(\frac{\omega_m}{\omega_{m0}}\right) X_{s0}\hat{I}_{a0}$$

or

$$\left(\frac{I_f}{I_{f0}}\right)\hat{E}_{af0} = \hat{V}_{a0} - jX_{s0}\hat{I}_{a0}$$

Since the subscripted quantities correspond to the solution of part (a), they must satisfy

$$\hat{E}_{af0} = \hat{V}_{a0} - jX_{s0}\hat{I}_{a0}$$

and thus we see that we must have $I_f = I_{f0}$. In other words, the field current for this operating condition is equal to that found in part (a), or $I_f = 3.69$ A.

Practice Problem 11.7

Consider 50-Hz operation of the synchronous motor of Example 11.7, part (b). If the load torque is reduced to 75 percent of rated torque, calculate the field current required to achieve unity power factor.

Solution

3.35 A

Although during steady-state operation the speed of a synchronous motor is determined by the frequency of the drive, speed control by means of frequency control is of limited use in practice. This is due in most part to the fact that it is difficult for the rotor of a synchronous machine to track arbitrary changes in the frequency of the applied armature voltage. In addition, starting is a major problem, and, as a result, the rotors of synchronous motors are often equipped with a squirrel-cage winding known as an *amortisseur* or *damper winding* similar to the squirrel-cage winding in an induction motor, as shown in Fig. 5.3. Following the application of a polyphase voltage to the armature, the rotor will come up almost to synchronous speed by induction-motor action with the field winding unexcited. If the load and inertia are not too great, the motor will pull into synchronism when the field winding is subsequently energized.

Problems with changing speed result from the fact that, in order to develop torque, the rotor of a synchronous motor must remain in synchronism with the stator flux. Control of synchronous motors can be greatly enhanced by control algorithms in which the stator flux and its relationship to the rotor flux are controlled directly. Such control, which amounts to direct control of torque, is discussed in Section 11.2.2.

11.2.2 Torque Control

Direct torque control in an ac machine, which can be implemented in a number of different ways, is commonly referred to as *field-oriented control* or *vector control*. To facilitate our discussion of field-oriented control, it is helpful to return to the discussion of Section 5.6.1. Under this viewpoint, which is formalized in Appendix C, stator quantities (flux, current, voltage, etc.) are resolved into components which rotate in synchronism with the rotor. *Direct-axis* quantities represent those components which are aligned with the field-winding axis, and *quadrature-axis* components are aligned perpendicular to the field-winding axis.

Section C.2 of Appendix C derives the basic machine relations in dq0 variables for a synchronous machine consisting of a field winding and a three-phase stator winding. The transformed flux-current relationships are found to be

$$\lambda_d = L_d i_d + L_{af} i_f \tag{11.17}$$

$$\lambda_q = L_q i_q \tag{11.18}$$

$$\lambda_f = \frac{3}{2} L_{af} i_d + L_{ff} i_f \tag{11.19}$$

where the subscripts d, q, and f refer to armature direct-, quadrature-axis, and field-winding quantities respectively. Note that throughout this chapter we will assume balanced operating conditions, in which case zero-sequence quantities will be zero and hence will be ignored.

The corresponding transformed voltage equations are

$$v_d = R_a i_d + \frac{d\lambda_d}{dt} - \omega_{me} \lambda_q \tag{11.20}$$

$$v_q = R_a i_q + \frac{d\lambda_q}{dt} + \omega_{me} \lambda_d \tag{11.21}$$

$$v_f = R_f i_f + \frac{d\lambda_f}{dt} \tag{11.22}$$

where $\omega_{me} = (\text{poles}/2)\omega_m$ is the electrical angular velocity of the rotor.

Finally, the electromagnetic torque acting on the rotor of a synchronous motor is shown to be (Eq. C.31)

$$T_{mech} = \frac{3}{2} \left(\frac{poles}{2} \right) (\lambda_d i_q - \lambda_q i_d) \qquad (11.23)$$

Under steady-state, balanced-three-phase operating conditions, $\omega_{me} = \omega_e$ where ω_e is the electrical frequency of the armature voltage and current in rad/sec. Because the armature-produced mmf and flux waves rotate in synchronism with the rotor and hence with respect to the dq reference frame, under these conditions an observer in the dq reference frame will see constant fluxes, and hence one can set $d/dt = 0$.[1]

Letting the subscripts F, D, and Q indicate the corresponding constant values of field-, direct- and quadrature-axis quantities respectively, the flux-current relationships of Eqs. 11.17 through 11.19 then become

$$\lambda_D = L_d i_D + L_{af} i_F \qquad (11.24)$$

$$\lambda_Q = L_q i_Q \qquad (11.25)$$

$$\lambda_F = \frac{3}{2} L_{af} i_D + L_{ff} i_F \qquad (11.26)$$

Armature resistance is typically quite small, and, if we neglect it, the steady-state voltage equations (Eqs. 11.20 through 11.22) then become

$$v_D = -\omega_e \lambda_Q \qquad (11.27)$$

$$v_Q = \omega_e \lambda_D \qquad (11.28)$$

$$v_F = R_f i_F \qquad (11.29)$$

Finally, we can write Eq. 11.23 as

$$T_{mech} = \frac{3}{2} \left(\frac{poles}{2} \right) (\lambda_D i_Q - \lambda_Q i_D) \qquad (11.30)$$

From this point on, we will focus our attention on machines in which the effects of saliency can be neglected. In this case, the direct- and quadrature-axis synchronous inductances are equal and we can write

$$L_d = L_q = L_s \qquad (11.31)$$

where L_s is the synchronous inductance. Substitution into Eqs. 11.24 and 11.25 and then into Eq. 11.30 gives

$$T_{mech} = \frac{3}{2} \left(\frac{poles}{2} \right) [(L_s i_D + L_{af} i_F) i_Q - L_s i_Q i_D]$$

$$= \frac{3}{2} \left(\frac{poles}{2} \right) L_{af} i_F i_Q \qquad (11.32)$$

---

[1] This can easily be derived formally by substituting expressions for the balanced-three-phase armature currents and voltages into the transformation equations.

Equation 11.32 shows that torque is produced by the interaction of the field flux (proportional to the field current) and the quadrature-axis component of the armature current, in other words the component of armature current that is orthogonal to the field flux. By analogy, we see that the direct-axis component of armature current, which is aligned with the field flux, produces no torque.

This result is fully consistent with the generalized torque expressions which are derived in Chapter 4. Consider for example Eq. 4.73 which expresses the torque in terms of the product of the stator and rotor mmfs (F_s and F_r respectively) and the sine of the angle between them.

$$T = -\left(\frac{\text{poles}}{2}\right)\left(\frac{\mu_0 \pi D l}{2g}\right) F_s F_r \sin \delta_{sr} \tag{11.33}$$

where δ_r is the electrical space angle between the stator and rotor mmfs. This shows clearly that no torque will be produced by the direct-axis component of the armature mmf which, by definition, is that component of the stator mmf which is aligned with that of the field winding on the rotor.

Equation 11.32 shows the torque in a nonsalient synchronous motor is proportional to the product of the field current and the quadrature-axis component of the armature current. This is directly analogous to torque production in a dc machine for which Eqs. 7.10 and 7.13 can be combined to show that the torque is proportional to the product of the field current and the armature current.

The analogy between a nonsalient synchronous machine and dc machine can be further reinforced. Consider Eq. 5.21, which expresses the rms value of the line-to-neutral generated voltage of a synchronous generator as

$$E_{af} = \frac{\omega_e L_{af} i_F}{\sqrt{2}} \tag{11.34}$$

Substitution into Eq. 11.32 gives

$$T_{mech} = \frac{3}{2}\left(\frac{\text{poles}}{\sqrt{2}}\right)\frac{E_{af} i_Q}{\omega_e} \tag{11.35}$$

This is directly analogous to Eq. 7.16 ($T_{mech} = E_a I_a / \omega_m$) for a dc machine in which the torque is proportional to the product of the generated voltage and the armature current.

The brushes and commutator of a dc machine force the commutated armature current and armature flux along the quadrature axis such that $I_d = 0$ and it is the interaction of this quadrature-axis current with the direct-axis field flux that produces the torque.[2] A field-oriented controller which senses the position of the rotor and controls the quadrature-axis component of armature current produces the same effect in a synchronous machine.

---

[2] In a practical dc motor, the brushes may be adjusted away from this ideal condition somewhat to improve commutation. In this case, some direct-axis flux will be produced, corresponding to a small direct-axis component of armature current.

Although the direct-axis component of the armature current does not play a role in torque production, it does play a role in determining the resultant stator flux and hence the machine terminal voltage, as can be readily shown. Specifically, from the transformation equations of Appendix C,

$$v_a = v_D \cos(\omega_e t) - v_Q \sin(\omega_e t) \tag{11.36}$$

and thus the rms amplitude of the armature voltage is equal to[3]

$$V_a = \sqrt{\frac{v_D^2 + v_Q^2}{2}} = \omega_e \sqrt{\frac{\lambda_D^2 + \lambda_Q^2}{2}}$$

$$= \omega_e \sqrt{\frac{(L_s i_D + L_{af} i_F)^2 + (L_s i_Q)^2}{2}} \tag{11.37}$$

Dividing V_a by the electrical frequency ω_e, we get an expression for rms armature flux linkages

$$(\lambda_a)_{rms} = \sqrt{\frac{\lambda_D^2 + \lambda_Q^2}{2}} = \sqrt{\frac{(L_s i_D + L_{af} i_F)^2 + (L_s i_Q)^2}{2}} \tag{11.38}$$

Similarly, the transformation equations of Appendix C can be used to show that the rms amplitude of the armature current is equal to

$$I_a = \sqrt{\frac{i_D^2 + i_Q^2}{2}} \tag{11.39}$$

From Eq. 11.32 we see that torque is controlled by the product $i_F i_Q$ of the field current and the quadrature-axis component of the armature current. Thus, simply specifying a desired torque is not sufficient to uniquely determine either i_F or i_Q. In fact, under the field-oriented-control viewpoint presented here, there are actually three indpendent variables, i_F, i_Q and i_D, and, in general, three constraints will be required to uniquely determine them. In addition to specifying the desired torque, a typical controller will implement additional constraints on terminal flux-linkages and current using the basic relationships found in Eqs. 11.38 and 11.39.

Figure 11.13a shows a typical field-oriented torque-control system in block-diagram form. The torque-controller block has two inputs, T_{ref}, the reference value or set point for torque and $(i_F)_{ref}$, the reference value or set point for the field current, which is also supplied to the power supply which supplies the current i_F to the motor field winding. $(i_F)_{ref}$ is determined by an auxiliary controller which also determines

---

[3] Strictly speaking, armature resistance should be included in the voltage expression, in which case the rms amplitude of the armature voltage would be given by the expression

$$V_a = \sqrt{\frac{v_D^2 + v_Q^2}{2}} = \sqrt{\frac{(R_a i_D - \omega_e \lambda_Q)^2 + (R_a i_Q + \omega_e \lambda_D)^2}{2}}$$

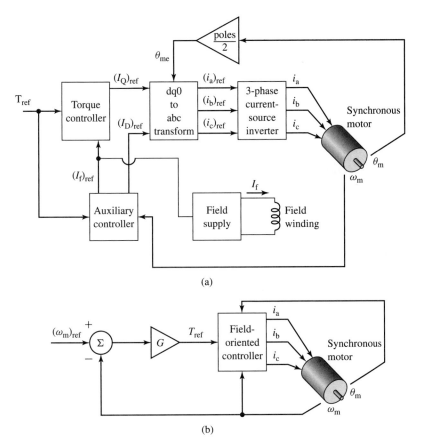

(a)

(b)

Figure 11.13 (a) Block diagram of a field-oriented torque-control system for a synchronous motor. (b) Block diagram of a synchronous-motor speed-control loop built around a field-oriented torque control system.

the reference value $(i_D)_{ref}$ of the direct-axis current, based upon desired values for the armature current and voltage. The torque controller calculates the desired quadrature-axis current from Eq. 11.32 based upon T_{ref} and $(i_F)_{ref}$

$$(i_Q)_{ref} = \frac{2}{3}\left(\frac{2}{poles}\right)\frac{T_{ref}}{L_{af}(i_F)_{ref}} \tag{11.40}$$

Note that a position sensor is required to determine the angular position of the rotor in order to implement the dq0 to abc transformation.

In a typical application, the ultimate control objective is not to control motor torque but to control speed or position. Figure 11.13b shows how the torque-control system of Fig. 11.13b can be used as a component of a speed-control loop, with speed feedback forming an outer control loop around the inner torque-control loop.

EXAMPLE 11.8

Consider again the 45-kVA, 220-V, six-pole synchronous motor of Example 11.7 operating at 60 Hz with a field current of 2.84 A. If the motor is loaded to rated torque and operating under a field-oriented control system such that $i_D = 0$, calculate (a) the phase currents $i_a(t)$, $i_b(t)$, and $i_c(t)$ as well as the per-unit value of the armature current and (b) the motor terminal voltage in per unit. Assume that $\theta_{me} = 0$ at time $t = 0$ (i.e., the rotor direct axis is aligned with phase a at $t = 0$).

■ **Solution**

a. We must first calculate L_{af}. From Example 11.7, we see that the motor produces rated open-circuit voltage (220-V rms, line-to-line) at a field current of 2.84 A. From Eq. 11.34

$$L_{af} = \frac{\sqrt{2}E_{af}}{\omega_e i_F}$$

where E_{af} is the rms, line-to-neutral generated voltage. Thus

$$L_{af} = \frac{\sqrt{2}(220/\sqrt{3})}{120\pi \times 2.84} = 0.168 \text{ H}$$

Rated torque for this six-pole motor is equal to

$$T_{rated} = \frac{P_{rated}}{(\omega_m)_{rated}} = \frac{P_{rated}}{(\omega_e)_{rated}(2/\text{poles})}$$

$$= \frac{45 \times 10^3}{120\pi(2/6)} = 358 \text{ N} \cdot \text{m}$$

Thus, setting $T_{ref} = T_{rated} = 358$ N·m and $(i_F)_{ref} = 2.84$ A, we can find i_Q from Eq. 11.40 as

$$i_Q = \frac{2}{3}\left(\frac{2}{\text{poles}}\right)\frac{T_{ref}}{L_{af}(i_F)_{ref}} = \frac{2}{3}\left(\frac{2}{6}\right)\frac{358}{0.168 \times 2.84} = 167 \text{ A}$$

Using the fact that $\theta_{me} = \omega_e t$ and setting $i_D = 0$, the transformation from dq0 variables to abc variables (Eq. C.2 of Appendix C) gives

$$i_a(t) = i_D \cos(\omega_e t) - i_Q \sin(\omega_e t) = -167\sin(\omega_e t) = -\sqrt{2}(118)\sin(\omega_e t) \text{ A}$$

where $\omega_e = 120\pi \approx 377$ rad/sec. Similarly

$$i_b(t) = -\sqrt{2}(118)\sin(\omega_e t - 120°) \text{ A}$$

and

$$i_c(t) = -\sqrt{2}(118)\sin(\omega_e t + 120°) \text{ A}$$

The rms armature current is 118 A and the machine base current is equal to

$$I_{base} = \frac{P_{base}}{\sqrt{3}V_{base}} = \frac{45 \times 10^3}{\sqrt{3}\,220} = 118 \text{ A}$$

Thus the per-unit machine terminal current is equal to $I_a = 118/118 = 1.0$ per unit.

b. The motor terminal voltage can most easily be found from the rms phasor relationship

$$\hat{V}_a = jX_s\hat{I}_a + \hat{E}_{af}$$

In part (a) we found that $i_a = -\sqrt{2}(118)\sin(\omega_e t)$ A and thus

$$\hat{I}_a = j118 \text{ A}$$

We can find E_{af} from Eq. 11.34 as

$$E_{af} = \frac{\omega_e L_{af} i_F}{\sqrt{2}} = \frac{120\pi \times 0.168 \times 2.84}{\sqrt{2}} = 127 \text{ V line-to-neutral}$$

and, thus, since $(\hat{E}_{af})_{rms}$ lies along the quadrature axis, as does $\hat{I}_a$,

$$\hat{E}_{af} = j127 \text{ V}$$

The base impedance of the machine is equal to

$$Z_{base} = \frac{V_{base}^2}{P_{base}} = \frac{220^2}{45 \times 10^3} = 1.076 \ \Omega$$

and the synchronous reactance of 0.836 pu is equal to $X_s = 0.836 \times 1.076 = 0.899 \ \Omega$. Thus the rms line-to-neutral terminal voltage

$$\hat{V}_a = jX_s\hat{I}_a + \hat{E}_{af} = j0.899(j118) + j127$$

$$= -106 + j127 = 165 \angle 129.9° \text{ V line-to-neutral}$$

or $V_a = 287$ V rms line-to-line $= 1.30$ per unit.

Note that the terminal voltage for this operating condition is considerably in excess of the rated voltage of this machine, and hence such operation would not be acceptable. As we shall now discuss, a control algorithm which takes advantage of the full capability to vary i_F, i_D, and i_Q can achieve rated torque while not exceeding rated terminal voltage.

Practice Problem 11.8

Calculate the per-unit terminal voltage and current of Example 11.8 if the field-oriented controller maintains $i_D = 0$ while reducing i_F to 2.6 A.

Solution

$V_a = 1.29$ per unit and $I_a = 1.09$ per unit.

As we have discussed, a practical field-oriented control must determine values for all three currents i_F, i_D, and i_Q. In Example 11.8 two of these values were chosen relatively arbitrarily ($i_F = 2.84$ and $i_D = 0$) and the result was a control that achieved the desired torque but with a terminal voltage 30 percent in excess of the motor-rated voltage. In a practical system, additional constraints are required to achieve an acceptable control algorithm.

One such algorithm would be to require that the motor operate at rated flux and at unity terminal power factor. Such an algorithm can be derived with reference to the

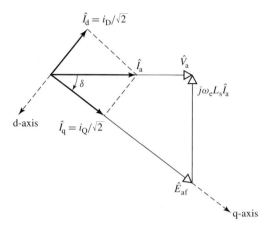

Figure 11.14 Phasor diagram for unity-power-
factor field-oriented-control algorithm.

phasor diagram of Fig 11.14 and can be implemented using the following steps:

Step 1. Calculate the line-to-neutral armature voltage corresponding to rated
flux as

$$V_a = (V_a)_{rated} \left(\frac{\omega_m}{(\omega_m)_{rated}} \right) \qquad (11.41)$$

where $(V_a)_{rated}$ is the rated line-to-neutral armature voltage at rated motor
speed, ω_m is the desired motor speed, and $(\omega_m)_{rated}$ is its rated speed.

Step 2. Calculate the rms armature current from the desired torque T_{ref} as

$$I_a = \frac{P_{ref}}{3V_a} = \frac{T_{ref}\,\omega_m}{3V_a} \qquad (11.42)$$

where P_{ref} is the mechanical power corresponding to the desired torque.

Step 3. Calculate the angle δ based upon the phasor diagram of Fig 11.14

$$\delta = -\tan^{-1}\left(\frac{\omega_e L_s I_a}{V_a} \right) \qquad (11.43)$$

where $\omega_e = \omega_{me} = (poles/2)\omega_m$ is the electrical frequency corresponding to
the desired motor speed.

Step 4. Calculate $(i_Q)_{ref}$ and $(i_D)_{ref}$

$$(i_Q)_{ref} = \sqrt{2}I_a \cos\delta \qquad (11.44)$$

$$(i_D)_{ref} = \sqrt{2}I_a \sin\delta \qquad (11.45)$$

Step 5. Calculate $(i_F)_{ref}$ from Eq. 11.32

$$(i_F)_{ref} = \frac{2}{3}\left(\frac{2}{poles} \right) \frac{T_{ref}}{L_{af}(i_Q)_{ref}} \qquad (11.46)$$

This algorithm is illustrated in Example 11.9.

EXAMPLE 11.9

The 45-kVA, 220-V synchronous motor of Example 11.8 is to be again operated at rated torque and speed from a field-oriented control system. Calculate the required motor field current and the per-unit terminal voltage and current if the unity-power-factor algorithm described above is implemented.

■ Solution

We will follow the individual steps outlined above.

Step 1. Since the motor is operating at rated speed, the desired terminal voltage will be the rated line-to-neutral voltage of the machine.

$$V_a = \frac{220}{\sqrt{3}} = 127 \text{ V} = 1.0 \text{ per unit}$$

Step 2. Setting $T_{ref} = 358$ N·m and $\omega_m = (2/\text{poles})\omega_e = 40\pi$, the rms armature current can be calculated from Eq. 11.42

$$I_a = \frac{T_{ref}\, \omega_m}{3(V_a)} = \frac{358 \times (40\pi)}{3 \times 127} = 118 \text{ A}$$

As calculated in Example 11.8, $I_{base} = 118$ A and thus $I_a = 1.0$ per unit. This is as expected, since we want the motor to operate at rated torque, speed and terminal voltage, and at unity power factor.

Step 3. Next calculate δ from Eq. 11.43. This calculation requires that we determine the synchronous inductance L_s.

$$L_s = \frac{X_s}{(\omega_e)_{rated}} = \frac{0.899}{120\pi} = 2.38 \text{ mH}$$

Thus

$$\delta = -\tan^{-1}\left(\frac{\omega_e L_s I_a}{V_a}\right)$$

$$= -\tan^{-1}\left(\frac{120\pi\, 2.38 \times 10^{-3} \times 118}{127}\right) = -0.695 \text{ rad} = -39.8°$$

Step 4. We can now calculate the desired values of i_Q and i_D from Eqs. 11.44 and 11.45.

$$(i_Q)_{ref} = \sqrt{2}I_a \cos\delta = 128 \text{ A}$$

and

$$(i_D)_{ref} = \sqrt{2}I_a \sin\delta = -107 \text{ A}$$

Step 5. $(i_F)_{ref}$ is found from Eq. 11.46

$$(i_F)_{ref} = \frac{2}{3}\left(\frac{2}{\text{poles}}\right)\frac{T_{ref}}{L_{af}(i_Q)_{ref}} = \frac{2}{3}\left(\frac{2}{6}\right)\frac{358}{0.168 \times 128} = 3.70 \text{ A}$$

Repeat Example 11.9 for the motor operating at rated torque and half of rated speed. Calculate (a) the desired motor field current, (b) the line-to-neutral terminal voltage (in volts), and (c) the armature current (in amperes).

Solution

    a. $(i_F)_{ref} = 3.70$ A

    b. $V_a = 63.5$ V line-to-neutral

    c. $I_a = 118$ A

The discussion of this section has focused upon synchronous machines with field windings and the corresponding capability to control the field excitation. The basic concept, of course, also applies to synchronous machines with permanent magnets on the rotor. However, in the case of permanent-magnet synchronous machines, the effective field excitation is fixed and, as a result, there is one less degree of freedom for the field-oriented control algorithm.

For a permanent-magnet synchronous machine, since the effective equivalent field current is fixed by the permanent magnet, the quadrature-axis current is determined directly by the desired torque. Consider a three-phase permanent-magnet motor whose rated rms, line-to-neutral open-circuit voltage is $(E_{af})_{rated}$ at electrical frequency $(\omega_e)_{rated}$. From Eq. 11.34 we see that the equivalent $L_{af}I_f$ product for this motor, which we will refer to by the symbol Λ_{PM}, is

$$\Lambda_{PM} = \frac{\sqrt{2}(E_{af})_{rated}}{(\omega_e)_{rated}} \tag{11.47}$$

Thus, the direct-axis flux-current relationship for this motor, corresponding to Eq. 11.24, becomes

$$\lambda_D = L_d i_D + \Lambda_{PM} \tag{11.48}$$

and the torque expression of Eq. 11.32 becomes

$$T_{mech} = \frac{3}{2}\left(\frac{poles}{2}\right)\Lambda_{PM}i_Q \tag{11.49}$$

From Eq. 11.49 we see that, for a permanent-magnet sychronous machine under field-oriented control, the quadrature-axis current is uniquely determined by the desired torque and Eq. 11.40 becomes

$$(i_Q)_{ref} = \frac{2}{3}\left(\frac{2}{poles}\right)\frac{T_{ref}}{\Lambda_{PM}} \tag{11.50}$$

Once $(i_Q)_{ref}$ has been specified, the only remaining control choice remains to determine the desired value for the direct-axis current, $(i_D)_{ref}$. One possibility is simply to set $(i_D)_{ref} = 0$. This will clearly result in the lowest possible armature current for a given torque. However, as we have seen in Example 11.8, this is likely to result in terminal voltages in excess of the rated voltage of the machine. As a

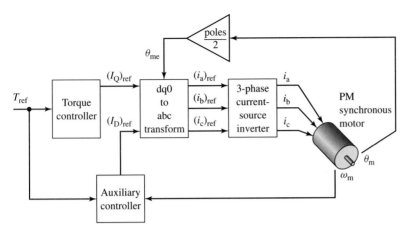

Figure 11.15 Block diagram of a field-oriented torque-control system for a permanent-magnet synchronous motor.

result, it is common to supply direct-axis current so as to reduce the direct-axis flux linkage of Eq. 11.48, which will in turn result in reduced terminal voltage. This technique is commonly referred to as *flux weakening* and comes at the expense of increased armature current.[4] In practice, the chosen operating point is determined by a trade-off between reducing the armature voltage and an increase in armature current. Figure 11.15 shows the block diagram for a field-oriented-control system for use with a permanent-magnet motor.

EXAMPLE 11.10

A 25-kW, 4000-rpm, 220-V, two-pole, three-phase permanent-magnet synchronous motor produces rated open-circuit voltage at a rotational speed of 3200 r/min and has a synchronous inductance of 1.75 mH. Assume the motor is to be operated under field-oriented control at 2800 r/min and 65 percent of rated torque.

a. Calculate the required quadrature-axis current.
b. If the controller is designed to minimize armature current by setting $i_D = 0$, calculate the resultant armature flux linkage in per-unit.
c. If the controller is designed to maintain the armature flux-linkage at its rated value, calculate the corresponding value of i_D and the corresponding rms and per-unit values of the armature current.

■ **Solution**

a. The rated speed of this machine is

$$(\omega_m)_{rated} = 4000 \left(\frac{\pi}{30} \right) = 419 \text{ rad/sec}$$

---

[4] See T. M. Jahns, "Flux-Weakening Regime Operation of an Interior Permanent Magnet Synchronous Motor Drive," *IEEE Transactions on Industry Applications,* Vol. 23, pp. 681–689.

and the rated torque is

$$T_{\text{rated}} = \frac{P_{\text{rated}}}{(\omega_m)_{\text{rated}}} = \frac{25 \times 10^3}{419} = 59.7 \text{ N} \cdot \text{m}$$

This motor achieves its rated-open-circuit voltage of $220/\sqrt{3} = 127$ V rms at a speed of $n = 3200$ r/min. The corresponding electrical frequency is

$$\omega_e = \left(\frac{\text{poles}}{2}\right)\left(\frac{\pi}{30}\right)n = \left(\frac{\pi}{30}\right)3200 = 335 \text{ rad/sec}$$

From Eq. 11.47,

$$\Lambda_{\text{PM}} = \frac{\sqrt{2}(E_{\text{af}})_{\text{rated}}}{\omega_e} = \frac{\sqrt{2}\,127}{335} = 0.536 \text{ Wb}$$

Thus, setting $T_{\text{ref}} = 0.65T_{\text{rated}} = 38.8$ N·m, from Eq. 11.50 we find that

$$(i_Q)_{\text{ref}} = \frac{2}{3}\left(\frac{2}{\text{poles}}\right)\frac{T_{\text{ref}}}{\Lambda_{\text{PM}}} = \frac{2}{3}\left(\frac{38.8}{0.536}\right) = 48.3 \text{ A}$$

b. With $(i_D)_{\text{ref}} = 0$,

$$\lambda_D = \Lambda_{\text{PM}} = 0.536 \text{ Wb}$$

and

$$\lambda_Q = L_s i_Q = (1.75 \times 10^{-3})48.3 = 0.0845 \text{ Wb}$$

Thus, the rms line-to-neutral armature flux is equal to

$$\lambda_a = \sqrt{\frac{\lambda_D^2 + \lambda_Q^2}{2}} = \sqrt{\frac{0.536^2 + 0.0845^2}{2}} = 0.384 \text{ Wb}$$

The base rms line-to-neutral armature flux can be determined from the base line-to-neutral voltage $(V_a)_{\text{base}} = 127$ V and the base frequency $(\omega_e)_{\text{base}} = 419$ rad/sec as

$$(\lambda_a)_{\text{base}} = \frac{(V_a)_{\text{base}}}{(\omega_e)_{\text{base}}} = 0.303 \text{ Wb}$$

Thus, the per-unit armature flux is equal to $0.384/0.303 = 1.27$ per unit. From this calculation we see that the motor is significantly saturated at this operating condition. In fact, the calculation is probably not accurate because such a degree of saturation will most likely give rise to a reduction in the synchronous inductance as well as the magnetic coupling between the rotor and the stator.

c. In order to maintain rated armature flux linkage, the control will have to produce a direct-axis component of armature current to reduce the direct-axis flux linkage such that the total armature flux linkage is equal to the rated value $(\lambda_a)_{\text{base}}$. Specifically, we must have

$$\lambda_D = \sqrt{2(\lambda_a)_{\text{base}}^2 - \lambda_Q^2} = \sqrt{2 \times 0.303^2 - 0.0844^2} = 0.420 \text{ Wb}$$

We can now find $(i_D)_{ref}$ from Eq. 11.48 (setting $L_d = L_s$)

$$(i_D)_{ref} = \frac{\lambda_D - \Lambda_{PM}}{L_s} = \frac{0.420 - 0.536}{1.75 \times 10^{-3}} = -66.3 \text{ A}$$

The corresponding rms armature current is

$$I_a = \sqrt{\frac{(i_D)_{ref}^2 + (i_Q)_{ref}^2}{2}} = \sqrt{\frac{66.3^2 + 48.3^2}{2}} = 58.0 \text{ A}$$

The base rms armature current for this machine is equal to

$$I_{base} = \frac{P_{base}}{\sqrt{3}V_{base}} = \frac{25 \times 10^3}{\sqrt{3}\ 220} = 65.6 \text{ A}$$

and hence the per-unit armature current is equal to $58.0/65.6 = 0.88$ per unit.

Comparing the results of parts (b) and (c) we see how flux weakening by the introduction of direct-axis current can be used to control the terminal voltage of a permanent-magnet synchronous motor under field-oriented control.

Practice Problem 11.10

Consider again the motor of Example 11.10. Repeat the calculations of parts (b) and (c) of Example 11.10 for the case in which the motor is operating at 80 percent of rated torque at a speed of 2500 r/min.

Solution

For part (b), $\lambda_a = 1.27$ per unit.

For part (c), $I_a = 0.98$ per unit.

11.3 CONTROL OF INDUCTION MOTORS

11.3.1 Speed Control

Induction motors supplied from a constant-frequency source admirably fulfill the requirements of substantially constant-speed drives. Many motor applications, however, require several speeds or even a continuously adjustable range of speeds. From the earliest days of ac power systems, engineers have been interested in the development of adjustable-speed ac motors.

The synchronous speed of an induction motor can be changed by (*a*) changing the number of poles or (*b*) varying the line frequency. The operating slip can be changed by (*c*) varying the line voltage, (*d*) varying the rotor resistance, or (*e*) applying voltages of the appropriate frequency to the rotor circuits. The salient features of speed-control methods based on these five possibilities are discussed in the following five sections.

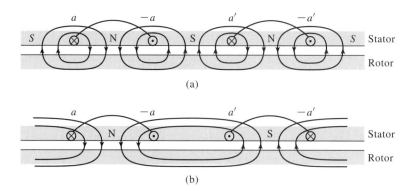

Figure 11.16 Principles of the pole-changing winding.

Pole-Changing Motors   In pole-changing motors, the stator winding is designed so that, by simple changes in coil connections, the number of poles can be changed in the ratio 2 to 1. Either of two synchronous speeds can then be selected. The rotor is almost always of the squirrel-cage type, which reacts by producing a rotor field having the same number of poles as the inducing stator field. With two independent sets of stator windings, each arranged for pole changing, as many as four synchronous speeds can be obtained in a squirrel-cage motor, for example, 600, 900, 1200, and 1800 r/min for 60-Hz operation.

The basic principles of the pole-changing winding are shown in Fig. 11.16, in which aa and $a'a'$ are two coils comprising part of the phase-a stator winding. An actual winding would, of course, consist of several coils in each group. The windings for the other stator phases (not shown in the figure) would be similarly arranged. In Fig. 11.16a the coils are connected to produce a four-pole field; in Fig. 11.16b the current in the $a'a'$ coil has been reversed by means of a controller, the result being a two-pole field.

Figure 11.17 shows the four possible arrangements of these two coils: they can be connected in series or in parallel, and with their currents either in the same direction (four-pole operation) or in the opposite direction (two-pole operation). Additionally, the machine phases can be connected either in Y or Δ, resulting in eight possible combinations.

Note that for a given phase voltage, the different connections will result in differing levels of air-gap flux density. For example, a change from a Δ to a Y connection will reduce the coil voltage (and hence the air-gap flux density) for a given coil arrangement by $\sqrt{3}$. Similarly, changing from a connection with two coils in series to two in parallel will double the voltage across each coil and therefore double the magnitude of the air-gap flux density. These changes in flux density can, of course, be compensated for by changes in the applied winding voltage. In any case, they must be considered, along with corresponding changes in motor torque, when the configuration to be used in a specific application is considered.

Armature-Frequency Control   The synchronous speed of an induction motor can be controlled by varying the frequency of the applied armature voltage. This method

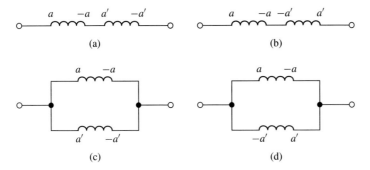

Figure 11.17 Four possible arrangements of phase-a stator coils in a pole-changing induction motor: (a) series-connected, four-pole; (b) series-connected, two-pole; (c) parallel-connected, four-pole; (d) parallel-connected, two-pole.

of speed control is identical to that discussed in Section 11.2.1 for synchronous machines. In fact, the same inverter configurations used for synchronous machines, such as the three-phase voltage-source inverter of Fig. 11.11, can be used to drive induction motors. As is the case with any ac motor, to maintain approximately constant flux density, the armature voltage should also be varied directly with the frequency (constant-volts-per-hertz).

The torque-speed curve of an induction motor for a given frequency can be calculated by using the methods of Chapter 6 within the accuracy of the motor parameters at that frequency. Consider the torque expression of Eq. 6.33 which is repeated here.

$$T_{\text{mech}} = \frac{1}{\omega_s} \left[\frac{n_{\text{ph}} V_{1,\text{eq}}^2 (R_2/s)}{(R_{1,\text{eq}} + (R_2/s))^2 + (X_{1,\text{eq}} + X_2)^2} \right] \tag{11.51}$$

where $\omega_s = (2/\text{poles})\omega_e$ and ω_e is the electrical excitation frequency of the motor in rad/sec,

$$\hat{V}_{1,\text{eq}} = \hat{V}_1 \left(\frac{jX_{\text{m}}}{R_1 + j(X_1 + X_{\text{m}})} \right) \tag{11.52}$$

and

$$R_{1,\text{eq}} + jX_{1,\text{eq}} = \frac{jX_{\text{m}}(R_1 + jX_1)}{R_1 + j(X_1 + X_{\text{m}})} \tag{11.53}$$

To investigate the effect of changing frequency, we will assume that R_1 is negligible. In this case,

$$\hat{V}_{1,\text{eq}} = \hat{V}_1 \left(\frac{X_{\text{m}}}{X_1 + X_{\text{m}}} \right) \tag{11.54}$$

$$R_{1,\text{eq}} = 0 \tag{11.55}$$

and

$$X_{1,eq} = \frac{X_m X_1}{X_1 + X_m} \qquad (11.56)$$

Let the subscript 0 represent rated-frequency values of each of the induction-motor parameters. As the electrical-excitation frequency is varied, we can then write

$$(X_{1,eq} + X_2) = \left(\frac{\omega_e}{\omega_{e0}}\right)(X_{1,eq} + X_2)_0 \qquad (11.57)$$

Under constant-volts-per-hertz control, we can also write the equivalent source voltage as

$$\hat{V}_1 = \left(\frac{\omega_e}{\omega_{e0}}\right)(\hat{V}_1)_0 \qquad (11.58)$$

and hence, since $\hat{V}_{1,eq}$ is equal to $\hat{V}_1$ multiplied by a reactance ratio which stays constant with changing frequency,

$$\hat{V}_{1,eq} = \left(\frac{\omega_e}{\omega_{e0}}\right)(\hat{V}_{1,eq})_0 \qquad (11.59)$$

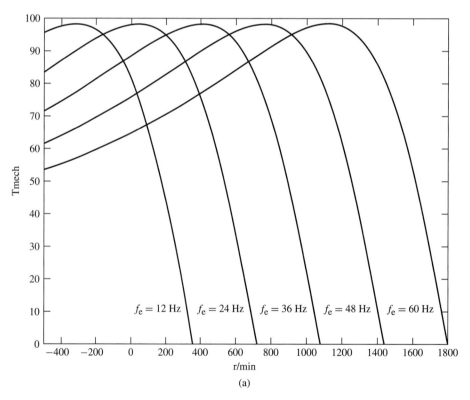

(a)

Figure 11.18 A family of typical induction-motor speed-torque curves for a four-pole motor for various values of the electrical supply frequency. (a) R_1 sufficiently small so that its effects are negligible. (b) R_1 not negligible.

Finally, we can write the motor slip as

$$s = \frac{\omega_s - \omega_m}{\omega_s} = \frac{poles}{2}\left(\frac{\Delta\omega_m}{\omega_e}\right) \tag{11.60}$$

where $\Delta\omega_m = \omega_s - \omega_m$ is the difference between the synchronous and mechanical angular velocities of the motor.

Substitution of Eqs. 11.57 through 11.60 into Eq. 11.51 gives

$$T_{mech} = \frac{n_{ph}[(V_{1,eq})_0]^2(R_2/\Delta\omega)}{\left[\left(\frac{2\omega_{e0}}{poles}\right)(R_2/\Delta\omega)\right]^2 + [(X_{1,eq} + X_2)_0]^2} \tag{11.61}$$

Equation 11.61 shows the general trend in which we see that the frequency dependence of the torque-speed characteristic of an induction motor appears only in the term $R_2/\Delta\omega$. Thus, under the assumption that R_1 is negligible, as the electrical supply frequency to an induction motor is changed, the shape of the speed-torque curve as a function of $\Delta\omega$ (the difference between the synchronous speed and the motor speed) will remain unchanged. As a result, the torque-speed characteristic will simply shift along the speed axis as $\omega_e(f_e)$ is varied.

A set of such curves is shown in Fig. 11.18a. Note that as the electrical frequency (and hence the synchronous speed) is decreased, a given value of $\Delta\omega$ corresponds to a

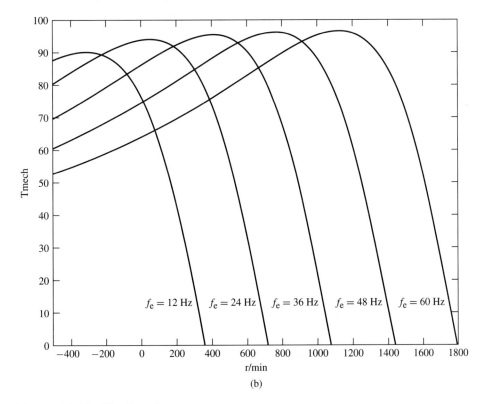

(b)

Figure 11.18 (Continued)

larger slip. Thus, for example, if the peak torque of a four-pole motor driven at 60 Hz occurs at 1638 r/min, corresponding to a slip of 9 percent, when driven at 30 Hz, the peak torque will occur at 738 r/min, corresponding to a slip of 18 percent.

In practice, the effects of R_1 may not be fully negligible, especially for large values of slip. If this is the case, the shape of the speed-torque curves will vary somewhat with the applied electrical frequency. Figure 11.18b shows a typical family of curves for this case.

EXAMPLE 11.11

The three-phase, 230-V, 60-Hz, 12-kW, four-pole induction motor of Example 6.4 (with $R_2 = 0.2\ \Omega$) is to be operated from a variable-frequency, constant-volts-per-hertz motor drive whose terminal voltage is 230 V at 60 Hz. The motor is driving a load whose power can be assumed to vary as

$$P_{load} = 10.5 \left(\frac{n}{1800} \right)^3 \text{ kW}$$

where n is the load speed in r/min. Motor rotational losses can be assumed to be negligible.

Write a MATLAB script to find the line-to-line terminal voltage, the motor speed in r/min, the slip and the motor load in kW for (a) a source frequency of 60 Hz and (b) a source frequency of 40 Hz.

■ Solution

As the electrical frequency f_e is varied, the motor reactances given in Example 6.4 must be varied as

$$X = X_0 \left(\frac{f_e}{60} \right)$$

where X_0 is the reactance value at 60 Hz. Similarly, the line-to-neutral armature voltage must be varied as

$$V_1 = \frac{220}{\sqrt{3}} \left(\frac{f_e}{60} \right) = 127 \left(\frac{f_e}{60} \right) \text{ V}$$

From Eq. 4.40, the synchronous angular velocity of the motor is equal to

$$\omega_s = \left(\frac{4\pi}{\text{poles}} \right) f_e = \pi f_e \text{ rad/sec}$$

and, at any given motor speed ω_m, the corresponding slip is given by

$$s = \frac{\omega_s - \omega_m}{\omega_s}$$

Using Eqs. 11.51 through 11.53, the motor speed can be found by searching over ω_m for that speed at which $P_{load} = \omega_m T_{mech}$. If this is done, the result is:

a. For $f_e = 60$ Hz:

    Terminal voltage $= 230$ V line-to-line

    Speed $= 1720$ r/min

Slip = 4.4%

$P_{load} = 9.17$ kW

b. For $f_e = 40$ Hz:

Terminal voltage = 153 V line-to-line

Speed = 1166 r/min

Slip = 2.8%

$P_{load} = 2.86$ kW

Here is the MATLAB script:

```
clc
clear

%Here are the 60-Hz motor parameters

V10 = 230/sqrt(3);
Nph = 3;
poles = 4;
fe0 = 60;

R1 = 0.095;
R2 = 0.2;
X10 = 0.680;
X20 = 0.672;
Xm0 = 18.7;

% Two frequency values

fe1 = 60;
fe2 = 40;

for m = 1:2,
   if m == 1
       fe = fe1;
   else
       fe = fe2;
   end

% Calculate the reactances and the voltage

   X1 = X10*(fe/fe0);
   X2 = X20*(fe/fe0);
   Xm = Xm0*(fe/fe0);
   V1 = V10*(fe/fe0);

%Calculate the synchronous speed

   omegas = 4*pi*fe/poles;
   ns = 120*fe/poles;

%Calculate stator Thevenin equivalent

   V1eq = abs(V1*j*Xm/(R1 + j*(X1+Xm)));
```

```
Z1eq = j*Xm*(R1+j*X1)/(R1 + j*(X1+Xm));
R1eq = real(Z1eq);
X1eq = imag(Z1eq);
```

%Search over the slip until the Pload = Pmech

```
    slip = 0.;
    error = 1;

    while error >= 0;
        slip = slip + 0.00001;
        rpm = ns*(1-slip);
        omegam = omegas*(1-slip);
        Tmech = (1/omegas)*Nph*V1eq^2*(R2/slip);
        Tmech = Tmech/((R1+R2/slip)^2 + (X1+X2)^2);
        Pmech = Tmech*omegam;
        Pload = 10.5e3*(rpm/1800)^3;

        error = Pload - Pmech;

    end   %End of while loop

    fprintf('\nFor fe = %g [Hz]:',fe)
    fprintf('\n   Terminal voltage = %g [V l-l]',V1*sqrt(3))
    fprintf('\n   rpm = %g',rpm)
    fprintf('\n   slip = %g [percent] ',100*slip)
    fprintf('\n   Pload = %g [kW]',Pload/1000)
    fprintf('\n\n')
```

end

Practice Problem 11.11

Repeat Example 11.11 for a source frequency of 50 Hz.

Solution

Terminal voltage = 192 V line-to-line

Speed = 1447 r/min

Slip = 3.6%

$P_{\text{load}} = 5.45$ kW

Line-Voltage Control   The internal torque developed by an induction motor is proportional to the square of the voltage applied to its primary terminals, as shown by the two torque-speed characteristics in Fig. 11.19. If the load has the torque-speed characteristic shown by the dashed line, the speed will be reduced from n_1 to n_2. This method of speed control is commonly used with small squirrel-cage motors driving fans, where cost is an issue and the inefficiency of high-slip operation can be tolerated. It is characterized by a rather limited range of speed control.

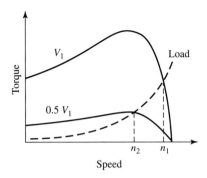

Figure 11.19 Speed control by means of line voltage.

Rotor-Resistance Control   The possibility of speed control of a wound-rotor motor by changing its rotor-circuit resistance has already been pointed out in Section 6.7.1. The torque-speed characteristics for three different values of rotor resistance are shown in Fig. 11.20. If the load has the torque-speed characteristic shown by the dashed line, the speeds corresponding to each of the values of rotor resistance are n_1, n_2, and n_3. This method of speed control has characteristics similar to those of dc shunt-motor speed control by means of resistance in series with the armature.

The principal disadvantages of both line-voltage and rotor-resistance control are low efficiency at reduced speeds and poor speed regulation with respect to change in load. In addition, the cost and maintenance requirements of wound-rotor induction motors are sufficiently high that squirrel-cage motors combined with solid-state drives have become the preferred option in most applications.

11.3.2 Torque Control

In Section 11.2.2 we developed the concept of field-oriented-control for synchronous machines. Under this viewpoint, the armature flux and current are resolved into two components which rotate synchronously with the rotor and with the air-gap flux wave.

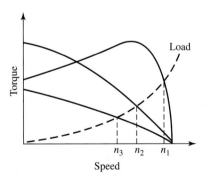

Figure 11.20 Speed control by means of rotor resistance.

The components of armature current and flux which are aligned with the field-winding are referred to as *direct-axis components* while those which are perpendicular to this axis are referred to as *quadrature-axis components*.

It turns out that the same viewpoint which we applied to synchronous machines can be applied to induction machines. As is discussed in Section 6.1, in the steady-state the mmf and flux waves produced by both the rotor and stator windings of an induction motor rotate at synchronous speed and in synchronism with each other. Thus, the torque-producing mechanism in an induction machine is equivalent to that of a synchronous machine. The difference between the two is that, in an induction machine, the rotor currents are not directly supplied but rather are induced as the induction-motor rotor slips with respect to the rotating flux wave produced by the stator currents.

To examine the application of field-oriented control to induction machines, we begin with the dq0 transformation of Section C.3 of Appendix C. This transformation transforms both the stator and rotor quantities into a synchronously rotating reference frame. Under balanced-three-phase, steady-state conditions, zero-sequence quantities will be zero and the remaining direct- and quadrature-axis quantites will be constant. Hence the flux-linkage current relationships of Eqs. C.52 through C.58 become

$$\lambda_D = L_S i_D + L_m i_{DR} \tag{11.62}$$

$$\lambda_Q = L_S i_Q + L_m i_{QR} \tag{11.63}$$

$$\lambda_{DR} = L_m i_D + L_R i_{DR} \tag{11.64}$$

$$\lambda_{QR} = L_m i_Q + L_R i_{QR} \tag{11.65}$$

In these equations, the subscripts D, Q, DR, and QR represent the constant values of the direct- and quadrature-axis components of the stator and rotor quantities respectively. It is a straight-forward matter to show that the inductance parameters can be determined from the equivalent-circuit parameters as

$$L_m = \frac{X_{m0}}{\omega_{e0}} \tag{11.66}$$

$$L_S = L_m + \frac{X_{10}}{\omega_{e0}} \tag{11.67}$$

$$L_R = L_m + \frac{X_{20}}{\omega_{e0}} \tag{11.68}$$

where the subscript 0 indicates the rated-frequency value.

The transformed voltage equations Eqs. C.63 through C.68 become

$$v_D = R_a i_D - \omega_e \lambda_Q \tag{11.69}$$

$$v_Q = R_a i_Q + \omega_e \lambda_D \tag{11.70}$$

$$0 = R_{aR} i_{DR} - (\omega_e - \omega_{me})\lambda_{QR} \tag{11.71}$$

$$0 = R_{aR} i_{QR} + (\omega_e - \omega_{me})\lambda_{DR} \tag{11.72}$$

where one can show that the resistances are related to those of the equivalent circuit as

$$R_a = R_1 \tag{11.73}$$

and

$$R_{aR} = R_2 \tag{11.74}$$

For the purposes of developing a field-oriented-control scheme, we will begin with the torque expression of Eq. C.70

$$T_{mech} = \frac{3}{2} \left(\frac{poles}{2} \right) \left(\frac{L_m}{L_R} \right) (\lambda_{DR} i_q - \lambda_{QR} i_d) \tag{11.75}$$

For the derivation of the dq0 transformation in Section C.3, the angular velocity of the reference frame was chosen to the synchronous speed as determined by the stator electrical frequency ω_e. It was not necessary for the purposes of the derivation to specify the absolute angular location of the reference frame. It is convenient at this point to choose the direct axis of the reference frame aligned with the rotor flux.

If this is done

$$\lambda_{QR} = 0 \tag{11.76}$$

and the torque expression of Eq. 11.75 becomes

$$T_{mech} = \frac{3}{2} \left(\frac{poles}{2} \right) \left(\frac{L_m}{L_R} \right) \lambda_{DR} i_Q \tag{11.77}$$

From Eq. 11.71 we see that

$$i_{DR} = 0 \tag{11.78}$$

and thus

$$\lambda_{DR} = L_m i_D \tag{11.79}$$

and

$$\lambda_D = L_s i_D \tag{11.80}$$

From Eqs. 11.79 and 11.80 we see that by choosing set $\lambda_{QR} = 0$ and thus aligning the synchronously rotating reference frame with the axis of the rotor flux, the direct-axis rotor flux (which is, indeed, the total rotor flux) as well as the direct-axis flux are determined by the direct-axis component of the armature current. Notice the direct analogy with a dc motor. In a dc motor, the field- and direct-axis armature fluxes are determined by the field current and in this field-oriented control scheme, the rotor and direct-axis armature fluxes are determined by the direct-axis armature current. In other words, in this field-oriented control scheme, the direct-axis component of armature current serves the same function as the field current in a dc machine.

The torque equation, Eq. 11.77, completes the analogy with the dc motor. We see that once the rotor direct-axis flux λ_{DR} is set by the direct-axis armature current, the torque is then determined by the quadrature-axis armature current just as the torque is determined by the armature current in a dc motor.

In a practical implementation of the technique which we have derived, the direct- and quadrature-axis currents i_D and i_Q must be transformed into the three motor phase currents $i_a(t)$, $i_b(t)$, and $i_c(t)$. This can be done using the inverse dq0 transformation of Eq. C.48 which requires knowledge of θ_S, the electrical angle between the axis of phase a, and the direct-axis of the synchronously rotating reference frame.

Since it is not possible to measure the axis of the rotor flux directly, it is necessary to calculate θ_S, where $\theta_S = \omega_e t + \theta_0$ as given by Eq. C.46. Solving Eq. 11.72 for ω_e gives

$$\omega_e = \omega_{me} - R_{aR}\left(\frac{i_{QR}}{\lambda_{DR}}\right) \tag{11.81}$$

From Eq. 11.65 with $\lambda_{QR} = 0$ we see that

$$i_{QR} = -\left(\frac{L_m}{L_R}\right)i_Q \tag{11.82}$$

Eq. 11.82 in combination with Eq. 11.79 then gives

$$\omega_e = \omega_{me} + \frac{R_{aR}}{L_R}\left(\frac{i_Q}{i_D}\right) = \omega_{me} + \frac{1}{\tau_R}\left(\frac{i_Q}{i_D}\right) \tag{11.83}$$

where $\tau_R = L_R/R_{aR}$ is the rotor time constant.

We can now integrate Eq. 11.83 to find

$$\hat{\theta}_S = \left[\omega_{me} + \frac{1}{\tau_R}\left(\frac{i_Q}{i_D}\right)\right]t + \theta_0 \tag{11.84}$$

where $\hat{\theta}_S$ indicates the calculated value of θ_S (often referred to as the *estimated value* of θ_S). In the more general dynamic sense

$$\hat{\theta}_S = \int_0^t \left[\omega_{me} + \frac{1}{\tau_R}\left(\frac{i_Q}{i_D}\right)\right]dt' + \theta_0 \tag{11.85}$$

Note that both Eqs. 11.84 and 11.85 require knowledge of θ_0, the value of θ_S at $t = 0$. Although we will not prove it here, it turns out that in a practical implementation, the effects of an error in this initial angle decay to zero with time, and hence it can be set to zero without any loss of generality.

Figure 11.21a shows a block diagram of a field-oriented torque-control system for an induction machine. The block labeled "Estimator" represents the calculation of Eq. 11.85 which calculates the estimate of θ_S required by the transformation from dq0 to abc variables.

Note that a speed sensor is required to provide the rotor speed measurement required by the estimator. Also notice that the estimator requires knowledge of the rotor time constant $\tau_R = L_R/R_{aR}$. In general, this will not be known exactly, both due to uncertainty in the machine parameters as well as due to the fact that the rotor

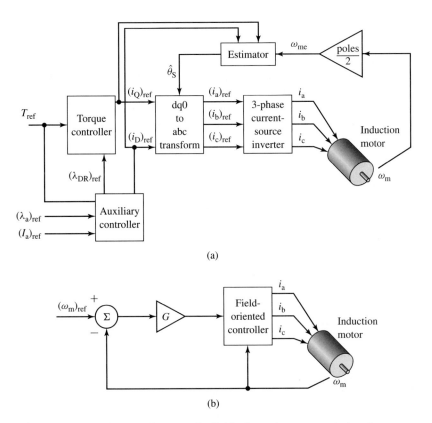

Figure 11.21 (a) Block diagram of a field-oriented torque-control system for an induction motor. (b) Block diagram of an induction-motor speed-control loop built around a field-oriented torque control system.

resistance R_{aR} will undoubtedly change with temperature as the motor is operated. It can be shown that errors in τ_R result in an offset in the estimate of θ_S, which in turn will result in an error in the estimate for the position of the rotor flux with the result that the applied armature currents will not be exactly aligned with the direct- and quadrature-axes. The torque controller will still work basically as expected, although there will be corresponding errors in the torque and rotor flux.

As with the synchronous motor, the rms armature flux-linkages can be found from Eq. 11.38 as

$$(\lambda_a)_{rms} = \sqrt{\frac{\lambda_D^2 + \lambda_Q^2}{2}} \tag{11.86}$$

Combining Eqs. 11.63 and 11.82 gives

$$\lambda_Q = L_S i_Q + L_m i_{QR} = \left(L_S - \frac{L_m^2}{L_R}\right) i_Q \tag{11.87}$$

Substituting Eqs. 11.80 and 11.87 into Eq. 11.86 gives

$$(\lambda_a)_{rms} = \sqrt{\frac{L_S^2 i_D^2 + \left(L_S - \frac{L_m^2}{L_R}\right)^2 i_Q^2}{2}} \tag{11.88}$$

Finally, as discussed in the footnote to Eq. 11.37, the rms line-to-neutral armature voltage can be found as

$$V_a = \sqrt{\frac{v_D^2 + v_Q^2}{2}} = \sqrt{\frac{(R_a i_D - \omega_e \lambda_Q)^2 + (R_a i_Q + \omega_e \lambda_D)^2}{2}}$$

$$= \sqrt{\frac{\left(R_a i_D - \omega_e \left(L_S - \frac{L_m^2}{L_R}\right) i_Q\right)^2 + (R_a i_Q + \omega_e L_S i_D)^2}{2}} \tag{11.89}$$

These equations show that the armature flux linkages and terminal voltage are determined by both the direct- and quadrature-axis components of the armature current. Thus, the block marked "Auxiliary Controller" in Fig. 11.21a, which calculates the reference values for the direct- and quadrature-axis currents, must calculate the reference currents $(i_D)_{ref}$ and $(i_Q)_{ref}$ which achieve the desired torque subject to constraints on armature flux linkages (to avoid saturation in the motor), armature current, $(I_a)_{rms} = \sqrt{(i_D^2 + i_Q^2)/2}$ (to avoid excessive armature heating) and armature voltage (to avoid potential insulation damage).

Note that, as we discussed with regard to synchronous machines in Section 11.2.2, the torque-control system of Fig. 11.21a is typically imbedded within a larger control loop. One such example is the speed-control loop of Fig. 11.21b.

EXAMPLE 11.12

The three-phase, 230-V, 60-Hz, 12-kW, four-pole induction motor of Example 6.7 and Example 11.11 is to be driven by a field-oriented speed-control system (similar to that of Fig. 11.21b) at a speed of 1740 r/min. Assuming the controller is programmed to set the rotor flux linkages λ_{DR} to the machine rated peak value, find the rms amplitude of the armature current, the electrical frequency, and the rms terminal voltage, if the electromagnetic power is 9.7 kW and the motor is operating at a speed of 1680 r/min.

■ Solution

We must first determine the parameters for this machine. From Eqs. 11.66 through Eq. 11.74

$$L_m = \frac{X_{m0}}{\omega_{e0}} = \frac{18.7}{120\pi} = 49.6 \text{ mH}$$

$$L_S = L_m + \frac{X_{10}}{\omega_{e0}} = 49.6 \text{ mH} + \frac{0.680}{120\pi} = 51.41 \text{ mH}$$

$$L_R = L_m + \frac{X_{20}}{\omega_{e0}} = 49.6 \text{ mH} + \frac{0.672}{120\pi} = 51.39 \text{ mH}$$

$$R_a = R_1 = 0.095 \ \Omega$$

$$R_{aR} = R_2 = 0.2 \ \Omega$$

The rated rms line-to-neutral terminal voltage for this machine is $230/\sqrt{3} = 132.8$ V and thus the peak rated flux for this machine is

$$(\lambda_{\text{rated}})_{\text{peak}} = \frac{\sqrt{2}(V_a)_{\text{rated}}}{\omega_e} = \frac{\sqrt{2} \times 132.8}{120\pi} = 0.498 \text{ Wb}$$

For the specified operating condition

$$\omega_m = n\left(\frac{\pi}{30}\right) = 1680\left(\frac{\pi}{30}\right) = 176 \text{ rad/sec}$$

and the mechanical torque is

$$T_{\text{mech}} = \frac{P_{\text{mech}}}{\omega_m} = \frac{9.7 \times 10^3}{176} = 55.1 \text{ N} \cdot \text{m}$$

From Eq. 11.77, with $\lambda_{\text{DR}} = \lambda_{\text{rated}} = 0.498$ Wb

$$\begin{aligned}
i_Q &= \frac{2}{3}\left(\frac{2}{\text{poles}}\right)\left(\frac{L_R}{L_m}\right)\left(\frac{T_{\text{mech}}}{\lambda_{\text{DR}}}\right) \\
&= \frac{2}{3}\left(\frac{2}{4}\right)\left(\frac{51.39 \times 10^{-3}}{49.6 \times 10^{-3}}\right)\left(\frac{55.1}{0.498}\right) = 38.2 \text{ A}
\end{aligned}$$

From Eq. 11.79,

$$i_D = \frac{\lambda_{\text{DR}}}{L_m} = \frac{0.498}{49.6 \times 10^{-3}} = 10.0 \text{ A}$$

The rms armature current is thus

$$I_a = \sqrt{\frac{i_D^2 + i_Q^2}{2}} = \sqrt{\frac{10.0^2 + 38.2^2}{2}} = 27.9 \text{ A}$$

The electrical frequency can be found from Eq. 11.81

$$\omega_e = \omega_{\text{me}} + \frac{R_{\text{aR}}}{L_R}\left(\frac{i_Q}{i_D}\right)$$

With $\omega_{\text{me}} = (\text{poles}/2)\omega_m = 2 \times 176 = 352$ rad/sec

$$\omega_e = 352 + \left(\frac{0.2}{51.39 \times 10^{-3}}\right)\left(\frac{38.2}{10.0}\right) = 367 \text{ rad/sec}$$

and $f_e = \omega_e/(2\pi) = 58.4$ Hz.

Finally, from Eq. 11.89, the rms line-to-neutral terminal voltage

$$V_a = \sqrt{\frac{\left(R_a i_D - \omega_e\left(L_S - \frac{L_m^2}{L_R}\right)i_Q\right)^2 + (R_a i_Q + \omega_e L_S i_D)^2}{2}}$$

$$= 140.6 \text{ V line-to-neutral} = 243.6 \text{ V line-to-line}$$

Consider again the induction motor and field-oriented-control system of Example 11.12. Assume that the speed is readjusted to 1700 r/min and that the electromagnetic power is known to increase to 10.0 kW. Find the rms amplitude of the armature current, the electrical frequency, and the rms terminal voltage for this new operating condition.

Solution

Armature current $= 28.4$ A

$f_e = 59.1$ Hz

Terminal voltage $= 142.5$ V line-to-neutral $= 246.9$ V line-to-line

The ability to independently control the rotor flux and the torque has important control implications. Consider, for example, the response of the direct-axis rotor flux to a change in direct-axis current. Equation C.66, with $\lambda_{qR} = 0$, becomes

$$0 = R_{aR} i_{dR} + \frac{d\lambda_{dR}}{dt} \tag{11.90}$$

Substituting for i_{dR} in terms of λ_{dR}

$$i_{dR} = \frac{\lambda_{dR} - L_m i_d}{L_R} \tag{11.91}$$

gives a differential equation for the rotor flux linkages λ_{DR}

$$\frac{d\lambda_{dR}}{dt} + \left(\frac{R_{aR}}{L_R}\right)\lambda_{dR} = \left(\frac{L_m}{L_R}\right) i_d \tag{11.92}$$

From Eq. 11.92 we see that the response of the rotor flux to a step change in direct-axis current i_d is relatively slow; λ_{dR} will change exponentially with the rotor time constant of $\tau_R = L_R/R_{aR}$. Since the torque is proportional to the product $\lambda_{dR} i_q$ we see that fast torque response will be obtained from changes in i_q. Thus, for example, to implement a step change in torque, a practical control algorithm might start with a step change in $(i_Q)_{ref}$ to achieve the desired torque change, followed by an adjustment in $(i_D)_{ref}$ (and hence λ_{dR}) to readjust the armature current or terminal voltage as desired. This adjustment in $(i_D)_{ref}$ would be coupled with a compensating adjustment in $(i_Q)_{ref}$ to maintain the torque at its desired value.

EXAMPLE 11.13

Consider again the induction motor of Example 11.12. Assuming the motor speed and electromagnetic power remain constant (at 1680 r/min and 9.7 kW), use MATLAB to plot the per-unit armature current I_a and terminal voltage V_a as a function of i_D as $(\lambda_{DR})_{ref}$ is varied between 0.8 and 1.2 per unit, where 1.0 per unit corresponds to the rated peak value.

■ Solution

The desired plot is given in Fig. 11.22. Note that the armature current decreases and the terminal voltage increases as λ_{DR} is increased. This clearly shows how i_D, which controls λ_{DR}, can be

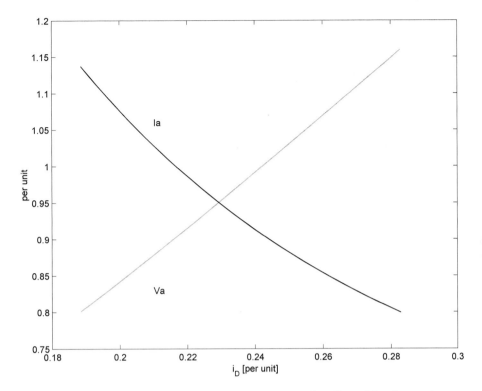

Figure 11.22 MATLAB plot for Example 11.13 showing the effect of the direct-axis current i_D on the armature voltage and current for an induction motor at constant speed and load.

chosen to optimize the tradeoff between such quantities as armature current, armature flux linkages, and terminal voltage.

Here is the MATLAB script:

```
clc
clear

%Motor rating and characteristics

Prated = 12e3;
Vrated = 230;
Varated = 230/sqrt(3);
ferated = 60;
omegaerated = 2*pi*ferated;
Lambdarated = sqrt(2)*Varated/omegaerated;
Irated = Prated/(sqrt(3)*Vrated);
Ipeakbase = sqrt(2)*Irated;
poles = 4;

%Here are the 60-Hz motor parameters
```

```
V10 = Vrated/sqrt(3);
X10 = 0.680;
X20 = 0.672;
Xm0 = 18.7;
R1 = 0.095;
R2 = 0.2;

%Calculate required dq0 parameters
Lm = Xm0/omegaerated;
LS = Lm + X10/omegaerated;
LR = Lm + X20/omegaerated;
Ra = R1;
RaR = R2;

% Operating point
n = 1680;
omegam = n*pi/30;
omegame = (poles/2)*omegam;
Pmech = 9.7e3;
Tmech = Pmech/omegam;

% Loop to plot over lambdaDR
for n = 1:41
    lambdaDR = (0.8 + (n-1)*0.4/40)*Lambdarated;
    lambdaDRpu(n) = lambdaDR/Lambdarated;
    iQ = (2/3)*(2/poles)*(LR/Lm)*(Tmech/lambdaDR);
    iD = (lambdaDR/Lm);
    iDpu(n) = iD/Ipeakbase;
    iQR = - (Lm/LR)*iQ;
    Ia = sqrt((iD^2 + iQ^2)/2);
    Iapu(n) = Ia/Irated;
    omegae = omegame - (RaR/LR)*(iQ/iD);
    fe(n) = omegae/(2*pi);
    Varms = sqrt(((Ra*iD-omegae*(LS-Lm^2/LR)*iQ)^2 + ...
        (Ra*iQ+ omegae*LS*iD)^2)/2);
    Vapu(n) = Varms/Varated;
end

%Now plot
plot(iDpu,Iapu)
hold
plot(iDpu,Vapu,':')
hold
xlabel('i_D [per unit]')
ylabel('per unit')
text(.21,1.06,'Ia')
text(.21,.83,'Va')
```

11.4 CONTROL OF VARIABLE-RELUCTANCE MOTORS

Unlike dc and ac (synchronous or induction) machines, VRMs cannot be simply "plugged in" to a dc or ac source and then be expected to run. As is dicussed in Chapter 8, the phases must be excited with (typically unipolar) currents, and the timing of these currents must be carefully correlated with the position of the rotor poles in order to produce a useful, time-averaged torque. The result is that although the VRM itself is perhaps the simplest of rotating machines, a practical VRM drive system is relatively complex.

VRM drive systems are competitive only because this complexity can be realized easily and inexpensively through power and microelectronic circuitry. These drive systems require a fairly sophisticated level of controllability for even the simplest modes of VRM operation. Once the capability to implement this control is available, fairly sophisticated control features can be added (typically in the form of additional software) at little additional cost, further increasing the competitive position of VRM drives.

In addition to the VRM itself, the basic VRM drive system consists of the following components: a rotor-position sensor, a controller, and an inverter. The function of the rotor-position sensor is to provide an indication of shaft position which can be used to control the timing and waveform of the phase excitation. This is directly analogous to the timing signal used to control the firing of the cylinders in an automobile engine.

The controller is typically implemented in software in microelectronic (microprocessor) circuitry. Its function is to determine the sequence and waveforms of the phase excitation required to achieve the desired motor speed-torque characteristics. In addition to set points of desired speed and/or torque and shaft position (from the shaft-position sensor), sophisticated controllers often employ additional inputs including shaft-speed and phase-current magnitude. Along with the basic control function of determining the desired torque for a given speed, the more sophisticated controllers attempt to provide excitations which are in some sense optimized (for maximum efficiency, stable transient behavior, etc.).

The control circuitry consists typically of low-level electronics which cannot be used to directly supply the currents required to excite the motor phases. Rather its output consists of signals which control an inverter which in turn supplies the phase currents. Control of the VRM is achieved by the application of an appropriate set of currents to the VRM phase windings.

Figures 11.23a to c show three common configurations found in inverter systems for driving VRMs. Note that these are simply H-bridge inverters of the type discussed in Section 10.3. Each inverter is shown in a two-phase configuration. As is clear from the figures, extension of each configuration to drive additional phases can be readily accomplished.

The configuration of Fig. 11.23a is perhaps the simplest. Closing switches S_{1a} and S_{1b} connects the phase-1 winding across the supply ($v_1 = V_0$) and causes the winding current to increase. Opening just one of the switches forces a short across

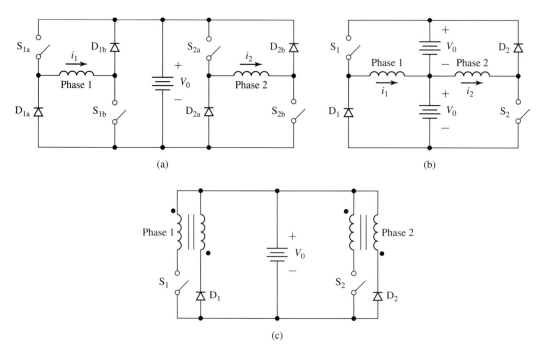

(a)

(b)

(c)

Figure 11.23 Inverter configurations. (a) Two-phase inverter which uses two switches per phase. (b) Two-phase inverter which uses a split supply and one switch per phase. (c) Two-phase inverter with bifilar phase windings and one switch per phase.

the winding and the current will decay, while opening both switches connects the winding across the supply with negative polarity through the diodes ($v_1 = -V_0$) and the winding current will decay more rapidly. Note that this configuration is capable of regeneration (returning energy to the supply), but not of supplying negative current to the phase winding. However, since the torque in a VRM is proportional to the square of the phase current, there is no need for negative winding current.

As discussed in Section 10.3.2, the process of pulse-width modulation, under which a series of switch configurations alternately charge and discharge a phase winding, can be used to control the average winding current. Using this technique, an inverter such as that of Fig. 11.23a can readily be made to supply the range of waveforms required to drive a VRM.

The inverter configuration of Fig. 11.23a is perhaps the simplest of H-bridge configurations which provide regeneration capability. Its main disadvantage is that it requires two switches per phase. In many applications, the cost of the switches (and their associated drive circuitry) dominates the cost of the inverter, and the result is that this configuration is less attractive in terms of cost when compared to other configurations which require one switch per phase.

Figure 11.23b shows one such configuration. This configuration requires a split supply (i.e., two supplies of voltage V_0) but only a single switch and diode per phase.

Closing switch S1 connects the phase-1 winding to the upper dc source. Opening the switch causes the phase current to transfer to diode D1, connecting the winding to the bottom dc source. Phase 1 is thus supplied by the upper dc source and regenerates through the bottom source. Note that to maintain symmetry and to balance the energy supplied from each source equally, phase 2 is connected oppositely so that it is supplied from the bottom source and regenerates into the top source.

The major disadvantages of the configuration of Fig. 11.23b are that it requires a split supply and that when the switch is opened, the switch must withstand a voltage of $2V_0$. This can be readily seen by recognizing that when diode D1 is forward-biased, the switch is connected across the two supplies. Such switches are likely to be more expensive than the switches required by the configuration of Fig. 11.23a. Both of these issues will tend to offset some of the economic advantage which can be gained by the elimination of one switch and one diode as compared with the inverter circuit of Fig. 11.23a.

A third inverter configuration is shown in Fig. 11.23c. This configuration requires only a single dc source and uses only a single switch and diode per phase. This configuration achieves regeneration through the use of *bifilar* phase windings. In a bifilar winding, each phase is wound with two separate windings which are closely coupled magnetically (this can be achieved by winding the two windings at the same time) and can be thought of as the primary and secondary windings of a transformer.

When switch S1 is closed, the primary winding of phase 1 is energized, exciting the phase winding. Opening the switch induces a voltage in the secondary winding (note the polarity indicated by the dots in Fig. 11.23c) in such a direction as to forward-bias D1. The result is that current is transferred from the primary to the secondary winding with a polarity such that the current in the phase decays to zero and energy is returned to the source.

Although this configuration requires only a single dc source, it requires a switch which must withstand a voltage in excess of $2V_0$ (the degree of excess being determined by the voltage developed across the primary leakage reactance as current is switched from the primary to the secondary windings) and requires the more complex bifilar winding in the machine. In addition, the switches in this configuration must include snubbing circuitry (typically consisting of a resistor-capacitor combination) to protect them from transient overvoltages. These overvoltages result from the fact that although the two windings of the bifilar winding are wound such that they are as closely coupled as possible, perfect coupling cannot be achieved. As a result, there will be energy stored in the leakage fields of the primary winding which must be dissipated when the switch is opened.

As is discussed in Section 10.3, VRM operation requires control of the current applied to each phase. For example, one control strategy for constant torque production is to apply constant current to each phase during the time that $dL/d\theta_m$ for that phase is constant. This results in constant torque proportional to the square of the phase-current magnitude. The magnitude of the torque can be controlled by changing the magnitude of the phase current.

The control required to drive the phase windings of a VRM is made more complex because the phase-winding inductances change both with rotor position and with current levels due to saturation effects in the magnetic material. As a result, it is not possible in general to implement an open-loop PWM scheme based on a precalculated algorithm. Rather, pulse-width-modulation is typically accomplished through the use of current feedback. The instantaneous phase current can be measured and a switching scheme can be devised such that the switch can be turned off when the current has been found to reach a desired maximum value and turned on when the current decays to a desired minimum value. In this manner the average phase current is controlled to a predetermined function of the rotor position and desired torque.

This section has provided a brief introduction to the topic of drive systems for variable-reluctance machines. In most cases, many additional issues must be considered before a practical drive system can be implemented. For example, accurate rotor-position sensing is required for proper control of the phase excitation, and the control loop must be properly compensated to ensure its stability. In addition, the finite rise and fall times of current buildup in the motor phase windings will ultimately limit the maximum achievable rotor torque and speed.

The performance of a complete VRM drive system is intricately tied to the performance of all its components, including the VRM, its controller, and its inverter. In this sense, the VRM is quite different from the induction, synchronous, and dc machines discussed earlier in this chapter. As a result, it is useful to design the complete drive system as an integrated package and not to design the individual components (VRM, inverter, controller, etc.) separately. The inverter configurations of Fig. 11.23 are representative of a number of possible inverter configurations which can be used in VRM drive systems. The choice of an inverter for a specific application must be made based on engineering and economic considerations as part of an integrated VRM drive system design.

11.5 SUMMARY

This chapter introduces various techniques for the control of electric machines. The broad topic of electric machine control requires a much more extensive discussion than is possible here so our objectives have been somewhat limited. Most noticeably, the discussion of this chapter focuses almost exclusively on steady-state behavior, and the issues of transient and dynamic behavior are not considered.

Much of the control flexibility that is now commonly associated with electric machinery comes from the capability of the power electronics that is used to drive these machines. This chapter builds therefore on the discussion of power electronics in Chapter 10.

The starting point is a discussion of dc motors for which it is convenient to subdivide the control techniques into two categories: speed and torque control. The algorithm for speed control in a dc motor is relatively straight forward. With the exception of a correction for voltage drop across the armature resistance, the steady-state speed is determined by the condition that the generated voltage must be equal to the applied armature voltage. Since the generated voltage is proportional to the field flux and motor speed, we see that the steady-state motor speed is proportional to the armature voltage and inversely proportional to the field flux.

An alternative viewpoint is that of torque control. Because the commutator/brush system maintains a constant angular relationship between the field and armature flux, the torque in a dc motor is simply proportional to the product of the armature current and the field flux. As a result, dc motor torque can be controlled directly by controlling the armature current as well as the field flux.

Because synchronous motors develop torque only at synchronous speed, the speed of a synchronous motor is simply determined by the electrical frequency of the applied armature excitation. Thus, steady-state speed control is simply a matter of armature frequency control. Torque control is also possible. By transforming the stator quantities into a reference frame rotating synchronously with the rotor (using the dq0 transformation of Appendix C), we found that torque is proportional to the field flux and the component of armature current in space quadrature with the field flux. This is directly analogous to the torque production in a dc motor. Control schemes which adopt this viewpoint are referred to as *vector* or *field-oriented* control.

Induction machines operate asynchronously; rotor currents are induced by the relative motion of the rotor with respect to the synchronously rotating stator-produced flux wave. When supplied by a constant-frequency source applied to the armature winding, the motor will operate at a speed somewhat lower than synchronous speed, with the motor speed decreasing as the load torque is increased. As a result, precise speed regulation is not a simple matter, although in most cases the speed will not vary from synchronous speed by an excessive amount.

Analogous to the situation in a synchronous motor, in spite of the fact that the rotor of an induction motor rotates at less than synchronous speed, the interaction between the rotor and stator flux waves is indeed synchronous. As a result, a transformation into a synchronously rotating reference frame results in rotor and stator flux waves which are constant. The torque can then be expressed in terms of the product of the rotor flux linkages and the component of armature current in quadrature with the rotor flux linkages (referred to as the *quadrature-axis component* of the armature current) in a fashion directly analogous to the field-oriented viewpoint of a synchronous motor. Furthermore, it can be shown that the rotor flux linkages are proportional to the direct-axis component of the armature current, and thus the direct-axis component of armature current behaves much like the field current in a synchronous motor. This field-oriented viewpoint of induction machine control, in combination with the power-electronic and control systems required to implement it, has led to the widespread applicability of induction machines to a wide range of variable-speed applications.

Finally, this chapter ends with a brief discussion of the control of variable-reluctance machines. To produce useful torque, these machines typically require relatively complex, nonsinusoidal current waveforms whose shape must be controlled as a function of rotor position. Typically, these waveforms are produced by pulse-width modulation combined with current feedback using an H-bridge inverter of the type discussed in Chapter 10. The details of these waveforms depend heavily upon the geometry and magnetic properties of the VRM and can vary significantly from motor to motor.

11.6 BIBLIOGRAPHY

Many excellent books are available which provide a much more thorough discussion of electric-machinery control than is possible in the introductory discussion presented here. This bibliography lists a few of the many textbooks available for readers who wish to study this topic in more depth.

Boldea, I., *Reluctance Synchronous Machines and Drives.* New York: Clarendon Press·Oxford, 1996.

Kenjo, T., *Stepping Motors and Their Microprocessor Controls.* New York: Clarendon Press·Oxford, 1984.

Leonhard, W., *Control of Electric Drives.* Berlin: Springer, 1996.

Miller, T. J. E., *Brushless Permanent-Magnet and Reluctance Motor Drives.* New York: Clarendon Press·Oxford, 1989.

Miller, T. J. E., *Switched Reluctance Motors and Their Controls.* New York: Magna Press Publishing and Clarendon Press·Oxford, 1993.

Mohan, N., *Advanced Electric Drives: Analysis, Control and Modeling Using Simulink.* Minneapolis: MNPERE (http://www.MNPERE.com), 2001.

Mohan, N., *Electric Drives: An Integrative Approach.* Minneapolis: MNPERE (http://www.MNPERE.com), 2001.

Murphy, J. M. D., and F. G. Turnbull, *Power Electronic Control of AC Motors.* New York: Pergamon Press, 1988.

Novotny, D. W., and T. A. Lipo, *Vector Control and Dynamics of AC Drives.* New York: Clarendon Press·Oxford, 1996.

Subrahmanyam, V., *Electric Drives: Concepts and Applications.* New York: McGraw-Hill, 1996.

Trzynadlowski, A. M., *Control of Induction Motors.* San Diego, California: Academic Press, 2001.

Vas, P., *Sensorless Vector and Direct Torque Control.* Oxford: Oxford University Press, 1998.

11.7 PROBLEMS

11.1 When operating at rated voltage, a 3-kW, 120-V, 1725 r/min separately excited dc motor achieves a no-load speed of 1718 r/min at a field current of 0.70 A. The motor has an armature resistance of 145 mA and a shunt-field resistance of 104 Ω. For the purposes of this problem you may assume the rotational losses to be negligible.

This motor will control the speed of a load whose torque is constant at 15.2 N·m over the speed range of 1500–1800 r/min. The motor will be operated at a constant armature voltage of 120 V. The field-winding will be supplied from the 120-V dc armature supply via a pulse-width modulation system, and the motor speed will be varied by varying the duty cycle of the pulse-width modulation.

   a. Calculate the field current required to achieve operation at 15.2 N·m torque and 1800 r/min. Calculate the corresponding PWM duty cycle D.

   b. Calculate the field current required to achieve operation at 15.2 N·m torque and 1500 r/min. Calculate the corresponding PWM duty cycle.

   c. Plot the required PWM duty cycle as a function of speed over the desired speed range of 1500 to 1800 r/min.

11.2 Repeat Problem 11.1 for a load whose torque is 15.2 N·m at 1600 r/min and which varies as the speed to the 1.8 power.

11.3 The dc motor of Problem 11.1 has a field-winding inductance $L_f = 3.7$ H and a moment of inertia $J = 0.081$ kg·m². The motor is operating at rated terminal voltage and an initial speed of 1300 r/min.

   a. Calculate the initial field current I_f and duty cycle D.

      At time $t = 0$, the PWM duty cycle is suddenly switched from the value found in part (a) to $D = 0.60$.

   b. Calculate the final values of the field current and motor speed after the transient has died out.

   c. Write an expression for the field-current transient as a function of time.

   d. Write a differential equation for the motor speed as a function of time during this transient.

11.4 A shunt-connected 240-V, 15-kW, 3000 r/min dc motor has the following parameters

Field resistance:	$R_f = 132\ \Omega$
Armature resistance:	$R_a = 0.168\ \Omega$
Geometric constant:	$K_f = 0.422$ V/(A·rad/sec)

When operating at rated voltage, no-load, the motor current is 1.56 A.

   a. Calculate the no-load speed and rotational loss.

   b. Assuming the rotational loss to be constant, use MATLAB to plot the motor output power as a function of speed. Limit your plot to a maximum power output of 15 kW.

   c. Armature-voltage control is to be used to maintain constant motor speed as the motor is loaded. For this operating condition, the shunt field voltage will be held constant at 240-V. Plot the armature voltage as a function of power output required to maintain the motor at a constant speed of 2950 r/min.

   d. Consider that the situation for armature-voltage control is applied to this motor while the field winding remains connected in shunt across the armature terminals. Repeat part (c) for this operating condition. Is such operation feasible? Why is the motor behavior significantly different from that in part (c)?

11.5 The data sheet for a small permanent-magnet dc motor provides the following parameters:

Rated voltage: $V_{rated} = 3$ V
Rated output power: $P_{rated} = 0.28$ W
No-load speed: $n_{nl} = 12{,}400$ r/min
Torque constant: $K_m = 0.218$ mV/(r/min)
Stall torque: $T_{stall} = 0.094$ oz·in

a. Calculate the motor armature resistance.
b. Calculate the no-load rotational loss.
c. Assume the motor to be connected to a load such that the total shaft power (actual load plus rotational loss) is equal 0.25 W at a speed of 12,000 r/min. Assuming this load to vary as the square of the motor speed, write a MATLAB script to plot the motor speed as a function of terminal voltage for 1.0 V $\leq V_a \leq 3.0$ V.

11.6 The data sheet for a 350-W permanent-magnet dc motor provides the following parameters:

Rated voltage: $V_{rated} = 24$ V
Armature resistance: $R_a = 97$ mΩ
No-load speed: $n_{nl} = 3580$ r/min
No-load current: $I_{a,nl} = 0.47$ A

a. Calculate the motor torque-constant K_m in V/(rad/sec).
b. Calculate the no-load rotational loss.
c. The motor is supplied from a 30-V dc supply through a PWM inverter. Table 11.1 gives the measured motor current as a function of the PWM duty cycle D.

Complete the table by calculating the motor speed and the load power for each value of D. Assume that the rotational losses vary as the square of the motor speed.

Table 11.1 Motor-performance data for Problem 11.6.

D	I_a (A)	r/min	P_{load} (W)
0.80	13.35		
0.75	12.70		
0.70	12.05		
0.65	11.40		
0.60	10.70		
0.55	10.05		
0.50	9.30		

11.7 The motor of Problem 11.5 has a moment of inertia of 6.4×10^{-7} oz·in·sec^2. Assuming it is unloaded and neglecting any effects of rotational loss, calculate the time required to achieve a speed of 12,000 r/min if the motor is supplied by a constant armature current of 100 mA.

11.8 An 1100-W, 150-V, 3000-r/min permanent-magnet dc motor is to be operated from a current-source inverter so as to provide direct control of the motor torque. The motor torque constant is $K_m = 0.465$ V/(rad/sec); its armature resistance is 1.37 Ω. The motor rotational loss at a speed of 3000 r/min is 87 W. Assume that the rotational loss can be represented by a constant loss torque as the motor speed varies.

     a. Calculate the rated armature current of this motor. What is the corresponding mechanical torque in N·m?

     b. The current source supplies a current of 6.2 A to the motor armature, and the motor speed is measured to be 2670 r/min. Calculate the load torque and power.

     c. Assume the load torque of part (b) to vary linearly with speed and the motor and load to have a combined inertia of 2.28×10^{-3} kg·m^2. Calculate the motor speed as a function of time if the armature current is suddenly increased to 7.0 A.

11.9 The permanent-magnet dc motor of Problem 11.8 is operating at its rated speed of 3000 r/min and no load. If rated current is suddenly applied to the motor armature in such a direction as to slow the motor down, how long will it take the motor to reach zero speed? The inertia of the motor alone is 1.86×10^{-3} kg·m^2. Ignore the effects of rotational loss.

11.10 A 1100-kVA, 4600-V, 60-Hz, three-phase, four-pole synchronous motor is to be driven from a variable-frequency, three-phase, constant V/Hz inverter rated at 1250-kVA. The synchronous motor has a synchronous reactance of 1.18 per unit and achieves rated open-circuit voltage at a field current of 85 A.

     a. Calculate the rated speed of the motor in r/min.

     b. Calculate the rated current of the motor.

     c. With the motor operating at rated voltage and speed and an input power of 1000-kW, calculate the field current required to achieve unity-power-factor operation.

     The load power of part (c) varies as the speed to the 2.5 power. With the motor field-current held fixed, the inverter frequency is reduced such that the motor is operating at a speed of 1300 r/min.

     d. Calculate the inverter frequency and the motor input power and power factor.

     e. Calculate the field current required to return the motor to unity power factor.

11.11 Consider a three-phase synchronous motor for which you are given the following data:

> Rated line-to-line voltage (V)
> Rated volt-amperes (VA)
> Rated frequency (Hz) and speed (r/min)
> Synchronous reactance in per unit
> Field current at rated open-circuit voltage (AFNL) (A)

The motor is to be operated from a variable-frequency, constant V/Hz inverter at speeds of up to 120 percent of the motor-rated speed.

a. Under the assumption that the motor terminal voltage and current cannot exceed their rated values, write a MATLAB script which calculates, for a given operating speed, the motor terminal voltage, the maximum possible motor input power, and the corresponding field current required to achieve this operating condition. You may consider the effects of saturation and armature resistance to be negligible.

b. Exercise your program on the synchronous motor of Problem 11.10 for motor speeds of 1500 and 2000 r/min.

11.12 For the purposes of performing field-oriented control calculations on non-salient synchronous motors, write a MATLAB script that will calculate the synchronous inductance L_s and armature-to-field mutual inductance L_{af}, both in henries, and the rated torque in N·m, given the following data:

> Rated line-to-line voltage (V)
> Rated (VA)
> Rated frequency (Hz)
> Number of poles
> Synchronous reactance in per unit
> Field current at rated open-circuit voltage (AFNL) (A)

11.13 A 100-kW, 460-V, 60-Hz, four-pole, three-phase synchronous machine is to be operated as a synchronous motor under field-oriented torque control using a system such as that shown in Fig. 11.13a. The machine has a synchronous reactance of 0.932 per unit and AFNL = 15.8 A. The motor is operating at rated speed, loaded to 50 percent of its rated torque at a field current of 14.0 A with the field-oriented controller set to maintain $i_D = 0$.

a. Calculate the synchronous inductance L_s and armature-to-field mutual inductance L_{af}, both in henries.

b. Find the quadrature-axis current i_Q and the corresponding rms magnitude of the armature current i_a.

c. Find the motor line-to-line terminal voltage.

11.14 The synchronous motor of Problem 11.13 is operating under field-oriented torque control such that $i_D = 0$. With the field current set equal to 14.5 A and

with the torque reference set equal to 0.75 of the motor rated torque, the
motor speed is observed to be 1475 r/min.

   a.   Calculate the motor output power.

   b.   Find the quadrature-axis current i_Q and the corresponding rms
magnitude of the armature current i_a.

   c.   Calculate the stator electrical frequency

   d.   Find the motor line-to-line terminal voltage.

11.15 Consider the case in which the load on the synchronous motor in the field-
oriented torque-control system of Problem 11.13 is increased, and the motor
begins to slow down. Based upon some knowledge of the load characteristic,
it is determined that it will be necessary to raise the torque set point T_{ref} from
50 percent to 80 percent of the motor-rated torque to return the motor to its
rated speed.

   a.   If the field current were left unchanged at 14.0 A, calculate the values of
quadrature-axis current, rms armature current, and motor line-to-line
terminal voltage (in V and in per unit) which would result in response to
this change in reference torque.

   b.   To achieve this operating condition with reasonable armature terminal
voltage, the field-oriented control algorithm is changed to the unity-
power-factor algorithm described in the text prior to Example 11.9.
Based upon that algorithm, calculate

      (i)   the motor terminal line-to-line terminal voltage (in V and in
per unit).

      (ii)   the rms armature current.

      (iii)   the direct- and quadrature-axis currents, i_D and i_Q.

      (iv)   the motor field current.

11.16 Consider a 500-kW, 2300-V, 50-Hz, eight-pole synchronous motor with a
synchronous reactance of 1.18 per unit and AFNL = 94 A. It is to be
operated under field-oriented torque control using the unity-power-factor
algorithm described in the text following Example 11.8. It will be used to
drive a load whose torque varies quadratically with speed and whose torque
at a speed of 750 r/min is 5900 N·m. The complete drive system will include
a speed-control loop such as that shown in Fig. 11.13b.

      Write a MATLAB script whose input is the desired motor speed (up to
750 r/min) and whose output is the motor torque, the field current, the direct-
and quadrature-axis currents, the armature current, and the line-to-line
terminal voltage. Exercise your script for a motor speed of 650 r/min.

11.17 A 2-kVA, 230-V, two-pole, three-phase permanent magnet synchronous
motor achieves rated open-circuit voltage at a speed of 3500 r/min. Its
synchronous inductance is 17.2 mH.

   a.   Calculate Λ_{PM} for this motor.

   b.   If the motor is operating at rated voltage and rated current at a speed of
3600 r/min, calculate the motor power in kW and the peak direct- and

quadrature-axis components of the armature current, i_D and i_Q respectively.

11.18 Field-oriented torque control is to be applied to the permanent-magnet synchronous motor of Problem 11.18. If the motor is to be operated at 4000 r/min at rated terminal voltage, calculate the maximum torque and power which the motor can supply and the corresponding values of i_D and i_Q.

11.19 A 15-kVA, 230-V, two-pole, three-phase permanent-magnet synchronous motor has a maximum speed of 10,000 r/min and produces rated open-circuit voltage at a speed of 7620 r/min. It has a synchronous inductance of 1.92 mH. The motor is to be operated under field-oriented torque control.

    a. Calculate the maximum torque the motor can produce without exceeding rated armature current.

    b. Assuming the motor to be operated with the torque controller adjusted to produce maximum torque (as found in part (a)) and $i_D = 0$, calculate the maximum speed at which it can be operated without exceeding rated armature voltage.

    c. To operate at speeds in excess of that found in part (b), flux weakening will be employed to maintain the armature voltage at its rated value. Assuming the motor to be operating at 10,000 r/min with rated armature voltage and current, calculate

        (i)   the direct-axis current i_D.

        (ii) the quadrature-axis current i_Q.

        (iii) the motor torque.

        (iv) the motor power and power factor.

11.20 The permanent magnet motor of Problem 11.17 is to be operated under vector control using the following algorithm.

    Terminal voltage not to exceed rated value

    Terminal current not to exceed rated value

    $i_D = 0$ unless flux weakening is required to avoid excessive armature voltage

    Write a MATLAB script to produce plots of the maximum power and torque which this system can produce as a function of motor speed for speeds up to 10,000 r/min.

11.21 Consider a 460-V, 25-kW, four-pole, 60-Hz induction motor which has the following equivalent-circuit parameters in ohms per phase referred to the stator:

$$R_1 = 0.103 \quad R_2 = 0.225 \quad X_1 = 1.10 \quad X_2 = 1.13 \quad X_m = 59.4$$

The motor is to be operated from a variable frequency, constant-V/Hz drive whose output is 460-V at 60-Hz. Neglect any effects of rotational loss. The motor drive is initially adjusted to a frequency of 60 Hz.

a.   Calculate the peak torque and the corresponding slip and motor speed in r/min.

b.   Calculate the motor torque at a slip of 2.9 percent and the corresponding output power.

c.   The drive frequency is now reduced to 35 Hz. If the load torque remains constant, estimate the resultant motor speed in r/min. Find the resultant motor slip, speed in r/min, and output power.

11.22 Consider the 460-V, 250-kW, four-pole induction motor and drive system of Problem 11.21.

a.   Write a MATLAB script to plot the speed-torque characteristic of the motor at drive frequencies of 20, 40, and 60 Hz for speeds ranging from −200 r/min to the synchronous speed at each frequency.

b.   Determine the drive frequency required to maximize the starting torque and calculate the corresponding torque in N·m.

11.23 A 550-kW, 2400-V, six-pole, 60-Hz three-phase induction motor has the following equivalent-circuit parameters in ohms-per-phase-Y referred to the stator:

$$R_1 = 0.108 \quad R_2 = 0.296 \quad X_1 = 1.18 \quad X_2 = 1.32 \quad X_m = 48.4$$

The motor will be driven by a constant-V/Hz drive whose voltage is 2400 V at a frequency of 60 Hz.

   The motor is used to drive a load whose power is 525 kW at a speed of 1138 r/min and which varies as the cube of speed. Using MATLAB, plot the motor speed as a function of frequency as the drive frequency is varied between 20 and 60 Hz.

11.24 A 150-kW, 60-Hz, six-pole, 460-V three-phase wound-rotor induction motor develops full-load torque at a speed of 1157 r/min with the rotor short-circuited. An external noninductive resistance of 870 mΩ is placed in series with each phase of the rotor, and the motor is observed to develop its rated torque at a speed of 1072 r/min. Calculate the resistance per phase of the original motor.

11.25 The wound rotor of Problem 11.24 will be used to drive a constant-torque load equal to the rated full-load torque of the motor. Using the results of Problem 11.24, calculate the external rotor resistance required to adjust the motor speed to 850 r/min.

11.26 A 75-kW, 460-V, three-phase, four-pole, 60-Hz, wound-rotor induction motor develops a maximum internal torque of 212 percent at a slip of 16.5 percent when operated at rated voltage and frequency with its rotor short-circuited directly at the slip rings. Stator resistance and rotational losses may be neglected, and the rotor resistance may be assumed to be constant, independent of rotor frequency. Determine

a.   the slip at full load in percent.

b.   the rotor I^2R loss at full load in watts.

c. the starting torque at rated voltage and frequency N·m.

If the rotor resistance is now doubled (by inserting external series resistance at the slip rings), determine

d. the torque in N·m when the stator current is at its full-load value.

e. the corresponding slip.

11.27 A 35-kW, three-phase, 440-V, six-pole wound-rotor induction motor develops its rated full-load output at a speed of 1169 r/min when operated at rated voltage and frequency with its slip rings short-circuited. The maximum torque it can develop at rated voltage and frequency is 245 percent of full-load torque. The resistance of the rotor winding is 0.23 Ω/phase Y. Neglect rotational and stray-load losses and stator resistance.

a. Compute the rotor $I^2 R$ loss at full load.

b. Compute the speed at maximum torque.

c. How much resistance must be inserted in series with the rotor to produce maximum starting torque?

The motor is now run from a 50-Hz supply with the applied voltage adjusted so that, for any given torque, the air-gap flux wave has the same amplitude as it does when operated 60 Hz at the same torque level.

d. Compute the 50-Hz applied voltage.

e. Compute the speed at which the motor will develop a torque equal to its rated value at 60-Hz with its slip rings short-circuited.

11.28 The three-phase, 2400-V, 550-kW, six-pole induction motor of Problem 11.23 is to be driven from a field-oriented speed-control system whose controller is programmed to set the rotor flux linkages λ_{DR} equal to the machine rated peak value. The machine is operating at 1148 r/min driving a load which is known to be 400 kW at this speed. Find:

a. the value of the peak direct- and quadrature-axis components of the armature currents i_D and i_Q.

b. the rms armature current under this operating condition.

c. the electrical frequency of the drive in Hz.

d. the rms line-to-line armature voltage.

11.29 A field-oriented drive system will be applied to a 230-V, 20-kW, four-pole, 60-Hz induction motor which has the following equivalent-circuit parameters in ohms per phase referred to the stator:

$$R_1 = 0.0322 \quad R_2 = 0.0703 \quad X_1 = 0.344 \quad X_2 = 0.353 \quad X_m = 18.6$$

The motor is connected to a load whose torque can be assumed proportional to speed as $T_{load} = 85(n/1800)$ N·m, where n is the motor speed in r/min.

The field-oriented controller is adjusted such that the rotor flux linkages λ_{DR} are equal to the machine's rated peak flux linkages, and the motor speed is 1300 r/min. Find

a. the electrical frequency in Hz.

b. the rms armature current and line-to-line voltage.

   c. the motor input kVA.

     If the field-oriented controller is set to maintain the motor speed at 1300 r/min, write a MATLAB script to plot the rms armature V/Hz as a percentage of the rated V/Hz as a function of λ_{DR} as λ_{DR} is varied between 80 and 120 percent of the machine's rated peak flux linkages.

11.30 The 20-kW induction motor-drive and load of Problem 11.29 is operating at a speed of 1450 r/min with the field-oriented controller adjusted to maintain the rotor flux linkages λ_{DR} equal to the machine's rated peak value.

   a. Calculate the corresponding values of the direct- and quadrature-axis components of the armature current, i_D and i_Q, and the rms armature current.

   b. Calculate the corresponding line-to-line terminal voltage drive electrical frequency.

     The quadrature-axis current i_Q is now increased by 10 percent while the direct-axis current is held constant.

   c. Calculate the resultant motor speed and power output.

   d. Calculate the terminal voltage and drive frequency.

   e. Calculate the total kVA input into the motor.

   f. With the controller set to maintain constant speed, determine the set point for λ_{DR}, as a percentage of rated peak flux linkages, that sets the terminal V/Hz equal to the rated machine rated V/Hz. (Hint: This solution is most easily found using a MATLAB script to search for the desired result.)

11.31 A three-phase, eight-pole, 60-Hz, 4160-V, 1250-kW squirrel-cage induction motor has the following equivalent-circuit parameters in ohms-per-phase-Y referred to the stator:

$$R_1 = 0.212 \quad R_2 = 0.348 \quad X_1 = 1.87 \quad X_2 = 2.27 \quad X_m = 44.6$$

It is operating from a field-oriented drive system at a speed of 805 r/min and a power output of 1050 kW. The field-oriented controller is set to maintain the rotor flux linkages λ_{DR} equal to the machine's rated peak flux linkages.

   a. Calculate the motor rms line-to-line terminal voltage, rms armature current, and electrical frequency.

   b. Show that steady-state induction-motor equivalent circuit and corresponding calculations of Chapter 6 give the same output power and terminal current when the induction motor speed is 828 r/min and the terminal voltage and frequency are equal to those found in part (a).

APPENDIX

Three-Phase Circuits

G eneration, transmission, and heavy-power utilization of ac electric energy almost invariably involve a type of system or circuit called a *polyphase system* or *polyphase circuit*. In such a system, each voltage source consists of a group of voltages having related magnitudes and phase angles. Thus, an *n*-phase system employs voltage sources which typically consist of *n* voltages substantially equal in magnitude and successively displaced by a phase angle of $360°/n$. A *three-phase system* employs voltage sources which typically consist of three voltages substantially equal in magnitude and displaced by phase angles of 120°. Because it possesses definite economic and operating advantages, the three-phase system is by far the most common, and consequently emphasis is placed on three-phase circuits in this appendix.

The three individual voltages of a three-phase source may each be connected to its own independent circuit. We would then have three separate *single-phase systems*. Alternatively, as will be shown in Section A.1, symmetrical electric connections can be made between the three voltages and the associated circuitry to form a three-phase system. It is the latter alternative that we are concerned with in this appendix. Note that the word *phase* now has two distinct meanings. It may refer to a portion of a polyphase system or circuit, or, as in the familiar steady-state circuit theory, it may be used in reference to the angular displacement between voltage or current phasors. There is very little possibility of confusing the two.

A.1 GENERATION OF THREE-PHASE VOLTAGES

Consider the elementary two-pole, three-phase generator of Fig. A.1. On the armature are three coils aa', bb', and cc' whose axes are displaced 120° in space from each other. This winding can be represented schematically as shown in Fig. A.2. When the field is excited and rotated, voltages will be generated in the three phases in accordance with Faraday's law. If the field structure is designed so that the flux is

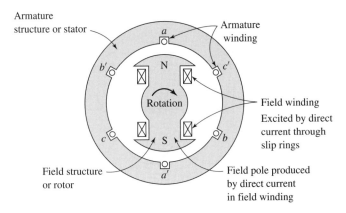

Figure A.1 Elementary two-pole, three-phase generator.

distributed sinusoidally over the poles, the flux linking any phase will vary sinusoidally with time, and sinusoidal voltages will be induced in the three phases. As shown in Fig. A.3, these three voltages will be displaced 120° electrical degrees in time as a result of the phases being displaced 120° in space. The corresponding phasor diagram is shown in Fig. A.4. In general, the time origin and the reference axis in diagrams such as Figs. A.3 and A.4 are chosen on the basis of analytical convenience.

There are two possibilities for the utilization of voltages generated in this manner. The six terminals a, a', b, b', c, and c' of the three-phase winding may be connected to three independent single-phase systems, or the three phases of the winding may be interconnected and used to supply a three-phase system. The latter procedure is adopted almost universally. The three phases of the winding may be interconnected in two possible ways, as shown in Fig. A.5. Terminals a', b', and c' may be joined to form the neutral o, yielding a Y *connection,* or terminals a and b', b and c', and c and a' may be joined individually, yielding a Δ *connection.* In the Y connection, a neutral conductor, shown dashed in Fig. A.5a, may or may not be brought out. If a neutral conductor exists, the system is a four-wire, three-phase system; if not, it is a three-wire, three-phase system. In the Δ connection (Fig. A.5b), no neutral exists and only a three-wire, three-phase system can be formed.

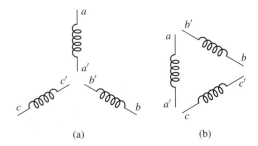

(a)   (b)

Figure A.2 Schematic representation of the windings of Fig. A.1.

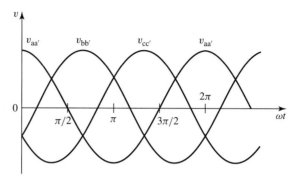

Figure A.3   Voltages generated in the windings of Figs. A.1 and A.2.

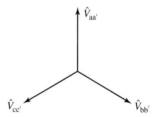

Figure A.4   Phasor diagram of generated voltages.

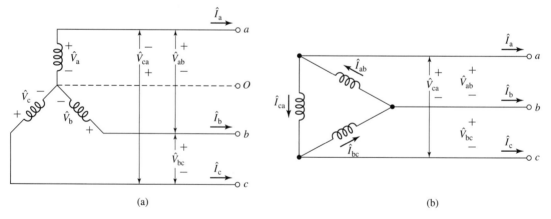

(a)   (b)

Figure A.5   Three-phase connections: (a) Y connection and (b) Δ connection.

The three phase voltages of Figs. A.3 and A.4, are equal and displaced in phase by 120 degrees, a general characteristic of a *balanced three-phase system*. Furthermore, in a balanced three-phase system the impedance in any one phase is equal to that in either of the other two phases, so that the resulting phase currents are also equal and displaced in phase from each other by 120 degrees. Likewise, equal power and equal reactive power flow in each phase. An *unbalanced three-phase system,* however, may be unbalanced in one or more of many ways; the source voltages may be unbalanced, either in magnitude or in phase, or the phase impedances may not be equal. Note that *only balanced systems are treated in this appendix, and none of the methods developed or conclusions reached apply to unbalanced systems.* Most practical analyses are conducted under the assumption of a balanced system. Many industrial loads are three-phase loads and therefore inherently balanced, and in supplying single-phase loads from a three-phase source definite efforts are made to keep the three-phase system balanced by assigning approximately equal single-phase loads to each of the three phases.

A.2 THREE-PHASE VOLTAGES, CURRENTS, AND POWER

When the three phases of the winding in Fig. A.1 are Y-connected, as in Fig. A.5a, the phasor diagram of voltages is that of Fig. A.6. The *phase order* or *phase sequence* in Fig. A.6 is *abc*; that is, the voltage of phase *a* reaches its maximum 120° before that of phase *b*.

The three-phase voltages $\hat{V}_a$, $\hat{V}_b$, and $\hat{V}_c$ are called *line-to-neutral voltages.* The three voltages $\hat{V}_{ab}$, $\hat{V}_{bc}$, and $\hat{V}_{ca}$ are called *line-to-line voltages.* The use of

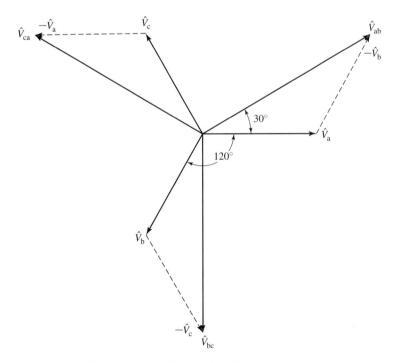

Figure A.6 Voltage phasor diagram for a Y-connected system.

double-subscript notation in Fig. A.6 greatly simplifies the task of drawing the complete diagram. The subscripts indicate the points between which the voltage is determined; for example, the voltage $\hat{V}_{ab}$ is calculated as $\hat{V}_{ab} = \hat{V}_a - \hat{V}_b$.

By Kirchhoff's voltage law, the line-to-line voltage $\hat{V}_{ab}$ is

$$\hat{V}_{ab} = \hat{V}_a - \hat{V}_b = \sqrt{3}\,\hat{V}_a \angle 30° \tag{A.1}$$

as shown in Fig. A.6. Similarly,

$$\hat{V}_{bc} = \sqrt{3}\,\hat{V}_b \angle 30° \tag{A.2}$$

and

$$\hat{V}_{ca} = \sqrt{3}\,\hat{V}_c \angle 30° \tag{A.3}$$

These equations show that *the magnitude of the line-to-line voltage is $\sqrt{3}$ times the line-to-neutral voltage*.

When the three phases are Δ-connected, the corresponding phasor diagram of currents is given in Fig. A.7. The Δ currents are $\hat{I}_{ab}$, $\hat{I}_{bc}$, and $\hat{I}_{ca}$. By Kirchhoff's current law, the line current $\hat{I}_a$ is

$$\hat{I}_a = \hat{I}_{ab} - \hat{I}_{ca} = \sqrt{3}\,\hat{I}_{ab} \angle -30° \tag{A.4}$$

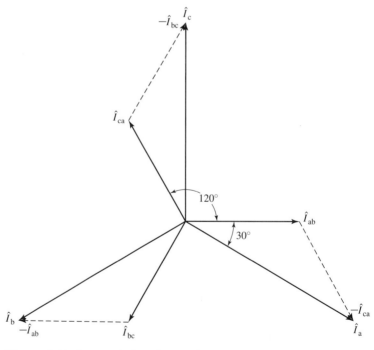

Figure A.7 Current phasor diagram for Δ connection.

as can be seen from the phasor diagram of Fig. A.7. Similarly,

$$\hat{I}_b = \sqrt{3}\,\hat{I}_{bc} \angle -30° \tag{A.5}$$

and

$$\hat{I}_c = \sqrt{3}\,\hat{I}_{ca} \angle -30° \tag{A.6}$$

Stated in words, Eqs. A.4 to A.6 show that for a Δ connection, *the magnitude of the line current is $\sqrt{3}$ times that of the Δ current*. Evidently, the relations between Δ currents and line currents of a Δ connection are similar to those between the line-to-neutral and line-to-line voltages of a Y connection.

With the time origin taken at the maximum positive point of the phase-*a* voltage wave, the instantaneous voltages of the three phases are

$$v_a(t) = \sqrt{2}\,V_{rms}\cos\omega t \tag{A.7}$$

$$v_b(t) = \sqrt{2}\,V_{rms}\cos(\omega t - 120°) \tag{A.8}$$

$$v_c(t) = \sqrt{2}\,V_{rms}\cos(\omega t + 120°) \tag{A.9}$$

where V_{rms} is the rms value of the phase-to-neutral voltage. When the phase currents are displaced from the corresponding phase voltages by the angle θ, the instantaneous phase currents are

$$i_a(t) = \sqrt{2}\,I_{rms}\cos(\omega t + \theta) \tag{A.10}$$

$$i_b(t) = \sqrt{2}\,I_{rms}\cos(\omega t + \theta - 120°) \tag{A.11}$$

$$i_c(t) = \sqrt{2}\,I_{rms}\cos(\omega t + \theta + 120°) \tag{A.12}$$

where I_{rms} is the rms value of the phase current.

The instantaneous power in each phase then becomes

$$p_a(t) = v_a(t)i_a(t) = V_{rms}I_{rms}[\cos(2\omega t + \theta) + \cos\theta] \tag{A.13}$$

$$p_b(t) = v_b(t)i_b(t) = V_{rms}I_{rms}[\cos(2\omega t + \theta - 240°) + \cos\theta] \tag{A.14}$$

$$p_c(t) = v_c(t)i_c(t) = V_{rms}I_{rms}[\cos(2\omega t + \theta + 240°) + \cos\theta] \tag{A.15}$$

Note that the average power of each phase is equal

$$<p_a(t)> = <p_b(t)> = <p_c(t)> = V_{rms}I_{rms}\cos\theta \tag{A.16}$$

The phase angle θ between the voltage and current is referred to as the *power-factor angle* and $\cos\theta$ is referred to as the *power factor*. If θ is negative, then the power factor is said to be *lagging;* if θ is power, then the power factor is said to be *leading*.

The total instantaneous power for all three phases is

$$p(t) = p_a(t) + p_b(t) + p_c(t) = 3V_{rms}I_{rms}\cos\theta \tag{A.17}$$

Notice that the sum of the cosine terms which involve time in Eqs. A.13 to A.15 (the first terms in the brackets) is zero. We have shown that *the total of the instantaneous power for the three phases of a balanced three-phase circuit is constant and*

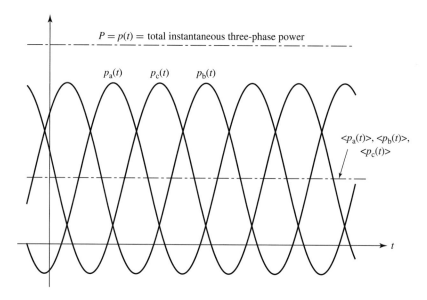

Figure A.8   Instantaneous power in a three-phase system.

does not vary with time. This situation is depicted graphically in Fig. A.8. Instantaneous powers for the three phases are plotted, together with the total instantaneous power, which is the sum of the three individual waves. *The total instantaneous power for a balanced three-phase system is equal to 3 times the average power per phase.* This is one of the outstanding advantages of polyphase systems. It is of particular advantage in the operation of polyphase motors since it means that the shaft-power output is constant and that torque pulsations, with the consequent tendency toward vibration, do not result.

On the basis of single-phase considerations, the average power per phase P_p for either a Y- or Δ-connected system connected to a balanced three-phase load of impedance $Z_p = R_p + jX_p$ Ω/phase is

$$P_p = V_{rms}I_{rms}\cos\theta = I_p^2 R_p \tag{A.18}$$

Here R_p is the resistance per phase. The total three-phase power P is

$$P = 3P_p \tag{A.19}$$

Similarly, for reactive power per phase Q_p and total three-phase reactive power Q,

$$Q_p = V_{rms}I_{rms}\sin\theta = I_p^2 X_p \tag{A.20}$$

and

$$Q = 3Q_p \tag{A.21}$$

where X_p is the reactance per phase.

The voltamperes per phase $(VA)_p$ and total three-phase voltamperes VA are

$$(VA)_p = V_{rms} I_{rms} = I_{rms}^2 Z_p \tag{A.22}$$

$$VA = 3(VA)_p \tag{A.23}$$

In Eqs. A.18 and A.20, θ is the angle between phase voltage and phase current. As in the single-phase case, it is given by

$$\theta = \tan^{-1} \frac{X_p}{R_p} = \cos^{-1} \frac{R_p}{Z_p} = \sin^{-1} \frac{X_p}{Z_p} \tag{A.24}$$

The power factor of a balanced three-phase system is therefore equal to that of any one phase.

A.3   Y- AND Δ-CONNECTED CIRCUITS

Three specific examples are given to illustrate the computational details of Y- and Δ-connected circuits. Explanatory remarks which are generally applicable are incorporated into the solutions.

EXAMPLE A.1

In Fig. A.9 is shown a 60-Hz transmission system consisting of a line having the impedance $Z_l = 0.05 + j0.20\ \Omega$, at the receiving end of which is a load of equivalent impedance $Z_L = 10.0 + j3.00\ \Omega$. The impedance of the return conductor should be considered zero.

a. Compute the line current I; the load voltage V_L; the power, reactive power, and voltamperes taken by the load; and the power and reactive-power loss in the line.
   Suppose now that three such identical systems are to be constructed to supply three such identical loads. Instead of drawing the diagrams one below the other, let them be drawn in the fashion shown in Fig. A.10, which is, of course, the same electrically.
b. For Fig. A.10 give the current in each line; the voltage at each load; the power, reactive power, and voltamperes supplied to each load; the power and reactive-power loss in each of the three transmission systems; the total power, reactive power, and voltamperes supplied to the loads; and the total power and reactive-power loss in the three transmission systems.

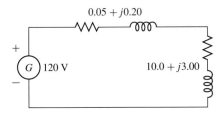

Figure A.9   Circuit for Example A.1, part (a).

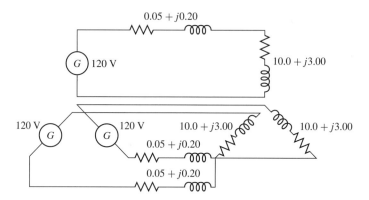

Figure A.10   Circuit for Example A.1, part (b).

     Next consider that the three return conductors are combined into one and that the phase relationship of the voltage sources is such that a balanced four-wire, three-phase system results, as in Fig. A.11.

c. For Fig. A.11 give the line current; the load voltage, both line-to-line and line-to-neutral; the power, reactive power, and voltamperes taken by each phase of the load; the power and reactive-power loss in each line; the total three-phase power, reactive power, and voltamperes taken by the load; and the total power and reactive-power loss in the lines.

d. In Fig. A.11 what is the current in the combined return or neutral conductor?

e. Can this conductor be dispensed with in Fig. A.11 if desired?

     Assume now that this neutral conductor is omitted. This results in the three-wire, three-phase system of Fig. A.12.

f. Repeat part (c) for Fig. A.12.

g. On the basis of the results of this example, outline briefly the method of reducing a balanced three-phase Y-connected circuit problem to its equivalent single-phase problem. Be careful to distinguish between the use of line-to-line and line-to-neutral voltages.

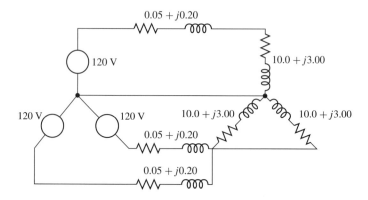

Figure A.11   Circuit for Example A.1, parts (c) to (e).

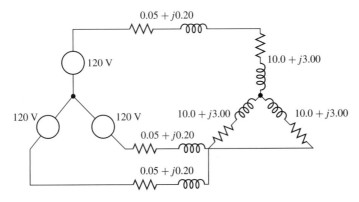

Figure A.12   Circuit for Example A.1 part (f).

■ **Solution**

a.

$$I = \frac{120}{\sqrt{(0.05 + 10.0)^2 + (0.20 + 3.00)^2}} = 11.4 \text{ A}$$

$$V_L = I|Z_L| = 11.4\sqrt{(10.0)^2 + (3.00)^2} = 119 \text{ V}$$

$$P_L = I^2 R_L = (11.4)^2(3.00) = 1200 \text{ W}$$

$$Q_L = I^2 X_L = (11.4)^2(3.00) = 390 \text{ VA reactive}$$

$$(VA)_L = I^2|Z_L| = (11.4)^2\sqrt{(10.0)^2 + (3.00)^2} = 1360 \text{ VA}$$

$$P_1 = I^2 R_1 = (11.4)^2(0.05) = 6.5 \text{ W}$$

$$Q_1 = I^2 X_1 = (11.4)^2(0.20) = 26 \text{ VA reactive}$$

b. The first four obviously have the same values as in part (a).

    Total power $= 3P_L = 3(1300) = 3900$ W

    Total reactive power $= 3Q_L = 3(390) = 1170$ VA reactive

    Total VA $= 3(VA)_L = 3(1360) = 4080$ VA

    Total power loss $= 3P_1 = 3(6.5) = 19.5$ W

    Total reactive-power loss $= 3Q_1 = 3(26) = 78$ VA reactive

c. The results obtained in part (b) are unaffected by this change. The voltage in parts (a) and (b) is now the line-to-neutral voltage. The line-to-line voltage is

$$\sqrt{3}(119) = 206 \text{ V}$$

d. By Kirchhoff's current law, the neutral current is the phasor sum of the three line currents. These line currents are equal and displaced in phase by 120°. Since the phasor sum of three equal phasors 120° apart is zero, the neutral current is zero.

e. The neutral current being zero, the neutral conductor can be dispensed with if desired.

f. Since the presence or absence of the neutral conductor does not affect conditions, the values are the same as in part (c).

g. A neutral conductor can be assumed, regardless of whether one is physically present. Since the neutral conductor in a balanced three-phase circuit carries no current and hence has no voltage drop across it, the neutral conductor should be considered to have zero impedance. Then one phase of the Y, together with the neutral conductor, can be removed for study. Since this phase is uprooted at the neutral, *line-to-neutral voltages must be used*. This procedure yields the single-phase equivalent circuit, in which all quantities correspond to those in one phase of the three-phase circuit. Conditions in the other two phases being the same (except for the 120° phase displacements in the currents and voltages), there is no need for investigating them individually. Line currents in the three-phase system are the same as in the single-phase circuit, and total three-phase power, reactive power, and voltamperes are three times the corresponding quantities in the single-phase circuit. If line-to-line voltages are desired, they must be obtained by multiplying voltages in the single-phase circuit by $\sqrt{3}$.

EXAMPLE A.2

Three impedances of value $Z_Y = 4.00 + j3.00 = 5.00\angle36.9°$ Ω are connected in Y, as shown in Fig. A.13. For balanced line-to-line voltages of 208 V, find the line current, the power factor, and the total power, reactive power, and voltamperes.

■ Solution

The rms line-to-neutral voltage V on any one phase, such as phase a, is

$$V = \frac{208}{\sqrt{3}} = 120 \text{ V}$$

Hence, the line current

$$\hat{I} = \frac{V}{Z_Y} = \frac{120}{5.00\angle36.9°} = 24.0 \angle-36.9° \text{ A}$$

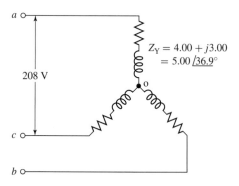

$$Z_Y = 4.00 + j3.00$$
$$= 5.00 \underline{/36.9°}$$

208 V

Figure A.13   Circuit for Example A.2.

and the power factor is equal to

$$\text{Power factor} = \cos\theta = \cos(-36.9°) = 0.80 \text{ lagging}$$

Thus

$$P = 3I^2 R_\text{Y} = 3(24.0)^2(4.00) = 6910 \text{ W}$$

$$Q = 3I^2 X_\text{Y} = 3(24.0)^2(3.00) = 5180 \text{ VA reactive}$$

$$\text{VA} = 3VI = 3(120)(24.0) = 8640 \text{ VA}$$

Note that phases a and c (Fig. A.13) do not form a simple series circuit. Consequently, the current cannot be found by dividing 208 V by the sum of the phase-a and -c impedances. To be sure, an equation can be written for voltage between points a and c by Kirchhoff's voltage law, but this must be a phasor equation taking account of the 120° phase displacement between the phase-a and phase-c currents. As a result, the method of thought outlined in Example A.1 leads to the simplest solution.

EXAMPLE A.3

Three impedances of value $Z_\Delta = 12.00 + j9.00 = 15.00∠36.9°\ \Omega$ are connected in Δ, as shown in Fig. A.14. For balanced line-to-line voltages of 208 V, find the line current, the power factor, and the total power, reactive power, and voltamperes.

■ Solution
The voltage across any one leg of the Δ, V_Δ is equal to the line-to-line voltage V_{l-l}, which is equal to $\sqrt{3}$ times the line-to-neutral voltage V. Consequently,

$$V = \frac{V_{l-l}}{\sqrt{3}} = \frac{208}{\sqrt{3}} = 120 \text{ V}$$

and the current in the Δ is given by the line-to-line voltage divided by the Δ impedance

$$\hat{I}_\Delta = \frac{V_{l-l}}{Z_\Delta} = \frac{208}{15.00 ∠36.9°} = 13.87 ∠-36.9° \text{ A}$$

$$\text{Power factor} = \cos\theta = \cos(-36.9°) = 0.80 \text{ lagging}$$

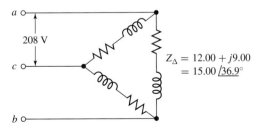

Figure A.14 Circuit for Example A.3.

From Eq. A.4 the phase current is equal to

$$I = \sqrt{3}I_\Delta = \sqrt{3}(13.87) = 24.0 \text{ A}$$

Also

$$P = 3P_\Delta = 3I_\Delta^2 R_\Delta = 3(13.87)^2(12.00) = 6910 \text{ W}$$

$$Q = 3Q_\Delta = 3I_\Delta^2 X_\Delta = 3(13.87)^2(9.00) = 5180 \text{ VA reactive}$$

and

$$\text{VA} = 3(\text{VA})_\Delta = 3V_{l-l}I_\Delta = 3(208)(13.87) = 8640 \text{ VA}$$

Note that phases *ab* and *bc* do not form a simple series circuit, nor does the path *cba* form a simple parallel combination with the direct path through the phase *ca*. Consequently, the line current cannot be found by dividing 208 V by the equivalent impedance of Z_{ca} in parallel with $Z_{ab} + Z_{bc}$. Kirchhoff's-law equations involving quantities in more than one phase can be written, but they must be phasor quantities taking account of the 120° phase displacement between phase currents and phase voltages. As a result, the method outlined above leads to the simplest solution.

Comparison of the results of Examples A.2 and A.3 leads to a valuable and interesting conclusion. Note that the line-to-line voltage, line current, power factor, total power, reactive power, and voltamperes are precisely equal in the two cases; in other words, conditions viewed from the terminals *A*, *B*, and *C* are identical, and one cannot distinguish between the two circuits from their terminal quantities. It will also be seen that the impedance, resistance, and reactance per phase of the Y connection (Fig. A.13) are exactly one-third of the corresponding values per phase of the Δ connection (Fig. A.14). Consequently, a balanced Δ connection can be replaced by a balanced Y connection providing that the circuit constants per phase obey the relation

$$Z_Y = \frac{1}{3}Z_\Delta \tag{A.25}$$

Conversely, a Y connection can be replaced by a Δ connection provided Eq. A.25 is satisfied. The concept of this Y-Δ equivalence stems from the general Y-Δ transformation and is not the accidental result of a specific numerical case.

Two important corollaries follow from this equivalence: (1) A general computational scheme for balanced circuits can be based entirely on Y-connected circuits or entirely on Δ-connected circuits, whichever one prefers. Since it is frequently more convenient to handle a Y connection, the former scheme is usually adopted. (2) In the frequently occurring problems in which the connection is not specified and is not pertinent to the solution, either a Y or a Δ connection may be assumed. Again the Y connection is more commonly selected. In analyzing three-phase motor performance, for example, the actual winding connections need not be known unless the investigation is to include detailed conditions within the windings themselves. The entire analysis can then be based on an assumed Y connection.

A.4 ANALYSIS OF BALANCED THREE-PHASE CIRCUITS; SINGLE-LINE DIAGRAMS

By combining the principle of Δ-Y equivalence with the technique revealed by Example A.1, a simple method of reducing a balanced three-phase-circuit problem to its corresponding single-phase problem can be developed. All the methods of single-phase-circuit analysis thus become available for its solution. The end results of the single-phase analysis are then translated back into three-phase terms to give the final results.

In carrying out this procedure, phasor diagrams need be drawn for only one phase of the Y connection, the diagrams for the other two phases being unnecessary repetition. Furthermore, circuit diagrams can be simplified by drawing only one phase. Examples of such *single-line diagrams* are given in Fig. A.15, showing two three-phase generators with their associated lines or cables supplying a common substation load. Specific connections of apparatus can be indicated if desired. Thus, Fig. A.15b shows that G_1 is Y-connected and G_2 is Δ-connected. Impedances are given in ohms per phase.

When one is dealing with power, reactive power, and voltamperes, it is sometimes more convenient to deal with the entire three-phase circuit at once instead of concentrating on one phase. This possibility arises because simple expressions for three-phase power, reactive power, and voltamperes can be written in terms of line-to-line voltage and line current regardless of whether the circuit is Y- or Δ-connected. Thus, from Eqs. A.18 and A.19, three-phase power is

$$P = 3P_p = 3V_p I_p \cos\theta \tag{A.26}$$

Since $V_{l-l} = \sqrt{3}V_p$, Eq. A.26 becomes

$$P = \sqrt{3}V_{l-l}I_p \cos\theta \tag{A.27}$$

Similarly,

$$Q = \sqrt{3}V_{l-l}I_p \sin\theta \tag{A.28}$$

and

$$VA = \sqrt{3}V_{l-l}I_p \tag{A.29}$$

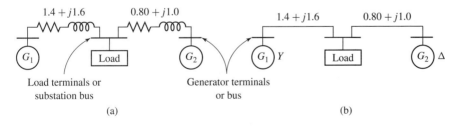

Figure A.15 Examples of single-line diagrams.

It should be borne in mind, however, that the power-factor angle θ, given by Eq. A.24, is the angle between $\hat{V}_p$ and $\hat{I}_p$ and not that between $\hat{V}_{l-l}$ and $\hat{I}_p$.

EXAMPLE A.4

Figure A.15 is the equivalent circuit of a load supplied from two three-phase generating stations over lines having the impedances per phase given on the diagram. The load requires 30 kW at 0.80 power factor lagging. Generator G_1 operates at a terminal voltage of 797 V line-to-line and supplies 15 kW at 0.80 power factor lagging. Find the load voltage and the terminal voltage and power and reactive-power output of G_2.

■ Solution

Let I, P, and Q, respectively, denote line current and three-phase active and reactive power. The subscripts 1 and 2 denote the respective branches of the system; the subscript r denotes a quantity measured at the receiving end of the line. We then have

$$I_1 = \frac{P_1}{\sqrt{3}E_1 \cos\theta_1} = \frac{15,000}{\sqrt{3}(797)(0.80)} = 13.6 \text{ A}$$

$$P_{r1} = P_1 - 3I_1^2 R_1 = 15,000 - 3(13.6)^2(1.4) = 14,220 \text{ W}$$

$$Q_{r1} = Q_1 - 3I_1^2 X_1 = 15,000 \tan(\cos^{-1}0.80) - 3(13.6)^2(1.6) = 10,350 \text{ VA reactive}$$

The factor 3 appears before $I_1^2 R_1$ and $I_1^2 X_1$ in the last two equations because the current I_1 is the phase current. The load voltage is

$$V_L = \frac{VA}{\sqrt{3}(\text{current})} = \frac{\sqrt{(14,220)^2 + (10,350)^2}}{\sqrt{3}(13.6)}$$

$$= 748 \text{ V line-to-line}$$

Since the load requires 30,000 W of real power and $30,000 \tan(\cos^{-1}0.80) = 22,500$ VA of reactive power,

$$P_{r2} = 30,000 - 14,220 = 15,780 \text{ W}$$

and

$$Q_{r2} = 22,500 - 10,350 = 12,150 \text{ VA reactive}$$

$$I_2 = \frac{VA}{\sqrt{3}V_{l-l}} = \frac{\sqrt{(15,780)^2 + (12,150)^2}}{\sqrt{3}(748)} = 15.4 \text{ A}$$

$$P_2 = P_{r2} + 3I_2^2 R_2 = 15,780 + 3(15.4)^2(0.80) = 16,350 \text{ W}$$

$$Q_2 = Q_{r2} + 3I_2^2 X_2 = 12,150 + 3(15.4)^2(1.0) = 12,870 \text{ VA reactive}$$

$$V_2 = \frac{VA}{\sqrt{3}I_2} = \frac{\sqrt{(16,350)^2 + (12,870)^2}}{\sqrt{3}(15.4)}$$

$$= 780 \text{ V (l-l)}$$

A.5 OTHER POLYPHASE SYSTEMS

Although three-phase systems are by far the most common of all polyphase systems, other numbers of phases are used for specialized purposes. The five-wire, four-phase system (Fig. A.16) is sometimes used for low-voltage distribution. It has the advantage that for a phase voltage of 115 V, single-phase voltages of 115 (between a, b, c, or d and o, Fig. A.16) and 230 V (between a and c or b and d) are available as well as a system of polyphase voltages. Essentially the same advantages are possessed by four-wire, three-phase systems having a line-to-neutral voltage of 120 V and a line-to-line voltage of 208 V, however.

Four-phase systems are obtained from three-phase systems by means of special transformer connections. Half of the four-phase system—the part aob (Fig. A.16), for example—constitutes a two-phase system. In some rectifier circuits, 6-, 12-, 18-, and 36-phase connections are used for the conversion of alternating to direct current. These systems are also obtained by transformation from three-phase systems.

When the loads and voltages are balanced, the methods of analysis for three-phase systems can be adapted to any of the other polyphase systems by considering one phase of that polyphase system. Of course, the basic voltage, current, and power relations must be modified to suit the particular polyphase system.

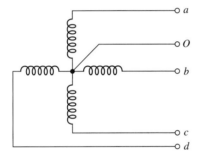

Figure A.16 A five-wire, four-phase system.

B APPENDIX

Voltages, Magnetic Fields, and Inductances of Distributed AC Windings

Both amplitude and waveform of the generated voltage and armature mmf's in machines are determined by the winding arrangements and general machine geometry. These configurations in turn are dictated by economic use of space and materials in the machine and by suitability for the intended service. In this appendix we supplement the introductory discussion of these considerations in Chapter 4 by analytical treatment of ac voltages and mmf's in the balanced steady state. Attention is confined to the time-fundamental component of voltages and the space-fundamental component of mmf's.

B.1 GENERATED VOLTAGES

In accordance with Eq. 4.50, the rms generated voltage per phase for a concentrated winding having N_{ph} turns per phase is

$$E = \sqrt{2}\,\pi f N_{ph}\Phi \tag{B.1}$$

where f is the frequency and Φ the fundamental flux per pole.

A more complex and practical winding will have coil sides for each phase distributed in several slots per pole. Equation B.1 can then be used to compute the voltage distribution of individual coils. To determine the voltage of an entire phase group, the voltages of the component coils must be added as phasors. Such addition of fundamental-frequency voltages is the subject of this article.

B.1.1 Distributed Fractional-Pitch Windings

A simple example of a distributed winding is illustrated in Fig. B.1 for a three-phase, two-pole machine. This case retains all the features of a more general one with any integral number of phases, poles, and slots per pole per phase. At the same time, a *double-layer winding* is shown. Double-layer windings usually lead to simpler end connections and to a machine which is more economical to manufacture and are found in all machines except some small motors below 10 kW. Generally, one side of a coil, such as a_1, is placed in the bottom of a slot, and the other side, $-a_1$, is placed in the top of another slot. Coil sides such as a_1 and a_3 or a_2 and a_4 which are in adjacent slots and associated with the same phase constitute a phase belt. All phase belts are alike when an integral number of slots per pole per phase are used, and for the normal machine the peripheral angle subtended by a phase belt is 60 electrical degrees for a three-phase machine and 90 electrical degrees for a two-phase machine.

Individual coils in Fig. B.1 all span a full pole pitch, or 180 electrical degrees; accordingly, the winding is a full-pitch winding. Suppose now that all coil sides in the tops of the slots are shifted one slot counterclockwise, as in Fig. B.2. Any coil, such as $a_1, -a_1$, then spans only five-sixths of a pole pitch or $\frac{5}{6}(180) = 150$ electrical degrees, and the winding is a fractional-pitch, or chorded, winding. Similar shifting by two slots yields a $\frac{2}{3}$-pitch winding, and so forth. Phase groupings are now intermingled, for some slots contain coil sides in phases a and b, a and c, and b and c. Individual phase

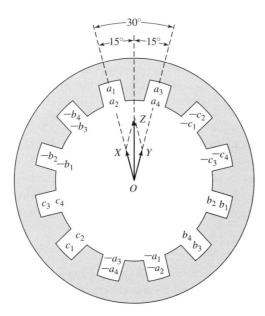

Figure B.1 Distributed two-pole, three-phase full-pitch armature winding with voltage phasor diagram.

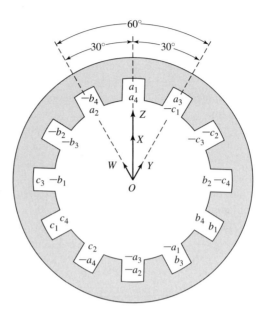

Figure B.2 Distributed two-pole, three-phase fractional-pitch armature winding with voltage phasor diagram.

groups, such as that formed by a_1, a_2, a_3, a_4 on one side and $-a_1, -a_2, -a_3, -a_4$ on the other, are still displaced by 120 electrical degrees from the groups in other phases so that three-phase voltages are produced. Besides the minor feature of shortening the end connections, fractional-pitch windings can be shown to decrease the harmonic content of both voltage and mmf waves.

The end connections between the coil sides are normally in a region of negligible flux density, and hence altering them does not significantly affect the mutual flux linkages of the winding. Allocation of coil sides in slots is then the factor determining the generated voltages, and only that allocation need be specified in Figs. B.1 and B.2. The only requisite is that all coil sides in a phase be included in the interconnection in such a manner that individual voltages make a positive contribution to the total. The practical consequence is that end connections can be made according to the dictates of manufacturing simplicity; the theoretical consequence is that when computational advantages result, the coil sides in a phase can be combined in an arbitrary fashion to form equivalent coils.

One sacrifice is made in using the distributed and fractional-pitch windings of Figs. B.1 and B.2 compared with a concentrated full-pitch winding: for the same number of turns per phase, the fundamental-frequency generated voltage is lower. The harmonics are, in general, lowered by an appreciably greater factor, however, and the total number of turns which can be accommodated on a fixed iron geometry is increased. The effect of distributing the winding in Fig. B.1 is that the voltages of coils a_1 and a_2 are not in phase with those of coils a_3 and a_4. Thus, the voltage of coils a_1 and a_2 can be represented by phasor OX in Fig. B.1, and that of coils a_3 and a_4 by the phasor OY. The time-phase displacement between these two voltages is the same as the electrical angle between adjacent slots, so that OX and OY coincide with the centerlines of adjacent slots. The resultant phasor OZ for phase a is obviously smaller than the arithmetic sum of OX and OY.

In addition, the effect of fractional pitch in Fig. B.2 is that a coil links a smaller portion of the total pole flux than if it were a full-pitch coil. The effect can be superimposed on that of distributing the winding by regarding coil sides a_2 and $-a_1$ as an equivalent coil with the phasor voltage OW (Fig. B.2), coil sides a_1, a_4, $-a_2$, and $-a_3$ as two equivalent coils with the phasor voltage OX (twice the length of OW), and coil sides a_3 and $-a_4$ as an equivalent coil with phasor voltage OY. The resultant phasor OZ for phase a is obviously smaller than the arithmetic sum of OW, OX, and OY and is also smaller than OZ in Fig. B.1.

The combination of these two effects can be included in a *winding factor* k_w to be used as a reduction factor in Eq. B.1. Thus, the generated voltage per phase is

$$E = \sqrt{2}\pi k_w f N_{ph} \Phi \qquad (B.2)$$

where N_{ph} is the total turns in series per phase and k_w accounts for the departure from the concentrated full-pitch case. For a three-phase machine, Eq. B.2 yields the line-to-line voltage for a Δ-connected winding and the line-to-neutral voltage for a Y-connected winding. As in any balanced Y connection, the line-to-line voltage of the latter winding is $\sqrt{3}$ times the line-to-neutral voltage.

B.1.2 Breadth and Pitch Factors

By separately considering the effects of distributing and of chording the winding, reduction factors can be obtained in generalized form convenient for quantitative analysis. The effect of distributing the winding in n slots per phase belt is to yield n voltage phasors displaced in phase by the electrical angle γ between slots, γ being equal to 180 electrical degrees divided by the number of slots per pole. Such a group of phasors is shown in Fig. B.3a and, in a more convenient form for addition, again in Fig. B.3b. Each phasor AB, BC, and CD is the chord of a circle with center at O and subtends the angle γ at the center. The phasor sum AD subtends the angle $n\gamma$, which, as noted previously, is 60 electrical degrees for the normal, uniformly distributed three-phase machine and 90 electrical degrees for the corresponding two-phase machine. From triangles OAa and OAd, respectively,

$$OA = \frac{Aa}{\sin(\gamma/2)} = \frac{AB}{2\sin(\gamma/2)} \qquad (B.3)$$

$$OA = \frac{Ad}{\sin(n\gamma/2)} = \frac{AD}{2\sin(n\gamma/2)} \qquad (B.4)$$

Equating these two values of OA yields

$$AD = AB \frac{\sin(n\gamma/2)}{\sin(\gamma/2)} \qquad (B.5)$$

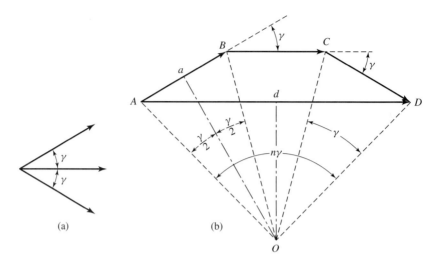

Figure B.3   (a) Coil voltage phasors and (b) phasor sum.

But the arithmetic sum of the phasors is $n(AB)$. Consequently, the reduction factor arising from distributing the winding is

$$k_b = \frac{AD}{nAB} = \frac{\sin(n\gamma/2)}{n\sin(\gamma/2)} \tag{B.6}$$

The factor k_b is called the *breadth factor* of the winding.

The effect of chording on the coil voltage can be obtained by first determining the flux linkages with the fractional-pitch coil. Since there are n coils per phase and N_{ph} total series turns per phase, each coil will have $N = N_{ph}/n$ turns per coil. From Fig. B.4 coil side $-a$ is only ρ electrical degrees from side a instead of the full 180°. The flux linkages with the N-turn coil are

$$\lambda = N B_{peak} l r \left(\frac{2}{poles}\right) \int_{\rho+\alpha}^{\alpha} \sin\theta \, d\theta \tag{B.7}$$

$$\lambda = N B_{peak} l r \left(\frac{2}{poles}\right) [\cos(\alpha + \rho) - \cos\alpha] \tag{B.8}$$

where

$$l = \text{axial length of coil side}$$
$$r = \text{coil radius}$$
$$poles = \text{number of poles}$$

With α replaced by ωt to indicate rotation at ω electrical radians per second, Eq. B.8 becomes

$$\lambda = N B_{peak} l r \left(\frac{2}{poles}\right) [\cos(\omega t + \rho) - \cos\omega t] \tag{B.9}$$

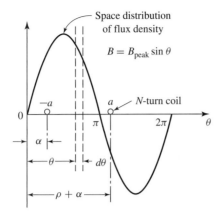

Figure B.4 Fractional-pitch coil in sinusoidal field.

The addition of cosine terms required in the brackets of Eq. B.9 may be performed by a phasor diagram as indicated in Fig. B.5, from which it follows that

$$\cos(\omega t + \rho) - \cos \omega t = -2 \cos \left(\frac{\pi - \rho}{2} \right) \cos \left(\omega t - \left(\frac{\pi - \rho}{2} \right) \right) \quad \text{(B.10)}$$

a result which can also be obtained directly from the terms in Eq. B.9 by the appropriate trigonometric transformations.

The flux linkages are then

$$\lambda = -N B_{peak} l r \left(\frac{4}{poles} \right) \cos \left(\frac{\pi - \rho}{2} \right) \cos \left(\omega t - \left(\frac{\pi - \rho}{2} \right) \right) \quad \text{(B.11)}$$

and the instantaneous voltage is

$$e = \frac{d\lambda}{dt} = \omega N B_{peak} l r \left(\frac{4}{poles} \right) \cos \left(\frac{\pi - \rho}{2} \right) \sin \left(\omega t - \left(\frac{\pi - \rho}{2} \right) \right) \quad \text{(B.12)}$$

The phase angle $(\pi - \rho)/2$ in Eq. B.12 merely indicates that the instantaneous voltage is no longer zero when α in Fig. B.4 is zero. The factor $\cos[(\pi - \rho)/2]$ is an

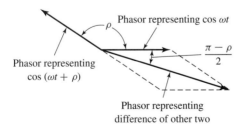

Figure B.5 Phasor addition for fractional-pitch coil.

amplitude-reduction factor, however, so that the rms voltage of Eq. B.1 is modified to

$$E = \sqrt{2}\pi k_{\text{p}} f N_{\text{ph}} \Phi \tag{B.13}$$

where the pitch factor k_{p} is

$$k_{\text{p}} = \cos\left(\frac{\pi - \rho}{2}\right) = \sin\left(\frac{\rho}{2}\right) \tag{B.14}$$

When both the breadth and pitch factors apply, the rms voltage is

$$E = \sqrt{2}\pi k_{\text{b}} k_{\text{p}} f N_{\text{ph}} \Phi = \sqrt{2}\,\pi k_{\text{w}} f N_{\text{ph}} \Phi \tag{B.15}$$

which is an alternate form of Eq. B.2; the winding factor k_{w} is seen to be the product of the pitch and breadth factors.

$$k_{\text{w}} = k_{\text{b}} k_{\text{p}} \tag{B.16}$$

EXAMPLE B.1

Calculate the breadth, pitch, and winding factors for the distributed fractional-pitch winding of Fig. B.2.

■ Solution

The winding of Fig. B.2 has two coils per phase belt, separated by an electrical angle of 30°. From Eq. B.6 the breadth factor is

$$k_{\text{b}} = \frac{\sin(n\gamma/2)}{n \sin(\gamma/2)} = \frac{\sin[2(30°)/2]}{2 \sin(30°/2)} = 0.966$$

The fractional-pitch coils span $150° = 5\pi/6$ rad, and from Eq. B.14 pitch factor is

$$k_{\text{p}} = \sin\left(\frac{\rho}{2}\right) = \sin\left(\frac{5\pi}{12}\right) = 0.966$$

The winding factor is

$$k_{\text{w}} = k_{\text{b}} k_{\text{p}} = 0.933$$

B.2   ARMATURE MMF WAVES

Distribution of a winding in several slots per pole per phase and the use of fractional-pitch coils influence not only the emf generated in the winding but also the magnetic field produced by it. Space-fundamental components of the mmf distributions are examined in this article.

B.2.1   Concentrated Full-Pitch Windings

We have seen in Section 4.3 that a concentrated polyphase winding of N_{ph} turns in a multipole machine produces a rectangular mmf wave around the air-gap circumference. With excitation by a sinusoidal current of amplitude I, the time-maximum amplitude of the space-fundamental component of the wave is, in accordance with Eq. 4.6,

$$(F_{ag1})_{peak} = \frac{4}{\pi} \frac{N_{ph}}{poles}(\sqrt{2}I) \text{ A} \cdot \text{turns/pole} \tag{B.17}$$

where the winding factor k_w of Eq. 4.6 has been set equal to unity since in this case we are discussing the mmf wave of a concentrated winding.

Each phase of a polyphase concentrated winding creates such a time-varying standing mmf wave in space. This situation forms the basis of the analysis leading to Eq. 4.39. For concentrated windings, Eq. 4.39 can be rewritten as

$$\mathcal{F}(\theta_{ae}, t) = \frac{3}{2} \frac{4}{\pi} \left(\frac{N_{ph}}{poles} \right)(\sqrt{2}I) \cos{(\theta_{ae} - \omega t)} = \frac{6}{\pi} \left(\frac{N_{ph}}{poles} \right)(\sqrt{2}I) \cos{(\theta_{ae} - \omega t)} \tag{B.18}$$

The amplitude of the resultant mmf wave in a three-phase machine in ampere-turns per pole is then

$$F_A = \frac{6}{\pi} \left(\frac{N_{ph}}{poles} \right)(\sqrt{2}I) \text{ A} \cdot \text{turns/pole} \tag{B.19}$$

Similarly, for a n_{ph}-phase machine, the amplitude is

$$F_A = \frac{2n_{ph}}{\pi} \left(\frac{N_{ph}}{poles} \right)(\sqrt{2}I) \text{ A} \cdot \text{turns/pole} \tag{B.20}$$

In Eqs. B.19 and B.20, I is the rms current per phase. The equations include only the fundamental component of the actual distribution and apply to concentrated full-pitch windings with balanced excitation.

B.2.2 Distributed Fractional-Pitch Winding

When the coils in each phase of a winding are distributed among several slots per pole, the resultant space-fundamental mmf can be obtained by superposition from the preceding simpler considerations for a concentrated winding. The effect of distribution can be seen from Fig. B.6, which is a reproduction of the two-pole, three-phase, full-pitch winding with two slots per pole per phase given in Fig. B.1. Coils a_1 and a_2, b_1 and b_2, and c_1 and c_2 by themselves constitute the equivalent of a three-phase, two-pole concentrated winding because they form three sets of coils excited by polyphase currents and mechanically displaced 120° from each other. They therefore produce a rotating space-fundamental mmf; the amplitude of this contribution is given by Eq. B.19 when N_{ph} is taken as the sum of the series turns in coils a_1 and a_2 only. Similarly, coils a_3 and a_4, b_3 and b_4, and c_3 and c_4 produce another identical mmf wave, but one which is phase-displaced in space by the slot angle γ from the former wave. The resultant space-fundamental mmf wave for the winding can be obtained by adding these two sinusoidal contributions.

The mmf contribution from the $a_1a_2b_1b_2c_1c_2$ coils can be represented by the phasor OX in Fig. B.6. Such phasor representation is appropriate because the waveforms concerned are sinusoidal, and phasor diagrams are simply convenient means for

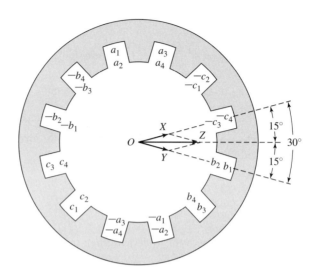

Figure B.6 Distributed two-pole, three-phase, full-pitch armature winding with mmf phasor diagram.

adding sine waves. These are space sinusoids, however, not time sinusoids. Phasor OX is drawn in the space position of the mmf peak for an instant of time when the current in phase a is a maximum. The length of OX is proportional to the number of turns in the associated coils. Similarly, the mmf contribution from the $a_3a_4b_3b_4c_3c_4$ coils may be represented by the phasor OY. Accordingly, the phasor OZ represents the resultant mmf wave. Just as in the corresponding voltage diagram, the resultant mmf is seen to be smaller than if the same number of turns per phase were concentrated in one slot per pole.

In like manner, mmf phasors can be drawn for fractional-pitch windings as illustrated in Fig. B.7, which is a reproduction of the two-pole, three-phase, fractional-pitch winding with two slots per pole per phase given in Fig. B.2. Phasor OW represents the contribution for the equivalent coils formed by conductors a_2 and $-a_1$, b_2 and $-b_1$, and c_2 and $-c_1$; OX for a_1a_4 and $-a_3 -a_2$, b_1b_4 and $-b_3 -b_2$, and c_1c_4 and $-c_3 -c_2$; and OY for a_3 and $-a_4$, b_3 and $-b_4$, and c_3 and $-c_4$. The resultant phasor OZ is, of course, smaller than the algebraic sum of the individual contributions and is also smaller than OZ in Fig. B.6.

By comparison with Figs. B.1 and B.2, these phasor diagrams can be seen to be identical with those for generated voltages. It therefore follows that pitch and breadth factors previously developed can be applied directly to the determination of resultant mmf. Thus, for a distributed, fractional-pitch, polyphase winding, the amplitude of the space-fundamental component of mmf can be obtained by using $k_b k_p N_{ph} = k_w N_{ph}$ instead of simply N_{ph} in Eqs. B.19 and B.20. These equations then become

$$F_A = \frac{6}{\pi}\left(\frac{k_b k_p N_{ph}}{\text{poles}}\right)(\sqrt{2}I) = \frac{6}{\pi}\left(\frac{k_w N_{ph}}{\text{poles}}\right)(\sqrt{2}I) \qquad \text{(B.21)}$$

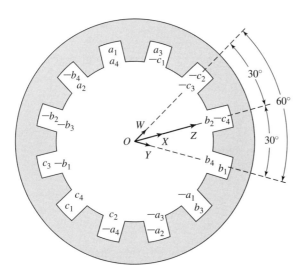

Figure B.7 Distributed two-pole, three-phase, fractional-pitch armature winding with mmf phasor diagram.

for a three-phase machine and

$$F_A = \frac{2n_{ph}}{\pi} \left(\frac{k_b k_p N_{ph}}{poles} \right) (\sqrt{2}I) = \frac{2n_{ph}}{\pi} \left(\frac{k_w N_{ph}}{poles} \right) (\sqrt{2}I) \qquad \text{(B.22)}$$

for a n_{ph}-phase machine, where F_A is in ampere-turns per pole.

B.3 AIR-GAP INDUCTANCES OF DISTRIBUTED WINDINGS

Figure B.8a shows an N-turn, full-pitch, concentrated armature winding in a two-pole magnetic structure with a concentric cylindrical rotor. The mmf of this configuration is shown in Fig. B.8b. Since the air-gap length g is much smaller than the average air-gap radius r, the air-gap radial magnetic field can be considered uniform and equal to the mmf divided by g. From Eq. 4.3 the space-fundamental mmf is given by

$$\mathcal{F}_{ag1} = \frac{4}{\pi} \frac{Ni}{2} \cos \theta_a \qquad \text{(B.23)}$$

and the corresponding air-gap flux density is

$$\mathcal{B}_{ag1} = \mu_0 \frac{\mathcal{F}_{ag1}}{g} = \frac{2\mu_0 Ni}{\pi g} \cos \theta_a \qquad \text{(B.24)}$$

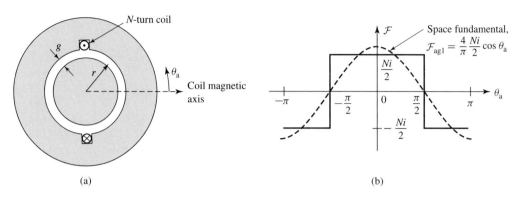

Figure B.8 (a) N-turn concentrated coil and (b) resultant mmf.

Equation B.24 can be integrated to find the fundamental air-gap flux per pole (Eq. 4.44)

$$\Phi = l \int_{-\pi/2}^{\pi/2} B_{ag1} r \, d\theta = \frac{4\mu_0 N l r}{\pi g} i \tag{B.25}$$

where l is the axial length of the air gap. The air-gap inductance of the coil can be found from Eq. 1.29

$$L = \frac{\lambda}{i} = \frac{N\Phi}{i} = \frac{4\mu_0 N^2 l r}{\pi g} \tag{B.26}$$

For a distributed multipole winding with N_{ph} series turns and a winding factor $k_w = k_b k_p$, the air-gap inductance can be found from Eq. B.26 by substituting for N the effective turns per pole pair $(2k_w N_{ph}/\text{poles})$

$$L = \frac{4\mu_0 l r}{\pi g}\left(\frac{2k_w N_{ph}}{\text{poles}}\right)^2 = \frac{16\mu_0 l r}{\pi g}\left(\frac{k_w N_{ph}}{\text{poles}}\right)^2 \tag{B.27}$$

Finally, Fig. B.9 shows schematically two coils (labeled 1 and 2) with winding factors k_{w1} and k_{w2} and with $2N_1/\text{poles}$ and $2N_2/\text{poles}$ turns per pole pair, respectively; their magnetic axes are separated by an electrical angle α (equal to poles/2 times their spatial angular displacement). The mutual inductance between these two windings is given by

$$L_{12} = \frac{4\mu_0}{\pi}\left(\frac{2k_{w1}N_1}{\text{poles}}\right)\left(\frac{2k_{w2}N_2}{\text{poles}}\right)\frac{l r}{g}\cos\alpha$$

$$= \frac{16\mu_0(k_{w1}N_1)(k_{w2}N_2)l r}{\pi g (\text{poles})^2}\cos\alpha \tag{B.28}$$

Although the figure shows one winding on the rotor and the second on the stator, Eq. B.28 is equally valid for the case where both windings are on the same member.

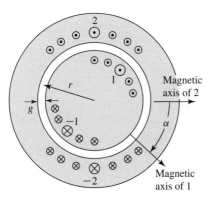

Figure B.9 Two distributed windings separated by electrical angle α.

The two-pole stator-winding distribution of Fig. B.2 is found on an induction motor with an air-gap length of 0.381 mm, an average rotor radius of 6.35 cm, and an axial length of 20.3 cm. Each stator coil has 15 turns, and the coil phase connections are as shown in Fig. B.10. Calculate the phase-a air-gap inductance L_{aa0} and the phase-a to phase-b mutual inductance L_{ab}.

■ **Solution**

Note that the placement of the coils around the stator is such that the flux linkages of each of the two parallel paths are equal. In addition, the air-gap flux distribution is unchanged if, rather than dividing equally between the two legs, as actually occurs, one path were disconnected and all the current were to flow in the remaining path. Thus, the phase inductances can be found by calculating the inductances associated with only one of the parallel paths.

This result may appear to be somewhat puzzling because the two paths are connected in parallel, and thus it would appear that the parallel inductance should be one-half that of the single-path inductance. However, the inductances share a common magnetic circuit, and their

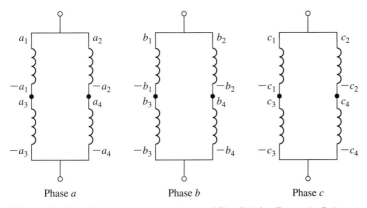

Figure B.10 Coil phase connections of Fig. B.2 for Example B.2.

combined inductance must reflect this fact. It should be pointed out, however, that the phase resistance is one-half that of each of the paths.

The winding factor has been calculated in Example B.1. Thus, from Eq. B.27,

$$L_{\text{aa0}} = \frac{16\mu_0 lr}{\pi g} \left(\frac{k_w N_{\text{ph}}}{\text{poles}}\right)^2$$

$$= \frac{16(4\pi \times 10^{-7}) \times 0.203 \times 0.0635}{\pi(3.81 \times 10^{-4})} \left(\frac{0.933 \times 30}{2}\right)^2$$

$$= 42.4 \text{ mH}$$

The winding axes are separated by $\alpha = 120°$, and thus from Eq. B.28

$$L_{\text{ab}} = \frac{16\mu_0 (k_w N_{\text{ph}})^2 lr}{\pi g (\text{poles})^2} \cos \alpha = -21.2 \text{ mH}$$

APPENDIX C

The dq0 Transformation

I n this appendix, the direct- and quadrature-axis (dq0) theory introduced in Section 5.6 is formalized. The formal mathematical transformation from three-phase stator quantities to their direct- and quadrature-axis components is presented. These transformations are then used to express the governing equations for a synchronous machine in terms of the dq0 quantities.

C.1 TRANSFORMATION TO DIRECT- AND QUADRATURE-AXIS VARIABLES

In Section 5.6 the concept of resolving synchronous-machine armature quantities into two rotating components, one aligned with the field-winding axis, the direct-axis component, and one in quadrature with the field-winding axis, the quadrature-axis component, was introduced as a means of facilitating analysis of salient-pole machines. The usefulness of this concept stems from the fact that although each of the stator phases sees a time-varying inductance due to the saliency of the rotor, the transformed quantities rotate with the rotor and hence see constant magnetic paths. Although not discussed here, additional saliency effects are present under transient conditions, due to the different conducting paths in the rotor, rendering the concept of this transformation all the more useful.

Similarly, this transformation is useful from the point of view of analyzing the interaction of the rotor and stator flux- and mmf-waves, independent of whether or not saliency effects are present. By transforming the stator quantities into equivalent quantities which rotate in synchronism with the rotor, under steady-state conditions these interactions become those of constant mmf- and flux-waves separated by a constant spatial angle. This indeed is the point of view which corresponds to that of an observer in the rotor reference frame.

The idea behind the transformation is an old one, stemming from the work of Andre Blondel in France, and the technique is sometimes referred to as the *Blondel two-reaction method*. Much of the development in the form used here was carried out

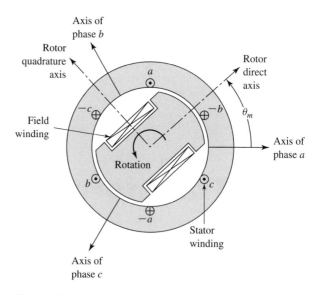

Figure C.1   Idealized synchronous machine

by R. E. Doherty, C. A. Nickle, R. H. Park, and their associates in the United States. The transformation itself, known as the *dq0 transformation,* can be represented in a straightforward fashion in terms of the electrical angle θ_{me} (equal to poles/2 times the spatial angle θ_m) between the rotor direct axis and the stator phase-a axis, as defined by Eq. 4.1 and shown in Fig. C.1.

Letting S represent a stator quantity to be transformed (current, voltage, or flux), we can write the transformation in matrix form as

$$\begin{bmatrix} S_d \\ S_q \\ S_0 \end{bmatrix} = \frac{2}{3} \begin{bmatrix} \cos(\theta_{me}) & \cos(\theta_{me} - 120°) & \cos(\theta_{me} + 120°) \\ -\sin(\theta_{me}) & -\sin(\theta_{me} - 120°) & -\sin(\theta_{me} + 120°) \\ \frac{1}{2} & \frac{1}{2} & \frac{1}{2} \end{bmatrix} \begin{bmatrix} S_a \\ S_b \\ S_c \end{bmatrix} \quad \text{(C.1)}$$

and the inverse transformation as

$$\begin{bmatrix} S_a \\ S_b \\ S_c \end{bmatrix} = \begin{bmatrix} \cos(\theta_{me}) & -\sin(\theta_{me}) & 1 \\ \cos(\theta_{me} - 120°) & -\sin(\theta_{me} - 120°) & 1 \\ \cos(\theta_{me} + 120°) & -\sin(\theta_{me} + 120°) & 1 \end{bmatrix} \begin{bmatrix} S_d \\ S_q \\ S_0 \end{bmatrix} \quad \text{(C.2)}$$

Here the letter S refers to the quantity to be transformed and the subscripts d and q represent the direct and quadrature axes, respectively. A third component, the *zero-sequence component,* indicated by the subscript 0, is also included. This component is required to yield a unique transformation of the three stator-phase quantities; it corresponds to components of armature current which produce no net air-gap flux and hence no net flux linking the rotor circuits. As can be seen from Eq. C.1, under balanced-three-phase conditions, there are no zero-sequence components. Only balanced-three-phase conditions are considered in this book, and hence zero-sequence components are not discussed in any detail.

Note that the dq0 transformation applies to the instantaneous values of the quantities to be transformed, not rms values. Thus when applying the formal instantaneous
transformations as presented here, one must be careful to avoid the use of rms values,
as are frequently used in phasor analyses such as are found in Chapter 5.

EXAMPLE C.1

A two-pole synchronous machine is carrying balanced three-phase armature currents

$$i_a = \sqrt{2}I_a \cos \omega t \qquad i_b = \sqrt{2}I_a \cos (\omega t - 120°) \qquad i_c = \sqrt{2}I_a \cos (\omega t + 120°)$$

The rotor is rotating at synchronous speed ω, and the rotor direct axis is aligned with the stator
phase-a axis at time $t = 0$. Find the direct- and quadrature-axis current components.

■ **Solution**

The angle between the rotor direct axis and the stator phase-a axis can be expressed as

$$\theta_{me} = \omega t$$

From Eq. C.1

$$i_d = \frac{2}{3}[i_a \cos \omega t + i_b \cos (\omega t - 120°) + i_c \cos (\omega t + 120°)]$$

$$= \frac{2}{3}\sqrt{2}I_a[\cos^2 \omega t + \cos^2 (\omega t - 120°) + \cos^2 (\omega t + 120°)]$$

Using the trigonometric identity $\cos^2 \alpha = \frac{1}{2}(1 + \cos 2\alpha)$ gives

$$i_d = \sqrt{2}I_a$$

Similarly,

$$i_q = -\frac{2}{3}[i_a \sin \omega t + i_b \sin (\omega t - 120°) + i_c \sin (\omega t + 120°)]$$

$$= -\frac{2}{3}\sqrt{2}I_a[\cos \omega t \sin \omega t + \cos (\omega t - 120°) \sin (\omega t - 120°)$$

$$+ \cos (\omega t + 120°) \sin (\omega t + 120°)]$$

and using the trigonometric identity $\cos \alpha \sin \alpha = \frac{1}{2} \sin 2\alpha$ gives

$$i_q = 0$$

This result corresponds directly to our physical picture of the dq0 transformation. From
the discussion of Section 4.5 we recognize that the balanced three-phase currents applied to
this machine produce a synchronously rotating mmf wave which produces flux along the stator
phase-a axis at time $t = 0$. This flux wave is thus aligned with the rotor direct axis at $t = 0$
and remains so since the rotor is rotating at the same speed. Hence the stator current produces
only direct-axis flux and thus consists only of a direct-axis component.

C.2 BASIC SYNCHRONOUS-MACHINE RELATIONS IN dq0 VARIABLES

Equations 5.2 to 5.5 give the flux-linkage current relationships for a synchronous machine consisting of a field winding and a three-phase stator winding. This simple machine is sufficient to demonstrate the basic features of the machine representation in dq0 variables; the effects of additional rotor circuits such as damper windings can be introduced in a straightforward fashion.

The flux-linkage current relationships in terms of phase variables (Eqs. 5.2 to 5.5) are repeated here for convenience

$$
\begin{bmatrix} \lambda_a \\ \lambda_b \\ \lambda_c \\ \lambda_f \end{bmatrix} = \begin{bmatrix} \mathcal{L}_{aa} & \mathcal{L}_{ab} & \mathcal{L}_{ac} & \mathcal{L}_{af} \\ \mathcal{L}_{ba} & \mathcal{L}_{bb} & \mathcal{L}_{bc} & \mathcal{L}_{bf} \\ \mathcal{L}_{ca} & \mathcal{L}_{cb} & \mathcal{L}_{cc} & \mathcal{L}_{cf} \\ \mathcal{L}_{fa} & \mathcal{L}_{fb} & \mathcal{L}_{fc} & \mathcal{L}_{ff} \end{bmatrix} \begin{bmatrix} i_a \\ i_b \\ i_c \\ i_f \end{bmatrix}
\tag{C.3}
$$

Unlike the analysis of Section 5.2, this analysis will include the effects of saliency, which causes the stator self and mutual inductances to vary with rotor position.

For the purposes of this analysis the idealized synchronous machine of Fig. C.1 is assumed to satisfy two conditions: (1) the air-gap permeance has a constant component as well as a smaller component which varies cosinusoidally with rotor angle as measured from the direct axis, and (2) the effects of space harmonics in the air-gap flux can be ignored. Although these approximations may appear somewhat restrictive, they form the basis of classical dq0 machine analysis and give excellent results in a wide variety of applications. Essentially they involve neglecting effects which result in time-harmonic stator voltages and currents and are thus consistent with our previous assumptions neglecting harmonics produced by discrete windings.

The various machine inductances can then be written in terms of the electrical rotor angle θ_{me} (between the rotor direct axis and the stator phase-a axis), using the notation of Section 5.2, as follows. For the stator self-inductances

$$
\mathcal{L}_{aa} = L_{aa0} + L_{al} + L_{g2} \cos 2\theta_{me}
\tag{C.4}
$$

$$
\mathcal{L}_{bb} = L_{aa0} + L_{al} + L_{g2} \cos (2\theta_{me} + 120°)
\tag{C.5}
$$

$$
\mathcal{L}_{cc} = L_{aa0} + L_{al} + L_{g2} \cos (2\theta_{me} - 120°)
\tag{C.6}
$$

For the stator-to-stator mutual inductances

$$
\mathcal{L}_{ab} = \mathcal{L}_{ba} = -\frac{1}{2} L_{aa0} + L_{g2} \cos (2\theta_{me} - 120°)
\tag{C.7}
$$

$$
\mathcal{L}_{bc} = \mathcal{L}_{cb} = -\frac{1}{2} L_{aa0} + L_{g2} \cos 2\theta_{me}
\tag{C.8}
$$

$$
\mathcal{L}_{ac} = \mathcal{L}_{ca} = -\frac{1}{2} L_{aa0} + L_{g2} \cos (2\theta_{me} + 120°)
\tag{C.9}
$$

For the field-winding self-inductance

$$\mathcal{L}_{ff} = L_{ff} \tag{C.10}$$

and for the stator-to-rotor mutual inductances

$$\mathcal{L}_{af} = \mathcal{L}_{fa} = L_{af} \cos \theta_{me} \tag{C.11}$$

$$\mathcal{L}_{bf} = \mathcal{L}_{fb} = L_{af} \cos (\theta_{me} - 120°) \tag{C.12}$$

$$\mathcal{L}_{cf} = \mathcal{L}_{fc} = L_{af} \cos (\theta_{me} + 120°) \tag{C.13}$$

Comparison with Section 5.2 shows that the effects of saliency appear only in the stator self- and mutual-inductance terms as an inductance term which varies with $2\theta_{me}$. This twice-angle variation can be understood with reference to Fig. C.1, where it can be seen that rotation of the rotor through 180° reproduces the original geometry of the magnetic circuit. Notice that the self-inductance of each stator phase is a maximum when the rotor direct axis is aligned with the axis of that phase and that the phase-phase mutual inductance is maximum when the rotor direct axis is aligned midway between the two phases. This is the expected result since the rotor direct axis is the path of lowest reluctance (maximum permeance) for air-gap flux.

The flux-linkage expressions of Eq. C.3 become much simpler when they are expressed in terms of dq0 variables. This can be done by application of the transformation of Eq. C.1 to both the flux linkages and the currents of Eq. C.3. The manipulations are somewhat laborious and are omitted here because they are simply algebraic. The results are

$$\lambda_d = L_d i_d + L_{af} i_f \tag{C.14}$$

$$\lambda_q = L_q i_q \tag{C.15}$$

$$\lambda_f = \frac{3}{2} L_{af} i_d + L_{ff} i_f \tag{C.16}$$

$$\lambda_0 = L_0 i_0 \tag{C.17}$$

In these equations, new inductance terms appear:

$$L_d = L_{al} + \frac{3}{2}(L_{aa0} + L_{g2}) \tag{C.18}$$

$$L_q = L_{al} + \frac{3}{2}(L_{aa0} - L_{g2}) \tag{C.19}$$

$$L_0 = L_{al} \tag{C.20}$$

The quantities L_d and L_q are the *direct-axis* and *quadrature-axis synchronous inductances,* respectively, corresponding directly to the direct- and quadrature-axis synchronous reactances discussed in Section 5.6 (i.e., $X_d = \omega_e L_d$ and $X_q = \omega_e L_q$). The inductance L_0 is the *zero-sequence inductance.* Notice that the transformed flux-linkage current relationships expressed in Eqs. C.14 to C.17 no longer contain inductances which are functions of rotor position. This feature is responsible for the usefulness of the dq0 transformation.

Transformation of the voltage equations

$$v_a = R_a i_a + \frac{d\lambda_a}{dt} \tag{C.21}$$

$$v_b = R_a i_b + \frac{d\lambda_b}{dt} \tag{C.22}$$

$$v_c = R_a i_c + \frac{d\lambda_c}{dt} \tag{C.23}$$

$$v_f = R_f i_f + \frac{d\lambda_f}{dt} \tag{C.24}$$

results in

$$v_d = R_a i_d + \frac{d\lambda_d}{dt} - \omega_{me}\lambda_q \tag{C.25}$$

$$v_q = R_a i_q + \frac{d\lambda_q}{dt} + \omega_{me}\lambda_d \tag{C.26}$$

$$v_f = R_f i_f + \frac{d\lambda_f}{dt} \tag{C.27}$$

$$v_0 = R_a i_0 + \frac{d\lambda_0}{dt} \tag{C.28}$$

(algebraic details are again omitted), where $\omega_{me} = d\theta_{me}/dt$ is the rotor electrical angular velocity.

In Eqs. C.25 and C.26 the terms $\omega_{me}\lambda_q$ and $\omega_{me}\lambda_d$ are speed-voltage terms which come as a result of the fact that we have chosen to define our variables in a reference frame rotating at the electrical angular velocity ω_{me}. These speed voltage terms are directly analogous to the speed-voltage terms found in the dc machine analysis of Chapter 9. In a dc machine, the commutator/brush system performs the transformation which transforms armature (rotor) voltages to the field-winding (stator) reference frame.

We now have the basic relations for analysis of our simple synchronous machine. They consist of the flux-linkage current equations C.14 to C.17, the voltage equations C.25 to C.28, and the transformation equations C.1 and C.2. When the rotor electrical angular velocity ω_{me} is constant, the differential equations are linear with constant coefficients. In addition, the transformer terms $d\lambda_d/dt$ and $d\lambda_q/dt$ in Eqs. C.25 and C.26 are often negligible with respect to the speed-voltage terms $\omega_{me}\lambda_q$ and $\omega_{me}\lambda_d$, providing further simplification. Omission of these terms corresponds to neglecting the harmonics and dc component in the transient solution for stator voltages and currents. In any case, the transformed equations are generally much easier to solve, both analytically and by computer simulation, than the equations expressed directly in terms of the phase variables.

In using these equations and the corresponding equations in the machinery literature, careful note should be made of the sign convention and units employed. Here we have chosen the motor-reference convention for armature currents, with positive armature current flowing into the machine terminals. Also, SI units (volts, amperes, ohms, henrys, etc.) are used here; often in the literature one of several per-unit systems is used to provide numerical simplifications.[1]

To complete the useful set of equations, expressions for power and torque are needed. The instantaneous power into the three-phase stator is

$$p_s = v_a i_a + v_b i_b + v_c i_c \tag{C.29}$$

Phase quantities can be eliminated from Eq. C.29 by using Eq. C.2 written for voltages and currents. The result is

$$p_s = \frac{3}{2}(v_d i_d + v_q i_q + 2 v_0 i_0) \tag{C.30}$$

The electromagnetic torque, T_{mech}, is readily obtained by using the techniques of Chapter 3 as the power output corresponding to the speed voltages divided by the shaft speed (in mechanical radians per second). From Eq. C.30 with the speed-voltage terms from Eqs. C.25 and C.26, and by recognizing ω_{me} as the rotor speed in electrical radians per second, we get

$$T_{mech} = \frac{3}{2}\left(\frac{poles}{2}\right)(\lambda_d i_q - \lambda_q i_d) \tag{C.31}$$

A word about sign conventions. When, as is the case in the derivation of this appendix, the motor-reference convention for currents is chosen (i.e., the positive reference direction for currents is into the machine), the torque of Eq. C.31 is the torque acting to accelerate the rotor. Alternatively, if the generator-reference convention is chosen, the torque of Eq. C.31 is the torque acting to decelerate the rotor. This result is, in general, conformity with torque production from interacting magnetic fields as expressed in Eq. 4.81. In Eq. C.31 we see the superposition of the interaction of components: the direct-axis magnetic flux produces torque via its interaction with the quadrature-axis mmf and the quadrature-axis magnetic flux produces torque via its interaction with the direct-axis mmf. Note that, for both of these interactions, the flux and interacting mmf's are 90 electrical degrees apart; hence the sine of the interacting angle (see Eq. 4.81) is unity which in turns leads to the simple form of Eq. C.31.

As a final cautionary note, the reader is again reminded that the currents, fluxes, and voltages in Eqs. C.29 through Eq. C.31 are instantaneous values. Thus, the reader is urged to avoid the use of rms values in these and the other transformation equations found in this appendix.

---

[1] See A. W. Rankin, "Per-Unit Impedances of Synchronous Machines," *Trans. AIEE* 64:569–573, 839–841 (1945).

C.3 BASIC INDUCTION-MACHINE RELATIONS IN dq0 VARIABLES

In this derivation we will assume that the induction machine includes three-phase windings on both the rotor and the stator and that there are no saliency effects. In this case, the flux-linkage current relationships can be written as

$$
\begin{bmatrix}
\lambda_a \\
\lambda_b \\
\lambda_c \\
\lambda_{aR} \\
\lambda_{bR} \\
\lambda_{cR}
\end{bmatrix}
=
\begin{bmatrix}
\mathcal{L}_{aa} & \mathcal{L}_{ab} & \mathcal{L}_{ac} & \mathcal{L}_{aaR} & \mathcal{L}_{abR} & \mathcal{L}_{acR} \\
\mathcal{L}_{ba} & \mathcal{L}_{bb} & \mathcal{L}_{bc} & \mathcal{L}_{baR} & \mathcal{L}_{bbR} & \mathcal{L}_{bcR} \\
\mathcal{L}_{ca} & \mathcal{L}_{cb} & \mathcal{L}_{cc} & \mathcal{L}_{caR} & \mathcal{L}_{cbR} & \mathcal{L}_{ccR} \\
\mathcal{L}_{Aa} & \mathcal{L}_{aRb} & \mathcal{L}_{aRc} & \mathcal{L}_{aRaR} & \mathcal{L}_{aRbR} & \mathcal{L}_{aRC} \\
\mathcal{L}_{bRa} & \mathcal{L}_{bRb} & \mathcal{L}_{bRc} & \mathcal{L}_{bRaR} & \mathcal{L}_{bRbR} & \mathcal{L}_{bRcR} \\
\mathcal{L}_{cRa} & \mathcal{L}_{cRb} & \mathcal{L}_{cRc} & \mathcal{L}_{cRaR} & \mathcal{L}_{cRbR} & \mathcal{L}_{cRcR}
\end{bmatrix}
\begin{bmatrix}
i_a \\
i_b \\
i_c \\
i_{aR} \\
i_{bR} \\
i_{cR}
\end{bmatrix}
\tag{C.32}
$$

where the subscripts (a, b, c) refer to stator quantities while the subscripts (aR, bR, cR) refer to rotor quantities.

The various machine inductances can then be written in terms of the electrical rotor angle θ_{me} (defined in this case as between the rotor phase-aR and the stator phase-a axes), as follows. For the stator self-inductances

$$
\mathcal{L}_{aa} = \mathcal{L}_{bb} = \mathcal{L}_{cc} = L_{aa0} + L_{al}
\tag{C.33}
$$

where L_{aa0} is the air-gap component of the stator self-inductance and L_{al} is the leakage component.

For the rotor self-inductances

$$
\mathcal{L}_{aRaR} = \mathcal{L}_{bRbR} = \mathcal{L}_{cRcR} = L_{aRaR0} + L_{aRl}
\tag{C.34}
$$

where L_{aRaR0} is the air-gap component of the rotor self-inductance and L_{aRl} is the leakage component.

For the stator-to-stator mutual inductances

$$
\mathcal{L}_{ab} = \mathcal{L}_{ba} = \mathcal{L}_{ac} = \mathcal{L}_{ca} = \mathcal{L}_{bc} = \mathcal{L}_{cb} = -\frac{1}{2}L_{aa0}
\tag{C.35}
$$

For the rotor-to-rotor mutual inductances

$$
\mathcal{L}_{aRbR} = \mathcal{L}_{bRaR} = \mathcal{L}_{aRcR} = \mathcal{L}_{cRaR} = \mathcal{L}_{bRcR} = \mathcal{L}_{cRbR} = -\frac{1}{2}L_{aRaR0}
\tag{C.36}
$$

and for the stator-to-rotor mutual inductances

$$
\mathcal{L}_{aaR} = \mathcal{L}_{aRa} = \mathcal{L}_{bbR} = \mathcal{L}_{bRb} = \mathcal{L}_{ccR} = \mathcal{L}_{cRc} = L_{aaR}\cos\theta_{me}
\tag{C.37}
$$

$$
\mathcal{L}_{baR} = \mathcal{L}_{aRb} = \mathcal{L}_{cbR} = \mathcal{L}_{bRc} = \mathcal{L}_{acR} = \mathcal{L}_{cRa} = L_{aaR}\cos(\theta_{me} - 120^\circ)
\tag{C.38}
$$

$$
\mathcal{L}_{caR} = \mathcal{L}_{aRc} = \mathcal{L}_{abR} = \mathcal{L}_{bRa} = \mathcal{L}_{bcR} = \mathcal{L}_{cRb} = L_{aaR}\cos(\theta_{me} + 120^\circ)
\tag{C.39}
$$

The corresponding voltage equations become

$$
v_a = R_a i_a + \frac{d\lambda_a}{dt}
\tag{C.40}
$$

$$v_b = R_a i_b + \frac{d\lambda_b}{dt} \tag{C.41}$$

$$v_c = R_a i_c + \frac{d\lambda_c}{dt} \tag{C.42}$$

$$v_{aR} = 0 = R_{aR} i_{aR} + \frac{d\lambda_{aR}}{dt} \tag{C.43}$$

$$v_{bR} = 0 = R_{aR} i_{bR} + \frac{d\lambda_{bR}}{dt} \tag{C.44}$$

$$v_{cR} = 0 = R_{aR} i_{cR} + \frac{d\lambda_{cR}}{dt} \tag{C.45}$$

where the voltages v_{aR}, v_{bR}, and v_{cR} are set equal to zero because the rotor windings are shorted at their terminals.

In the case of a synchronous machine in which the stator flux wave and rotor rotate in synchronism (at least in the steady state), the choice of reference frame for the dq0 transformation is relatively obvious. Specifically, the most useful transformation is to a reference frame fixed to the rotor.

The choice is not so obvious in the case of an induction machine. For example, one might choose a reference frame fixed to the rotor and apply the transformation of Eqs. C.1 and C.2 directly. If this is done, because the rotor of an induction motor does not rotate at synchronous speed, the flux linkages seen in the rotor reference frame will not be constant, and hence the time derivatives in the transformed voltage equations will not be equal to zero. Correspondingly, the direct- and quadrature-axis flux linkages, currents, and voltages will be found to be time-varying, with the result that the transformation turns out to be of little practical value.

An alternative choice is to choose a reference frame rotating at the synchronous angular velocity. In this case, both the stator and rotor quantities will have to be transformed. In the case of the stator quantities, the rotor angle θ_{me} in Eqs. C.1 and C.2 would be replaced by θ_S where

$$\theta_S = \omega_e t + \theta_0 \tag{C.46}$$

is the angle between the axis of phase a and that of the synchronously-rotating dq0 reference frame and θ_0.

The transformation equations for the stator quantities then become

$$\begin{bmatrix} S_d \\ S_q \\ S_0 \end{bmatrix} = \frac{2}{3} \begin{bmatrix} \cos(\theta_S) & \cos(\theta_S - 120°) & \cos(\theta_S + 120°) \\ -\sin(\theta_S) & -\sin(\theta_S - 120°) & -\sin(\theta_S + 120°) \\ \frac{1}{2} & \frac{1}{2} & \frac{1}{2} \end{bmatrix} \begin{bmatrix} S_a \\ S_b \\ S_c \end{bmatrix} \tag{C.47}$$

and the inverse transformation

$$\begin{bmatrix} S_a \\ S_b \\ S_c \end{bmatrix} = \begin{bmatrix} \cos(\theta_S) & -\sin(\theta_S) & 1 \\ \cos(\theta_S - 120°) & -\sin(\theta_S - 120°) & 1 \\ \cos(\theta_S + 120°) & -\sin(\theta_S + 120°) & 1 \end{bmatrix} \begin{bmatrix} S_d \\ S_q \\ S_0 \end{bmatrix} \tag{C.48}$$

Similarly, in the case of the rotor, θ_S would be replaced by θ_R where

$$\theta_R = (\omega_e - \omega_{me})t + \theta_0 \tag{C.49}$$

is the angle between the axis of rotor phase aR and that of the synchronously-rotating dq0 reference frame and $(\omega_e - \omega_{me})$ is the electrical angular velocity of the synchronously rotating reference frame as seen from the rotor frame.

The transformation equations for the rotor quantities then become

$$
\begin{bmatrix} S_{dR} \\ S_{qR} \\ S_{0R} \end{bmatrix} = \frac{2}{3} \begin{bmatrix} \cos(\theta_R) & \cos(\theta_R - 120°) & \cos(\theta_R + 120°) \\ -\sin(\theta_R) & -\sin(\theta_R - 120°) & -\sin(\theta_R + 120°) \\ \frac{1}{2} & \frac{1}{2} & \frac{1}{2} \end{bmatrix} \begin{bmatrix} S_{aR} \\ S_{bR} \\ S_{cR} \end{bmatrix} \tag{C.50}
$$

and the inverse transformation

$$
\begin{bmatrix} S_{aR} \\ S_{bR} \\ S_{cR} \end{bmatrix} = \begin{bmatrix} \cos(\theta_R) & -\sin(\theta_R) & 1 \\ \cos(\theta_R - 120°) & -\sin(\theta_R - 120°) & 1 \\ \cos(\theta_R + 120°) & -\sin(\theta_R + 120°) & 1 \end{bmatrix} \begin{bmatrix} S_{dR} \\ S_{qR} \\ S_{0R} \end{bmatrix} \tag{C.51}
$$

Using this set of transformations for the rotor and stator quantities, the transformed flux-linkage current relationships become

$$\lambda_d = L_S i_d + L_m i_{dR} \tag{C.52}$$

$$\lambda_q = L_S i_q + L_m i_{qR} \tag{C.53}$$

$$\lambda_0 = L_0 i_0 \tag{C.54}$$

for the stator and

$$\lambda_{dR} = L_m i_d + L_R i_{dR} \tag{C.55}$$

$$\lambda_{qR} = L_m i_q + L_R i_{qR} \tag{C.56}$$

$$\lambda_{0R} = L_{0R} i_{0R} \tag{C.57}$$

for the rotor
Here we have defined a set of new inductances

$$L_S = \frac{3}{2} L_{aa0} + L_{al} \tag{C.58}$$

$$L_m = \frac{3}{2} L_{aaR0} \tag{C.59}$$

$$L_0 = L_{al} \tag{C.60}$$

$$L_R = \frac{3}{2} L_{aRaR0} + L_{aRl} \tag{C.61}$$

$$L_{0R} = L_{aRl} \tag{C.62}$$

The transformed stator-voltage equations are

$$v_\mathrm{d} = R_\mathrm{a}i_\mathrm{d} + \frac{d\lambda_\mathrm{d}}{dt} - \omega_\mathrm{e}\lambda_\mathrm{q} \tag{C.63}$$

$$v_\mathrm{q} = R_\mathrm{a}i_\mathrm{q} + \frac{d\lambda_\mathrm{q}}{dt} + \omega_\mathrm{e}\lambda_\mathrm{d} \tag{C.64}$$

$$v_0 = R_\mathrm{a}i_0 + \frac{d\lambda_0}{dt} \tag{C.65}$$

and those for the rotor are

$$0 = R_\mathrm{aR}i_\mathrm{dR} + \frac{d\lambda_\mathrm{dR}}{dt} - (\omega_\mathrm{e} - \omega_\mathrm{me})\lambda_\mathrm{qR} \tag{C.66}$$

$$0 = R_\mathrm{aR}i_\mathrm{qR} + \frac{d\lambda_\mathrm{qR}}{dt} + (\omega_\mathrm{e} - \omega_\mathrm{me})\lambda_\mathrm{dR} \tag{C.67}$$

$$0 = R_\mathrm{aR}i_\mathrm{0R} + \frac{d\lambda_\mathrm{0R}}{dt} \tag{C.68}$$

Finally, using the techniques of Chapter 3, the torque can be expressed in a number of equivalent ways including

$$T_\mathrm{mech} = \frac{3}{2}\left(\frac{\mathrm{poles}}{2}\right)(\lambda_\mathrm{d}i_\mathrm{q} - \lambda_\mathrm{q}i_\mathrm{d}) \tag{C.69}$$

and

$$T_\mathrm{mech} = \frac{3}{2}\left(\frac{\mathrm{poles}}{2}\right)\left(\frac{L_\mathrm{m}}{L_\mathrm{R}}\right)(\lambda_\mathrm{dR}i_\mathrm{q} - \lambda_\mathrm{qR}i_\mathrm{d}) \tag{C.70}$$

D APPENDIX

Engineering Aspects of Practical Electric Machine Performance and Operation

In this book the basic essential features of electric machinery have been discussed; this material forms the basis for understanding the behavior of electric machinery of all types. In this appendix our objective is to introduce practical issues associated with the engineering implementation of the machinery concepts which have been developed. Issues common to all electric machine types such as losses, cooling, and rating are discussed.

D.1 LOSSES

Consideration of machine losses is important for three reasons: (1) Losses determine the efficiency of the machine and appreciably influence its operating cost; (2) losses determine the heating of the machine and hence the rating or power output that can be obtained without undue deterioration of the insulation; and (3) the voltage drops or current components associated with supplying the losses must be properly accounted for in a machine representation. Machine efficiency, like that of transformers or any energy-transforming device, is given by

$$\text{Efficiency} = \frac{\text{output}}{\text{input}} \tag{D.1}$$

which can also be expressed as

$$\text{Efficiency} = \frac{\text{input} - \text{losses}}{\text{input}} = 1 - \frac{\text{losses}}{\text{input}} \tag{D.2}$$

$$\text{Efficiency} = \frac{\text{output}}{\text{output} + \text{losses}} \tag{D.3}$$

Rotating machines in general operate efficiently except at light loads. For example, the full-load efficiency of average motors ranges from 80 to 90 percent for motors on the order of 1 to 10 kW, 90 to 95 percent for motors up to a few hundred kW, and up to a few percent higher for larger motors.

The forms given by Eqs. D.2 and D.3 are often used for electric machines, since their efficiency is most commonly determined by measurement of losses instead of by directly measuring the input and output under load. Efficiencies determined from loss measurements can be used in comparing competing machines if exactly the same methods of measurement and computation are used in each case. For this reason, the various losses and the conditions for their measurement are precisely defined by the American National Standards Institute (ANSI), the Institute of Electrical and Electronics Engineers (IEEE), and the National Electrical Manufacturers Association (NEMA). The following discussion summarizes some of the various commonly considered loss mechanisms.

Ohmic Losses *Ohmic,* or $I^2 R$ *losses,* are found in all windings of a machine. By convention, these losses are computed on the basis of the dc resistances of the winding at 75°C. Actually the $I^2 R$ loss depends on the effective resistance of the winding under the operating frequency and flux conditions. The increment in loss represented by the difference between dc and effective resistances is included with stray load losses, discussed below. In the field windings of synchronous and dc machines, only the losses in the field winding are charged against the machine; the losses in external sources supplying the excitation are charged against the plant of which the machine is a part. Closely associated with $I^2 R$ loss is the *brush-contact loss* at slip rings and commutators. By convention, this loss is normally neglected for induction and synchronous machines. For industrial-type dc machines the voltage drop at the brushes is regarded as constant at 2 V total when carbon and graphite brushes with shunts (pigtails) are used.

Mechanical Losses *Mechanical losses* consist of brush and bearing friction, windage, and the power required to circulate air through the machine and ventilating system, if one is provided, whether by self-contained or external fans (except for the power required to force air through long or restricted ducts external to the machine). Friction and windage losses can be measured by determining the input to the machine running at the proper speed but unloaded and unexcited. Frequently they are lumped with core loss and determined at the same time.

Open-Circuit, or No-Load, Core Loss *Open-circuit core loss* consists of the hysteresis and eddy-current losses arising from changing flux densities in the iron of the machine with only the main exciting winding energized. In dc and synchronous machines, these losses are confined largely to the armature iron, although the flux variations arising from slot openings will cause losses in the field iron as well, particularly in the pole shoes or surfaces of the field iron. In induction machines the losses are confined largely to the stator iron. Open-circuit core loss can be found by measuring the input to the machine when it is operating unloaded at rated speed or frequency and under the appropriate flux or voltage conditions, and then deducting the friction and windage loss and, if the machine is self-driven during the test, the no-load armature I^2R loss (no-load stator I^2R loss for an induction motor). Usually, data are taken for a curve of core loss as a function of armature voltage in the neighborhood of rated voltage. The core loss under load is then considered to be the value at a voltage equal to rated voltage corrected for the armature resistance drop under load (a phasor correction for an ac machine). For induction motors, however, this correction is dispensed with, and the core loss at rated voltage is used. For efficiency determination alone, there is no need to segregate open-circuit core loss and friction and windage loss; the sum of these two losses is termed the *no-load rotational loss*.

   Eddy-current loss varies with the square of the flux density, the frequency, and the thickness of laminations. Under normal machine conditions it can be expressed to a sufficiently close approximation as

$$P_e = K_e(B_{max} f \delta)^2 \tag{D.4}$$

where

$$\delta = \text{lamination thickness}$$
$$B_{max} = \text{maximum flux density}$$
$$f = \text{frequency}$$
$$K_e = \text{proportionality constant}$$

The value of K_e depends on the units used, volume of iron, and resistivity of the iron.
   Variation of *hysteresis loss* can be expressed in equation form only on an empirical basis. The most commonly used relation is

$$P_h = K_h f B_{max}^n \tag{D.5}$$

where K_h is a proportionality constant dependent on the characteristics and volume of iron and the units used and the exponent n ranges from 1.5 to 2.5, a value of 2.0 often being used for estimating purposes in machines. In both Eqs. D.4 and D.5, frequency can be replaced by speed and flux density by the appropriate voltage, with the proportionality constants changed accordingly.

   When the machine is loaded, the space distribution of flux density is significantly changed by the mmf of the load currents. The actual core losses may increase noticeably. For example, mmf harmonics cause appreciable losses in the iron near the air-gap surfaces. The total increment in core loss is classified as part of the stray load loss.

Stray Load Loss   *Stray load loss* consists of the losses arising from nonuniform current distribution in the copper and the additional core losses produced in the iron by distortion of the magnetic flux by the load current. It is a difficult loss to determine accurately. By convention it is taken as 1.0 percent of the output for dc machines. For synchronous and induction machines it can be found by test.

D.2   RATING AND HEATING

The rating of electrical devices such as machines and transformers is often determined by mechanical and thermal considerations. For example, the maximum winding current is typically determined by the maximum operating temperature which the insulation can withstand without damage or excessive loss of life. Similarly the maximum speed of a motor or generator is typically determined by mechanical considerations related to the structural integrity of the rotor or the performance of the bearings. The temperature rise resulting from the losses considered in Section D.1 is therefore a major factor in the rating of a machine.

The operating temperature of a machine is closely associated with its life expectancy because deterioration of the insulation is a function of both time and temperature. Such deterioration is a chemical phenomenon involving slow oxidation and brittle hardening and leading to loss of mechanical durability and dielectric strength. In many cases the deterioration rate is such that the life of the insulation can be represented as an exponential

$$\text{Life} = Ae^{B/T} \tag{D.6}$$

where A and B are constants and T is the absolute temperature. Thus, according to Eq. D.6, when life is plotted to a logarithmic scale against the reciprocal of absolute temperature on a uniform scale, a straight line should result. Such plots form valuable guides in the thermal evaluation of insulating materials and systems. A very rough idea of the life-temperature relation can be obtained from the old and more or less obsolete rule of thumb that the time to failure of organic insulation is halved for each 8 to 10°C rise.

The evaluation of insulating materials and insulation systems (which may include widely different materials and techniques in combination) is to a large extent a functional one based on accelerated life tests. Both normal life expectancy and service conditions will vary widely for different classes of electric equipment. Life expectancy, for example, may be a matter of minutes in some military and missile applications, may be 500 to 1000 h in certain aircraft and electronic equipment, and may range from 10 to 30 years or more in large industrial equipment. The test procedures will accordingly vary with the type of equipment. Accelerated life tests on models, called *motorettes,* are commonly used in insulation evaluation. Such tests, however, cannot be easily applied to all equipment, especially the insulation systems of large machines.

Insulation life tests generally attempt to simulate service conditions. They usually include the following elements:

- Thermal shock resulting from heating to the test temperature.
- Sustained heating at that temperature.
- Thermal shock resulting from cooling to room temperature or below.
- Vibration and mechanical stress such as may be encountered in actual service.
- Exposure to moisture.
- Dielectric testing to determine the condition of the insulation.

Enough samples must be tested to permit statistical methods to be applied in analyzing the results. The life-temperature relations obtained from these tests lead to the classification of the insulation or insulating system in the appropriate temperature class.

For the allowable temperature limits of insulating systems used commercially, the latest standards of ANSI, IEEE, and NEMA should be consulted. The three NEMA insulation-system classes of chief interest for industrial machines are class B, class F, and class H. Class B insulation includes mica, glass fiber, asbestos, and similar materials with suitable bonding substances. Class F insulation also includes mica, glass fiber, and synthetic substances similar to those in class B, but the system must be capable of withstanding higher temperatures. Class H insulation, intended for still higher temperatures, may consist of materials such as silicone elastomer and combinations including mica, glass fiber, asbestos, and so on, with bonding substances such as appropriate silicone resins. Experience and tests showing the material or system to be capable of operation at the recommended temperature form the important classifying criteria.

When the temperature class of the insulation is established, the permissible observable temperature rises for the various parts of industrial-type machines can be found by consulting the appropriate standards. Reasonably detailed distinctions are made with respect to type of machine, method of temperature measurement, machine part involved, whether the machine is enclosed, and the type of cooling (air-cooled, fan-cooled, hydrogen-cooled, etc.). Distinctions are also made between general-purpose machines and definite- or special-purpose machines. The term *general-purpose motor* refers to one of standard rating "up to 200 hp with standard operating characteristics and mechanical construction for use under usual service conditions without restriction to a particular application or type of application." In contrast a *special-purpose motor* is "designed with either operating characteristics or mechanical construction, or both, for a particular application." For the same class of insulation, the permissible rise of temperature is lower for a general-purpose motor than for a special-purpose motor, largely to allow a greater factor of safety where service conditions are unknown. Partially compensating the lower rise, however, is the fact that general-purpose motors are allowed a *service factor* of 1.15 when operated at rated voltage; the service factor is a multiplier which, applied to the rated output, indicates a permissible loading which may be carried continuously under the conditions specified for that service factor.

Examples of allowable temperature rises can be seen from Table D.1. The table applies to integral-horsepower induction motors, is based on 40°C ambient temperature, and assumes measurement of temperature rise by determining the increase of winding resistances.

The most common machine rating is the *continuous rating* defining the output (in kilowatts for dc generators, kilovoltamperes at a specified power factor for ac generators, and horsepower or kilowatts for motors) which can be carried indefinitely

Table D.1 Allowable temperature rise, °C[†].

Motor type	Class B	Class F	Class H
1.15 service factor	90	115	
1.00 service factor, encapsulated windings	85	110	
Totally enclosed, fan-cooled	80	105	125
Totally enclosed, nonventilated	85	110	135

[†]Excerpted from NEMA standards.

without exceeding established limitations. For intermittent, periodic, or varying duty, a machine may be given a short-time rating defining the load which can be carried for a specific time. Standard periods for short-time ratings are 5, 15, 30, and 60 minutes. Speeds, voltages, and frequencies are also specified in machine ratings, and provision is made for possible variations in voltage and frequency. Motors, for example, must operate successfully at voltages 10 percent above and below rated voltage and, for ac motors, at frequencies 5 percent above and below rated frequency; the combined variation of voltage and frequency may not exceed 10 percent. Other performance conditions are so established that reasonable short-time overloads can be carried. Thus, the user of a motor can expect to be able to apply for a short time an overload of, say, 25 percent at 90 percent of normal voltage with an ample margin of safety.

The converse problem to the rating of machinery, that of choosing the size of machine for a particular application, is a relatively simple one when the load requirements remain substantially constant. For many motor applications, however, the load requirements vary more or less cyclically and over a wide range. The duty cycle of a typical crane or hoist motor offers a good example. From the thermal viewpoint, the average heating of the motor must be found by detailed study of the motor losses during the various parts of the cycle. Account must be taken of changes in ventilation with motor speed for open and semiclosed motors. Judicious selection is based on a large amount of experimental data and considerable experience with the motors involved. For estimating the required size of motors operating at substantially constant speeds, it is sometimes assumed that the heating of the insulation varies as the square of the load, an assumption which obviously overemphasizes the role of armature I^2R loss at the expense of the core loss. The rms ordinate of the power-time curve representing the duty cycle is obtained by the same technique used to find the rms value of periodically varying currents, and a motor rating is chosen on the basis of the result; i.e.,

$$\text{rms kW} = \sqrt{\frac{\Sigma(\text{kW})^2 \times \text{time}}{\text{running time} + (\text{standstill time}/k)}} \qquad \text{(D.7)}$$

where the constant k accounts for the poorer ventilation at standstill and equals approximately 4 for an open motor. The time for a complete cycle must be short compared with the time for the motor to reach a steady temperature.

Although crude, the rms-kW method is used fairly often. The necessity for rounding the result to a commercially available motor size[1] obviates the need for precise computations. Special consideration must be given to motors that are frequently started or reversed, for such operations are thermally equivalent to heavy overloads. Consideration must also be given to duty cycles having such high torque

---

[1] Commercially available motors are generally found in standard sizes as defined by NEMA. NEMA Standards on Motors and Generators specify motor rating as well as the type and dimensions of the motor frame.

peaks that motors with continuous ratings chosen on purely thermal bases would be unable to furnish the torques required. It is to such duty cycles that special-purpose motors with short-time ratings are often applied. Short-time-rated motors in general have better torque-producing capability than motors rated to produce the same power output continuously, although, of course, they have a lower thermal capacity. Both these properties follow from the fact that a short-time-rated motor is designed for high flux densities in the iron and high current densities in the copper. In general, the ratio of torque capacity to thermal capacity increases as the period of the short-time rating decreases. Higher temperature rises are allowed in short-time-rated motors than for general-purpose motors. A motor with a 150-kW, 1-hr, 50°C rating, for example, may have the torque ability of a 200-kW continuously rated motor; it will be able to carry only about 0.8 times its rated output, or 120 kW continuously, however. In many cases it will be the economical solution for a drive requiring a continuous thermal capacity of 120 kW but having torque peaks which require the capability of a 200-kW continuously rated motor.

D.3   COOLING MEANS FOR ELECTRIC MACHINES

The cooling problem in electric apparatus in general increases in difficulty with increasing size. The surface area from which the heat must be carried away increases roughly as the square of the dimensions, whereas the heat developed by the losses is roughly proportional to the volume and therefore increases approximately as the cube of the dimensions. This problem is a particularly serious one in large turbine generators, where economy, mechanical requirements, shipping, and erection all demand compactness, especially for the rotor forging. Even in moderate size machines, for example, above a few thousand kVA for generators, a closed ventilating system is commonly used. Rather elaborate systems of cooling ducts must be provided to ensure that the cooling medium will effectively remove the heat arising from the losses.

For turbine generators, hydrogen is commonly used as the cooling medium in the totally enclosed ventilating system. Hydrogen has the following properties which make it well suited to the purpose:

■ Its density is only about 0.07 times that of air at the same temperature and pressure, and therefore windage and ventilating losses are much less.

■ Its specific heat on an equal-weight basis is about 14.5 times that of air. This means that, for the same temperature and pressure, hydrogen and air are about equally effective in their heat-storing capacity per unit volume, but the heat transfer by forced convection between the hot parts of the machine and the cooling gas is considerably greater with hydrogen than with air.

■ The life of the insulation is increased and maintenance expenses decreased because of the absence of dirt, moisture, and oxygen.

■ The fire hazard is minimized. A hydrogen-air mixture will not explode if the hydrogen content is above about 70 percent.

The result of the first two properties is that for the same operating conditions the heat which must be dissipated is reduced and at the same time the ease with which it can be carried off is increased.

The machine and its water-cooled heat exchanger for cooling the hydrogen must be sealed in a gas-tight envelope. The crux of the problem is in sealing the bearings. The system is maintained at a slight pressure (at least 0.5 psi) above atmospheric so that gas leakage is outward and an explosive mixture cannot accumulate in the machine. At this pressure, the rating of the machine can be increased by about 30 percent above its aircooled rating, and the full-load efficiency increased by about 0.5 percent. The trend is toward the use of higher pressures (15 to 60 psi). Increasing the hydrogen pressure from 0.5 to 15 psi increases the output for the same temperature rise by about 15 percent; a further increase to 30 psi provides about an additional 10 percent.

An important step which has made it possible almost to double the output of a hydrogen-cooled turbine-generator of given physical size is the development of *conductor cooling,* also called *inner cooling.* Here the coolant (liquid or gas) is forced through hollow ducts inside the conductor or conductor strands. Examples of such conductors can be seen in Fig. D.1. Thus, the thermal barrier presented by the electric insulation is largely circumvented, and the conductor losses can be absorbed directly by the coolant. Hydrogen is usually the cooling medium for the rotor conductors. Either gas or liquid cooling may be used for the stator conductors. Hydrogen is the coolant in the former case, and transit oil or water is commonly used in the latter. A sectional view of a conductor-cooled turbine generator is given in Fig. D.2. A large hydroelectric generator in which both stator and rotor are water-cooled is shown in Figs. 4.1 and 4.9.

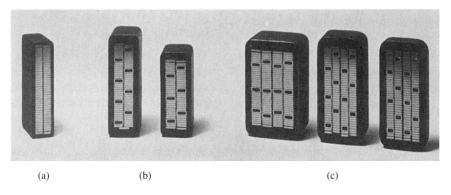

(a)   (b)   (c)

Figure D.1 Cross sections of bars for two-layer stator windings of turbine-generators. Insulation system consists of synthetic resin with vacuum impregnation. (a) Indirectly cooled bar with tubular strands; (b) water-cooled bars, two-wire-wide mixed strands, (c) water-cooled bars, four-wire-wide mixed strands. (*Brown Boveri Corporation.*)

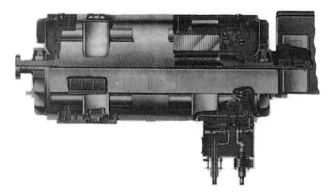

Figure D.2 Cutaway view of a two-pole 3600 r/min turbine rated at 500 MVA, 0.90 power factor, 22 kV, 60 Hz, 45 psig H_2 pressure. Stator winding is water-cooled; rotor winding is hydrogen-cooled. (*General Electric Company.*)

D.4 EXCITATION

The resultant flux in the magnetic circuit of a machine is established by the combined mmf of all the windings on the machine. For the conventional dc machine, the bulk of the effective mmf is furnished by the field windings. For the transformer, the net excitation may be furnished by either the primary or the secondary winding, or a portion may be furnished by each. A similar situation exists in ac machines. Furnishing excitation to ac machines has two different operational aspects which are of economic importance in the application of the machines.

D.4.1 Power Factor in AC Machines

The power factor at which ac machines operate is an economically important feature because of the cost of reactive kilovoltamperes. Low power factor adversely affects system operation in three principal ways. (1) Generators, transformers, and transmission equipment are rated in terms of kVA rather than kW because their losses and heating are very nearly determined by voltage and current regardless of power factor. The physical size and cost of ac apparatus are roughly proportional to kVA rating. The investment in generators, transformers, and transmission equipment for supplying a given useful amount of active power therefore is roughly inversely proportional to the power factor. (2) Low power factor means more current and greater I^2R losses in the generating and transmitting equipment. (3) A further disadvantage is poor voltage regulation.

Factors influencing reactive-kVA requirements in motors can be visualized readily in terms of the relationship of these requirements to the establishment of magnetic flux. As in any electromagnetic device, the resultant flux necessary for motor operation must be established by a magnetizing component of current. It makes no difference either in the magnetic circuit or in the fundamental energy conversion process whether this magnetizing current be carried by the rotor or stator winding, just as it makes no basic difference in a transformer which winding carries the exciting current. In some cases, part of it is supplied from each winding. If all or part of the magnetizing current is supplied by an ac winding, the input to that winding must include lagging reactive kVA, because magnetizing current lags voltage drop by 90°. In effect, the lagging reactive kVA set up flux in the motor.

The only possible source of excitation in an induction motor is the stator input. The induction motor therefore must operate at a lagging power factor. This power factor is very low at no load and increases to about 85 to 90 percent at full load, the improvement being caused by the increased real-power requirements with increasing load.

With a synchronous motor, there are two possible sources of excitation: alternating current in the armature or direct current in the field winding. If the field current is just sufficient to supply the necessary mmf, no magnetizing-current component or reactive kVA are needed in the armature and the motor operates at unity power factor. If the field current is less, i.e., the motor is *underexcited,* the deficit in mmf must be made up by the armature and the motor operates at a lagging power factor. If the field current is greater, i.e., the motor is *overexcited,* the excess mmf must be counterbalanced in the armature and a leading component of current is present; the motor then operates at a leading power factor.

Because magnetizing current must be supplied to inductive loads such as transformers and induction motors, the ability of overexcited synchronous motors to supply lagging current is a highly desirable feature which may have considerable economic importance. In effect, overexcited synchronous motors act as generators of lagging reactive kilovoltamperes and thereby relieve the power source of the necessity for supplying this component. They thus may perform the same function as a local capacitor installation. Sometimes unloaded synchronous machines are installed in power systems solely for power-factor correction or for control of reactive-kVA flow. Such machines, called *synchronous condensers,* may be more economical in the larger sizes than static capacitors.

Both synchronous and induction machines may become self-excited when a sufficiently heavy capacitive load is present in their stator circuits. The capacitive current then furnishes the excitation and may cause serious overvoltage or excessive transient torques. Because of the inherent capacitance of transmission lines, the problem may arise when synchronous generators are energizing long unloaded or lightly loaded lines. The use of shunt reactors at the sending end of the line to compensate the capacitive current is sometimes necessary. For induction motors, it is normal practice to avoid self-excitation by limiting the size of any parallel capacitor when the motor and capacitor are switched as a unit.

D.4.2 Turbine-Generator Excitation Systems

As the available ratings of turbine-generators have increased, the problems of supplying the dc field excitation (amounting to 4000 A or more in the larger units) have grown progressively more difficult. A common excitation source is a shaft-driven dc generator whose output is supplied to the alternator field through brushes and slip rings. Alternatively, excitation may be supplied from a shaft-driven alternator of conventional design as the main exciter. This alternator has a stationary armature and a rotating-field winding. Its frequency may be 180 or 240 Hz. Its output is fed to a stationary solid-state rectifier, which in turn supplies the turbine-generator field through slip rings.

Cooling and maintenance problems are inevitably associated with slip rings, commutators, and brushes. Many modern excitation systems have minimized these problems by minimizing the use of sliding contacts and brushes. As a result, some excitation systems employ shaft-driven ac alternators whose field windings are stationary and whose ac windings rotate. By the use of rotating rectifiers, dc excitation can be applied directly to the generator field winding without the use of slip rings.

Excitation systems of the latest design are being built without any sort of rotating exciter-alternator. In these systems, the excitation power is obtained from a special auxiliary transformer fed from the local power system. Alternatively it may be obtained directly from the main generator terminals; in one system a special armature winding is included in the main generator to supply the excitation power. In each of these systems the power is rectified using phase-controlled silicon controlled rectifiers (SCRs). These types of excitation system, which have been made possible by the development of reliable, high-power SCRs, are relatively simple in design and provide the fast response characteristics required in many modern applications.

D.5 ENERGY EFFICIENCY OF ELECTRIC MACHINERY

With increasing concern for both the supply and cost of energy comes a corresponding concern for efficiency in its use. Although electric energy can be converted to mechanical energy with great efficiency, achieving maximum efficiency requires both careful design of the electric machinery and proper matching of machine and intended application.

Clearly, one means to maximize the efficiency of an electric machine is to minimize its internal losses, such as those described in Section D.1. For example, the winding I^2R losses can be reduced by increasing the slot area so that more copper can be used, thus increasing the cross-sectional area of the windings and reducing the resistance.

Core loss can be reduced by decreasing the magnetic flux density in the iron of the machine. This can be done by increasing the volume of iron, but although the loss goes down in terms of watts per pound, the total volume of material (and hence the mass) is increased; depending on how the machine design is changed, there may be a point beyond which the losses actually begin to increase. Similarly, for a given flux density, eddy-current losses can be reduced by using thinner iron laminations.

One can see that there are trade-offs involved here; machines of more efficient design generally require more material and thus are bigger and more costly. Users will generally choose the "lowest-cost" solution to a particular requirement; if the increased capital cost of a high-efficiency motor can be expected to be offset by energy savings over the expected lifetime of the machine, they will probably select the high-efficiency machine. If not, users are very unlikely to select this option in spite of the increased efficiency.

Similarly, some types of electric machines are inherently more efficient than others. For example, single-phase capacitor-start induction motors (Section 9.2) are relatively inexpensive and highly reliable, finding use in all sorts of small appliances, e.g., refrigerators, air conditioners, and fans. Yet they are inherently less efficient than their three-phase counterparts. Modifications such as a capacitor-run feature can lead to greater efficiency in the single-phase induction motor, but they are expensive and often not economically justifiable.

To optimize the efficiency of use of electric machinery the machine must be properly matched to the application, both in terms of size and performance. Since typical induction motors tend to draw nearly constant reactive power, independent of load, and since this causes resistive losses in the supply lines, it is wise to pick the smallest-rating induction motor which can properly satisfy the requirements of a specific application. Alternatively, capacitative power-factor correction may be used. Proper application of modern solid-state control technology can also play an important role in optimizing both performance and efficiency.

There are, of course, practical limitations which affect the selection of the motor for any particular application. Chief among them is that motors are generally available only in certain standard sizes. For example, a typical manufacturer might make fractional-horsepower ac motors rated at $\frac{1}{8}$, $\frac{1}{6}$, $\frac{1}{4}$, $\frac{1}{3}$, $\frac{1}{2}$, $\frac{3}{4}$, and 1 hp (NEMA standard ratings). This discrete selection thus limits the ability to fine-tune a particular application; if the need is 0.8 hp, the user will undoubtedly end up buying a 1-hp device and settling for a somewhat lower than optimum efficiency. A custom-designed and manufactured 0.8-hp motor can be economically justified only if it is needed in large quantities.

It should be pointed out that an extremely common source of inefficiency in electric motor applications is the mismatch of the motor to its application. Even the most efficient 50-kW motors will be somewhat inefficient when driving a 20-kW load. Yet mismatches of this type often occur in practice, due in great extent to the difficulty in characterizing operating loads and a tendency on the part of application engineers to be conservative to make sure that the system in question is guaranteed to operate in the face of design uncertainties. More careful attention to this issue can go a long way toward increasing the efficiency of energy use in electric machine applications.

E APPENDIX

Table of Constants and Conversion Factors for SI Units

Constants

Permeability of free space $\mu_0 = 4\pi \times 10^{-7}$ H/m

Permittivity of free space $\epsilon_0 = 8.854 \times 10^{-12}$ F/m

Conversion Factors

Length	1 m = 3.281 ft = 39.37 in
Mass	1 kg = 0.0685 slug = 2.205 lb (mass) = 35.27 oz
Force	1 N = 0.225 lbf = 7.23 poundals
Torque	1 N·m = 0.738 lbf·ft = 141.6 oz·in
Pressure	1 Pa (N/m^2) = 1.45×10^{-4} lbf/in^2 = 9.86×10^{-6} atm
Energy	1 J (W·sec) = 9.48×10^{-4} BTU = 0.239 calories
Power	1 W = 1.341×10^{-3} hp = 3.412 BTU/hr
Moment of intertia	1 kg·m^2 = 0.738 slug·ft^2 = 23.7 lb·ft^2 = 141.6 oz·in·sec^2
Magnetic flux	1 Wb = 10^8 lines (maxwells)
Magnetic flux density	1 T (Wb/m^2) = 10,000 gauss = 64.5 kilolines/in^2
Magnetizing force	1 A·turn/m = 0.0254 A·turn/in = 0.0126 oersted

INDEX